STUDENT SOLUTIONS MANUAL

SEVENTH EDITION

CALCULUS

FOR BUSINESS, ECONOMICS, LIFE SCIENCES, AND SOCIAL SCIENCES

RAYMOND A. BARNETT

MICHAEL R. ZIEGLER

PRENTICE HALL, Upper Saddle River, New Jersey 07458

Acquisition Editor: *George Lobell*
Production Editor: *Alison Aquino*
Page Layout Designer: *Jeanne Wallace*
Production Supervisor: *Joan Eurell*
Supplement Acquisitions Editor: *Audra Walsh*
Production Coordinator: *Alan Fischer*

 © 1996 by Prentice-Hall, Inc.
Simon & Schuster / A Viacom Company
Upper Saddle River, NJ 07458

Printed in the United States of America

10 9 8 7 6 5 4 3

ISBN 0-13-232752-X

Prentice-Hall International (UK) Limited, *London*
Prentice-Hall of Australia Pty. Limited, *Sydney*
Prentice-Hall Canada, Inc., *Toronto*
Prentice-Hall Hispanoamericana, S.A., *Mexico*
Prentice-Hall of India Private Limited, *New Delhi*
Prentice-Hall of Japan, Inc., *Tokyo*
Simon & Schuster Asia Pte. Ltd., *Singapore*
Editora Prentice-Hall do Brasil, Ltda., *Rio de Janeiro*

CONTENTS

1 A BEGINNING LIBRARY OF ELEMENTARY FUNCTIONS

Things to remember:

<u>1</u>. A FUNCTION is a rule (process or method) that produces a correspondence between one set of elements, called a DOMAIN, and a second set of elements, called the RANGE, such that to each element in the domain there corresponds one and only one element in the range.

<u>2</u>. EQUATIONS AND FUNCTIONS:

Given an equation in two variables. If there corresponds exactly one value of the dependent variable (output) to each value of the independent variable (input), then the equation defines a function. If there is more than one output for at least one input, then the equation does not define a function.

<u>3</u>. VERTICAL LINE TEST FOR A FUNCTION

An equation defines a function if each vertical line in the coordinate system passes through at most one point on the graph of the equation. If any vertical line passes through two or more points on the graph of an equation, then the equation does not define a function.

<u>4</u>. AGREEMENT ON DOMAINS AND RANGES

If a function is specified by an equation and the domain is not given explicitly, then assume that the domain is the set of all real number replacements of the independent variable (inputs) that produce real values for the dependent variable (outputs). The range is the set of all outputs corresponding to input values.

In many applied problems, the domain is determined by practical considerations within the problem.

<u>5</u>. FUNCTION NOTATION—THE SYMBOL $f(x)$

For any element x in the domain of the function f, the symbol $f(x)$ represents the element in the range of f corresponding to x in the domain of f. If x is an input value, then $f(x)$ is the corresponding output value. If x is an element which is not in the domain of f, then f is NOT DEFINED at x and $f(x)$ DOES NOT EXIST.

1. The table specifies a function, since for each domain value there corresponds one and only one range value.

3. The table does not specify a function, since more than one range value corresponds to a given domain value. (Range values 5, 6 correspond to domain value 3; range values 6, 7 correspond to domain value 4.)

5. This is a function.

7. The graph specifies a function; each vertical line in the plane intersects the graph in at most one point.

9. The graph does not specify a function. There are vertical lines which intersect the graph in more than one point. For example, the y-axis intersects the graph in three points.

11. The graph specifies a function.

13. $f(x) = 3x - 2$
$f(2) = 3(2) - 2 = 4$

15. $f(-1) = 3(-1) - 2$
$= -5$

17. $g(x) = x - x^2$
$g(3) = 3 - 3^2 = -6$

19. $f(0) = 3(0) - 2$
$= -2$

21. $g(-3) = -3 - (-3)^2$
$= -12$

23. $f(1) + g(2)$
$= [3(1) - 2] + (2 - 2^2) = -1$

25. $g(2) - f(2)$
$= (2 - 2^2) - [3(2) - 2]$
$= -2 - 4 = -6$

27. $g(3) \cdot f(0) = (3 - 3^2)[3(0) - 2]$
$= (-6)(-2)$
$= 12$

29. $\dfrac{g(-2)}{f(-2)} = \dfrac{-2 - (-2)^2}{3(-2) - 2} = \dfrac{-6}{-8} = \dfrac{3}{4}$

31. $y = f(-5) = 0$

33. $y = f(5) = 4$

35. $f(x) = 0$ at $x = -5, 0, 4$

37. $f(x) = -4$ at $x = 6$

39. domain: all real numbers or $(-\infty, \infty)$

41. domain: all real numbers except -4

43. $x^2 + 3x - 4 = (x + 4)(x - 1)$; domain: all real numbers except -4 and 1.

45. $x^2 + 6x + 9 = (x + 3)^2$; domain: all real numbers except -3.

47. $7 - x \geq 0$ for $x \leq 7$; domain: $x \leq 7$ or $(-\infty, 7]$

49. $7 - x > 0$ for $x < 7$; domain: $x < 7$ or $(-\infty, 7)$

51. f is not defined at the values of x where $x^2 - 9 = 0$, that is, at 3 and -3; f is defined at $x = 2$, $f(2) = \dfrac{0}{-5} = 0$.

53. $g(x) = 2x^3 - 5$

55. $G(x) = 2\sqrt{x} - x^2$

57. Function f multiplies the domain element by 2 and subtracts 3 from the result.

59. Function F multiplies the cube of the domain element by 3 and subtracts twice the square root of the domain element from the result.

61. Given $4x - 5y = 20$. Solving for y, we have:

$$-5y = -4x + 20$$

$$y = \frac{4}{5}x - 4$$

Since each input value x determines a unique output value y, the equation specifies a function. The domain is R, the set of real numbers.

63. Given $x^2 - y = 1$. Solving for y, we have:

$$-y = -x^2 + 1 \quad \text{or} \quad y = x^2 - 1$$

This equation specifies a function. The domain is R, the set of real numbers.

65. Given $x + y^2 = 10$. Solving for y, we have:

$$y^2 = 10 - x$$

$$y = \pm\sqrt{10 - x}$$

This equation does not specify a function since each value of x, $x \leq 10$, determines two values of y. For example, corresponding to $x = 1$, we have $y = 3$ and $y = -3$; corresponding to $x = 6$, we have $y = 2$ and $y = -2$.

67. Given $xy - 4y = 1$. Solving for y, we have:

$$(x - 4)y = 1 \quad \text{or} \quad y = \frac{1}{x - 4}$$

This equation specifies a function. The domain is all real numbers except $x = 4$.

69. Given $x^2 + y^2 = 25$. Solving for y, we have:

$$y^2 = 25 - x^2 \quad \text{or} \quad y = \pm\sqrt{25 - x^2}$$

Thus, the equation does not specify a function since, for $x = 0$, we have $y = \pm 5$, when $x = 4$, $y = \pm 3$, and so on.

71. Given $F(t) = 4t + 7$. Then:

$$\frac{F(3 + h) - F(3)}{h} = \frac{4(3 + h) + 7 - (4 \cdot 3 + 7)}{h}$$

$$= \frac{12 + 4h + 7 - 19}{h} = \frac{4h}{h} = 4$$

73. Given $g(w) = w^2 - 4$. Then:

$$\frac{g(1 + h) - g(1)}{h} = \frac{(1 + h)^2 - 4 - (1^2 - 4)}{h} = \frac{1 + 2h + h^2 - 4 + 3}{h}$$

$$= \frac{2h + h^2}{h} = \frac{h(2 + h)}{h} = 2 + h$$

75. Given $Q(x) = x^2 - 5x + 1$. Then:

$$\frac{Q(2 + h) - Q(2)}{h} = \frac{(2 + h)^2 - 5(2 + h) + 1 - (2^2 - 5 \cdot 2 + 1)}{h}$$

$$= \frac{4 + 4h + h^2 - 10 - 5h + 1 - (-5)}{h} = \frac{h^2 - h - 5 + 5}{h}$$

$$= \frac{h(h - 1)}{h} = h - 1$$

77. Given $f(x) = 4x - 3$. Then:

$$\frac{f(a + h) - f(a)}{h} = \frac{4(a + h) - 3 - (4a - 3)}{h}$$

$$= \frac{4a + 4h - 3 - 4a + 3}{h} = \frac{4h}{h} = 4$$

79. Given $f(x) = 4x^2 - 7x + 6$. Then:

$$\frac{f(a + h) - f(a)}{h} = \frac{4(a + h)^2 - 7(a + h) + 6 - (4a^2 - 7a - 6)}{h}$$

$$= \frac{4(a^2 + 2ah + h^2) - 7a - 7h + 6 - 4a^2 + 7a - 6}{h}$$

$$= \frac{4a^2 + 8ah + 4h^2 - 7h - 4a^2}{h} = \frac{8ah + 4h^2 - 7h}{h}$$

$$= \frac{h(8a + 4h - 7)}{h} = 8a + 4h - 7$$

81. Given $f(x) = x^3$. Then:

$$\frac{f(a + h) - f(a)}{h} = \frac{(a + h)^3 - a^3}{h} = \frac{a^3 + 3a^2h + 3ah^2 + h^3 - a^3}{h}$$

$$= \frac{h(3a^2 + 3ah + h^2)}{h} = 3a^2 + 3ah + h^2$$

83. Given $f(x) = \sqrt{x}$. Then:

$$\frac{f(a + h) - f(a)}{h} = \frac{\sqrt{a + h} - \sqrt{a}}{h}$$

$$= \frac{\sqrt{a + h} - \sqrt{a}}{h} \cdot \frac{\sqrt{a + h} + \sqrt{a}}{\sqrt{a + h} + \sqrt{a}} \quad \text{(rationalizing the numerator)}$$

$$= \frac{a + h - a}{h(\sqrt{a + h} + \sqrt{a})} = \frac{h}{h(\sqrt{a + h} + \sqrt{a})} = \frac{1}{\sqrt{a + h} + \sqrt{a}}$$

85. Given $A = \ell w = 25$.

Thus, $\ell = \dfrac{25}{w}$. Now $P = 2\ell + 2w$

$$= 2\left(\frac{25}{w}\right) + 2w = \frac{50}{w} + 2w.$$

The domain is $w > 0$.

87. Given $P = 2\ell w + 2w = 100$ or $\ell + w = 50$ and $w = 50 - \ell$.

Now $A = \ell w = \ell(50 - \ell)$ and $A = 50\ell - \ell^2$.

The domain is $0 \leq \ell \leq 50$. [<u>Note</u>: $\ell \leq 50$ since $\ell > 50$ implies $w < 0$.]

89.

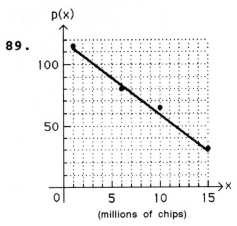

(millions of chips)

$p(8) = 71$ dollars per chip
$p(11) = 53$ dollars per chip

91. (A) $R(x) = xp(x) = x(119 - 6x)$
Domain: $1 \leq x \leq 15$

(C)

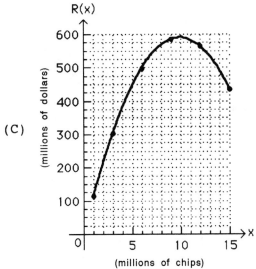

(millions of chips)

(B) Table 10 Revenue

x(millions)	$R(x)$(millions)
1	$113
3	303
6	498
9	585
12	564
15	435

93. (A) $P(x) = R(x) - C(x)$
$= x(119 - 6x) - (234 + 23x)$
$= -6x^2 + 96x - 234$ million dollars
Domain: $1 \leq x \leq 15$

(B) Table 12 Profit

x(millions)	$P(x)$(millions)
1	-$144
3	0
6	126
9	144
12	54
15	-144

(C)

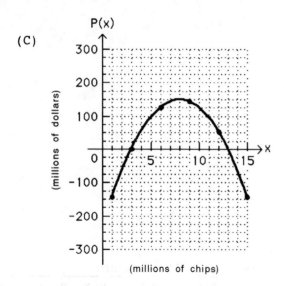

95.

```
        12
   ┌──────────┐
   │☐        ☐│ x
 8 │          │
 x │☐        ☐│
   └──────────┘
   x          x
```

(A) V = (length)(width)(height)
   $V(x) = (12 - 2x)(8 - 2x)x$
   $= x(8 - 2x)(12 - 2x)$

(B) Domain: $0 \le x \le 4$

(C) $V(1) = (12 - 2)(8 - 2)(1)$
        $= (10)(6)(1) = 60$
   $V(2) = (12 - 4)(8 - 4)(2)$
        $= (8)(4)(2) = 64$
   $V(3) = (12 - 6)(8 - 6)(3)$
        $= (6)(2)(3) = 36$

Thus,

Volume

x	$V(x)$
1	60
2	64
3	36

(D)

97. Given $(w + a)(v + b) = c$. Let $a = 15$, $b = 1$, and $c = 90$. Then:

$(w + 15)(v + 1) = 90$

Solving for v, we have

$v + 1 = \dfrac{90}{w + 15}$ and $v = \dfrac{90}{w + 15} - 1 = \dfrac{90 - (w + 15)}{w + 15}$, so that $v = \dfrac{75 - w}{w + 15}$.

If $w = 16$, then $v = \dfrac{75 - 16}{16 + 15} = \dfrac{59}{31} \approx 1.9032$ cm/sec.

EXERCISE 1-2

Things to remember:

1. **LIBRARY OF ELEMENTARY FUNCTIONS**

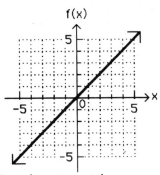

Identity Function
$f(x) = x$
Domain: All real numbers
Range: All real numbers
(a)

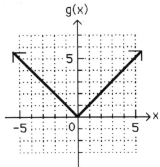

Absolute Value Function
$g(x) = |x|$
Domain: All real numbers
Range: $[0, \infty)$
(b)

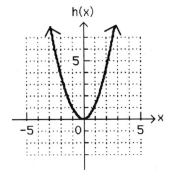

Square Function
$h(x) = x^2$
Domain: All real numbers
Range: $[0, \infty)$
(c)

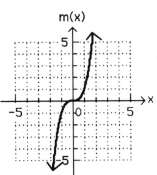

Cube Function
$m(x) = x^3$
Domain: All real numbers
Range: All real numbers
(d)

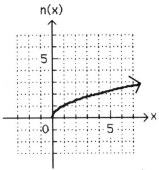

Square-Root Function
$n(x) = \sqrt{x}$
Domain: $[0, \infty)$
Range: $[0, \infty)$
(e)

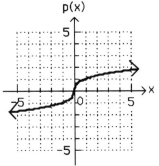

Cube-Root Function
$p(x) = \sqrt[3]{x}$
Domain: All real numbers
Range: All real numbers
(f)

NOTE: Letters used to designate the above functions may vary from context to context.

<u>2</u>. GRAPH TRANSFORMATIONS SUMMARY

<u>Vertical Translation:</u>

$y = f(x) + k$ $\begin{cases} k > 0 & \text{Shift graph of } y = f(x) \text{ up } k \text{ units} \\ k < 0 & \text{Shift graph of } y = f(x) \text{ down } |k| \text{ units} \end{cases}$

<u>Horizontal Translation:</u>

$y = f(x + h)$ $\begin{cases} h > 0 & \text{Shift graph of } y = f(x) \text{ left } h \text{ units} \\ h < 0 & \text{Shift graph of } y = f(x) \text{ right } |h| \text{ units} \end{cases}$

<u>Reflection:</u>

$y = -f(x)$ Reflect the graph of $y = f(x)$ in the x axis

<u>Vertical Expansion and Contraction:</u>

$y = Af(x)$ $\begin{cases} A > 1 & \text{Vertically expand graph of } y = f(x) \\ & \text{by multiplying each ordinate value by } A \\ \\ 0 < A < 1 & \text{Vertically contract graph of } y = f(x) \\ & \text{by multiplying each ordinate value by } A \end{cases}$

---

1. Domain: all real numbers; Range: all real numbers

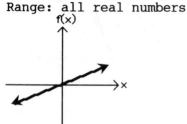

3. Domain: all real numbers; Range: $(-\infty, 0]$

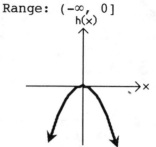

5. Domain: $[0, \infty)$; Range: $(-\infty, 0]$

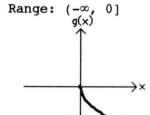

7. Domain: all real numbers; Range: all real numbers

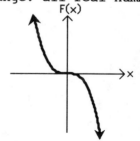

9.

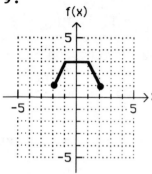

11.

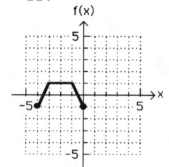

13.

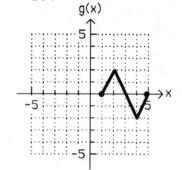

15.

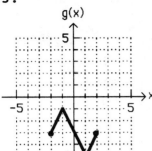

17.

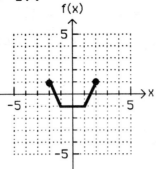

19.

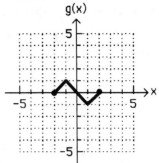

21. The graph of $g(x) = -|x + 3|$ is the graph of $y = |x|$ reflected in the x axis and shifted 3 units to the left.

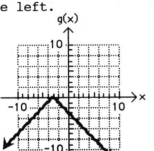

23. The graph of $f(x) = (x - 4)^2 - 3$ is the graph of $y = x^2$ shifted 4 units to the right and 3 units down.

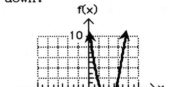

25. The graph of $f(x) = 7 - \sqrt{x}$ is the graph of $y = \sqrt{x}$ reflected in the x axis and shifted 7 units up.

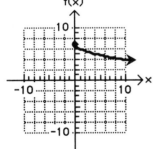

27. The graph of $h(x) = -3|x|$ is the graph of $y = |x|$ reflected in the x axis and vertically expanded by a factor of 3.

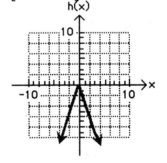

29. The graph of the basic function $y = x^2$ is shifted 2 units to the left and 3 units down. Equation: $y = (x + 2)^2 - 3$.

31. The graph of the basic function $y = x^2$ is reflected in the x axis, shifted 3 units to the right and 2 units up. Equation: $y = 2 - (x - 3)^2$.

33. The graph of the basic function $y = \sqrt{x}$ is reflected in the x axis and shifted 4 units up. Equation: $y = 4 - \sqrt{x}$.

35. The graph of the basic function $y = x^3$ is shifted 2 units to the left and 1 unit down. Equation: $y = (x + 2)^3 - 1$.

37. $g(x) = \sqrt{x - 2} - 3$ **39.** $g(x) = -|x + 3|$ **41.** $g(x) = -(x - 2)^3 - 1$

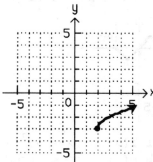 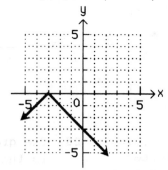 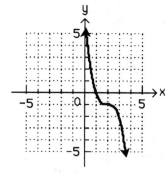

43. The graph of the basic function: $y = |x|$ is reflected in the x axis and has a vertical contraction by the factor 0.5. Equation: $y = -0.5|x|$.

45. The graph of the basic function $y = x^2$ is reflected in the x axis and is vertically expanded by the factor 2. Equation: $y = -2x^2$.

47. The graph of the basic function $y = \sqrt[3]{x}$ is reflected in the x axis and is vertically expanded by the factor 3. Equation: $y = -3\sqrt[3]{x}$.

49. (A) The graph of the basic function $y = \sqrt{x}$ is reflected in the x axis, vertically expanded by a factor of 4, and shifted up 115 units.

    (B)

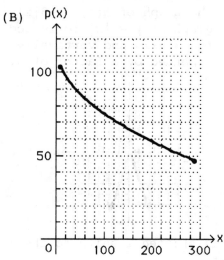

51. (A) The graph of the basic function $y = x^3$ is vertically contracted by a factor of 0.00048 and shifted right 500 units and up 60,000 units.

(B)

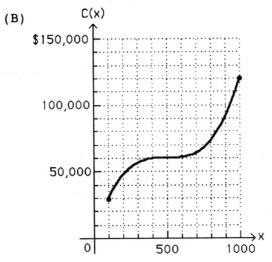

53. (A) The graph of the basic function $y = x$ is vertically expanded by a factor of 5.5 and shifted down 220 units.

(B)

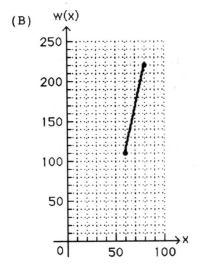

55. (A) The graph of the basic function $y = \sqrt{x}$ is vertically expanded by a factor of 7.08.

(B)

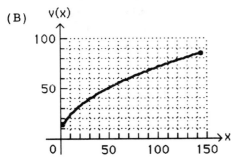

Things to remember:

1. INTERCEPTS

    If the graph of a function f crosses the x axis at a point with
    x coordinate a, then a is called an **x intercept** of f. If the
    graph of f crosses the y axis at a point with y coordinate b,
    then b is called the **y intercept**. The x intercepts are the
    real solutions or roots of $f(x) = 0$; if f is defined at 0, then
    $f(0)$ is the y intercept.

2. LINEAR AND CONSTANT FUNCTIONS

    A function f is a LINEAR FUNCTION if

    $$f(x) = mx + b \qquad m \neq 0$$

    where m and b are real numbers. The DOMAIN is the set of all
    real numbers and the RANGE is the set of all real numbers.
    If $m = 0$, then f is called a CONSTANT FUNCTION

    $$f(x) = b$$

    which has the set of all real numbers as its DOMAIN and the
    constant b as its RANGE.

    THE GRAPH OF A LINEAR FUNCTION IS A STRAIGHT LINE THAT IS
    NEITHER HORIZONTAL NOR VERTICAL. THE GRAPH OF A CONSTANT
    FUNCTION IS A HORIZONTAL STRAIGHT LINE.

3. GRAPH OF A LINEAR EQUATION IN TWO VARIABLES

    The graph of any equation of the form

    $$AX + By = C \qquad \text{Standard Form} \tag{5}$$

    where A, B, and C are real constants (A and B not both 0) is a
    straight line. Every straight line in a Cartesian coordinate
    system is the graph of an equation of this type. Vertical and
    horizontal lines have particularly simple equations, which are
    special cases of equation (5):

    Horizontal line with y intercept b: $y = b$
    Vertical line with x intercept a: $\quad x = a$

<u>**4.**</u> SLOPE OF A LINE

If a line passes through two distinct points $P_1(x_1, y_1)$ and $P_2(x_2, y_2)$, then its slope is given by the formula

$$m = \frac{y_2 - y_1}{x_2 - x_1} \qquad x_1 \neq x_2$$

$$= \frac{\text{Vertical change (rise)}}{\text{Horizontal change (run)}}$$

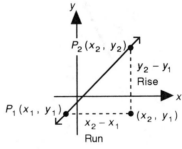

GEOMETRIC INTERPRETATION OF SLOPE

Line	Slope	Example
Rising as x moves from left to right	Positive	
Falling as x moves from left to right	Negative	
Horizontal	0	
Vertical	Not defined	

<u>**5.**</u> The equation
$$y = mx + b \qquad m = \text{slope, } b = y \text{ intercept}$$
is called the SLOPE-INTERCEPT FORM of an equation of a line.

<u>**6.**</u> An equation of the line with slope m that passes through (x_1, y_1) is:
$$y - y_1 = m(x - x_1)$$
This equation is called the POINT-SLOPE FORM of an equation of a line.

1. (d)

3. (c); The slope is 0.

5. $y = 2x - 3$

x	y
0	-3
1	-1
4	5

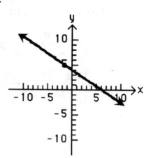

7. $2x + 3y = 12$

x	y
0	4
6	0
9	-2

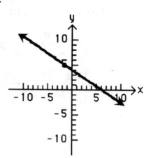

9. Slope $m = 2$

y intercept $b = -3$

11. Slope $m = -\dfrac{2}{3}$

y intercept $b = 2$

13. $m = -2$

$b = 4$

Using $\underline{5}$, $y = -2x + 4$.

15. $m = -\dfrac{3}{5}$

$b = 3$

Using $\underline{5}$, $y = -\dfrac{3}{5}x + 3$.

17. $y = -\dfrac{2}{3}x - 2$

$m = -\dfrac{2}{3}$, $b = -2$

x	y
0	-2
3	-4
-3	0

19. $3x - 2y = 10$

x	y
0	-5
10	10
-4	-11

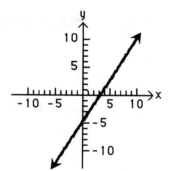

21.

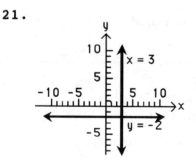

23. $3x + y = 5$

$y = -3x + 5$

$m = -3$ (using $\underline{5}$)

25. $2x + 3y = 12$

$3y = -2x + 12$

Divide both sides by 3:

$y = -\dfrac{2}{3}x + \dfrac{12}{3} = -\dfrac{2}{3}x + 4$

$m = -\dfrac{2}{3}$ (using $\underline{5}$)

27. (A)

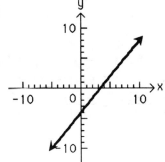

(B) x intercept--set $f(x) = 0$: $1.2x - 4.2 = 0$

$x = 3.5$

y intercept--set $x = 0$: $y = -4.2$

(C)

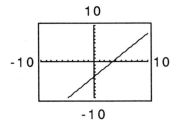

(D) x intercept: 3.5; y intercept: -4.2

(E) $x > 3.5$ or $(3.5, \infty)$

29. Using $\underline{3}$ with $a = 3$ for the vertical line and $b = -5$ for the horizontal line, we find that the equation of the vertical line is $x = 3$ and the equation of the horizontal line is $y = -5$.

31. Using $\underline{3}$ with $a = -1$ for the vertical line and $b = -3$ for the horizontal line, we find that the equation of the vertical line is $x = -1$ and the equation of the horizontal line is $y = -3$.

33. $m = -3$
For the point $(4, -1)$, $x_1 = 4$ and $y_1 = -1$. Using $\underline{6}$, we get:
$$y - (-1) = -3(x - 4)$$
$$y + 1 = -3x + 12$$
$$y = -3x + 11$$

35. $m = \dfrac{2}{3}$
For the point $(-6, -5)$, $x_1 = -6$ and $y_1 = -5$. Using $\underline{6}$, we get:
$$y - (-5) = \frac{2}{3}[x - (-6)]$$
$$y + 5 = \frac{2}{3}(x + 6)$$
$$y + 5 = \frac{2}{3}x + 4$$
$$y = \frac{2}{3}x - 1$$

37. $y - (-5) = 0(x - 3)$

   $y + 5 = 0$ or $y = -5$

   $y = 0x - 5$

39. The points are $(1, 3)$ and $(7, 5)$. Let $x_1 = 1$, $y_1 = 3$, $x_2 = 7$, and $y_2 = 5$. Using $\underline{4}$, we get:

$$m = \frac{5 - 3}{7 - 1} = \frac{2}{6} = \frac{1}{3}$$

41. Let $x_1 = -5$, $y_1 = -2$, $x_2 = 5$, and $y_2 = -4$. Using $\underline{4}$, we get:

$$m = \frac{-4 - (-2)}{5 - (-5)} = \frac{-4 + 2}{5 + 5} = \frac{-2}{10} = -\frac{1}{5}$$

43. $m = \dfrac{-3 - 7}{2 - 2} = \dfrac{-10}{0}$, the slope is not defined; the line through $(2, 7)$ and $(2, -3)$ is vertical.

45. $m = \dfrac{3 - 3}{-5 - 2} = \dfrac{0}{-7} = 0$

47. First, find the slope using $\underline{4}$:

$$m = \frac{y_2 - y_1}{x_2 - x_1} = \frac{5 - 3}{7 - 1} = \frac{2}{6} = \frac{1}{3}$$

Then, by using $\underline{6}$, $y - y_1 = m(x - x_1)$, where $m = \dfrac{1}{3}$ and $(x_1, y_1) = (1, 3)$ or $(7, 5)$, we get:

$$y - 3 = \frac{1}{3}(x - 1) \quad \text{or} \quad y - 5 = \frac{1}{3}(x - 7)$$

These two equations are equivalent. After simplifying either one of these, we obtain:

$-x + 3y = 8$ or $x - 3y = -8$

49. First, find the slope using $\underline{4}$:

$$m = \frac{-4 - (-2)}{5 - (-5)} = \frac{-4 + 2}{5 + 5} = \frac{-2}{10} = -\frac{1}{5}$$

By using $\underline{6}$, and either one of these points, we obtain:

$$y - (-2) = -\frac{1}{5}[x - (-5)] \quad \text{[using } (-5, -2)\text{]}$$

$$y + 2 = -\frac{1}{5}(x + 5)$$

$5(y + 2) = -x - 5$

$5y + 10 = -x - 5$

$x + 5y = -15$

51. $(2, 7)$ and $(2, -3)$

Since each point has the same x coordinate, the graph of the line formed by these two points will be a *vertical line*. Then, using $\underline{3}$, with $a = 2$, we have $x = 2$ as the equation of the line.

53. $(2, 3)$ and $(-5, 3)$

Since each point has the same y coordinate, the graph of the line formed by these two points will be a *horizontal line*. Then, using $\underline{3}$, with $b = 3$, we have $y = 3$ as the equation of the line.

55. A linear function

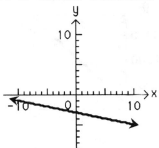

57. Not a function

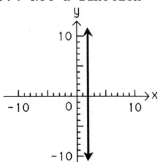

59. A constant function

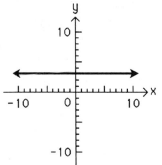

61. (A)

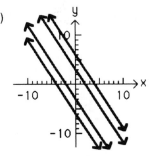

(B) Varying C produces a family of parallel lines. This is verified by observing that varying C does not change the slope of the lines but changes the intercepts.

63. $A = Prt + P$ (1)
Rate $r = 0.06$
Principal $P = 100$
Substituting in (1), we get:
$A = 6t + 100$ (2)

(A) Let $t = 5$ and $t = 20$ and substitute in (2). We get:
$A = 6(5) + 100 = \$130$
$A = 6(20) + 100 = \$220$

(B)

t	A
0	100
10	160
20	220

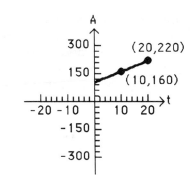

(C) Consider two points $(10, 160)$ and $(20, 220)$. Using $\underline{4}$, we have:
$$m = \frac{220 - 160}{20 - 10} = \frac{60}{10} = 6$$

65. (A) We find an equation $C(x) = mx + b$ for the line passing through $(0, 200)$ and $(20, 3800)$.
$$m = \frac{3800 - 200}{20 - 0} = \frac{3600}{20} = 180$$
Also, since $C(x) = 200$ when $x = 0$, it follows that $b = 200$.
Thus, $C(x) = 180x + 200$.

(B) The total costs at 12 boards per day
are:
$$C(x) = 180(12) + 200 = 2,360 \text{ or } \$2,360$$

(C)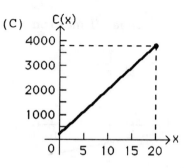

67. (A)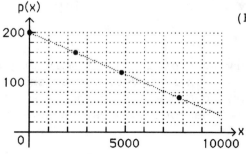

(B) slope: $m = \dfrac{160 - 200}{2,400 - 0} = \dfrac{-40}{2400} = -\dfrac{1}{60}$

y intercept: 200

equation: $p(x) = -\dfrac{1}{60}x + 200$

(C) $p(3000) = -\dfrac{1}{60}(3000) + 200 = -50 + 200 = \$150$

69. Mix A contains 20% protein. Mix B contains 10% protein. Let x be the amount of A used, and let y be the amount of B used. Then $0.2x$ is the amount of protein from mix A and $0.1y$ is the amount of protein from mix B. Thus, the linear equation is:

$0.2x + 0.1y = 20$

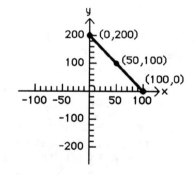

The table shows different combinations of mix A and mix B to provide 20 grams of protein.

[Note: We can get many more combinations. In fact, each point on the graph indicates a combination of mix A and mix B.]

Mix A	Mix B
x	y
100	0
0	200
50	100
10	180

71. $p = -\dfrac{1}{5}d + 70$, $30 \le d \le 175$, where d = distance in centimeters and p = pull in grams

(A) $d = 30$ $d = 175$

$p = -\dfrac{1}{5}(30) + 70 = 64$ grams $p = -\dfrac{1}{5}(175) + 70 = 35$ grams

(B)

d	p
30	64
50	60
175	35

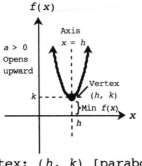

(C) Select two points $(30, 64)$ and $(50, 60)$ as (x_1, y_1) and (x_2, y_2), respectively, from part (B). Using $\underline{2}$:

$$\text{Slope } m = \frac{y_2 - y_1}{x_2 - x_1} = \frac{60 - 64}{50 - 30}$$

$$= -\frac{4}{20} = -\frac{1}{5}$$

EXERCISE 1-4

Things to remember:

$\underline{1}$.   QUADRATIC FUNCTION

A function f is a QUADRATIC FUNCTION if
$$f(x) = ax^2 + bx + c \qquad a \neq 0$$
where a, b, and c are real numbers. The domain of a quadratic function is the set of all real numbers.

$\underline{2}$.   PROPERTIES OF A QUADRATIC FUNCTION AND ITS GRAPH

Given a quadratic function
$$f(x) = ax^2 + bx + c \qquad a \neq 0$$
and the form obtained by completing the square
$$f(x) = a(x - h)^2 + k$$
we summarize general properties as follows:

a. The graph of f is a parabola:

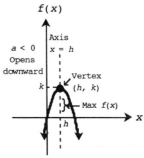

b. Vertex: (h, k) [parabola increases on one side of the vertex and decreases on the other]

c. Axis (of symmetry): $x = h$ (parallel to y axis)

d. $f(h) = k$ is the minimum if $a > 0$ and the maximum if $a < 0$

e. Domain: All real numbers
   Range: $(-\infty, k]$ if $a < 0$ or $[k, \infty)$ if $a > 0$

f. The graph of f is the graph of $g(x) = ax^2$ translated horizontally h units and vertically k units.

1. (a), (c), (e), (f)   **3.** (A) m   (B) g   (C) f   (D) n

5. (A) x intercepts: 1, 3; y intercept: -3   (B) Vertex: (2, 1)
(C) Maximum: 1   (D) Range: $y \le 1$ or $(-\infty, 1]$
(E) Increasing interval: $x \le 2$ or $(-\infty, 2]$
(F) Decreasing interval: $x \ge 2$ or $[2, \infty)$

7. (A) x intercepts: -3, -1; y intercept: 3   (B) Vertex: (-2, -1)
(C) Minimum: -1   (D) Range: $y \ge -1$ or $[-1, \infty)$
(E) Increasing interval: $x \ge -2$ or $[-2, \infty)$
(F) Decreasing interval: $x \le -2$ or $(-\infty, -2]$

9. $f(x) = -(x - 2)^2 + 1 = -x^2 + 4x - 4 + 1 = -x^2 + 4x - 3 = -(x - 3)(x - 1)$
(A) x intercepts: 1, 3; y intercepts: -3   (B) Vertex: (2, 1)
(C) Maximum: 1   (D) Range: $y \le 1$ or $(-\infty, 1]$

11. $M(x) = (x + 2)^2 - 1 = x^2 + 4x + 4 - 1 = x^2 + 4x + 3 = (x + 3)(x + 1)$
(A) x intercepts: -3, -1; y intercept 3   (B) Vertex: (-2, -1)
(C) Minimum: -1   (D) Range: $[-1, \infty)$

13. $y = -[x - (-2)]^2 + 5 = -(x + 2)^2 + 5$

15. $y = (x - 1)^2 - 3$

17. $f(x) = x^2 - 8x + 13 = x^2 - 8x + 16 - 3 = (x - 4)^2 - 3$
(A) x intercepts: $(x - 4)^2 - 3 = 0$
$$(x - 4)^2 = 3$$
$$x - 4 = \pm\sqrt{3}$$
$$x = 4 + \sqrt{3} \approx 5.7, \; 4 - \sqrt{3} \approx 2.3$$
y intercept: 13
(B) Vertex: (4, -3)   (C) Minimum: -3   (D) Range: $y \ge -3$ or $[-3, \infty)$

19. $M(x) = 1 - 6x - x^2 = -(x^2 + 6x + 9) + 1 + 9 = -(x + 3)^2 + 10$
(A) x intercepts: $-(x + 3)^2 + 10 = 0$
$$(x + 3)^2 = 10$$
$$x + 3 = \pm\sqrt{10}$$
$$x = -3 + \sqrt{10} \approx 0.2, \; -3 - \sqrt{10} = -6.2$$
y intercept: 1
(B) Vertex: (-3, 10)   (C) Maximum: 10   (D) Range: $y \le 10$ or $(-\infty, 10]$

21. $G(x) = 0.5x^2 - 4x + 10 = \frac{1}{2}(x^2 - 8x + 16) + 2$
$$= \frac{1}{2}(x - 4)^2 + 2$$

(A) x intercepts: none, since $G(x) = \frac{1}{2}(x - 4)^2 + 2 \ge 2$ for all x;
y intercept: 10

(B) Vertex: (4, 2)   (C) Minimum: 2   (D) Range: $y \ge 2$ or $[2, \infty)$

23. The vertex of the parabola is on the x axis.

25. $g(x) = 0.25x^2 - 1.5x - 7 = 0.25(x^2 - 6x + 9) - 2.25 - 7$
$$= 0.25(x - 3)^2 - 9.25$$

(A) x intercepts: $0.25(x - 3)^2 - 9.25 = 0$
$$(x - 3)^2 = 37$$
$$x - 3 = \pm\sqrt{37}$$
$$x = 3 + \sqrt{37} \approx 9.1, \ 3 - \sqrt{37} \approx -3.1$$

y intercept: -7
(B) Vertex: $(3, -9.25)$ (C) Minimum: -9.25
(D) Range: $y \geq -9.25$ or $[-9.25, \infty)$

27. $f(x) = -0.12x^2 + 0.96x + 1.2$
$$= -0.12(x^2 - 8x + 16) + 1.92 + 1.2$$
$$= -0.12(x - 4)^2 + 3.12$$

(A) x intercepts: $-0.12(x - 4)^2 + 3.12 = 0$
$$(x - 4)^2 = 26$$
$$x - 4 = \pm\sqrt{26}$$
$$x = 4 + \sqrt{26} \approx 9.1, \ 4 - \sqrt{26} \approx -1.1$$

y intercept: 1.2
(B) Vertex: $(4, 3.12)$ (C) Maximum: 3.12
(D) Range: $y \leq 3.12$ or $(-\infty, 3.12]$

29. 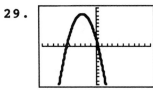 $x = -5.37, \ 0.37$

$-10 \leq x \leq 10$ xscl = 1
$-10 \leq y \leq 10$ yscl = 1

31. 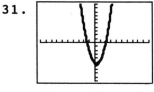 $-1.37 < x < 2.16$

$-10 \leq x \leq 10$ xscl = 1
$-10 \leq y \leq 10$ yscl = 1

33. 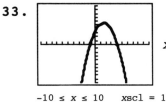 $x \leq -0.74$ or $x \geq 4.19$

$-10 \leq x \leq 10$ xscl = 1
$-10 \leq y \leq 10$ yscl = 1

35. (A)

(B) $f(x) = g(x)$

$$-0.4x(x - 10) = 0.3x + 5$$
$$-0.4x^2 + 4x = 0.3x + 5$$
$$-0.4x^2 + 3.7x = 5$$
$$-0.4x^2 + 3.7x - 5 = 0$$

$$x = \frac{-3.7 \pm \sqrt{3.7^2 - 4(-0.4)(-5)}}{2(-0.4)}$$

$$x = \frac{-3.7 \pm \sqrt{5.69}}{-0.8} \approx 1.64, \ 7.61$$

(C) $f(x) > g(x)$ for $1.64 < x < 7.61$
(D) $f(x) < g(x)$ for $0 \le x < 1.64$ or $7.61 < x \le 10$

37. (A)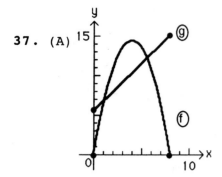

(B) $f(x) = g(x)$

$$-0.9x^2 + 7.2x = 1.2x + 5.5$$
$$-0.9x^2 + 6x = 5.5$$
$$-0.9x^2 + 6x - 5.5 = 0$$

$$x = \frac{-6 \pm \sqrt{36 - 4(-0.9)(-5.5)}}{2(-0.9)}$$

$$x = \frac{-6 \pm \sqrt{16.2}}{-1.8} \approx 1.1, \ 5.57$$

(C) $f(x) > g(x)$ for $1.10 < x < 5.57$
(D) $f(x) < g(x)$ for $0 \le x < 1.10$ or $5.57 < x \le 8$

39. $f(x) = x^2 + 1$ and $g(x) = -(x - 4)^2 - 1$ are two examples. Their graphs are:

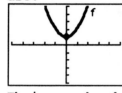 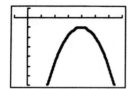

Their graphs do not intersect the x axis.

41. (A)

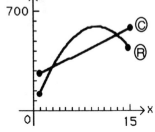

(B) $R(x) = x(119 - 6x) = -6x^2 + 119x$

$$= -6\left(x^2 - \frac{119}{6}x\right)$$

$$= -6\left(x^2 - 19.833x + 98.340\right) + 590.042$$

$$= -6(x - 9.917)^2 + 590.042$$

Output: 9.917 million chips, i.e., 9,917,000 chips
Maximum revenue: 590.042 million dollars, i.e. $590,042,000

(C)

(D) 9.917 million chips (9,917,000 chips)
590.042 million dollars ($590,042,000)

(E) $p(9.917) = 119 - 6(9.917) \approx \$59$

43. (A)

(B) $R(x) = C(x)$

$$x(119 - 6x) = 234 + 23x$$
$$-6x^2 + 96x = 234$$
$$x^2 - 16x = -39$$
$$x^2 - 16x + 39 = 0$$
$$(x - 13)(x - 3) = 0$$
$$x = 13, \ 3$$

Break-even at 3 million 3,000,000 and 13 million (13,000,000) chip
production levels.

(C)

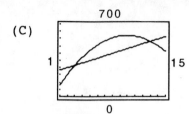

(D) Break-even at 3 million and 13 million chip production levels
(E) Loss: $1 \le x < 3$ or $13 < x \le 15$ Profit: $3 < x < 13$

45. (A) Solve: $f(x) = 1{,}000(0.04 - x^2) = 20$
$$40 - 1000x^2 = 20$$
$$1000x^2 = 20$$
$$x^2 = 0.02$$
$$x = 0.14 \text{ or } -0.14$$

Since we are measuring distance, we take the positive solution:
$x = 0.14$ cm

(B) $x = 0.14$ cm

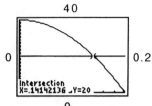

CHAPTER 1 REVIEW

1.

(1-1)

2. (A) Not a function; fails vertical line test   (B) A function
(C) A function   (D) Not a function; fails vertical line test   (1-1)

3. $f(x) = 2x - 1$, $g(x) = x^2 - 2x$
(A) $f(-2) + g(-1) = 2(-2) - 1 + (-1)^2 - 2(-1) = -2$
(B) $f(0) \cdot g(4) = (2 \cdot 0 - 1)(4^2 - 2 \cdot 4) = -8$
(C) $\dfrac{g(2)}{f(3)} = \dfrac{2^2 - 2 \cdot 2}{2 \cdot 3 - 1} = 0$
(D) $\dfrac{f(3)}{g(2)}$ not defined because $g(2) = 0$ (1-1)

4. (A) $y = 4$ (B) $x = 0$ (C) $y = 1$ (D) $x = -1$ or 1
(E) $y = -2$ (F) $x = -5$ or 5 (1-1)

5. (A) 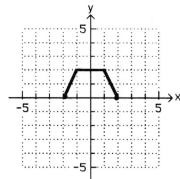 (B)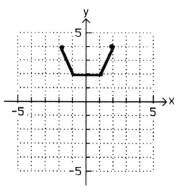

(C) 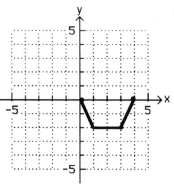 (D)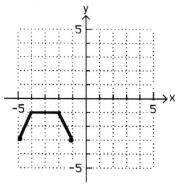

(1-2)

6. (A) (n) (B) (g) (C) (m); slope is zero
(D) (f); slope is not defined (1-3)

7. $y = -\dfrac{2}{3}x + 6$ (1-3)

8. vertical line: $x = -6$; horizontal line: $y = 5$ (1-3)

9. x intercept: $2x = 18$, $x = 9$;
y intercept: $-3y = 18$, $x = -6$;
slope-intercept form: $y = \dfrac{2}{3}x - 6$; slope $= \dfrac{2}{3}$
graph:

(1-3)

10. (b), (c), (d), (f) (1-4)

11. (A) g (B) m (C) n (D) f (1-2, 1-4)

12. $y = f(x) = (x + 2)^2 - 4$

    (A) x intercepts: $(x + 2)^2 - 4 = 0$

$$(x + 2)^2 = 4$$
$$x + 2 = -2 \text{ or } 2$$
$$x = -4, \ 0$$

       y intercept: 0

    (B) Vertex: $(-2, -4)$ (C) Minimum: -4 (D) Range: $y \geq -4$ or $[-4, \infty)$

    (E) Increasing interval $[-2, \infty)$ (F) Decreasing interval $(-\infty, -2]$

                                                                                      (1-4)

13. Linear function: (a), (c), (e), (f); Constant function: (d)   (1-3)

14. (A) $x^2 - x - 6 = 0$ at $x = -2, \ 3$

        Domain: all real numbers except $x = -2, \ 3$

    (B) $5 - x > 0$ for $x < 5$

        Domain: $x < 5$ or $(-\infty, 5)$ (1-1)

15. Function g multiplies a domain element by 2 and then subtracts three times the square root of the domain element from the result.   (1-1)

16. The graph of $x = -3$ is a vertical line 3 units to the *left* of the y axis; $y = 2$ is a horizontal line 2 units *above* the x axis.

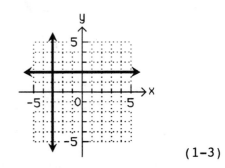

                                                                         (1-3)

17.

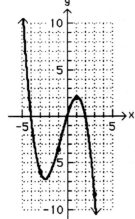

                    (1-1)

18. $f(x) = 3 - 2x$

$$\frac{f(2 + h) - f(2)}{h} = \frac{3 - 2(2 + h) - (3 - 2 \cdot 2)}{h}$$

$$= \frac{3 - 4 - 2h - 3 + 4}{h}$$

$$= \frac{-2h}{h}$$

$$= -2 \qquad\qquad (1-1)$$

19. $f(x) = x^2 - 3x + 1$

$$\frac{f(a + h) - f(a)}{h} = \frac{(a + h)^2 - 3(a + h) + 1 - (a^2 - 3a + 1)}{h}$$

$$= \frac{a^2 + 2ah + h^2 - 3a - 3h + 1 - a^2 + 3a - 1}{h}$$

$$= \frac{2ah + h^2 - 3h}{h}$$

$$= \frac{(2a + h - 3)h}{h}$$

$$= 2a + h - 3 \qquad (1\text{-}1)$$

20. The graph of m is the graph of $y = |x|$ reflected on the x axis and shifted 4 units to the right. $\qquad (1\text{-}2)$

21. The graph of g is the graph of $y = x^3$ vertically contracted by a factor of 0.3 and shifted up 3 units. $\qquad (1\text{-}2)$

22. The graph of $y = x^2$ is vertically expanded by a factor of 2, reflected in the x axis and shifted to the left 3 units. Equation: $y = -2(x + 3)^2$ $\qquad (1\text{-}2)$

23. Equation: $f(x) = 2\sqrt{x + 3} - 1$

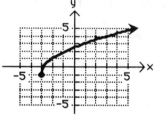

$\qquad (1\text{-}2)$

24. Use the point-slope form:

(A) $y - 2 = -\frac{2}{3}[x - (-3)]$ \qquad (B) $y - 3 = 0(x - 3)$

$\qquad\quad y - 2 = -\frac{2}{3}(x + 3)$ \qquad\qquad\quad $y = 3$

$\qquad\qquad\quad y = -\frac{2}{3}x$ $\qquad\qquad\qquad\qquad\qquad (1\text{-}3)$

25. (A) Slope: $\dfrac{-1 - 5}{1 - (-3)} = -\dfrac{3}{2}$ \qquad (B) Slope: $\dfrac{5 - 5}{4 - (-1)} = 0$

$\qquad\qquad\quad y - 5 = -\frac{3}{2}(x + 3)$ \qquad\qquad\qquad $y - 5 = 0(x - 1)$

$\qquad\qquad\quad 3x + 2y = 1$ \qquad\qquad\qquad\qquad\quad $y = 5$

(C) Slope: $\dfrac{-2 - 7}{-2 - (-2)}$ not defined since $2 - (-2) = 0$

$\qquad x = -2$ $\qquad\qquad\qquad\qquad\qquad\qquad\qquad (1\text{-}3)$

26. $y = -(x - 4)^2 + 3$ $\qquad\qquad\qquad\qquad\qquad\qquad (1\text{-}2, 1\text{-}4)$

27. $f(x) = -0.4x^2 + 3.2x - 1.2 = -0.4(x^2 - 8x + 16) + 7.6$
$$= -0.4(x - 4)^2 + 7.6$$

(A) y intercept: 1.2
   x intercepts: $-0.4(x - 4)^2 + 7.6 = 0$
$$(x - 4)^2 = 19$$
$$x = 4 + \sqrt{19} \approx 8.4, \ 4 - \sqrt{19} \approx -0.4$$

(B) Vertex: (4.0, 7.6)   (C) Maximum: 7.6

(D) Range: $x \leq 7.6$ or $(-\infty, 7.6]$ (1-4)

28.

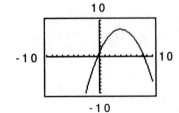

(A) y intercept: 1.2
   x intercepts: -0.4, 8.4

(B) Vertex: (4.0, 7.6)

(C) Maximum: 7.6

(D) Range: $x \leq 7.6$ or $(-\infty, 7.6]$ (1-4)

29. The graph of $y = \sqrt[3]{x}$ is vertically expanded by a factor of 2, reflected in the x axis, shifted 1 unit to the left and 1 unit down.

Equation: $y = -2\sqrt[3]{x + 1} - 1$ (1-2)

30. The graphs of the pairs $\{y = 2x, \ y = -\frac{1}{2}x\}$ and

$\{y = \frac{2}{3}x + 2, \ y = -\frac{3}{2}x + 2\}$ are shown below:

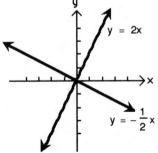 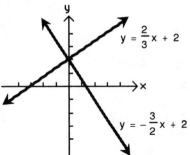

In each case, the graphs appear to be perpendicular to each other. It can be shown that two slant lines are perpendicular if and only if their slopes are negative reciprocals.   (1-3)

31. (A) $f(x) = \sqrt{x}$

$$\frac{f(x + h) - f(x)}{h} = \frac{\sqrt{x + h} - \sqrt{x}}{h} = \frac{\sqrt{x + h} - \sqrt{x}}{h} \cdot \frac{\sqrt{x + h} + \sqrt{x}}{\sqrt{x + h} + \sqrt{x}} \ \text{rationalize}$$
$$\text{the numerator}$$

$$= \frac{x + h - x}{h[\sqrt{x + h} + \sqrt{x}]} = \frac{1}{\sqrt{x + h} + \sqrt{x}}$$

(B) $f(x) = \dfrac{1}{x}$

$$\dfrac{f(x+h) - f(x)}{h} = \dfrac{\dfrac{1}{x+h} - \dfrac{1}{x}}{h} = \dfrac{\dfrac{x - (x+h)}{x(x+h)}}{h}$$

$$= \dfrac{-h}{hx(x+h)} = \dfrac{-1}{x(x+h)} \qquad (1\text{-}2)$$

32. $G(x) = 0.3x^2 + 1.2x - 6.9 = 0.3(x^2 + 4x + 4) - 8.1$
$$= 0.3(x+2)^2 - 8.1$$

(A) *y* intercept: -6.9
   x intercepts: $0.3(x+2)^2 - 8.1 = 0$
$$(x+2)^2 = 27$$
$$x = -2 + \sqrt{27} \approx 3.2, \ -2 - \sqrt{27} \approx -7.2$$

(B) Vertex: $(-2, -8.1)$ (C) Minimum: -8.1
(D) Range: $x \geq -8.1$ or $[-8.1, \infty)$
(E) Decreasing: $(-\infty, -2]$; Increasing: $[-2, \infty)$ (1-4)

33.

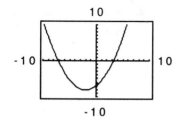

(A) *y* intercept: -6.9
   x intercept: -7.2, 3.2
(B) Vertex: $(-2, -8.1)$
(C) Minimum: -8.1
(D) Range: $x \geq -8.1$ or $[-8.1, \infty)$
(E) Decreasing: $(-\infty, -2]$
   Increasing: $[-2, \infty)$ (1-4)

34. (A) $V(0) = 12,000$, $V(8) = 2,000$
   Slope: $\dfrac{2,000 - 12,000}{8 - 0} = \dfrac{-10,000}{8} = -1,250$
   V intercept: 12,000
   Equation: $V(t) = -1,250t + 12,000$

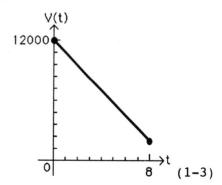

(B) $V(5) = -1,250(5) + 12,000 = \$5,750$

                (1-3)

35. (A)

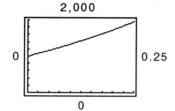

(B) $r = 0.1447$ or 14.7% compounded annually
   Alternative algebraic solution:
$$1000(1 + r)^3 = 1500$$
$$(1 + r)^3 = 1.5$$
$$1 + r = \sqrt[3]{1.5} \approx 1.1447$$
$$r = 0.1447$$

                (1-1, 1-2)

36. (A) $R(130) = 208$, $R(50) = 80$

Slope: $\dfrac{208 - 80}{130 - 50} = \dfrac{128}{80} = 1.6$

Equation: $R - 80 = 1.6(C - 50)$ or $R = 1.6C$

(B) $R(120) = 1.6(120) = \$192$

(C) $176 = 1.6C$; $C = \$110$

(D) 1.6; The slope gives the change in retail price per unit change in the cost. (1-3)

37. (A) Let x = number of video tapes produced.

$C(x) = 84{,}000 + 15x$

$R(x) = 50x$

(B) $R(x) = C(x)$

$50x = 84{,}000 + 15x$

$35x = 84{,}000$

$x = 2{,}400$ units

$R < C$ for $0 \le x < 2{,}400$; $R > C$ for $x > 2{,}400$

(C) $R = C$ at $x = 2{,}400$ units

$R < C$ for $0 \le x < 2{,}400$; $R > C$ for $x > 2{,}400$ (1-3)

38. $p(x) = 50 - 1.25x$ Price-demand function

$C(x) = 160 + 10x$ Cost function

$R(x) = xp(x)$

$\qquad = x(50 - 1.25x)$ Revenue function

(A)

(B) $R = C$

$$x(50 - 1.25x) = 160 + 10x$$
$$-1.25x^2 + 50x = 160 + 10x$$
$$-1.25x^2 + 40x = 160$$
$$-1.25(x^2 - 32x + 256) = 160 - 320$$
$$-1.25(x - 16)^2 = -160$$
$$(x - 16)^2 = 128$$
$$x = 16 + \sqrt{128} \approx 27.314, \ 16 - \sqrt{128} \approx 4.686$$

$R = C$ at $x = 4.686$ thousand units (4,686 units) and

$x = 27.314$ thousand units (27,314 units)

$R < C$ for $1 \le x < 4.686$ or $27.314 < x \le 40$

$R > C$ for $4.686 < x < 27.314$

(C) Max Rev: $50x - 1.25x^2 = R$

$-1.25(x^2 - 40x + 400) + 500 = R$

$-1.25(x - 20)^2 + 500 = R$

Vertex at (20, 500)

Max. Rev. = 500 thousand ($500,000) occurs when <u>output</u> is 20 thousand (20,000 units)

<u>Wholesale price</u> at this output: $p(x) = 50 - 1.25x$

$$p(20) = 50 - 1.25(20)$$
$$= \$25 \qquad (1\text{-}3, \ 1\text{-}4)$$

39. (A) $P(x) = R(x) - C(x) = x(50 - 1.25x) - (160 + 10x)$

$$= -1.25x^2 + 40x - 160$$

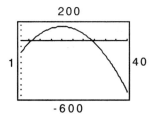

(B) $P = 0$ for $x = 4.686$ thousand uits (4,686 units) and $x = 27.314$ thousand units (27,314 units)

$P < 0$ for $1 \leq x < 4.686$ or $27.314 < x \leq 40$

$P > 0$ for $4.686 < x < 27.314$

(C) Maximum profit is 160 thousand dollars ($160,000), and this occurs at $x = 16$ thousand units (16,000 units). The wholesale price at this output is $p(16) = 50 - 1.25(16) = \$30$, which is $5 greater than the $25 found in 38(C). $\qquad (1\text{-}4)$

40. (A) The area A enclosed by the pens is given by

$$A = (2y)x$$

Now, $3x + 4y = 840$

and $\qquad y = 210 - \dfrac{3}{4}x$

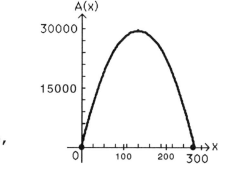

Thus $\qquad A(x) = 2\left(210 - \dfrac{3}{4}x\right)x$

$$= 420x - \dfrac{3}{2}x^2$$

(B) Since x and y must both be nonnegative,

$$210 - \dfrac{3}{4}x \geq 0$$

$$-\dfrac{3}{4}x \geq -210$$

$$x \leq 280$$

Domain: $0 \leq x \leq 280$

(C) Maximum combined area is 29,400 feet. This occurs at $x = 140$ feet, $y = 105$ feet. $\qquad (1\text{-}4)$

41. (A) We are given $P(0) = 20$ and $m = 15$. Thus, $P(x) = 15x + 20$

(B) 1 PM is 5 hours after 8 AM
$$P(5) = 15(5) + 20 = 95$$

(C)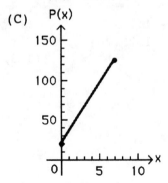

(D) Slope = 15   (1-4)

42. $\frac{\Delta s}{s} = k$.   For $k = \frac{1}{30}$, $\frac{\Delta s}{s} = \frac{1}{30}$ or $\Delta s = \frac{1}{30}s$

(A) When $s = 30$, $\Delta s = \frac{1}{30}(30) = 1$ pound.   (B) $\Delta s = \frac{1}{30}s$

   When $s = 90$, $\Delta s = \frac{1}{30}(90) = 3$ pounds.   Slope $m = \frac{1}{30}$

   y intercept $b = 0$

(C) Slope $m = \frac{1}{30}$

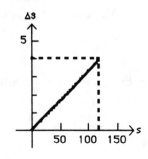

(1-4)

2 ADDITIONAL ELEMENTARY FUNCTIONS

Things to remember:

1. POLYNOMIAL FUNCTION

   A POLYNOMIAL FUNCTION is a function of the form
   $$f(x) = a_n x^n + a_{n-1} x^{n-1} + \ldots + a_1 x + a_0$$
   for n a nonnegative integer, called the DEGREE of the polynomial. The coefficients a_0, a_1, ..., a_n are real numbers with $a_n \neq 0$. The DOMAIN of a polynomial function is the set of all real numbers.

2. TURNING POINT

   A TURNING POINT on a graph is a point that separates an increasing portion from a decreasing portion, or vice versa. The graph of a polynomial function of degree $n \geq 1$ can have at most $n-1$ turning points and can cross the x axis at most n times.

3. A RATIONAL FUNCTION is any function of the form
   $$f(x) = \frac{n(x)}{d(x)} \qquad d(x) \neq 0$$
   where $n(x)$ and $d(x)$ are polynomials. The DOMAIN is the set of all real numbers such that $d(x) \neq 0$. We assume $n(x)/d(x)$ is reduced to lowest terms.

4. ASYMPTOTES OF RATIONAL FUNCTIONS

   Given the rational function
   $$f(x) = \frac{n(x)}{d(x)}$$
   where $n(x)$ and $d(x)$ are polynomials without common factors.

   (a) If a is a real number such that $d(a) = 0$, then the line $x = a$ is a VERTICAL ASYMPTOTE of the graph of $y = f(x)$.

   (b) HORIZONTAL ASYMPTOTES, if any exists, can be found by dividing each term of the numerator $n(x)$ and denominator $d(x)$ by the highest power of x that appears in the numerator and denominator.

---

1. (A) 2   (B) 1   (C) 2   (D) 0   (E) 1   (F) 1

3. (A) 5   (B) 4   (C) 5   (D) 1   (E) 1   (F) 1

5. (A) 6   (B) 5   (C) 6   (D) 0   (E) 1   (F) 1

7. (A) 3   (B) 4   (C) negative   **9.** (A) 4   (B) 5   (C) negative

11. (A) 0   (B) 1   (C) negative   **13.** (A) 5   (B) 6   (C) positive

15. $f(x) = \dfrac{x + 2}{x - 2}$

    (A) *Intercepts:*

        x intercepts: $f(x) = 0$ only if $x + 2 = 0$ or $x = -2$.
                     The x intercept is -2.

        y intercept: $f(0) = \dfrac{0 + 2}{0 - 2} = -1$
                     The y intercept is -1.

    (B) *Domain:* The denominator is 0 at $x = 2$. Thus, the domain is the set of all real numbers except 2.

    (C) *Asymptotes:*

        Vertical asymptotes: $f(x) = \dfrac{x + 2}{x - 2}$

                        The denominator is 0 at $x = 2$. Therefore, the line $x = 2$ is a vertical asymptote.

        Horizontal asymptotes: $f(x) = \dfrac{x + 2}{x - 2} = \dfrac{1 + \dfrac{2}{x}}{1 - \dfrac{2}{x}}$

                        As x increases or decreases without bound, the numerator tends to 1 and the denominator tends to 1. Therefore, the line $y = 1$ is a horizontal asymptote.

    (D)

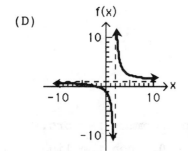

    (E)

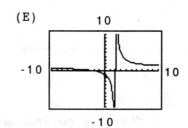

17. $f(x) = \dfrac{3x}{x + 2}$

    (A) *Intercepts:*

        x intercepts: $f(x) = 0$ only if $3x = 0$ or $x = 0$.
                     The x intercept is 0.

        y intercept: $f(0) = \dfrac{3 \cdot 0}{0 + 2} = 0$
                     The y intercept is 0.

(B) *Domain:* The denominator is 0 at $x = -2$. Thus, the domain is the set of all real numbers except -2.

(C) *Asymptotes:*

Vertical asymptotes: $f(x) = \dfrac{3x}{x + 2}$

The denominator is 0 at $x = -2$. Therefore, the line $x = -2$ is a vertical asymptote.

Horizontal asymptotes: $f(x) = \dfrac{3x}{x + 2} = \dfrac{3}{1 + \dfrac{2}{x}}$

As x increases or decreases without bound, the numerator is 3 and the denominator tends to 1. Therefore, the line $y = 3$ is a horizontal asymptote.

(D)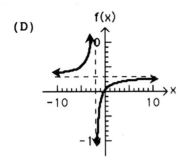

(E)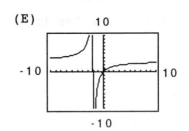

19. $f(x) = \dfrac{4 - 2x}{x - 4}$

(A) *Intercepts:*

x intercepts: $f(x) = 0$ only if $4 - 2x = 0$ or $x = 2$.
The x intercept is 2.

y intercept: $f(0) = \dfrac{4 - 2 \cdot 0}{0 - 4} = -1$
The y intercept is -1.

(B) *Domain:* The denominator is 0 at $x = 4$. Thus, the domain is the set of all real numbers except 4.

(C) *Asymptotes:*

Vertical asymptotes: $f(x) = \dfrac{4 - 2x}{x - 4}$

The denominator is 0 at $x = 4$. Therefore, the line $x = 4$ is a vertical asymptote.

Horizontal asymptotes: $f(x) = \dfrac{4 - 2x}{x - 4} = \dfrac{\dfrac{4}{x} - 2}{1 - \dfrac{4}{x}}$

As x increases or decreases without bound, the numerator tends to -2 and the denominator tends to 1. Therefore, the line $y = -2$ is a horizontal asymptote.

(D)

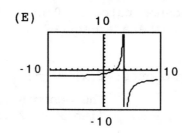

(E)

21. The graph of $f(x) = 2x^4 - 5x^2 + x + 2 = 2x^4\left(1 - \dfrac{5}{2x^2} + \dfrac{1}{2x^3} + \dfrac{1}{x^4}\right)$ will "look like" the graph of $y = 2x^4$. For large x, $f(x) \approx 2x^4$.

23. The graph of $f(x) = -x^5 + 4x^3 - 4x + 1 = -x^5\left(1 - \dfrac{4}{x^2} + \dfrac{4}{x^4} - \dfrac{1}{x^5}\right)$ will "look like" the graph of $y = -x^5$. For large x, $f(x) \approx -x^5$.

25. (A)

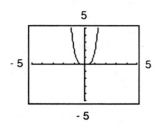

$$y = 2x^4 \qquad\qquad y = 2x^4 - 5x^2 + x + 2$$

(B)

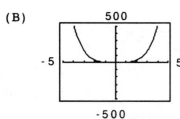
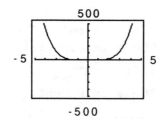

$$y = 2x^4 \qquad\qquad y = 2x^4 - 5x^2 + x + 2$$

27. (A)

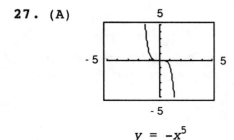
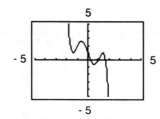

$$y = -x^5 \qquad\qquad y = -x^5 + 4x^3 - 4x + 1$$

(B)

$$y = -x^5$$

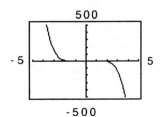

$$y = -x^5 + 4x^3 - 4x + 1$$

29. $f(x) = \dfrac{2x^2}{x^2 - x - 6}$

(A) *Intercepts:*

x intercepts: $f(x) = 0$ only if $2x^2 = 0$ or $x = 0$.
The x intercept is 0.

y intercept: $f(0) = \dfrac{2 \cdot 0^2}{0^2 - 0 - 6} = 0$
The y intercept is 0.

(B) *Asymptotes:*

Vertical asymptotes: $f(x) = \dfrac{2x^2}{x^2 - x - 6} = \dfrac{2x^2}{(x - 3)(x + 2)}$

The denominator is 0 at $x = -2$ and $x = 3$.
Thus, the lines $x = -2$ and $x = 3$ are vertical asymptotes.

Horizontal asymptotes: $f(x) = \dfrac{2x^2}{x^2 - x - 6} = \dfrac{2}{1 - \dfrac{1}{x} - \dfrac{6}{x^2}}$

As x increases or decreases without bound, the numerator is 2 and the denominator tends to 1. Therefore, the line $y = 2$ is a horizontal asymptote.

(C)

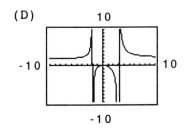

(D)

31. $f(x) = \dfrac{6 - 2x^2}{x^2 - 9}$

(A) *Intercepts:*

x intercepts: $f(x) = 0$ only if $6 - 2x^2 = 0$

$$2x^2 = 6$$
$$x^2 = 3$$
$$x = \pm\sqrt{3}$$

The x intercepts are $\pm\sqrt{3}$.

y intercept: $f(0) = \dfrac{6 - 2 \cdot 0^2}{0^2 - 9} = -\dfrac{2}{3}$

The y intercept is $-\dfrac{2}{3}$.

(B) *Asymptotes:*

Vertical asymptotes: $f(x) = \dfrac{6 - 2x^2}{x^2 - 9} = \dfrac{6 - 2x^2}{(x - 3)(x + 3)}$

The denominator is 0 at $x = -3$ and $x = 3$. Thus, the lines $x = -3$ and $x = 3$ are vertical asymptotes.

Horizontal asymptotes: $f(x) = \dfrac{6 - 2x^2}{x^2 - 9} = \dfrac{\dfrac{6}{x^2} - 2}{1 - \dfrac{9}{x^2}}$

As x increases or decreases without bound, the numerator tends to -2 and the denominator tends to 1. Therefore, the line $y = -2$ is a horizontal asymptote.

(C)

(D)

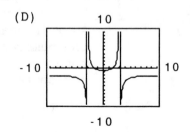

33. $f(x) = \dfrac{-4x}{x^2 + x - 6}$

(A) *Intercepts:*

x intercepts: $f(x) = 0$ only if $-4x = 0$ or $x = 0$.
The x intercept is 0.

y intercept: $f(0) = \dfrac{-4 \cdot 0}{0^2 + 0 - 6} = 0$

The y intercept is 0.

(B) *Asymptotes:*

Vertical asymptotes: $f(x) = \dfrac{-4x}{x^2 + x - 6} = \dfrac{-4x}{(x + 3)(x - 2)}$

The denominator is 0 at $x = -3$ and $x = 2$. Thus, the lines $x = -3$ and $x = 2$ are vertical asymptotes.

Horizontal asymptotes: $f(x) = \dfrac{-4x}{x^2 + x - 6} = \dfrac{-\dfrac{4}{x}}{1 + \dfrac{1}{x} - \dfrac{6}{x^2}}$

As x increases or decreases without bound, the numerator tends to 0 and the denominator tends to 1. Therefore, the line $y = 0$ (the x axis) is a horizontal asymptote.

(C)

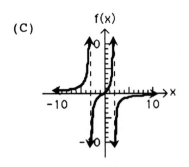

(D)

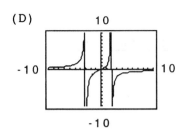

35. The graph has 1 turning point which implies degree $n = 2$. The x intercepts are $x = -1$ and $x = 2$. Thus, $f(x) = (x + 1)(x - 2) = x^2 - x - 2$.

37. The graph has 2 turning points which implies degree $n = 3$. The x intercepts are $x = -2$, $x = 0$, and $x = 2$. The direction of the graph indicates that leading coefficient is negative $f(x) = -(x + 2)(x)(x - 2) = 4x - x^3$.

39. (A) Since $C(x)$ is a linear function of x, it can be written in the form
$$C(x) = mx + b$$
Since the fixed costs are \$200, $b = 200$.
Also, $C(20) = 3800$, so
$$3800 = m(20) + 200$$
$$20m = 3600$$
$$m = 180$$
Therefore, $C(x) = 180x + 200$

(B) $\overline{C}(x) = \dfrac{C(x)}{x} = \dfrac{180x + 200}{x}$

(C)

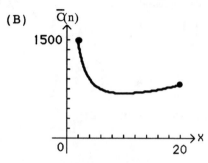

(D) $\overline{C}(x) = \dfrac{180x + 200}{x} = \dfrac{180 + \dfrac{200}{x}}{1}$

As x increases, the numerator tends to 180 and the denominator is 1. Therefore, $\overline{C}(x)$ tends to 180 or $180 per board.

41. (A) $\overline{C}(n) = \dfrac{2500 + 175n + 25n^2}{n}$

(B)

(C) Using the graph, we calculate

$C(8) = \dfrac{2500 + 175(8) + 25(8)^2}{8} = 687.50$

$C(9) = \dfrac{2500 + 175(9) + 25(9)^2}{9} = 677.78$

$C(10) = \dfrac{2500 + 175(10) + 25(10)^2}{10} = 675.00$

$C(11) = \dfrac{2500 + 175(11) + 25(11)^2}{11} = 677.27$

$C(12) = \dfrac{2500 + 175(12) + 25(12)^2}{12} = 683.33$

Thus, it appears that the average cost per year is a minimum at $n = 10$ years; at 10 years, the average minimum cost is $675.00 per year.

(D) 10 years; $675.00 per year

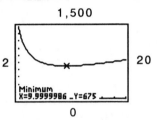

43. (A) $\overline{C}(x) = \dfrac{0.00048(x - 500)^3 + 60,000}{x}$

(B)

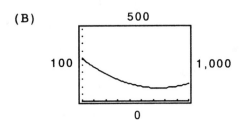

(C) The caseload which yields the minimum average cost per case is 750 cases per month. At 750 cases per month, the average cost per case is $90.

45. (A) $v(x) = \dfrac{26 + 0.06x}{x} = \dfrac{\dfrac{26}{x} + 0.06}{1}$

As x increases, the numerator tends to 0.06 and the denominator is 1. Therefore, $v(x)$ approaches 0.06 centimeters per second as x increases.

(B)

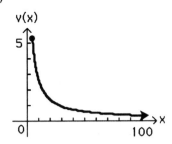

Things to remember:

<u>1</u>. EXPONENTIAL FUNCTION

The equation

$$f(x) = b^x, \; b > 0, \; b \neq 1$$

defines an EXPONENTIAL FUNCTION for each different constant b, called the BASE. The DOMAIN of f is all real numbers, and the RANGE of f is the set of positive real numbers.

<u>2</u>. BASIC PROPERTIES OF THE GRAPH OF $f(x) = b^x, \; b > 0, \; b \neq 1$

a. All graphs pass through $(0,1)$; $b^0 = 1$ for any base b.

b. All graphs are continuous curves; there are no holes or jumps.

c. The x-axis is a horizontal asymptote.

d. If $b > 1$, then b^x increases as x increases.

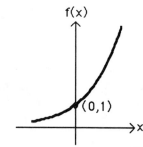

Graph of $f(x) = b^x, \; b > 1$

e. If $0 < b < 1$, then b^x decreases as x increases.

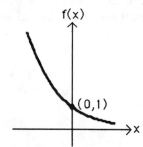

Graph of $f(x) = b^x, \; 0 < b < 1$

<u>3</u>. EXPONENTIAL FUNCTION PROPERTIES
For $a, \; b > 0, \; a \neq 1, \; b \neq 1$, and $x, \; y$ real numbers:

a. EXPONENT LAWS

(i) $a^x a^y = a^{x+y}$ (iv) $(ab)^x = a^x b^x$

(ii) $\dfrac{a^x}{a^y} = a^{x-y}$ (v) $\left(\dfrac{a}{b}\right)^x = \dfrac{a^x}{b^x}$

(iii) $(a^x)^y = a^{xy}$

b. $a^x = a^y$ if and only if $x = y$.

c. For $x \neq 0$, $a^x = b^x$ if and only if $a = b$.

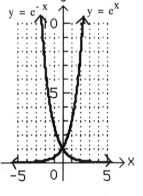

$y = e^{-x}$ $y = e^x$

4. EXPONENTIAL FUNCTION WITH BASE $e = 2.71828...$

Exponential functions with base e and base $1/e$ are respectively defined by $y = e^x$ and $y = e^{-x}$.

    Domain: $(-\infty, \infty)$

    Range: $(0, \infty)$

5. COMPOUND INTEREST

If a principal P (present value) is invested at an annual rate r (expressed as a decimal) compounded m times per year, then the amount A (future value) in the account at the end of t years is given by:

$$A = P\left(1 + \frac{r}{m}\right)^{mt}$$

6. CONTINUOUS COMPOUND INTEREST FORMULA

If a principal P (present value) is invested at an annual rate r (expressed as a decimal) compounded continuously, then the amount A (future value) in the account at the end of t years is given by

$$A = Pe^{rt}$$

7. INTEREST FORMULAS

(a) $A = P(1 + rt)$ Simple interest

(b) $A = P\left(1 + \frac{r}{m}\right)^{mt}$ Compound interest

(c) $A = Pe^{rt}$ Continuous compound interest

---

1. $y = 5^x$, $-2 \leq x \leq 2$

x	y
-2	$\frac{1}{25}$
-1	$\frac{1}{5}$
0	1
1	5
2	25

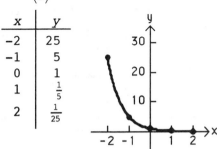

3. $y = \left(\frac{1}{5}\right)^x = 5^{-x}$, $-2 \leq x \leq 2$

x	y
-2	25
-1	5
0	1
1	$\frac{1}{5}$
2	$\frac{1}{25}$

5. $f(x) = -5^x$, $-2 \le x \le 2$

x	$f(x)$
-2	$-\frac{1}{25}$
-1	$-\frac{1}{5}$
0	-1
1	-5
2	-25

7. $y = -e^{-x}$, $-3 \le x \le 3$

x	y
-3	≈ -20
-2	≈ -7.4
-1	≈ -2.7
0	-1
1	≈ -0.4
2	≈ -0.1
3	≈ -0.05

9. $y = 100e^{0.1x}$, $-5 \le x \le 5$

x	y
-5	≈ 60
-3	≈ 74
-1	≈ 90
0	100
1	≈ 111
3	≈ 135
5	≈ 165

11. $g(t) = 10e^{-0.2t}$, $-5 \le t \le 5$

g	$g(t)$
-5	≈ 27.2
-3	≈ 18.2
-1	≈ 12.2
0	10
1	≈ 8.2
3	≈ 5.5
5	≈ 3.7

13. $(4^{3x})^{2y} = 4^{6xy}$ [see 3a(iii)]

15. $\dfrac{e^{x-3}}{e^{x-4}} = e^{(x-3)-(x-4)} = e^{x-3-x+4} = e$ [See 3a(ii)]

17. $(2e^{1.2t})^3 = 2^3 e^{3(1.2t)} = 8e^{3.6t}$ [see 3a (iv)]

19. $g(x) = -f(x)$; the graph of g is the graph of f reflected in the x axis.

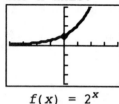

$f(x) = 2^x$

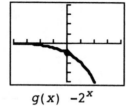

$g(x) \ -2^x$

21. $g(x) = f(x + 1)$; the graph of g is the graph of f shifted one unit to the left.

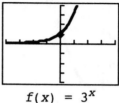

$f(x) = 3^x$

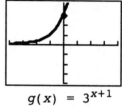

$g(x) = 3^{x+1}$

23. $g(x) = f(x) + 1$; the graph of g is the graph of f shifted one unit up.

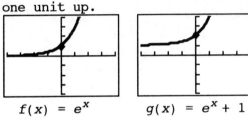

$f(x) = e^x$ $g(x) = e^x + 1$

25. $g(x) = 2f(x + 2)$; the graph of g is the graph of f vertically expanded by a factor of 2 and shifted to the left 2 units.

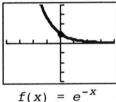   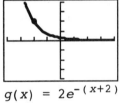

$f(x) = e^{-x}$ $g(x) = 2e^{-(x+2)}$

27. $f(t) = 2^{t/10}$, $-30 \le t \le 30$

t	f(t)
-30	$\frac{1}{8}$
-20	$\frac{1}{4}$
-10	$\frac{1}{2}$
0	1
10	2
20	4
30	8

29. $y = -3 + e^{1+x}$, $-4 \le x \le 2$

x	y
-4	≈ -3
-2	≈ -2.6
-1	-2
0	≈ -0.3
1	≈ 4.4
2	≈ 17.1

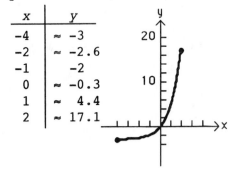

31. $y = e^{|x|}$, $-3 \le x \le 3$

x	y
-3	≈ 20.1
-1	≈ 2.7
0	1
1	≈ 2.7
3	≈ 20.1

33. $C(x) = \dfrac{e^x + e^{-x}}{2}$, $-5 \le x \le 5$

x	C(x)
-5	≈ 74
-3	≈ 10
0	1
3	≈ 10
5	≈ 74

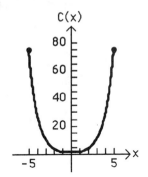

35. $y = e^{-x^2}$, $-3 \le x \le 3$

x	y
-3	0.0001
-2	0.0183
-1	0.3679
0	1
1	0.3679
2	0.0183
3	0.0001

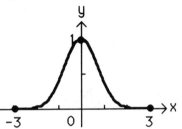

37. The top curve is the graph of $f(x) = 2^x$, the bottom curve is the graph of $g(x) = e^x$; e^x approaches 0 more rapidly than 2^x as $x \to -\infty$.

39. The top curve is the graph of $g(x) = e^{-x}$, the bottom curve is the graph of $f(x) = 2^{-x}$; e^{-x} grows more rapidly than 2^{-x} as $x \to -\infty$.

41. $10^{2-3x} = 10^{5x-6}$ implies (see $\underline{3}$b)
$$2 - 3x = 5x - 6$$
$$-8x = -8$$
$$x = 1$$

43. $4^{5x-x^2} = 4^{-6}$ implies
$$5x - x^2 = -6$$
$$\text{or} \quad -x^2 + 5x + 6 = 0$$
$$x^2 - 5x - 6 = 0$$
$$(x - 6)(x + 1) = 0$$
$$x = 6, -1$$

45. $5^3 = (x + 2)^3$ implies (by property $\underline{3}$c)
$$5 = x + 2$$
Thus, $x = 3$.

47. $(x - 3)e^x = 0$
$$x - 3 = 0 \quad \text{(since } e^x \neq 0\text{)}$$
$$x = 3$$

49. $3xe^{-x} + x^2e^{-x} = 0$
$$e^{-x}(3x + x^2) = 0$$
$$3x + x^2 = 0 \quad \text{(since } e^{-x} \neq 0\text{)}$$
$$x(3 + x) = 0$$
$$x = 0, -3$$

51. $h(x) = x2^x$, $-5 \le x \le 0$

x	$h(x)$
-5	$-\frac{5}{32}$
-4	$-\frac{1}{4}$
-3	$-\frac{3}{8}$
-2	$-\frac{1}{2}$
-1	$-\frac{1}{2}$
0	0

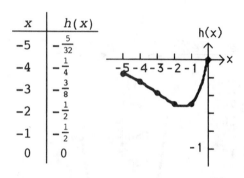

53. $N = \dfrac{100}{1 + e^{-t}}$, $0 \le t \le 5$

t	N
0	50
1	≈ 73.1
2	≈ 88.1
3	≈ 95.3
5	≈ 99.3

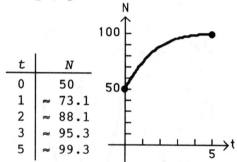

55. Using $\underline{4}$, $A = P\left(1 + \dfrac{r}{m}\right)^{mt}$, we have:

(A) $P = 2,500$, $r = 0.07$, $m = 4$, $t = \dfrac{3}{4}$
$$A = 2,500\left(1 + \frac{0.07}{4}\right)^{4 \cdot 3/4} = 2,500(1 + 0.0175)^3 = 2,633.56$$
Thus, $A = \$2,633.56$.

(B) $A = 2,500\left(1 + \dfrac{0.07}{4}\right)^{4 \cdot 15} = 2,500(1 + 0.0175)^{60} = 7079.54$
Thus, $A = \$7,079.54$.

57. Using $\underline{6}$ with $P = 7,500$ and $r = 0.0835$, we have:
$$A = 7,500e^{0.0835t}$$

(A) $A = 7,500e^{(0.0835)5.5} = 7,500e^{0.45925} \approx 11,871.65$
Thus, there will be $\$11,871.65$ in the account after 5.5 years.

(B) $A = 7,500e^{(0.0835)12} = 7,500e^{1.002} \approx 20,427.93$
Thus, there will be \$20,427.93 in the account after 12 years.

59. Using $A = P\left(1 + \dfrac{r}{m}\right)^{mt}$, we have:

$A = 15,000$, $r = 0.0975$, $m = 52$, $t = 5$

Thus, $15,000 = P\left(1 + \dfrac{0.0975}{52}\right)^{52 \cdot 5} = P(1 + 0.001875)^{260} \approx P(1.6275)$ and

$P = \dfrac{15,000}{1.6275} \approx 9,217.$ Therefore, $P \approx \$9,217.$

61. Alamo Savings:

From Section 2-1, $A = P\left(1 + \dfrac{r}{m}\right)^{mt}$, where P is the principal, r is the annual rate, and m is the number of compounding periods per year. Thus:

$A = 10,000\left(1 + \dfrac{0.0825}{4}\right)^{4} = 10,000(1.020625)^{4} \approx \$10,850.88$

Lamar Savings:
$A = 10,000e^{0.0805} \approx \$10,838.29$

63. In $A = Pe^{rt}$, we are given $A = 50,000$, $r = 0.1$, and $t = 5.5$. Thus:

$50,000 = Pe^{(0.1)5.5}$ or $P = \dfrac{50,000}{e^{0.55}} \approx 28,847.49$

You should be willing to pay \$28,847.49 for the note.

65. Given $N = 2(1 - e^{-0.037t})$, $0 \le t \le 50$

t	N
0	0
10	≈ 0.62
30	≈ 1.34
50	≈ 1.69

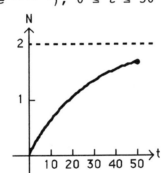

N approaches 2 as t increases without bound.

67. Given $I = I_0 e^{-0.23d}$
(A) $I = I_0 e^{-0.23(10)} = I_0 e^{-2.3} \approx I_0(0.10)$
Thus, about 10% of the surface light will reach a depth of 10 feet.
(B) $I = I_0 e^{-0.23(20)} = I_0 e^{-4.6} \approx I_0(0.010)$
Thus, about 1% of the surface light will reach a depth of 20 feet.

69. (A) Using $\underline{6}$ with $N_0 = 40,000$ and $r = 0.21$, we have $N = 40,000e^{0.21t}$.

(B) At the end of the year 2,000, $t = 8$ years and
$N(8) = 40,000e^{0.21(8)} = 40,000e^{1.6} \approx 215,000$

At the end of the year 2005, $t = 13$ years and
$$N(13) = 40,000e^{0.21(13)} = 40,000e^{2.73} \approx 613,000$$

(C)

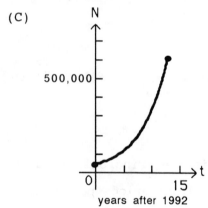

years after 1992

71. (A) Using $\underline{6}$ with $P_0 = 5.7$ and $r = 0.0114$, we have
$$P = 5.7e^{0.0114t}$$

(B) In the year 2010, $t = 15$ and
$$P = 5.7e^{0.0114(15)} = 5.7e^{0.171} \approx 6.8 \text{ billion}$$
In the year 2030, $t = 35$ and
$$P = 5.7e^{0.0114(35)} = 5.7e^{1.490} \approx 8.5 \text{ billion}$$

(C)

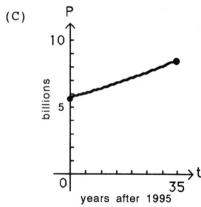

years after 1995

EXERCISE 2-3

Things to remember:

$\underline{1}$. ONE-TO-ONE FUNCTIONS

A function f is said to be ONE-TO-ONE if each range value corresponds to exactly one domain value.

2. INVERSE OF A FUNCTION

If f is a one-to-one function, then the INVERSE of f is the function formed by interchanging the independent and dependent variables for f. Thus, if (a, b) is a point on the graph of f, then (b, a) is a point on the graph of the inverse of f.

Note: If f is not one-to-one, then f DOES NOT HAVE AN INVERSE.

3. LOGARITHMIC FUNCTIONS

The inverse of an exponential function is called a LOGARITHMIC FUNCTION. For $b > 0$ and $b \neq 1$,

Logarithmic form		Exponential form
$y = \log_b x$	is equivalent to	$x = b^y$

The LOG TO THE BASE b OF x is the exponent to which b must be raised to obtain x. [Remember: A logarithm is an exponent.] The DOMAIN of the logarithmic function is the range of the corresponding exponential function, and the RANGE of the logarithmic function is the domain of the corresponding exponential function. Typical graphs of an exponential function and its inverse, a logarithmic function, for $b > 1$, are shown in the figure below:

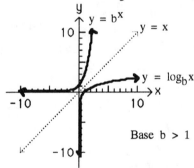

4. PROPERTIES OF LOGARITHMIC FUNCTIONS

If b, M, and N are positive real numbers, $b \neq 1$, and p and x are real numbers, then:

a. $\log_b 1 = 0$

b. $\log_b b = 1$

c. $\log_b b^x = x$

d. $b^{\log_b x} = x,\ x > 0$

e. $\log_b MN = \log_b M + \log_b N$

f. $\log_b \frac{M}{N} = \log_b M - \log_b N$

g. $\log_b M^p = p \log_b M$

h. $\log_b M = \log_b N$ if and only if $M = N$

$\underline{5}$. LOGARITHMIC NOTATION; LOGARITHMIC-EXPONENTIAL RELATIONSHIPS

Common logarithm $\log x = \log_{10} x$

Natural logarithm $\ln x = \log_e x$

$\log x = y$ is equivalent to $x = 10^y$

$\ln x = y$ is equivalent to $x = e^y$

---

1. $27 = 3^3$ (using $\underline{3}$)   3. $1 = 10^0$ 5. $8 = 4^{3/2}$

7. $\log_7 49 = 2$ 9. $\log_4 8 = \frac{3}{2}$ 11. $\log_b A = u$

13. $\log_{10} 1 = y$ is equivalent to $10^y = 1$; $y = 0$.

15. $\log_e e = y$ is equivalent to $e^y = e$; $y = 1$.

17. $\log_{0.2} 0.2 = y$ is equivalent to $(0.2)^y = 0.2$; $y = 1$.

19. $\log_{10} 10^3 = 3$ 21. $\log_2 2^{-3} = -3$ 23. $\log_{10} 1,000 = \log_{10} 10^3 = 3$
    (using $\underline{2}$a)

25. $\log_b \frac{P}{Q} = \log_b P - \log_b Q$ (using $\underline{4}$f)   27. $\log_b L^5 = 5 \log_b L$ (using $\underline{4}$g)

29. $\log_b \frac{p}{qrs} = \log_b p - \log_b qrs$ (using $\underline{4}$f)

$\qquad = \log_b p - (\log_b q + \log_b r + \log_b s)$ (using $\underline{4}$e)

$\qquad = \log_b p - \log_b q - \log_b r - \log_b s$

31. $\log_3 x = 2$ 33. $\log_7 49 = y$ 35. $\log_b 10^{-4} = -4$

$\quad x = 3^2$ (using $\underline{3}$) $\quad \log_7 7^2 = y$ $\quad 10^{-4} = b^{-4}$

$\quad x = 9$ $\qquad 2 = y$ This equality implies

$\qquad\qquad$ Thus, $y = 2$. $b = 10$ (since the
exponents are the same).

37. $\log_4 x = \frac{1}{2}$ 39. $\log_{1/3} 9 = y$ 41. $\log_b 1,000 = \frac{3}{2}$

$\quad x = 4^{1/2}$ $\quad 9 = \left(\frac{1}{3}\right)^y$ $\quad \log_b 10^3 = \frac{3}{2}$

$\quad x = 2$ $\quad 3^2 = (3^{-1})^y$ $\quad 3 \log_b 10 = \frac{3}{2}$

$\qquad\qquad 3^2 = 3^{-y}$ $\quad \log_b 10 = \frac{1}{2}$

This inequality $\qquad 10 = b^{1/2}$
implies that   Square both sides:
$2 = -y$ or $y = -2$. $100 = b$, i.e., $b = 100$.

43. $\log_b \dfrac{x^5}{y^3}$

$= \log_b x^5 - \log_b y^3$

$= 5\log_b x - 3\log_b y$

45. $\log_b \sqrt[3]{N} = \log_b N^{1/3}$

$= \dfrac{1}{3}\log_b N$

47. $\log_b(x^2 \sqrt[3]{y}) = \log_b x^2 + \log_b y^{1/3} = 2\log_b x + \dfrac{1}{3}\log_b y$

49. $\log_b(50 \cdot 2^{-0.2t}) = \log_b 50 + \log_b 2^{-0.2t} = \log_b 50 - 0.2t\log_b 2$

51. $\log_b P(1+r)^t = \log_b P + \log_b(1+r)^t = \log_b P + t\log_b(1+r)$

53. $\log_e 100e^{-0.01t} = \log_e 100 + \log_e e^{-0.01t}$

$= \log_e 100 - 0.01t\log_e e = \log_e 100 - 0.01t$

55. $\log_b x = \dfrac{2}{3}\log_b 8 + \dfrac{1}{2}\log_b 9 - \log_b 6 = \log_b 8^{2/3} + \log_b 9^{1/2} - \log_b 6$

$= \log_b 4 + \log_b 3 - \log_b 6 = \log_b \dfrac{4 \cdot 3}{6}$

$\log_b x = \log_b 2$

$x = 2$ (using $\underline{2}$e)

57. $\log_b x = \dfrac{3}{2}\log_b 4 - \dfrac{2}{3}\log_b 8 + 2\log_b 2 = \log_b 4^{3/2} - \log_b 8^{2/3} + \log_b 2^2$

$= \log_b 8 - \log_b 4 + \log_b 4 = \log_b 8$

$\log_b x = \log_b 8$

$x = 8$ (using $\underline{2}$e)

59. $\log_b x + \log_b(x-4) = \log_b 21$

$\log_b x(x-4) = \log_b 21$

Therefore, $x(x-4) = 21$

$x^2 - 4x - 21 = 0$

$(x-7)(x+3) = 0$

Thus, $x = 7$.

[Note: $x = -3$ is not a solution since $\log_b(-3)$ is not defined.]

61. $\log_{10}(x-1) - \log_{10}(x+1) = 1$

$\log_{10}\!\left(\dfrac{x-1}{x+1}\right) = 1$

Therefore, $\dfrac{x-1}{x+1} = 10^1 = 10$

$x - 1 = 10(x+1)$

$x - 1 = 10x + 10$

$-9x = 11$

$x = -\dfrac{11}{9}$

There is *no solution*, since

$\log_{10}\!\left(-\dfrac{11}{9} - 1\right) = \log_{10}\!\left(-\dfrac{20}{9}\right)$

is not defined. Similarly,

$\log_{10}\!\left(-\dfrac{11}{9} + 1\right) = \log_{10}\!\left(-\dfrac{2}{9}\right)$

is not defined.

63. $y = \log_2(x - 2)$
$x - 2 = 2^y$
$x = 2^y + 2$

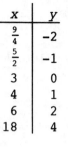

x	y
$\frac{9}{4}$	-2
$\frac{5}{2}$	-1
3	0
4	1
6	2
18	4

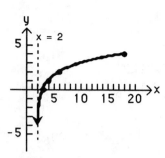

65. The graph of $y = \log_2(x - 2)$ is the graph of $y = \log_2 x$ shifted to the right 2 units.

67. Since logarithmic functions are defined only for positive "inputs", we must have $x + 1 > 0$ or $x > -1$; domain: $(-1, \infty)$. The range of $y = 1 + \ln(x + 1)$ is the set of all real numbers.

69. (A) 3.54743
(B) −2.16032
(C) 5.62629
(D) −3.19704

71. (A) $\log x = 1.1285$
$x = 13.4431$
(B) $\log x = -2.0497$
$x = 0.0089$
(C) $\ln x = 2.7763$
$x = 16.0595$
(D) $\ln x = -1.8879$
$x = 0.1514$

73. $10^x = 12$ (Take common logarithms of both sides)
$\log 10^x = \log 12 \approx 1.0792$
$x \approx 1.0792$ ($\log 10^x = x \log 10 = x$; $\log 10 = 1$)

75. $e^x = 4.304$ (Take natural logarithms of both sides)
$\ln e^x = \ln 4.304 \approx 1.4595$
$x \approx 1.4595$ ($\ln e^x = x \ln e = x$; $\ln e = 1$)

77. $1.03^x = 2.475$ (Take either common or natural logarithms of both sides; we use common logarithms)
$\log(1.03)^x = \log 2.475$
$x = \dfrac{\log 2.475}{\log 1.03} \approx 30.6589$

79. $1.005^{12t} = 3$ (Take either common or natural logarithms of both sides; here we'll use natural logarithms.)
$\ln 1.005^{12t} = \ln 3$
$12t = \dfrac{\ln 3}{\ln 1.005} \approx 220.2713$
$t = 18.3559$

81. $y = \ln x$, $x > 0$

x	y
0.5	≈ -0.69
1	0
2	≈ 0.69
4	≈ 1.39
5	≈ 1.61

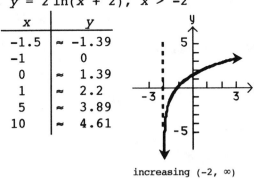

increasing $(0, \infty)$

83. $y = |\ln x|$, $x > 0$

x	y
0.5	≈ 0.69
1	0
2	≈ 0.69
4	≈ 1.39
5	≈ 1.6

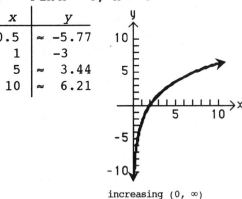

decreasing $(0, 1]$
increasing $[1, \infty)$

85. $y = 2\ln(x + 2)$, $x > -2$

x	y
-1.5	≈ -1.39
-1	0
0	≈ 1.39
1	≈ 2.2
5	≈ 3.89
10	≈ 4.61

increasing $(-2, \infty)$

87. $y = 4\ln x - 3$, $x > 0$

x	y
0.5	≈ -5.77
1	-3
5	≈ 3.44
10	≈ 6.21

increasing $(0, \infty)$

89. The calculator interprets $\log \dfrac{13}{7}$ as $\dfrac{\log 13}{7}$ not as $\log\left(\dfrac{13}{7}\right)$. To find $\log\left(\dfrac{13}{7}\right)$, calculate $\dfrac{13}{7}$ and take the common logarithm of the result:

$$\log\left(\frac{13}{7}\right) = \log(1.8571...) \approx 0.2688453123$$

or calculate $\log 13 - \log 7$ to get the same result.

91. For any number b, $b > 0$, $b \neq 1$, $\log_b 1 = y$ is equivalent to $b^y = 1$ which implies $y = 0$. Thus, $\log_b 1 = 0$ for any permissible base b.

93. $\log_{10} y - \log_{10} c = 0.8x$

$\log_{10} \dfrac{y}{c} = 0.8x$

Therefore, $\dfrac{y}{c} = 10^{0.8x}$ (using $\underline{1}$)

and $y = c \cdot 10^{0.8x}$.

95.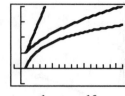

$1 \le x \le 16$

A function f is "larger than" a function g on an interval $[a, b]$ if $f(x) > g(x)$ for $a \le x \le b$. $r(x) > q(x) > p(x)$ for $1 \le x \le 16$, that is $x > \sqrt{x} > \ln x$ for $1 < x \le 16$

97. From the compound interest formula $A = P(1 + r)^t$, we have:

$2P = P(1 + .06)^t$ or $(1.06)^t = 2$

Take the natural log of both sides of this equation:

$\ln(1.06)^t = \ln 2$ [Note: The common log could have been used instead of the natural log.]

$t \ln(1.06) = \ln 2$

$$t = \frac{\ln 2}{\ln(1.06)} \approx \frac{.69315}{.05827} = 11.90 \approx 12 \text{ years}$$

99. (A) $A = P\left(1 + \frac{r}{m}\right)^{mt}$, $r = 0.06$, $m = 4$, $P = 1000$, $A = 1800$.

$$1800 = 1000\left(1 + \frac{0.06}{4}\right)^{4t} = 1000(1.015)^{4t}$$

$$(1.015)^{4t} = \frac{1800}{1000} = 1.8$$

$$4t \ln(1.015) = \ln(1.8)$$

$$t = \frac{\ln(1.8)}{4 \ln(1.015)} \approx 9.87$$

$1000 at 6% compounded quarterly will grow to $1800 in 9.87 years.

(B) $A = Pe^{rt}$, $r = 0.06$, $P = 1000$, $A = 1800$

$1000e^{0.06t} = 1800$

$e^{0.06t} = 1.8$

$0.06t = \ln 1.8$

$$t = \frac{\ln 1.8}{0.06} \approx 9.80$$

$1000 at 6% compounded continuously will grow to $1800 in 9.80 years.

101. $A = Pe^{rt}$, $P = 10,000$, $A = 20,000$, $t = 8$

$20,000 = 10,000e^{8r}$

$e^{8r} = 2$

$8r = \ln 2$

$$r = \frac{\ln 2}{8} \approx 0.08664$$

$10,000 invested at an annual interest rate of 8.664% compounded continuously will yield $20,000 after 8 years.

103. $I = I_0 10^{N/10}$

Take the common log of both sides of this equation. Then:
$$\log I = \log(I_0 10^{N/10}) = \log I_0 + \log 10^{N/10}$$
$$= \log I_0 + \frac{N}{10} \log 10 = \log I_0 + \frac{N}{10} \text{ (since } \log 10 = 1)$$

So, $\frac{N}{10} = \log I - \log I_0 = \log\left(\frac{I}{I_0}\right)$ and $N = 10 \log\left(\frac{I}{I_0}\right)$.

105. Assuming that the world population is currently 5.8 billion and that it will grow at the rate of 1.14% compounded continuously, the population will be
$$P = 5.8e^{0.0114t}$$
after t years.

Given that there are $1.68 \times 10^{14} = 168,000$ billion square yards of land, we solve
$$168,000 = 5.8e^{0.0114t}$$
for t:
$$e^{0.0114t} \approx 28,966$$
$$0.0114t = \ln(28,966) \approx 10.2739$$
$$t \approx 901$$

It will take approximately 901 years.

CHAPTER 2 REVIEW

1. $u = e^v$
$v = \ln u$ (2-3)

2. $x = 10^y$
$y = \log x$ (2-3)

3. $\ln M = N$
$M = e^N$ (2-3)

4. $\log u = v$
$u = 10^v$ (2-3)

5. $\dfrac{5^{x+4}}{5^{4-x}} = 5^{x+4-(4-x)} = 5^{2x}$ (2-2)

6. $\left(\dfrac{e^u}{e^{-u}}\right)^u = (e^{u+u})^u = (e^{2u})^u = e^{2u^2}$ (2-2)

7. $\log_3 x = 2$
$x = 3^2$
$x = 9$ (2-3)

8. $\log_x 36 = 2$
$x^2 = 36$
$x = 6$ (2-3)

9. $\log_2 16 = x$
$2^x = 16$
$x = 4$ (2-3)

10. $10^x = 143.7$
$x = \log 143.7$
$x \approx 2.157$ (2-3)

11. $e^x = 503,000$
$x = \ln 503,000 \approx 13.128$ (2-3)

12. $\log x = 3.105$
$x = 10^{3.105} \approx 1273.503$ (2-3)

13. $\ln x = -1.147$
$x = e^{-1.147} \approx 0.318$ (2-3)

14. (A) 3   (B) 2   (C) 3   (D) 1   (E) 1   (F) 1   (2-1)

15. (A) 4   (B) 3   (C) 4   (D) 0   (E) 1   (F) 1   (2-1)

16. (A) 2   (B) 3   (C) positive   (2-1)

17. (A) 3   (B) 4   (C) negative   (2-1)

18. $f(x) = \dfrac{x + 4}{x - 2}$

(A) *Intercepts:*

x intercepts: $f(x) = 0$ only if $x + 4 = 0$ or $x = -4$.
The x intercept is -4.

y intercepts: $f(0) = \dfrac{0 + 4}{0 - 2} = -2$
The y intercept is -2.

(B) *Domain:* The denominator is 0 at $x = 2$. Thus, the domain is the set of all real numbers except 2.

(C) *Asymptotes:*

Vertical asymptotes: $f(x) = \dfrac{x + 4}{x - 2}$

The denominator is 0 at $x = 2$. Therefore, the line $x = 2$ is a vertical asymptote.

Horizontal asymptotes: $f(x) = \dfrac{x + 4}{x - 2} = \dfrac{1 + \dfrac{4}{x}}{1 - \dfrac{2}{x}}$

As x increases or decreases without bound, the numerator tends to 1 and the denominator tends to 1. Therefore, the line $y = 1$ is a horizontal asymptote.

(D)

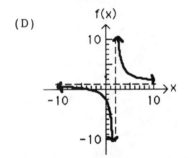

(E)

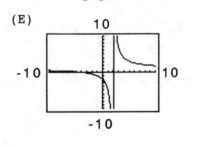

(2-1)

19. $f(x) = \dfrac{3x - 4}{2 + x}$

(A) *Intercepts:*

x intercepts: $f(x) = 0$ only if $3x - 4 = 0$ or $x = \dfrac{4}{3}$.

The x intercept is $\dfrac{4}{3}$.

y intercepts: $f(0) = \dfrac{3 \cdot 0 - 4}{2 + 0} = -2$
The y intercept is -2.

(B) *Domain:* The denominator is 0 at $x = -2$. Thus, the domain is the set of all real numbers except -2.

(C) *Asymptotes:*

Vertical asymptotes: $f(x) = \dfrac{3x - 4}{x + 2}$

The denominator is 0 at $x = -2$. Therefore, the line $x = -2$ is a vertical asymptote.

Horizontal asymptotes: $f(x) = \dfrac{3x - 4}{x + 2} = \dfrac{3 - \dfrac{4}{x}}{1 + \dfrac{2}{x}}$

As x increases or decreases without bound, the numerator tends to 3 and the denominator tends to 1. Therefore, the line $y = 3$ is a horizontal asymptote.

(D)

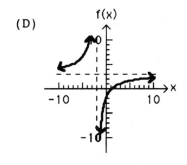

(E)

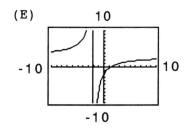

(2-1)

20. $\log(x + 5) = \log(2x - 3)$
$$x + 5 = 2x - 3$$
$$-x = -8$$
$$x = 8 \quad (2\text{-}3)$$

21. $2 \ln(x - 1) = \ln(x^2 - 5)$
$$\ln(x - 1)^2 = \ln(x^2 - 5)$$
$$(x - 1)^2 = x^2 - 5$$
$$x^2 - 2x + 1 = x^2 - 5$$
$$-2x = -6$$
$$x = 3 \quad (2\text{-}3)$$

22. $\quad 9^{x-1} = 3^{1+x}$
$$(3^2)^{x-1} = 3^{1+x}$$
$$3^{2x-2} = 3^{1+x}$$
$$2x - 2 = 1 + x$$
$$x = 3 \quad (2\text{-}2)$$

23. $\quad e^{2x} = e^{x^2-3}$
$$2x = x^2 - 3$$
$$x^2 - 2x - 3 = 0$$
$$(x - 3)(x + 1) = 0$$
$$x = 3, -1 \quad (2\text{-}2)$$

24. $\quad 2x^2 e^x = 3xe^x$
$$2x^2 = 3x \text{ (divide both sides}$$
$$2x^2 - 3x = 0 \quad \text{by } e^x)$$
$$x(2x - 3) = 0$$
$$x = 0, \frac{3}{2} \quad (2\text{-}2)$$

25. $\log_{1/3} 9 = x$

$\left(\dfrac{1}{3}\right)^x = 9$

$\dfrac{1}{3^x} = 9$

$3^x = \dfrac{1}{9}$

$x = -2$ (2-3)

26. $\log_x 8 = -3$

$x^{-3} = 8$

$\dfrac{1}{x^3} = 8$

$x^3 = \dfrac{1}{8}$

$x = \dfrac{1}{2}$ (2-3)

27. $\log_9 x = \dfrac{3}{2}$

$9^{3/2} = x$

$x = 27$ (2-3)

28. $x = 3(e^{1.49}) \approx 13.3113$ (2-3)

29. $x = 230(10^{-0.161}) \approx 158.7552$ (2-3)

30. $\log x = -2.0144$
$x \approx 0.0097$ (2-3)

31. $\ln x = 0.3618$
$x \approx 1.4359$ (2-3)

32. $35 = 7(3^x)$

$3^x = 5$

$\ln 3^x = \ln 5$

$x \ln 3 = \ln 5$

$x = \dfrac{\ln 5}{\ln 3} \approx 1.4650$ (2-3)

33. $0.01 = e^{-0.05x}$

$\ln(0.01) = \ln(e^{-0.05x}) = -0.05x$

Thus, $x = \dfrac{\ln(0.01)}{-0.05} \approx 92.1034$ (2-3)

34. $8{,}000 = 4{,}000(1.08)^x$

$(1.08)^x = 2$

$\ln(1.08)^x = \ln 2$

$x \ln 1.08 = \ln 2$

$x = \dfrac{\ln 2}{\ln 1.08} \approx 9.0065$

(2-3)

35. $5^{2x-3} = 7.08$

$\ln(5^{2x-3}) = \ln 7.08$

$(2x - 3)\ln 5 = \ln 7.08$

$2x \ln 5 - 3 \ln 5 = \ln 7.08$

$x = \dfrac{\ln 7.08 + 3 \ln 5}{2 \ln 5}$

$ = \dfrac{\ln 7.08 + \ln 5^3}{2 \ln 5}$

$ = \dfrac{\ln[7.08(125)]}{2 \ln 5} \approx 2.1081$ (2-3)

36. $x = \log_2 7 = \dfrac{\log 7}{\log 2} \approx 2.8074$

or $x = \log_2 7 = \dfrac{\ln 7}{\ln 2} \approx 2.8074$

(2-3)

37. $x = \log_{0.2} 5.321 = \dfrac{\log 5.321}{\log 0.2} \approx -1.0387$

or $x = \log_{0.2} 5.321 = \dfrac{\ln 5.321}{\ln 0.2} \approx -1.0387$

(2-3)

38. The graph of $f(x) = x^4 - 4x^2 + 1 = x^4\left(1 - \dfrac{4}{x^2} + \dfrac{1}{x^4}\right)$ will "look like" the
graph of $y = x^4$; for large x, $f(x) \approx x^4$.   (2-1)

39. (A) 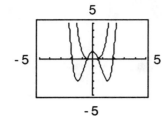 (B)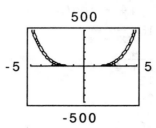

(2-1)

40. $e^x(e^{-x} + 1) - (e^x + 1)(e^{-x} - 1) = 1 + e^x - (1 - e^x + e^{-x} - 1)$
$$= 1 + e^x + e^x - e^{-x}$$
$$= 1 + 2e^x - e^{-x}$$

(2-2)

41. $(e^x - e^{-x})^2 - (e^x + e^{-x})(e^x - e^{-x})$
$$= (e^x)^2 - 2(e^x)(e^{-x}) + (e^{-x})^2 - [(e^x)^2 - (e^{-x})^2]$$
$$= e^{2x} - 2 + e^{-2x} - [e^{2x} - e^{-2x}]$$
$$= 2e^{-2x} - 2$$

(2-2)

42. $y = 2^{x-1}$, $-2 \le x \le 4$

x	y
-2	$\frac{1}{8}$
-1	$\frac{1}{4}$
0	$\frac{1}{2}$
1	1
2	2
4	8

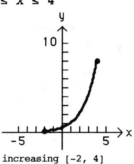

increasing [-2, 4]   (2-2)

43. $f(t) = 10e^{-0.08t}$, $t \ge 0$

t	f(t)
0	10
10	≈ 4.5
20	≈ 2
30	≈ 0.9
40	≈ 0.4

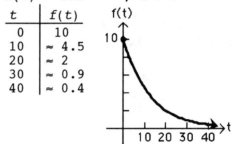

decreasing [0, ∞)   (2-2)

44. $y = \ln(x + 1)$, $-1 < x \le 10$

x	y
-0.5	≈ -0.7
0	0
4	≈ 1.6
8	≈ 2.2
10	≈ 2.4

increasing (-1, 10]   (2-2)

45. $\log 10^\pi = \pi \log 10 = \pi$ (see logarithm properties <u>4</u>.b & g, Section 2-3)
$10^{\log\sqrt{2}} = y$ is equivalent to $\log y = \log\sqrt{2}$
which implies $y = \sqrt{2}$
Similarly, $\ln e^\pi = \pi \ln e = \pi$ (Section 2-3, <u>4</u>.b & g) and $e^{\ln\sqrt{2}} = y$
implies $\ln y = \ln\sqrt{2}$ and $y = \sqrt{2}$.

(2-3)

46. $\log x - \log 3 = \log 4 - \log(x + 4)$

$$\log \frac{x}{3} = \log \frac{4}{x + 4}$$

$$\frac{x}{3} = \frac{4}{x + 4}$$

$$x(x + 4) = 12$$

$$x^2 + 4x - 12 = 0$$

$$(x + 6)(x - 2) = 0$$

$$x = -6, \ 2$$

Since $\log(-6)$ and $\log(-2)$ are not defined, -6 is not a solution. Therefore, the solution is $x = 2$.   (2-3)

47. $\ln(2x - 2) - \ln(x - 1) = \ln x$

$$\ln\left(\frac{2x - 2}{x - 1}\right) = \ln x$$

$$\ln\left[\frac{2(x - 1)}{x - 1}\right] = \ln x$$

$$\ln 2 = \ln x$$

$$x = 2 \quad (2-3)$$

48. $\ln(x + 3) - \ln x = 2 \ln 2$

$$\ln\left(\frac{x + 3}{x}\right) = \ln(2^2)$$

$$\frac{x + 3}{x} = 4$$

$$x + 3 = 4x$$

$$3x = 3$$

$$x = 1 \qquad (2-3)$$

49. $\log 3x^2 = 2 + \log 9x$

$$\log 3x^2 - \log 9x = 2$$

$$\log\left(\frac{3x^2}{9x}\right) = 2$$

$$\log\left(\frac{x}{3}\right) = 2$$

$$\frac{x}{3} = 10^2 = 100$$

$$x = 300 \qquad (2-3)$$

50. $\ln y = -5t + \ln c$

$$\ln y - \ln c = -5t$$

$$\ln \frac{y}{c} = -5t$$

$$\frac{y}{c} = e^{-5t}$$

$$y = ce^{-5t} \qquad (2-3)$$

51. Let x be *any* positive real number and suppose $\log_1 x = y$. Then $1^y = x$.

But, $1^y = 1$, so $x = 1$, i.e., $x = 1$ for all positive real numbers x. This is clearly impossible.   (2-3)

52. $A = P\left(1 + \frac{r}{m}\right)^{mt}$.

We let $P = 5,000$, $r = 0.12$, $m = 52$, and $t = 6$. Then we have:

$A = 5,000\left(1 + \frac{0.12}{52}\right)^{52(6)} \approx 5,000(1 + 0.0023)^{312} \approx 10,263.65$

Thus, there will be $10,263.65 in the account 6 years from now.   (2-2)

53. $A = Pe^{rt}$. We let $P = 5,000$, $r = 0.12$, and $t = 6$. Then:

$A = 5,000e^{(0.12)6} \approx 10,272.17$

Thus, there will be $10,272.17 in the account 6 years from now.   (2-2)

54. The compound interest formula for money invested at 15% compounded annually is:

$A = P(1 + 0.15)^t$

To find the tripling time, we set $A = 3P$ and solve for t:

$$3P = P(1.15)^t$$
$$(1.15)^t = 3$$
$$\ln(1.15)^t = \ln 3$$
$$t \ln 1.15 = \ln 3$$
$$t = \frac{\ln 3}{\ln 1.15} \approx 7.86$$

Thus, the tripling time (to the nearest year) is 8 years.   (2-2)

55. The compound interest formula for money invested at 10% compounded continuously is:

$A = Pe^{0.1t}$

To find the doubling time, we set $A = 2P$ and solve for t:

$$2P = Pe^{0.1t}$$
$$e^{0.1t} = 2$$
$$0.1t = \ln 2$$
$$t = \frac{\ln 2}{0.1} \approx 6.93 \text{ years}$$ (2-3)

56. (A) Since $C(x)$ is a linear function of x, it can be written in the form
$$C(x) = mx + b$$
Since the fixed costs are $300, $b = 300$.
Also, $C(100) = 4300$, so
$$4300 = 100m + 300$$
$$100m = 4000$$
$$m = 40$$
Therefore,
$$C(x) = 40x + 300$$
and $$\overline{C}(x) = \frac{40x + 300}{x}$$

(B)

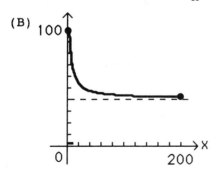

(C) $\overline{C}(x) = \dfrac{40x + 300}{x} = \dfrac{40 + \dfrac{300}{x}}{1}$, $5 \le x \le 200$

As x increases, the numerator tends to 40 and the denominator is 1. Therefore, $\overline{C}(x)$ approaches 40; The line $y = 40$ is a horizontal asymptote.

(D) $\overline{C}(x)$ approaches \$40 per pair as production increases. (2-1)

57. (A) $\overline{C}(x) = \dfrac{C(x)}{x} = \dfrac{20x^3 - 360x^2 + 2,300x - 1,000}{x}$

(B)

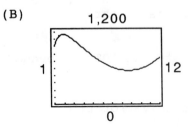

(C) From the graph, $\overline{C}(x)$ has a minimum at $x \approx 8.667$. Thus, the minimum average cost occurs when 8.667 thousand (8,667) cases are handled per year.

$$\overline{C}(8.667) = \dfrac{20(8.667)^3 - 360(8.667)^2 + 2300(8.667) - 1000}{8.667}$$

$\approx$ \$567 per case (2-1)

58. (A) $N(0) = 1$

$N\left(\dfrac{1}{2}\right) = 2$

$N(1) = 4 = 2^2$

$N\left(\dfrac{3}{2}\right) = 8 = 2^3$

$N(2) = 16 = 2^4$

$\vdots$

Thus, we conclude that
$N(t) = 2^{2t}$ or $N = 4^t$.

(B) We need to solve:

$2^{2t} = 10^9$

$\log 2^{2t} = \log 10^9 = 9$

$2t \log 2 = 9$

$t = \dfrac{9}{2 \log 2} \approx 14.95$

Thus, the mouse will die in 15 days.

(2-2, 2-3)

59. Given $I = I_0 e^{-kd}$. When $d = 73.6$, $I = \dfrac{1}{2} I_0$. Thus, we have:

$\dfrac{1}{2} I_0 = I_0 e^{-k(73.6)}$

$e^{-k(73.6)} = \dfrac{1}{2}$

$-k(73.6) = \ln \dfrac{1}{2}$

$k = \dfrac{\ln(0.5)}{-73.6} \approx 0.00942$

Thus, $k \approx 0.00942$.

To find the depth at which 1% of the surface light remains, we set $I = 0.01I_0$ and solve

$$0.01I_0 = I_0 e^{-0.00942d}$$

for d:

$$0.01 = e^{-0.00942d}$$

$$-0.00942d = \ln 0.01$$

$$d = \frac{\ln 0.01}{-0.00942} \approx 488.87$$

Thus, 1% of the surface light remains at approximately 489 feet.

(2-2, 2-3)

60. Using the model $P = P_0(1 + r)^t$, we must solve $2P_0 = P_0(1 + 0.03)^t$ for t:

$$2 = (1.03)^t$$

$$\ln(1.03)^t = \ln 2$$

$$t \ln(1.03) = \ln 2$$

$$t = \frac{\ln 2}{\ln 1.03} \approx 23.4$$

Thus, at a 3% growth rate, the population will double in approximately 23.4 years.

(2-2, 2-3)

61. Using the continuous compounding model, we have:

$$2P_0 = P_0 e^{0.03t}$$

$$2 = e^{0.03t}$$

$$0.03t = \ln 2$$

$$t = \frac{\ln 2}{0.03} \approx 23.1$$

Thus, the model predicts that the population will double in approximately 23.1 years.

(2-2, 2-3)

3 THE DERIVATIVE

Things to remember:

1. AVERAGE RATE OF CHANGE

   For $y = f(x)$, the AVERAGE RATE OF CHANGE FROM $x = a$ TO $x = a + h$ is

   $$\frac{f(a + h) - f(a)}{(a + h) - a} = \frac{f(a + h) - f(a)}{h} \qquad h \neq 0$$

   The expression $\dfrac{f(a + h) - f(a)}{h}$ is called the DIFFERENCE QUOTIENT.

2. INSTANTANEOUS RATE OF CHANGE

   For $y = f(x)$, the INSTANTANEOUS RATE OF CHANGE AT $x = a$ is

   $$\lim_{h \to 0} \frac{f(a + h) - f(a)}{h}$$

   if the limit exists.

3. SECANT LINE

   A line through two points on the graph of a function is called a SECANT LINE. If $(a, f(a))$ and $((a + h), f(a + h))$ are two points on the graph of $y = f(x)$, then

   $$\text{Slope of secant line} = \frac{f(a + h) - f(a)}{h} \quad \text{[Difference quotient]}$$

4. SLOPE OF A GRAPH

   For $y = f(x)$, the SLOPE OF THE GRAPH at the point $(a, f(a))$ is given by

   $$\lim_{h \to 0} \frac{f(a + h) - f(a)}{h}$$

   provided the limit exists. The slope of the graph is also the SLOPE OF THE TANGENT LINE at the point $(a, f(a))$.

---

1. $f(4) - f(1) = 3(4)^2 - 3(1)^2 = 48 - 3 = 45$

3. Average rate of change $= \dfrac{f(4) - f(1)}{4 - 1} = \dfrac{3(4)^2 - 3(1)^2}{3} = \dfrac{45}{3} = 15$

5. Slope of secant line $= \dfrac{f(4) - f(1)}{4 - 1} = \dfrac{3(4)^2 - 3(1)^2}{3} = \dfrac{45}{3} = 15$

7.

h	-0.1	-0.01	-0.001	$\rightarrow 0 \leftarrow$	0.001	0.01	0.1
$\dfrac{f(1 + h) - f(1)}{h}$	5.7	5.97	5.997	$\rightarrow 6 \leftarrow$	6.003	6.03	6.3

9. By Problem 7, instantaneous rate of change $= 6$.

11. By Problem 7, instantaneous velocity $= 6$ ft/sec.

13. By Problem 7, slope of the graph $= 6$.

15. Average velocity $= \dfrac{f(6) - f(4)}{6 - 4} = \dfrac{10(6)^2 - 10(4)^2}{2} = \dfrac{360 - 160}{2}$
$= 100$ ft/sec

17.

h	-0.1	-0.01	-0.001	$\rightarrow 0 \leftarrow$	0.001	0.01	0.1
$\dfrac{f(4 + h) - f(4)}{h}$	79	79.9	79.99	$\rightarrow 80 \leftarrow$	80.01	80.1	81

Instantaneous velocity at $x = 4$: 80 ft/sec

19. (A) Slope of secant line through $(2, f(2))$, $(4, f(4))$:

$\dfrac{f(4) - f(2)}{4 - 2} = \dfrac{[4^2 - 2(4) - 4] - [2^2 - 2(2) - 4]}{2} = \dfrac{4 - (-4)}{2} = 4$

(B) Slope of secant line through $(2, f(2))$, $(3, f(3))$:

$\dfrac{f(3) - f(2)}{3 - 2} = \dfrac{[3^2 - 2(3) - 4] - [2^2 - 2(2) - 4]}{1} = -1 - (-4) = 3$

(C)

h	-0.1	-0.01	-0.001	$\rightarrow 0 \leftarrow$	0.001	0.01	0.1
$\dfrac{f(2 + h) - f(2)}{h}$	1.9	1.99	1.999	$\rightarrow 2 \leftarrow$	2.001	2.01	2.1

(D)

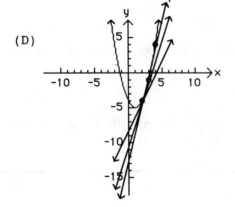

21. At $x = -1$, slope $= 1$; at $x = 3$, slope $= -2$

23. At $x = -3$, slope $= -5$; at $x = -1$, slope $= 0$; at $x = 1$, slope $= -1$;
at $x = 3$, slope $= 4$

25.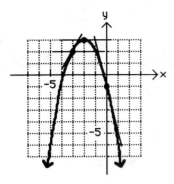

27.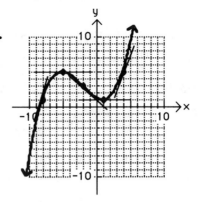

29. 8

31. 0.25 or $\frac{1}{4}$

33.

x	-3	-2	-1	0	1	2	3
slope	-3	-2	-1	0	1	2	3

slope function: $y = x$

35. The slope of the line is m. The slope of the graph of $f(x) = mx + b$ at any point on the graph is also m.

37.
$$\frac{f(1 + h) - f(1)}{h} = \frac{3(1 + h)^2 - 3(1)^2}{h} = \frac{3(1 + 2h + h^2) - 3}{h}$$
$$= \frac{3 + 6h + 3h^2 - 3}{h} = \frac{h(6 + 3h)}{h} = 6 + 3h;$$

As $h \to 0$, $6 + 3h \to 6$.
The slope of the graph at $x = 1$ is 6.

39.

h	-0.1	-0.01	-0.001	→	0	← 0.001	0.01	0.1
$\frac{f(0 + h) - f(0)}{h}$	-1	-1	-1	→ -1	≠ 1 ←	1	1	1

The slope of the graph is not defined at (0, 0).

41. (A) Average rate of change, 1975—1985: $\frac{34.9 - 36.1}{10} = -0.12$ hrs/yr

(B) Average rate of change, 1975—1990: $\frac{10.01 - 4.53}{15} = \$0.37$ per yr

43. (A) $R(1000) = 60(1000) - 0.025(1000)^2 = \$35,000$
$$\frac{R(1000 + h) - R(1000)}{h} = \frac{60(1000 + h) - 0.025(1000 + h)^2 - 35,000}{h}$$
$$= \frac{60,000 + 60h - 0.025(1,000,000 + 2000h + h^2) - 35,000}{h}$$
$$= \frac{10h - 0.025h^2}{h} = 10 - 0.025h \to 10 \text{ as } h \to 0$$

At a production level of 1,000 car seats, the revenue is \$35,000 and is INCREASING at the rate of \$10 per seat.

(B) $R(1300) = 60(1300) - 0.025(1300)^2 = \$35,750$

$$\frac{R(1300 + h) - R(1300)}{h} = \frac{60(1300 + h) - 0.025(1300 + h)^2 - 35,750}{h}$$

$$= \frac{78,000 + 60h - 0.025(1,690,000 + 2600h + h^2) - 35,750}{h}$$

$$= \frac{-5h - 0.025h^2}{h} = -5 - 0.025h \to -5 \text{ as } h \to 0$$

At a production level of 1,300 car seats, the revenue is \$35,750 and is DECREASING at the rate of \$5 per seat.

45. $f(3) = -150(3)^2 + 770(3) + 10,400 = 11,360.$

$$\frac{f(3 + h) - f(3)}{h} = \frac{-150(3 + h)^2 + 770(3 + h) + 10,400 - 11,360}{h}$$

$$= \frac{-150(9 + 6h + h^2) + 2310 + 770h + 10,400 - 11,360}{h}$$

$$= \frac{-130h - 150h^2}{h} = -130 - 150h \to -130 \text{ as } h \to 0.$$

In 1990, the annual production was 11,360,000 metric tons and was DECREASING at the rate of 130,000 metric tons per year.

47. (A) Average rate of change, 1988—1991: $\dfrac{728.6 - 526.2}{3} \approx \$67.5$ billion/yr

(B) Average rate of change, 1987—1989: $\dfrac{20.7 - 17.3}{2} = \$1.7$ billion/yr

49. $f(30) = 0.008(30)^2 - 0.9(30) + 29.6 = 9.8$

$$\frac{f(30 + h) - f(30)}{h} = \frac{0.008(30 + h)^2 - 0.9(30 + h) + 29.6 - 9.8}{h}$$

$$= \frac{0.008(900 + 60h + h^2) - 27 - 0.9h + 29.6 - 9.8}{h}$$

$$= \frac{-0.42h + 0.008h^2}{h} = -0.42 + 0.008h$$

In 1990, the number of male infant deaths per 100,000 births was 9.8 and was DECREASING at the rate of 0.42 deaths per 100,000 births per year.

EXERCISE 3-2

Things to remember:

<u>1</u>. LIMIT

We write

$$\lim_{x \to c} f(x) = L \text{ or } f(x) \to L \text{ as } x \to c$$

if the functional value $f(x)$ is close to the single real number L whenever x is close to but not equal to c (on either side of c).

[Note: The existence of a limit at c has nothing to do with the value of the function at c. In fact, c may not even be in the domain of f(see Examples 2 and 3). However, the function must be defined on both sides of c.]

2. ONE-SIDED LIMITS

We write $\lim\limits_{x \to c^-} f(x) = K$ [$x \to c^-$ is read "x approaches c from the left" and means $x \to c$ and $x < c$] and call K the LIMIT FROM THE LEFT or LEFT-HAND LIMIT if $f(x)$ is close to K whenever x is close to c, but to the left of c on the real number line.

We write $\lim\limits_{x \to c^+} f(x) = L$ [$x \to c^+$ is read "x approaches c from the right" and means $x \to c$ and $x > c$] and call L the LIMIT FROM THE RIGHT or RIGHT-HAND LIMIT if $f(x)$ is close to L whenever x is close to c, but to the right of c on the real number line.

3. EXISTENCE OF A LIMIT

In order for a limit to exist, the limit from the left and the limit from the right must both exist, and must be equal.

4. PROPERTIES OF LIMITS

Let f and g be two functions and assume that

$$\lim_{x \to c} f(x) = L \qquad \lim_{x \to c} g(x) = M$$

where L and M are real numbers (both limits exist). Then,

(a) $\lim\limits_{x \to c}[f(x) + g(x)] = \lim\limits_{x \to c} f(x) + \lim\limits_{x \to c} g(x) = L + M,$

(b) $\lim\limits_{x \to c}[f(x) - g(x)] = \lim\limits_{x \to c} f(x) - \lim\limits_{x \to c} g(x) = L - M,$

(c) $\lim\limits_{x \to c} kf(x) = k \lim\limits_{x \to c} f(x) = kL$ for any constant $k,$

(d) $\lim\limits_{x \to c}[f(x)g(x)] = \left(\lim\limits_{x \to c} f(x)\right)\left(\lim\limits_{x \to c} g(x)\right) = LM,$

(e) $\lim\limits_{x \to c} \dfrac{f(x)}{g(x)} = \dfrac{\lim\limits_{x \to c} f(x)}{\lim\limits_{x \to c} g(x)} = \dfrac{L}{M}$ if $M \neq 0,$

(f) $\lim\limits_{x \to c} \sqrt[n]{f(x)} = \sqrt[n]{\lim\limits_{x \to c} f(x)} = \sqrt[n]{L}$ ($L \geq 0$ for n even).

5. LIMIT OF A POLYNOMIAL FUNCTION

If $f(x)$ is a polynomial function and c is any real number, then

$$\lim_{x \to c} f(x) = f(c)$$

1.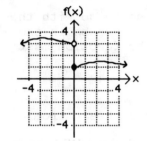

f(x)

3.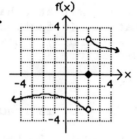

f(x)

5.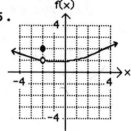

f(x)

7. (A) $\lim\limits_{x\to 0^-} f(x) = 2$ (B) $\lim\limits_{x\to 0^+} f(x) = 2$ (C) $\lim\limits_{x\to 0} f(x) = 2$ (D) $f(0) = 2$

9. (A) $\lim\limits_{x\to 2^-} f(x) = 1$ (B) $\lim\limits_{x\to 2^+} f(x) = 2$ (C) $\lim\limits_{x\to 2} f(x)$ does not exist

     (D) $f(2) = 2$

11. (A) $\lim\limits_{x\to 1^-} g(x) = 1$ (B) $\lim\limits_{x\to 1^+} g(x) = 2$ (C) $\lim\limits_{x\to 1} g(x) =$ does not exist

     (D) $g(1)$ does not exist

13. (A) $\lim\limits_{x\to 3^-} g(x) = 1$ (B) $\lim\limits_{x\to 3^+} g(x) = 1$ (C) $\lim\limits_{x\to 3} g(x) = 1$ (D) $g(3) = 3$

15. $\lim\limits_{x\to 3}[f(x) - g(x)] = \lim\limits_{x\to 3} f(x) - \lim\limits_{x\to 3} g(x)$ [Property $\underline{4}$(b)]

$$= 5 - 9 = -4$$

17. $\lim\limits_{x\to 3} 4g(x) = 4 \lim\limits_{x\to 3} g(x)$ [Property $\underline{4}$(c)]

$$= 4 \cdot 9 = 36$$

19. $\lim\limits_{x\to 3} \dfrac{f(x)}{g(x)} = \dfrac{\lim\limits_{x\to 3} f(x)}{\lim\limits_{x\to 3} g(x)}$ [since $\lim\limits_{x\to 3} g(x) \neq 0$, Property $\underline{4}$(e)]

$$= \frac{5}{9}$$

21. $\lim\limits_{x\to 3} \sqrt{f(x)} = \sqrt{\lim\limits_{x\to 3} f(x)} = \sqrt{5}$ [Property $\underline{4}$(f)]

23. $\lim\limits_{x\to 3} \dfrac{f(x) + g(x)}{2f(x)} = \dfrac{\lim\limits_{x\to 3}[f(x) + g(x)]}{\lim\limits_{x\to 3} 2f(x)}$ [since $\lim\limits_{x\to 3} 2f(x) \neq 0$]

$$= \frac{\lim\limits_{x\to 3} f(x) + \lim\limits_{x\to 3} g(x)}{2 \cdot \lim\limits_{x\to 3} f(x)} = \frac{5 + 9}{2 \cdot 5} = \frac{14}{10} = \frac{7}{5} = 1.4$$

25. $\lim\limits_{x\to 5}(2x^2 - 3) = 2(5)^2 - 3$ [since $f(x) = 2x^2 - 3$ is continuous at $x = 5$]

$$= 47$$

27. $\lim\limits_{x \to 2} \dfrac{5x}{2 + x^2} = \dfrac{5(2)}{2 + (2)^2}$ $\left[f(x) = \dfrac{5x}{2 + x^2} \text{ is continuous at } x = 2 \right]$

$\qquad\qquad\qquad = \dfrac{10}{6} = \dfrac{5}{3}$

29. $\lim\limits_{x \to 2} (x + 1)^3 (2x - 1)^2 = \lim\limits_{x \to 2} (x + 1)^3 \cdot \lim\limits_{x \to 2} (2x - 1)^2$

$\qquad\qquad\qquad\qquad = (2 + 1)^3 \cdot (2 \cdot 2 - 1)^2 = 3^3 \cdot 3^2 = 3^5 = 243$

31. $\lim\limits_{x \to -1} \sqrt{5 - 4x} = \sqrt{\lim\limits_{x \to -1} (5 - 4x)} = \sqrt{5 - 4(-1)} = \sqrt{9} = 3$

33. Since $\dfrac{x^2 - 9}{x + 3} = \dfrac{(x + 3)(x - 3)}{x + 3} = x - 3, \; x \neq -3,$

$\lim\limits_{x \to -3} \dfrac{x^2 - 9}{x + 3} = \lim\limits_{x \to -3} (x - 3) = -6.$

35. For $x > 1$, $|x - 1| = x - 1$. Thus, $\dfrac{|x - 1|}{x - 1} = \dfrac{x - 1}{x - 1} = 1$ for $x > 1$;

and $\lim\limits_{x \to 1^+} \dfrac{|x - 1|}{x - 1} = 1.$

37. For $x < 1$, $|x - 1| = -(x - 1)$. Thus, $\dfrac{|x - 1|}{x - 1} = \dfrac{-(x - 1)}{x - 1} = -1$

for $x < 1$, and $\lim\limits_{x \to 1^-} \dfrac{|x - 1|}{x - 1} = -1.$

39. It follows from Problems 35 and 37 that $\lim\limits_{x \to 1} \dfrac{|x - 1|}{x - 1}$ does not exist.

41. $\lim\limits_{x \to 1} \dfrac{x - 2}{x^2 - 2x} = \dfrac{1 - 2}{1 - 2} = 1$ $\left[\underline{\text{Note:}} \; f(x) = \dfrac{x - 2}{x^2 - 2x} \text{ is continuous at } x = 1. \right]$

43. $\lim\limits_{x \to 2} \dfrac{x - 2}{x^2 - 2x}$ is a 0/0 indeterminate form.

Thus, we try to manipulate the expression algebraically.

$\dfrac{x - 2}{x^2 - 2x} = \dfrac{x - 2}{x(x - 2)} = \dfrac{1}{x}, \; x \neq 2$

Now, $\lim\limits_{x \to 2} \dfrac{x - 2}{x^2 - 2x} = \lim\limits_{x \to 2} \dfrac{1}{x} = \dfrac{1}{2}.$

45. $\lim\limits_{x \to 2} \dfrac{x^2 - x - 6}{x + 2} = \dfrac{2^2 - 2 - 6}{2 + 2} = -1$

47. $\lim\limits_{x \to -2} \dfrac{x^2 - x - 6}{x + 2}$ is a 0/0 indeterminate form.

$\dfrac{x^2 - x - 6}{x + 2} = \dfrac{(x - 3)(x + 2)}{x + 2} = x - 3, \; x \neq -2$

Thus, $\lim\limits_{x \to -2} \dfrac{x^2 - x - 6}{x + 2} = \lim\limits_{x \to -2} (x - 3) = -5.$

49. $\lim\limits_{x\to 3}\left(\dfrac{x}{x+3} + \dfrac{x-3}{x^2-9}\right) = \lim\limits_{x\to 3}\dfrac{x}{x+3} + \lim\limits_{x\to 3}\dfrac{x-3}{x^2-9} = \dfrac{3}{6} + \lim\limits_{x\to 3}\dfrac{x-3}{x^2-9}$

Now, $\dfrac{x-3}{x^2-9} = \dfrac{x-3}{(x-3)(x+3)} = \dfrac{1}{x+3},\ x \neq 3.$

Thus, $\lim\limits_{x\to 3}\left(\dfrac{x}{x+3} + \dfrac{x-3}{x^2-9}\right) = \dfrac{1}{2} + \lim\limits_{x\to 3}\dfrac{1}{x+3} = \dfrac{1}{2} + \dfrac{1}{6} = \dfrac{2}{3}.$

51. $\lim\limits_{x\to 0}\left(\sqrt{x^2+9} - \dfrac{x^2+3x}{x}\right) = \lim\limits_{x\to 0}\sqrt{x^2+9} - \lim\limits_{x\to 0}\dfrac{x(x+3)}{x}$

$= \sqrt{\lim\limits_{x\to 0}(x^2+9)} - \lim\limits_{x\to 0}(x+3) = \sqrt{9} - 3 = 0$

53. $f(x) = 3x + 1$

$\lim\limits_{h\to 0}\dfrac{f(2+h)-f(2)}{h} = \lim\limits_{h\to 0}\dfrac{3(2+h)+1-(3\cdot 2+1)}{h}$

$= \lim\limits_{h\to 0}\dfrac{6+3h+1-7}{h} = \lim\limits_{h\to 0}\dfrac{3h}{h} = \lim\limits_{h\to 0}3 = 3$

55. $f(x) = x^2 + 1$

$\lim\limits_{h\to 0}\dfrac{f(2+h)-f(2)}{h} = \lim\limits_{h\to 0}\dfrac{(2+h)^2+1-(2^2+1)}{h}$

$= \lim\limits_{h\to 0}\dfrac{4+4h+h^2+1-5}{h} = \lim\limits_{h\to 0}\dfrac{4h+h^2}{h}$

$= \lim\limits_{h\to 0}(4+h) = 4$

57. $f(x) = 5$

$\lim\limits_{h\to 0}\dfrac{f(2+h)-f(2)}{h} = \lim\limits_{h\to 0}\dfrac{5-5}{h} = \lim\limits_{h\to 0}0 = 0$

59. $f(x) = \sqrt{x} - 2$

$\lim\limits_{h\to 0}\dfrac{f(2+h)-f(2)}{h} = \lim\limits_{h\to 0}\dfrac{\sqrt{2+h}-2-(\sqrt{2}-2)}{h} = \lim\limits_{h\to 0}\dfrac{\sqrt{2+h}-\sqrt{2}}{h}$

$= \lim\limits_{h\to 0}\dfrac{\sqrt{2+h}-\sqrt{2}}{h}\cdot\dfrac{\sqrt{2+h}+\sqrt{2}}{\sqrt{2+h}+\sqrt{2}} = \lim\limits_{h\to 0}\dfrac{2+h-2}{h(\sqrt{2+h}+\sqrt{2})}$

$= \lim\limits_{h\to 0}\dfrac{h}{h(\sqrt{2+h}+\sqrt{2})} = \lim\limits_{h\to 0}\dfrac{1}{\sqrt{2+h}+\sqrt{2}} = \dfrac{1}{2\sqrt{2}}$

61. $f(x) = |x-2| - 3$

$\lim\limits_{h\to 0}\dfrac{f(2+h)-f(2)}{h} = \lim\limits_{h\to 0}\dfrac{|(2+h)-2|-3-(|2-2|-3)}{h}$

$= \lim\limits_{h\to 0}\dfrac{|h|-3+3}{h} = \lim\limits_{h\to 0}\dfrac{|h|}{h}$ does not exist.

63. Slope of the graph of $y = x^2 - 3$ at $(2, 1)$:

$\lim\limits_{h\to 0}\dfrac{f(2+h)-f(2)}{h} = \lim\limits_{h\to 0}\dfrac{(2+h)^2-3-1}{h} = \lim\limits_{h\to 0}\dfrac{4+4h+h^2-4}{h}$

$= \lim\limits_{h\to 0}4+h = 4$

65. Slope of the graph of $y = \sqrt{x}$ at $(4, 2)$:

$$\lim_{h \to 0} \frac{\sqrt{4+h} - \sqrt{4}}{h} = \lim_{h \to 0} \frac{\sqrt{4+h} - 2}{h} \cdot \frac{\sqrt{4+h} + 2}{\sqrt{4+h} + 2} = \lim_{h \to 0} \frac{4 + h - 4}{h[\sqrt{4+h} + 2]}$$

$$= \lim_{h \to 0} \frac{1}{\sqrt{4+h} + 2} = \frac{1}{4}$$

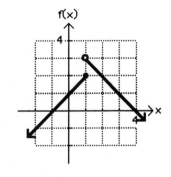

67. Instantaneous velocity at $x = 4$:

$$\lim_{h \to 0} \frac{f(4 + h) - f(4)}{h} = \lim_{h \to 0} \frac{10(4+h)^2 - 160}{h} = \lim_{h \to 0} \frac{10(16 + 8h + h^2) - 160}{h}$$

$$= \lim_{h \to 0} \frac{80h + 10h^2}{h} = \lim_{h \to 0} 80 + 10h = 80 \text{ ft/sec}$$

69. At $x = -3$, slope $= -1$; at $x = -1$, slope $= 0$; at $x = 3$, slope $= 2$.

71. (A)

x	-0.1	-0.01	-0.001	-0.0001	$\to 0$
$\frac{1}{x}$	-10	-100	-1000	-10,000	

The limit does not exist; the values of $\frac{1}{x}$ are increasingly large negative numbers as x approaches 0 from the left.

(B)

x	0	$\leftarrow$	0.0001	0.001	0.01	0.1
$\frac{1}{x}$			10,000	1000	100	10

The limit does not exist; the values of $\frac{1}{x}$ are increasingly large positive numbers as x approaches 0 from the right.

73. (A) $\lim\limits_{x \to 1^-} f(x) = \lim\limits_{x \to 1^-} (1 + x) = 2$

$\lim\limits_{x \to 1^+} f(x) = \lim\limits_{x \to 1^+} (4 - x) = 3$

(B) $\lim\limits_{x\to 1^-} f(x) = \lim\limits_{x\to 1^-} (1 + 2x) = 3$

   $\lim\limits_{x\to 1^+} f(x) = \lim\limits_{x\to 1^+} (4 - 2x) = 2$

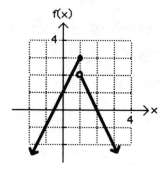

(C) $\lim\limits_{x\to 1^-} f(x) = \lim\limits_{x\to 1^-} (1 + mx) = 1 + m$

   $\lim\limits_{x\to 1^+} f(x) = \lim\limits_{x\to 1^+} (4 - mx) = 4 - m$

   $1 + m = 4 - m$

   $2m = 3$

   $m = \dfrac{3}{2}$

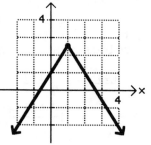

(D) The graph in (A) is broken at $x = 1$; it jumps up from $(1, 2)$ to $(1, 3)$.

   The graph in (B) is also broken at $x = 1$; it jumps down from $(1, 3)$ to $(1, 2)$.

   The graph in (C) is not broken; the two pieces meet at $\left(1, \dfrac{5}{2}\right)$.

75. $\lim\limits_{h\to 0} \dfrac{(a + h)^2 - a^2}{h} = \lim\limits_{h\to 0} \dfrac{a^2 + 2ah + h^2 - a^2}{h}$

$\qquad\qquad = \lim\limits_{h\to 0} \dfrac{2ah + h^2}{h} = \lim\limits_{h\to 0}(2a + h) = 2a$

77. $\lim\limits_{h\to 0} \dfrac{\sqrt{a + h} - \sqrt{a}}{h} = \lim\limits_{h\to 0} \dfrac{\sqrt{a + h} - \sqrt{a}}{h} \cdot \dfrac{\sqrt{a + h} + \sqrt{a}}{\sqrt{a + h} + \sqrt{a}} = \lim\limits_{h\to 0} \dfrac{(a + h) - a}{h(\sqrt{a + h} + \sqrt{a})}$

$\qquad\qquad = \lim\limits_{h\to 0} \dfrac{1}{\sqrt{a + h} + \sqrt{a}} = \dfrac{1}{2\sqrt{a}}$

79. (A) At $a = 1$,

$\qquad \lim\limits_{h\to 0} \dfrac{f(1 + h) - f(1)}{h} = \lim\limits_{h\to 0} \dfrac{(1 + h)^2 - 3(1 + h) + 1 - (-1)}{h}$

$\qquad\qquad\qquad = \lim\limits_{h\to 0} \dfrac{1 + 2h + h^2 - 3 - 3h + 2}{h}$

$\qquad\qquad\qquad = \lim\limits_{h\to 0} \dfrac{-h + h^2}{h} = \lim\limits_{h\to 0} (-1 + h) = -1$

At $a = 2$,
$$\lim_{h \to 0} \frac{f(2 + h) - f(2)}{h} = \lim_{h \to 0} \frac{(2 + h)^2 - 3(2 + h) + 1 - (-1)}{h}$$
$$= \lim_{h \to 0} \frac{4 + 4h + h^2 - 6 - 3h + 2}{h}$$
$$= \lim_{h \to 0} \frac{h + h^2}{h} = \lim_{h \to 0} (1 + h) = 1$$

At $a = 3$,
$$\lim_{h \to 0} \frac{f(3 + h) - f(3)}{h} = \lim_{h \to 0} \frac{(3 + h)^2 - 3(3 + h) + 1 - (1)}{h}$$
$$= \lim_{h \to 0} \frac{9 + 6h + h^2 - 9 - 3h}{h}$$
$$= \lim_{h \to 0} \frac{3h + h^2}{h} = \lim_{h \to 0} (3 + h) = 3$$

(B) $\lim\limits_{h \to 0} \dfrac{f(a + h) - f(a)}{h} = \lim\limits_{h \to 0} \dfrac{(a + h)^2 - 3(a + h) + 1 - [a^2 - 3a + 1]}{h}$
$$= \lim_{h \to 0} \frac{a^2 + 2ah + h^2 - 3a - 3h - a^2 + 3a}{h}$$
$$= \lim_{h \to 0} \frac{(2a - 3)h + h^2}{h}$$
$$= \lim_{h \to 0} (2a - 3 + h) = 2a - 3$$

(C) At $a = 1$, $2(1) - 3 = -1$; at $a = 2$, $2(2) - 3 = 1$;
at $a = 3$, $2(3) - 3 = 3$

The slopes are the same as in part (A). In part (A), each value required a new limit operation. In part (B) a single limit operation produces the slope for all values of a.

81.

x	0.9	0.99	0.999	→ 1 ←	1.001	1.01	1.1
$\dfrac{x^{10} - 1}{x - 1}$	6.513	9.562	9.955	→ 10 ←	10.045	10.462	15.937

$$\lim_{x \to 1} \frac{x^{10} - 1}{x - 1} = 10$$

83.

x	-0.1	-0.01	-0.001	→ 0	← 0.001	0.01	0.1
$\dfrac{2^x - 1}{x}$	0.670	0.691	0.693	→ 0.693	← 0.693	0.696	0.718

$$\lim_{x \to 0} \frac{2^x - 1}{x} \approx 0.693$$

85.

x	-0.1	-0.01	-0.001	$\to$ 0	$\leftarrow$ 0.001	0.01	0.1
$(1 + x)^{1/x}$	2.868	2.732	2.7196	$\to$ 2.718	$\leftarrow$ 2.7169	2.705	2.594

$\lim\limits_{x \to 0} (1 + x)^{1/x} \approx 2.718$

87. Typical values of n are 95 on a TI-81, 94 on a TI-82 and 126 on a TI-85.

89. $\lim\limits_{x \to -2^-} f(x) = \lim\limits_{x \to -2^+} f(x) = 2$

$\lim\limits_{x \to 2^-} f(x) = \lim\limits_{x \to 2^+} f(x) = 2$

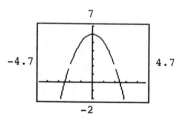

91. $\lim\limits_{x \to 2^-} f(x) = -4, \quad \lim\limits_{x \to 2^+} f(x) = 4$

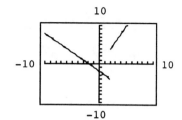

93. $\lim\limits_{x \to -3^-} f(x) = -3, \quad \lim\limits_{x \to -3^+} f(x) = 3,$
$\lim\limits_{x \to 3^-} f(x) = -3, \quad \lim\limits_{x \to 3^+} f(x) = 3$

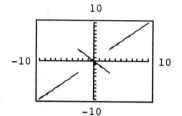

95. (A) $R(900) = 200(900) - 0.1(900)^2 = \$99{,}000$

$\lim\limits_{h \to 0} \dfrac{R(900 + h) - R(900)}{h} = \lim\limits_{h \to 0} \dfrac{200(900 + h) - 0.1(900 + h)^2 - 99{,}000}{h}$

$= \lim\limits_{h \to 0} \dfrac{180{,}000 + 200h - 0.1(810{,}000 + 1800h + h^2) - 99{,}000}{h}$

$= \lim\limits_{h \to 0} \dfrac{20h - 0.1h^2}{h} = \lim\limits_{h \to 0} (20 - 0.1h) = 20$

At a production level of 900 units, the revenue is \$99,000 and is INCREASING at the rate of \$20 per jigsaw.

(B) $R(1{,}200) = 200(1{,}200) - 0.1(1{,}200)^2 = \$96{,}000$

$\lim\limits_{h \to 0} \dfrac{R(1{,}200 + h) - R(1{,}200)}{h}$

$= \lim\limits_{h \to 0} \dfrac{200(1{,}200 + h) - 0.1(1{,}200 + h)^2 - 96{,}000}{h}$

$= \lim\limits_{h \to 0} \dfrac{-40h - 0.1h^2}{h} = \lim\limits_{h \to 0} (-40 - 0.1h) = -40$

At a production level of 1,200 units, the revenue is \$96,000 and is DECREASING at the rate of \$40 per jigsaw.

97. $f(20) = 0.62(20)^2 - 20 + 5.1 = \$233.1$ billion

$$\lim_{h \to 0} \frac{f(20 + h) - f(20)}{h} = \lim_{h \to 0} \frac{0.62(20 + h)^2 - (20 + h) + 5.1 - 233.1}{h}$$

$$= \lim_{h \to 0} \frac{0.62(400 + 40h + h^2) - 20 - h + 5.1 - 233.1}{h}$$

$$= \lim_{h \to 0} \frac{23.8h + 0.62h^2}{h} = \lim_{h \to 0} (23.8 + 0.62h) = 23.8$$

In 1990, the revolving credit debt was \$233.1 bilion and was INCREASING at the rate of \$23.8 billion per year.

99. At $x = 9$(thousand) $y = 0.04(9)^2 - 3.66(9) + 100 = 70.3\%$.
Let $f(x) = 0.04x^2 - 3.66x + 100$. Then

$$\lim_{h \to 0} \frac{f(9 + h) - f(9)}{h} = \lim_{h \to 0} \frac{0.04(9 + h)^2 - 3.66(9 + h) + 100 - 70.3}{h}$$

$$= \lim_{h \to 0} \frac{0.04(81 + 18h + h^2) - 32.94 - 3.66h + 29.7}{h}$$

$$= \lim_{h \to 0} \frac{-2.94h + 0.04h^2}{h} = \lim_{h \to 0} (-2.94 + 0.04h)$$

$$= -2.94$$

The aveolar pressure at 9,000 feet is 70.3% of that at sea level and is DECREASING at the rate of 2.94% per 1,000 feet.

101. Let $y = f(t) = -0.03t^2 + 1.5t + 32$.

(A) $f(20) = -0.03(20)^2 + 1.5(20) + 32 = 50$ million

$$\lim_{h \to 0} \frac{f(20 + h) - f(20)}{h} = \lim_{h \to 0} \frac{-0.03(20 + h)^2 + 1.5(20 + h) + 32 - 50}{h}$$

$$= \lim_{h \to 0} \frac{-0.03(400 + 40h + h^2) + 30 + 1.5h - 18}{h}$$

$$= \lim_{h \to 0} \frac{0.3h - 0.03h^2}{h} = \lim_{h \to 0} (0.3 - 0.03h)$$

$$= 0.3$$

The school age population in 1970 was 50 million and was INCREASING at the rate of 0.3 million per year.

(B) $f(40) = -0.03(40)^2 + 1.5(40) + 32 = 44$ million

$$\lim_{h \to 0} \frac{f(40 + h) - f(40)}{h} = \lim_{h \to 0} \frac{-0.03(40 + h)^2 + 1.5(40 + h) + 32 - 44}{h}$$

$$= \lim_{h \to 0} \frac{-0.03(1,600 + 80h + h^2) + 60 + 1.5h - 12}{h}$$

$$= \lim_{h \to 0} \frac{-0.9h - 0.03h^2}{h} = \lim_{h \to 0} (-0.9 - 0.03)h$$

$$= -0.9$$

The school age population in 1990 was 44 million and was DECREASING at the rate of 0.9 million per year.

Things to remember:

<u>1</u>. THE DERIVATIVE

For $y = f(x)$, we define THE DERIVATIVE OF f AT x, denoted by $f'(x)$, to be

$$f'(x) = \lim_{h \to 0} \frac{f(x + h) - f(x)}{h} \quad \text{if the limit exists.}$$

If $f'(x)$ exists for each x in the open interval (a, b), then f is said to be DIFFERENTIABLE OVER (a, b).

<u>2</u>. INTERPRETATIONS OF THE DERIVATIVE

The derivative of a function f is a new function f'. The domain of f' is a subset of the domain of f. Interpretations of the derivative are:

a. Slope of the tangent line. For each x in the domain of f', $f'(x)$ is the slope of the line tangent to the graph of f at the point $(x, f(x))$.

b. Instantaneous rate of change. For each x in the domain of f', $f'(x)$ is the instantaneous rate of change of $y = f(x)$ with respect to x.

1. (A) $\dfrac{f(1) - f(0)}{1 - 0} = \dfrac{0 - (-1)}{1} = 1$; slope of secant line through $(0, f(0))$ and $(1, f(1))$.

(B) $\dfrac{f(1 + h) - f(1)}{h} = \dfrac{(1 + h)^2 - 1 - 0}{h} = \dfrac{1 + 2h + h^2 - 1}{h} = \dfrac{h(2 + h)}{h}$
$= 2 + h$, $h \neq 0$, slope of secant line through $(1 + h, f(1 + h))$ and $(1, f(1))$.

(C) $\lim\limits_{h \to 0} \dfrac{f(1 + h) - f(1)}{h} = \lim\limits_{h \to 0} \dfrac{(1 + h)^2 - 1 - 0}{h} = \lim\limits_{h \to 0} (2 + h) = 2$;
slope of tangent line at $(1, f(1))$.

3. $\lim\limits_{h \to 0} \dfrac{f(x + h) - f(x)}{h} = \lim\limits_{h \to 0} \dfrac{4hx - 3h + 2h^2}{h} = \lim\limits_{h \to 0} \dfrac{h(4x - 3 + 2h)}{h}$
$$= \lim\limits_{h \to 0} (4x - 3 + 2h) = 4x - 3$$

5. $\lim\limits_{h \to 0} \dfrac{f(x + h) - f(x)}{h} = \lim\limits_{h \to 0} \dfrac{3hx^2 - 2xh + 3h^2x - h^2 + h^3}{h}$
$$= \lim\limits_{h \to 0} \dfrac{h(3x^2 - 2x + 3hx - h + h^2)}{h}$$
$$= \lim\limits_{h \to 0} (3x^2 - 2x + 3hx - h + h^2) = 3x^2 - 2x$$

7. $f(x) = 2x - 3$

    Step 1. Simplify $\dfrac{f(x + h) - f(x)}{h}$.

$$\frac{f(x + h) - f(x)}{h} = \frac{2(x + h) - 3 - (2x - 3)}{h}$$

$$= \frac{2x + 2h - 3 - 2x + 3}{h} = \frac{2h}{h} = 2$$

    Step 2. Evaluate $\lim\limits_{h \to 0} \dfrac{f(x + h) - f(x)}{h}$.

$$\lim_{h \to 0} \frac{f(x + h) - f(x)}{h} = \lim_{h \to 0} 2 = 2$$

    Thus, $f'(x) = 2$. Now $f'(1) = 2$, $f'(2) = 2$, $f'(3) = 2$.

9. $f(x) = 2 - x^2$

    Step 1. Simplify $\dfrac{f(x + h) - f(x)}{h}$.

$$\frac{f(x + h) - f(x)}{h} = \frac{2 - (x + h)^2 - (2 - x^2)}{h}$$

$$= \frac{2 - (x^2 + 2xh + h^2) - 2 + x^2}{h}$$

$$= \frac{-2xh - h^2}{h} = -2x - h$$

    Step 2. Evaluate $\lim\limits_{h \to 0} \dfrac{f(x + h) - f(x)}{h}$.

$$\lim_{h \to 0} \frac{f(x + h) - f(x)}{h} = \lim_{h \to 0}(-2x - h) = -2x$$

    Thus, $f'(x) = -2x$. Now $f'(1) = -2$, $f'(2) = -4$, $f'(3) = -6$.

11. At $x = -3$, $f'(-3) = -1$; at $x = 3$, $f'(3) = 2$; $f'(-1) = 0$

13. At $x = -5$, $f'(-5) = -3$; at $x = 1$, $f'(1) = 1$; at $x = 7$, $f'(7) = -3$; $f'(-2) = 0$, $f'(4) = 0$

15. $y = f(x) = x^2 + x$

    (A) $f(1) = 1^2 + 1 = 2$, $f(3) = 3^2 + 3 = 12$

        Slope of secant line: $\dfrac{f(3) - f(1)}{3 - 1} = \dfrac{12 - 2}{2} = 5$

    (B) $f(1) = 2$, $f(1 + h) = (1 + h)^2 + (1 + h) = 1 + 2h + h^2 + 1 + h$

                       $= 2 + 3h + h^2$

        Slope of secant line: $\dfrac{f(1 + h) - f(1)}{h} = \dfrac{2 + 3h + h^2 - 2}{h} = 3 + h$

    (C) Slope of tangent line at $(1, f(1))$:

$$\lim_{h \to 0} \frac{f(1 + h) - f(1)}{h} = \lim_{h \to 0}(3 + h) = 3$$

    (D) Equation of tangent line at $(1, f(1))$:

        $y - f(1) = f'(1)(x - 1)$ or $y - 2 = 3(x - 1)$ and $y = 3x - 1$.

17. $f(x) = x^2 + x$

(A) Average velocity: $\dfrac{f(3) - f(1)}{3 - 1} = \dfrac{3^2 + 3 - (1^2 + 1)}{2} = \dfrac{12 - 2}{2}$

$$= 5 \text{ meters/sec.}$$

(B) Average velocity: $\dfrac{f(1 + h) - f(1)}{h} = \dfrac{(1 + h)^2 + (1 + h) - (1^2 + 1)}{h}$

$$= \dfrac{1 + 2h + h^2 + 1 + h - 2}{h}$$

$$= \dfrac{3h + h^2}{h} = 3 + h \text{ meters/sec.}$$

(C) Instantaneous velocity: $\displaystyle\lim_{h \to 0} \dfrac{f(1 + h) - f(1)}{h}$

$$= \lim_{h \to 0}(3 + h) = 3 \text{ meters/sec.}$$

19. $f(x) = 6x - x^2$

Step 1. Simplify $\dfrac{f(x + h) - f(x)}{h}$.

$$\dfrac{f(x + h) - f(x)}{h} = \dfrac{6(x + h) - (x + h)^2 - (6x - x^2)}{h}$$

$$= \dfrac{6x + 6h - (x^2 + 2xh + h^2) - 6x + x^2}{h}$$

$$= \dfrac{6h - 2xh - h^2}{h} = 6 - 2x - h$$

Step 2. Evaluate $\displaystyle\lim_{h \to 0} \dfrac{f(x + h) - f(x)}{h}$.

$$\lim_{h \to 0} \dfrac{f(x + h) - f(x)}{h} = \lim_{h \to 0}(6 - 2x - h) = 6 - 2x$$

Therefore, $f'(x) = 6 - 2x$. $f'(1) = 6 - 2(1) = 4$, $f'(2) = 6 - 2(2) = 2$, $f'(3) = 6 - 2(3) = 0$.

21. $f(x) = \sqrt{x} - 3$

Step 1. Simplify $\dfrac{f(x + h) - f(x)}{h}$

$$\dfrac{f(x + h) - f(x)}{h} = \dfrac{\sqrt{x + h} - 3 - (\sqrt{x} - 3)}{h}$$

$$= \dfrac{\sqrt{x + h} - \sqrt{x}}{h} \cdot \dfrac{\sqrt{x + h} + \sqrt{x}}{\sqrt{x + h} + \sqrt{x}}$$

$$= \dfrac{x + h - x}{h(\sqrt{x + h} + \sqrt{x})} = \dfrac{1}{\sqrt{x + h} + \sqrt{x}}$$

Step 2. Evaluate $\displaystyle\lim_{h \to 0} \dfrac{f(x + h) - f(x)}{h}$.

$$\lim_{h \to 0} \dfrac{f(x + h) - f(x)}{h} = \lim_{h \to 0} \dfrac{1}{\sqrt{x + h} + \sqrt{x}} = \dfrac{1}{2\sqrt{x}}$$

Therefore, $f'(x) = \dfrac{1}{2\sqrt{x}}$. $f'(1) = \dfrac{1}{2\sqrt{1}} = \dfrac{1}{2}$, $f'(2) = \dfrac{1}{2\sqrt{2}}$, $f'(3) = \dfrac{1}{2\sqrt{3}}$.

23. $f(x) = -\dfrac{1}{x}$

Step 1. Simplify $\dfrac{f(x + h) - f(x)}{h}$.

$$\dfrac{f(x + h) - f(x)}{h} = \dfrac{-\dfrac{1}{x + h} - \left(-\dfrac{1}{x}\right)}{h} = \dfrac{-\dfrac{1}{x + h} + \dfrac{1}{x}}{h}$$

$$= \dfrac{\dfrac{-x + x + h}{x(x + h)}}{h} = \dfrac{1}{x(x + h)}$$

Step 2. Evaluate $\lim\limits_{h \to 0} \dfrac{f(x + h) - f(x)}{h}$.

$$\lim_{h \to 0} \dfrac{f(x + h) - f(x)}{h} = \lim_{h \to 0} \dfrac{1}{x(x + h)} = \dfrac{1}{x^2}$$

Therefore, $f'(x) = \dfrac{1}{x^2}$. $f'(1) = \dfrac{1}{1^2} = 1$, $f'(2) = \dfrac{1}{2^2} = \dfrac{1}{4}$, $f'(3) = \dfrac{1}{3^2} = \dfrac{1}{9}$.

25. $F'(x)$ does exist at $x = a$.

27. $F'(x)$ does not exist at $x = c$; the graph has a vertical tangent line at $(c, F(c))$.

29. $F'(x)$ does not exist at $x = e$; F is not defined at $x = e$.

31. $F'(x)$ does exist at $x = g$.

33. $f(x) = x^2 - 4x$

   (A) Step 1. Simplify $\dfrac{f(x + h) - f(x)}{h}$.

$$\dfrac{f(x + h) - f(x)}{h} = \dfrac{(x + h)^2 - 4(x + h) - (x^2 - 4x)}{h}$$

$$= \dfrac{x^2 + 2xh + h^2 - 4x - 4h - x^2 + 4x}{h}$$

$$= \dfrac{2xh + h^2 - 4h}{h} = 2x - 4 + h$$

        Step 2. Evaluate $\lim\limits_{h \to 0} \dfrac{f(x + h) - f(x)}{h}$.

$$\lim_{h \to 0} \dfrac{f(x + h) - f(x)}{h} = \lim_{h \to 0}(2x - 4 + h) = 2x - 4$$

Therefore, $f'(x) = 2x - 4$.

   (B) $f'(0) = -4$, $f'(2) = 0$,
$f'(4) = 4$

   (C) Since f is a quadratic
function, the graph of f is
a parabola.

y intercept: $y = 0$
x intercepts: $x = 0$, $x = 4$
Vertex: $(2, -4)$

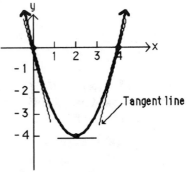

35. To find $v = f'(x)$, use the two-step process for the given distance function, $f(x) = 4x^2 - 2x$.

Step 1. $f(x + h) = 4(x + h)^2 - 2(x + h)$
$$= 4(x^2 + 2xh + h^2) - 2(x + h)$$
$$= 4x^2 + 8xh + 4h^2 - 2x - 2h$$

$f(x + h) - f(x) = (4x^2 + 8xh + 4h^2 - 2x - 2h) - (4x^2 - 2x)$
$$= 4x^2 + 8xh + 4h^2 - 2x - 2h - 4x^2 + 2x$$
$$= 8xh + 4h^2 - 2h$$
$$= h(8x + 4h - 2)$$

$$\frac{f(x + h) - f(x)}{h} = \frac{h(8x + 4h - 2)}{h}$$
$$= 8x + 4h - 2, \ h \neq 0$$

Step 2. $\lim\limits_{h \to 0} \dfrac{f(x + h) - f(x)}{h} = \lim\limits_{h \to 0}(8x + 4h - 2) = 8x - 2$

Thus, the velocity, $v = f'(x) = 8x - 2$
$$f'(1) = 8 \cdot 1 - 2 = 6 \text{ feet per second}$$
$$f'(3) = 8 \cdot 3 - 2 = 22 \text{ feet per second}$$
$$f'(5) = 8 \cdot 5 - 2 = 38 \text{ feet per second}$$

37.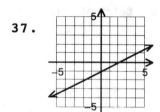

39.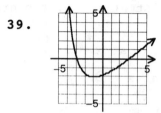

41. $f(x) = 2^x$; $f'(0) \approx 0.69$

43. $f(x) = \sqrt{2 + 2x - x^2}$; $f'(0) \approx 0.71$

45. (A) The graphs of g and h are vertical translations of the graph of f. All three functions should have the same derivative.

(B) Step 1. $\dfrac{m(x + h) - m(x)}{h} = \dfrac{(x + h)^2 + C - (x^2 + C)}{h}$

$$= \frac{x^2 + 2xh + h^2 + C - x^2 - C}{h}$$

$$= \frac{2xh + h^2}{h} = \frac{h(2x + h)}{h}$$

$$= 2x + h \quad h \neq 0$$

Step 2. $\lim\limits_{h \to 0} \dfrac{m(x + h) - m(x)}{h} = \lim\limits_{h \to 0}(2x + h) = 2x$; $m'(x) = 2x$

47. (A) The graph of $f(x) = C$, C a constant, is a horizontal line C units above or below the x axis depending on the sign of C. At any given point on the graph, the slope of the tangent line is 0.

(B) Step 1. $\dfrac{f(x + h) - f(x)}{h} = \dfrac{C - C}{h} = 0$

      Step 2. $\lim\limits_{h \to 0} \dfrac{f(x + h) - f(x)}{h} = \lim\limits_{h \to 0} 0 = 0$

49. The graph of $f(x) = \begin{cases} 2x, & x < 1 \\ 2, & x \ge 1 \end{cases}$ is:

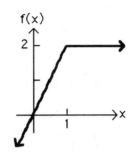

f is not differentiable at $x = 1$ because the graph of f has a sharp corner at this point.

51. $f(x) = |x|$

$\lim\limits_{h \to 0} \dfrac{f(0 + h) - f(0)}{h} = \lim\limits_{h \to 0} \dfrac{|0 + h| - |0|}{h} = \lim\limits_{h \to 0} \dfrac{|h|}{h}$

The limit does not exist. Thus, f is not differentiable at $x = 0$.

53. $f(x) = \sqrt[3]{x} = x^{1/3}$

$\lim\limits_{h \to 0} \dfrac{f(0 + h) - f(0)}{h} = \lim\limits_{h \to 0} \dfrac{(0 + h)^{1/3} - 0^{1/3}}{h} = \lim\limits_{h \to 0} \dfrac{h^{1/3}}{h} = \lim\limits_{h \to 0} \dfrac{1}{h^{2/3}}$

The limit does not exist. Thus, f is not differentiable at $x = 0$.

55. $f(x) = 2x - x^2,\ 0 \le x \le 2$

(A) For $0 < x < 2$:

$$\lim\limits_{h \to 0} \dfrac{f(x + h) - f(x)}{h} = \lim\limits_{h \to 0} \dfrac{2(x + h) - (x + h)^2 - (2x - x^2)}{h}$$

$$= \lim\limits_{h \to 0} \dfrac{2x + 2h - x^2 - 2xh - h^2 - 2x + x^2}{h}$$

$$= \lim\limits_{h \to 0} \dfrac{2h - 2xh - h^2}{h} = \lim\limits_{h \to 0} (2 - 2x - h) = 2 - 2x$$

Thus, $f'(x) = 2 - 2x,\ 0 < x < 2$.

(B) For $x = 0$:

$$\lim\limits_{h \to 0^+} \dfrac{f(0 + h) - f(0)}{h} = \lim\limits_{h \to 0^+} \dfrac{2(0 + h) - (0 + h)^2 - 0}{h}$$

$$= \lim\limits_{h \to 0^+} \dfrac{2h - h^2}{h} = 2$$

(C) For $x = 2$:

$$\lim_{h \to 0^-} \frac{f(2 + h) - f(2)}{h} = \lim_{h \to 0^-} \frac{2(2 + h) - (2 + h)^2 - (2 \cdot 2 - 2^2)}{h}$$

$$= \lim_{h \to 0^-} \frac{4 + 2h - 4 - 4h - h^2 - 0}{h}$$

$$= \lim_{h \to 0^-} -\frac{2h + h^2}{h} = -2$$

57. (A) $S(t) = 2\sqrt{t + 10}$

Step 1. $\dfrac{S(t + h) - S(t)}{h} = \dfrac{2\sqrt{t + h + 10} - 2\sqrt{t + 10}}{h}$

$$= \frac{2[\sqrt{t + h + 10} - \sqrt{t + 10}]}{h} \cdot \frac{\sqrt{t + h + 10} + \sqrt{t + 10}}{\sqrt{t + h + 10} + \sqrt{t + 10}}$$

$$= \frac{2[t + h + 10 - (t + 10)]}{h[\sqrt{t + h + 10} + \sqrt{t + 10}]} = \frac{2h}{h[\sqrt{t + h + 10} + \sqrt{t + 10}]}$$

$$= \frac{2}{\sqrt{t + h + 10} + \sqrt{t + 10}}$$

Step 2. $\displaystyle\lim_{h \to 0} \frac{S(t + h) - S(t)}{h} = \lim_{h \to 0} \frac{2}{\sqrt{t + h + 10} + \sqrt{t + 10}}$

$$= \frac{2}{2\sqrt{t + 10}} = \frac{1}{\sqrt{t + 10}};$$

$$S'(t) = \frac{1}{\sqrt{t + 10}}$$

(B) $S(15) = 2\sqrt{15 + 10} = 2\sqrt{25} = 10;$

$$S'(15) = \frac{1}{\sqrt{15 + 10}} = \frac{1}{\sqrt{25}} = \frac{1}{5} = 0.2$$

After 15 months, the total sales are $10 million and are INCREASING at the rate of $0.2 million = $200,000 per month.

(C) The estimated total sales are $10.2 million after 16 months and $10.4 million after 17 months.

59. (A) $A(t) = 100(1.06)^t;$ $A(5) = 100(1.06)^5 = 133.82$

h	-0.1	-0.01	-0.001 → 0	← 0.001	0.01	0.1
$\dfrac{A(5 + h) - A(5)}{h}$	7.78	7.80	7.80 → 7.80 ← 7.80	7.80	7.82

$A'(5) \approx 7.80$

(B) After 5 years, the original $100 investment has grown to $133.82 and is continuing to grow at the rate of $7.80 per year.

61. (A) In March, the price of the stock was $80 and was INCREASING at the rate of $5 per month.

(B) The stock reached its highest price in April. The rate of change at that point in time was 0.

63. (A) $P(t) = 80 + 12t - t^2$

Step 1. $\dfrac{P(t + h) - P(t)}{h} = \dfrac{80 + 12(t + h) - (t + h)^2 - [80 + 12t - t^2]}{h}$

$\qquad = \dfrac{80 + 12t + 12h - (t^2 + 2th + h^2) - 80 - 12t + t^2}{h}$

$\qquad = \dfrac{12h - 2th - h^2}{h} = \dfrac{h(12 - 2t - h)}{h} = 12 - 2t - h$

Step 2. $\displaystyle\lim_{h\to 0}\dfrac{P(t + h) - P(t)}{h} = \lim_{h\to 0}(12 - 2t - h) = 12 - 2t;$

$\qquad P'(t) = 12 - 2t$

(B) $P(3) = 80 + 12(3) - (3)^2 = 107;$ $\quad P'(3) = 12 - 2(3) = 6$

After 3 hours, the ozone level is 107 ppb and is INCREASING at the rate of 6 ppb per hour.

65. (A) $P(t) = 26(1.02)^t$

$\qquad P(30) = 26(1.02)^{30} = 47.1$

h	-0.1	-0.01	-0.001 → 0	← 0.001	0.01	0.1
$\dfrac{P(30 + h) - P(30)}{h}$	0.932	0.933	0.933 → 0.933	← 0.933	0.933	0.934

Therefore, $P'(30) \approx 0.9$

(B) In 2010 there will be 47.1 million people aged 65 or older, and the number of people in this group will be INCREASING at the rate of 900,000 per year.

(C) The estimated population is 48 million in 2011 and 48.9 million in 2012.

EXERCISE 3-4

Things to remember:

1. DERIVATIVE NOTATION

   Given $y = f(x)$, then

   $$f'(x), \quad y', \quad \dfrac{dy}{dx}$$

   all represent the derivative of f at x.

2. DERIVATIVE OF A CONSTANT FUNCTION RULE

   If $f(x) = C$, C a constant, then $f'(x) = 0$. Also

   $$y' = 0 \text{ and } \dfrac{dy}{dx} = 0.$$

<u>3</u>. POWER RULE

If $f(x) = x^n$, n any real number, then

$$f'(x) = nx^{n-1}.$$

Also, $y' = nx^{n-1}$ and $\dfrac{dy}{dx} = nx^{n-1}$

<u>4</u>. CONSTANT TIMES A FUNCTION RULE

If $y = f(x) = ku(x)$, where k is a constant, then

$$f'(x) = ku'(x).$$

Also,
$$y' = ku' \text{ and } \dfrac{dy}{dx} = k\dfrac{du}{dx}.$$

<u>5</u>. SUM AND DIFFERENCE RULE

If $y = f(x) = u(x) \pm v(x)$, then

$$f'(x) = u'(x) \pm v'(x).$$

Also,
$$y' = u' \pm v' \text{ and } \dfrac{dy}{dx} = \dfrac{du}{dx} \pm \dfrac{dv}{dx}$$

[<u>Note</u>: This rule generalizes to the sum and difference of any given number of functions.]

<u>6</u>. THE MARGINAL COST FUNCTION

If $C(x)$ is the total cost of producing x items, then the marginal cost function $C'(x)$ approximates the cost of producing one more item at a production level of x items.

---

1. $f(x) = 12$

$f'(x) = (12)' = 0$ (using <u>2</u>)

3. $\dfrac{d}{dx}(23) = 0$ (23 is a constant)

5. $y = x^{12}$
$\dfrac{dy}{dx} = 12x^{12-1}$ (using <u>3</u>)
$\quad = 12x^{11}$

7. $f(x) = x$
$f'(x) = 1x^{1-1}$ (using <u>3</u>)
$\quad\quad = x^0$ ($x^0 = 1$)
$\quad\quad = 1$

9. $y = x^{-7}$
$y' = -7x^{-7-1}$
$\quad = -7x^{-8}$

11. $y = x^{5/2}$
$\dfrac{dy}{dx} = \dfrac{5}{2}x^{5/2-1} = \dfrac{5}{2}x^{5/2-2/2} = \dfrac{5}{2}x^{3/2}$

13. $f(x) = \dfrac{1}{x^5} = x^{-5}$
$\dfrac{d}{dx}(x^{-5}) = -5x^{-5-1} = -5x^{-6}$

15. $f(x) = 2x^4$
$f'(x) = 2 \cdot 4x^3$ (using <u>4</u>)
$\quad\quad = 8x^3$

17. $\dfrac{d}{dx}\left(\dfrac{1}{3}x^6\right) = \dfrac{1}{3}\dfrac{d}{dx}(x^6)$ (using $\underline{4}$)

$\qquad\qquad\quad = \dfrac{1}{3}\cdot 6x^5 = 2x^5$

19. $y = \dfrac{x^5}{15} = \dfrac{1}{15}x^5$

$\qquad \dfrac{dy}{dx} = \dfrac{1}{15}\cdot 5x^4 = \dfrac{x^4}{3}$

21. $h(x) = 4f(x)$; $h'(2) = 4\cdot f'(2) = 4(3) = 12$

23. $h(x) = f(x) + g(x)$; $h'(2) = f'(2) + g'(2) = 3 + (-1) = 2$

25. $h(x) = 2f(x) - 3g(x) + 7$; $h'(2) = 2f'(2) - 3g'(2)$

$\qquad\qquad\qquad\qquad\qquad\qquad\quad = 2(3) - 3(-1) = 9$

27. $\dfrac{d}{dx}(2x^{-5}) = 2\dfrac{d}{dx}(x^{-5}) = 2(-5)x^{-6} = -10x^{-6}$

29. $f(x) = \dfrac{4}{x^4} = 4x^{-4}$

$\qquad f'(x) = 4(-4)x^{-5} = -16x^{-5}$

31. $\dfrac{d}{dx}\left(\dfrac{-1}{2x^2}\right) = \dfrac{d}{dx}\left(-\dfrac{1}{2}x^{-2}\right) = -\dfrac{1}{2}\dfrac{d}{dx}(x^{-2})$

$\qquad\qquad\qquad\qquad\qquad = -\dfrac{1}{2}(-2)x^{-3} = x^{-3}$

33. $f(x) = -3x^{1/3}$

$\qquad f'(x) = -3\left(\dfrac{1}{3}\right)x^{1/3-1} = -x^{-2/3}$

35. $\dfrac{d}{dx}(2x^2 - 3x + 4) = \dfrac{d}{dx}(2x^2) - \dfrac{d}{dx}(3x) + \dfrac{d}{dx}(4)$ (using $\underline{5}$)

$\qquad\qquad\qquad\qquad = 2\dfrac{d}{dx}(x^2) - 3\dfrac{d}{dx}(x) + \dfrac{d}{dx}(4)$

$\qquad\qquad\qquad\qquad = 2\cdot 2x - 3\cdot 1 = 4x - 3$

37. $y = 3x^5 - 2x^3 + 5$

$\qquad \dfrac{dy}{dx} = (3x^5)' - (2x^3)' + (5)' = 15x^4 - 6x^2$

39. $\dfrac{d}{dx}(3x^{-4} + 2x^{-2}) = \dfrac{d}{dx}(3x^{-4}) + \dfrac{d}{dx}(2x^{-2}) = -12x^{-5} - 4x^{-3}$

41. $y = \dfrac{1}{2x} - \dfrac{2}{3x^3} = \dfrac{1}{2}x^{-1} - \dfrac{2}{3}x^{-3}$

$\qquad \dfrac{dy}{dx} = -\dfrac{1}{2}x^{-2} - \dfrac{2}{3}(-3)x^{-4} = -\dfrac{1}{2}x^{-2} + 2x^{-4}$

43. $\dfrac{d}{dx}(3x^{2/3} - 5x^{1/3}) = \dfrac{d}{dx}(3x^{2/3}) - \dfrac{d}{dx}(5x^{1/3})$

$\qquad\qquad\qquad\qquad = 3\left(\dfrac{2}{3}\right)x^{-1/3} - 5\left(\dfrac{1}{3}\right)x^{-2/3} = 2x^{-1/3} - \dfrac{5}{3}x^{-2/3}$

45. $\dfrac{d}{dx}\left(\dfrac{3}{x^{3/5}} - \dfrac{6}{x^{1/2}}\right) = \dfrac{d}{dx}(3x^{-3/5} - 6x^{-1/2}) = \dfrac{d}{dx}(3x^{-3/5}) - \dfrac{d}{dx}(6x^{-1/2})$

$\qquad\qquad\qquad\qquad = 3\left(\dfrac{-3}{5}\right)x^{-8/5} - 6\left(-\dfrac{1}{2}\right)x^{-3/2} = \dfrac{-9}{5}x^{-8/5} + 3x^{-3/2}$

47. $\dfrac{d}{dx}\dfrac{1}{\sqrt[3]{x}} = \dfrac{d}{dx}(x^{-1/3}) = -\dfrac{1}{3}x^{-4/3}$

49. $y = \dfrac{12}{\sqrt{x}} - 3x^{-2} + x = 12x^{-1/2} - 3x^{-2} + x$

$\dfrac{dy}{dx} = 12\left(-\dfrac{1}{2}\right)x^{-3/2} - 3(-2)x^{-3} + 1 = -6x^{-3/2} + 6x^{-3} + 1$

51. $f(x) = 6x - x^2$

(A) $f'(x) = 6 - 2x$

(B) Slope of the graph of f at $x = 2$: $f'(2) = 6 - 2(2) = 2$
Slope of the graph of f at $x = 4$: $f'(4) = 6 - 2(4) = -2$

(C) Tangent line at $x = 2$: $y - y_1 = m(x - x_1)$
$x_1 = 2$
$y_1 = f(2) = 6(2) - 2^2 = 8$
$m = f'(2) = 2$
Thus, $y - 8 = 2(x - 2)$ or $y = 2x + 4$.
Tangent line at $x = 4$: $y - y_1 = m(x - x_1)$
$x_1 = 4$
$y_1 = f(4) = 6(4) - 4^2 = 8$
$m = f'(4) = -2$
Thus, $y - 8 = -2(x - 4)$ or $y = -2x + 16$

(D) The tangent line is horizontal at the values $x = c$ such that
$f'(c) = 0$. Thus, we must solve the following:
$f'(x) = 6 - 2x = 0$
$\qquad\qquad 2x = 6$
$\qquad\qquad\ \ x = 3$

53. $f(x) = 3x^4 - 6x^2 - 7$

(A) $f'(x) = 12x^3 - 12x$

(B) Slope of the graph of $x = 2$: $f'(2) = 12(2)^3 - 12(2) = 72$
Slope of the graph of $x = 4$: $f'(4) = 12(4)^3 - 12(4) = 720$

(C) Tangent line at $x = 2$: $y - y_1 = m(x - x_1)$, where $x_1 = 2$,
$y_1 = f(2) = 3(2)^4 - 6(2)^2 - 7 = 17$, $m = 72$.
$y - 17 = 72(x - 2)$ or $y = 72x - 127$

Tangent line at $x = 4$: $y - y_1 = m(x - x_1)$, where $x_1 = 4$,
$y_1 = f(4) = 3(4)^4 - 6(4)^2 - 7 = 665$, $m = 720$.
$y - 665 = 720(x - 4)$ or $y = 720x - 2215$

(D) Solve $f'(x) = 0$ for x:
$$12x^3 - 12x = 0$$
$$12x(x^2 - 1) = 0$$
$$12x(x - 1)(x + 1) = 0$$
$$x = -1, \; x = 0, \; x = 1$$

55. $f(x) = 176x - 16x^2$

(A) $v = f'(x) = 176 - 32x$

(B) $v\big|_{x=0} = f'(0) = 176$ feet/sec.

$v\big|_{x=3} = f'(3) = 176 - 32(3) = 80$ feet/sec.

(C) Solve $v = f'(x) = 0$ for x:
$$176 - 32x = 0$$
$$32x = 176$$
$$x = 5.5 \text{ seconds}$$

57. $f(x) = x^3 - 9x^2 + 15x$

(A) $v = f'(x) = 3x^2 - 18x + 15$

(B) $v\big|_{x=0} = f'(0) = 15$ feet/sec.

$v\big|_{x=3} = f'(3) = 3(3)^2 - 18(3) + 15 = -12$ feet/sec.

(C) Solve $v = f'(x) = 0$ for x:
$$3x^2 - 18x + 15 = 0$$
$$3(x^2 - 6x + 5) = 0$$
$$3(x - 5)(x - 1) = 0$$
$$x = 1, \; x = 5$$

59. $f(x) = x^2 - 3x - 4\sqrt{x} = x^2 - 3x - 4x^{1/2}$
$f'(x) = 2x - 3 - 2x^{-1/2}$
The graph of f has a horizontal tangent line at the value(s) of x where $f'(x) = 0$. Thus, we need to solve the equation
$$2x - 3 - 2x^{-1/2} = 0$$
By graphing the function $y = 2x - 3 - 2x^{-1/2}$, we see that there is one zero. To two decimal places, it is $x = 2.18$.

61. $f(x) = 3\sqrt[3]{x^4} - 1.5x^2 - 3x = 3x^{4/3} - 1.5x^2 - 3x$
$f'(x) = 4x^{1/3} - 3x - 3$
The graph of f has a horizontal tangent line at the value(s) of x where $f'(x) = 0$. Thus, we need to solve the equation
$$4x^{1/3} - 3x - 3 = 0$$
Graphing the function $y = 4x^{1/3} - 3x - 3$, we see that there is one zero. To two decimal places, it is $x = -2.90$.

63. $f(x) = 0.05x^4 - 0.1x^3 - 1.5x^2 - 1.6x + 3$

$f'(x) = 0.2x^3 + 0.3x^2 - 3x - 1.6$

The graph of f has a horizontal tangent line at the value(s) of x where $f'(x) = 0$. Thus, we need to solve the equation

$$0.2x^3 + 0.3x^2 - 3x - 1.6 = 0$$

By graphing the function $y = 0.2x^3 + 0.3x^2 - 3x - 1.6$ we see that there are three zeros. To two decimal places, they are

$$x_1 = -4.46, \ x_2 = -0.52, \ x_3 = 3.48$$

65. $f(x) = 0.2x^4 - 3.12x^3 + 16.25x^2 - 28.25x + 7.5$

$f'(x) = 0.8x^3 - 9.36x^2 + 32.5x - 28.25$

The graph of f has a horizontal tangent line at the value(s) of x where $f'(x) = 0$. Thus, we need to solve the equation

$$0.8x^3 - 9.36x^2 + 32.5x - 28.25 = 0$$

Graphing the function $y = 0.8x^3 - 9.36x^2 + 32.5x - 28.25$, we see that there is one zero. To two decimal places, it is $x = 1.30$.

67. $f(x) = ax^2 + bx + c; \ f'(x) = 2ax + b.$

The derivative is 0 at the vertex of the parabola:

$$2ax + b = 0$$

$$x = -\frac{b}{2a}$$

69. (A) $f(x) = x^3 + x$ (B) $f(x) = x^3$ (C) $f(x) = x^3 - x$

71. $f(x) = \dfrac{10x + 20}{x} = 10 + \dfrac{20}{x} = 10 + 20x^{-1}$

$f'(x) = -20x^{-2}$

73. $\dfrac{d}{dx}\left(\dfrac{x^4 - 3x^3 + 5}{x^2}\right) = \dfrac{d}{dx}\left(x^2 - 3x + \dfrac{5}{x^2}\right)$

$$= \dfrac{d}{dx}(x^2) - \dfrac{d}{dx}(3x) + \dfrac{d}{dx}(5x^{-2}) = 2x - 3 - 10x^{-3}$$

75. Let $f(x) = x^3$

Step 1. Simplify $\dfrac{f(x + h) - f(x)}{h}$.

$$\dfrac{f(x + h) - f(x)}{h} = \dfrac{(x + h)^3 - x^3}{h} = \dfrac{x^3 + 3x^2h + 3xh^2 + h^3 - x^3}{h}$$

$$= \dfrac{3x^2h + 3xh^2 + h^3}{h} = \dfrac{h(3x^2 + 3xh + h^2)}{h} = 3x^2 + 3xh + h^2 \quad (h \neq 0)$$

Step 2. Evaluate $\lim\limits_{h \to 0} \dfrac{f(x + h) - f(x)}{h}$.

$$\lim\limits_{h \to 0} \dfrac{f(x + h) - f(x)}{h} = \lim\limits_{h \to 0}(3x^2 + 3xh + h^2) = 3x^2$$

Therefore, $\dfrac{d}{dx}x^3 = 3x^2.$

77. $f(x) = x^{1/3}$; $f'(x) = \frac{1}{3}x^{-2/3} = \frac{1}{3x^{2/3}}$

The domain of f' is the set of all real numbers except $x = 0$. The graph of f is smooth, but it has a vertical tangent line at $(0, 0)$.

79. $C(x) = 800 + 60x - \frac{x^2}{4}$, $0 \le x \le 120$

(A) Marginal cost $= C'(x) = 60 - \frac{1}{2}x$

(B) $C'(60) = 60 - \frac{1}{2}(60) = 30$ or \$30 per racket

Interpretation: At a production level of 60 rackets, the rate of change of total cost with respect to production is \$30 per racket. Thus, the cost of producing one more racket at this level of production is approximately \$30.

(C) $C(61) - C(60) = 800 + 60(61) - \frac{(61)^2}{4} - \left[800 + 60(60) - \frac{(60)^2}{4}\right]$

$= 3529.75 - 3500 = 29.75$

The actual cost of producing the 61st racket is \$29.75.

(D) $C'(80) = 60 - \frac{1}{2}(80) = 20$ or \$20 per racket

Interpretation: At a production level of 80 rackets, the rate of change of total cost with respect to production is \$20 per racket. Thus, the cost of producing the 81st racket is approximately \$20.

81. The approximate cost of producing the 101st oven is greater than the approximate cost of producing the 401st oven. Since the marginal costs are decreasing, the manufacturing process is becoming more efficient.

83. (A) $N(x) = 1,000 - \frac{3,780}{x} = 1,000 - 3,780x^{-1}$

$N'(x) = 3,780x^{-2} = \frac{3,780}{x^2}$

(B) $N'(10) = \frac{3,780}{(10)^2} = 37.8$

At the \$10,000 level of advertising, sales are INCREASING at the rate of 37.8 boats per \$1000 spent on advertising.

$N'(20) = \frac{3,780}{(20)^2} = 9.45$

At the \$20,000 level of advertising, sales are INCREASING at the rate of 9.45 boats per \$1000 spent on advertising.

85. $y = 590x^{-1/2}$, $30 \le x \le 75$

First, find $\frac{dy}{dx} = \frac{d}{dx}590x^{-1/2} = -295x^{-3/2} = \frac{-295}{x^{3/2}}$, the instantaneous rate of change of pulse when a person is x inches tall.

(A) The instantaneous rate of change of pulse rate at $x = 36$ is:
$$\frac{-295}{(36)^{3/2}} = \frac{-295}{216} = -1.37 \text{ (1.37 decrease in pulse rate)}$$

(B) The instantaneous rate of change of pulse rate at $x = 64$ is:
$$\frac{-295}{(64)^{3/2}} = \frac{-295}{512} = -0.58 \text{ (0.58 decrease in pulse rate)}$$

87. $y = 50\sqrt{x}$, $0 \le x \le 9$

First, find $y' = (50\sqrt{x})' = (50x^{1/2})' = 25x^{-1/2}$
$$= \frac{25}{\sqrt{x}}, \text{ the rate of learning at the end of } x \text{ hours.}$$

(A) Rate of learning at the end of 1 hour:
$$\frac{25}{\sqrt{1}} = 25 \text{ items per hour}$$

(B) Rate of learning at the end of 9 hours:
$$\frac{25}{\sqrt{9}} = \frac{25}{3} = 8.33 \text{ items per hour}$$

EXERCISE 3-5

Things to remember:

1. PRODUCT RULE

If
$$y = f(x) = F(x)S(x)$$
and if $F'(x)$ and $S'(x)$ exist, then
$$f'(x) = F(x)S'(x) + S(x)F'(x).$$
Also,
$$y' = FS' + SF';$$
$$\frac{dy}{dx} = F\frac{dS}{dx} + S\frac{dF}{dx}.$$

2. QUOTIENT RULE

If
$$y = f(x) = \frac{T(x)}{B(x)}$$
and if $T'(x)$ and $B'(x)$ exist, then
$$f'(x) = \frac{B(x)T'(x) - T(x)B'(x)}{[B(x)]^2}.$$
Also,
$$y' = \frac{BT' - TB'}{B^2};$$
$$\frac{dy}{dx} = \frac{B\left(\frac{dT}{dx}\right) - T\left(\frac{dB}{dx}\right)}{B^2}.$$

1. $f(x) = 2x^3(x^2 - 2)$
   $f'(x) = 2x^3(x^2 - 2)' + (x^2 - 2)(2x^3)'$ [using $\underline{1}$ with $F(x) = 2x^3$,
   $ = 2x^3(2x) + (x^2 - 2)(6x^2)$ $S(x) = x^2 - 2]$
   $ = 4x^4 + 6x^4 - 12x^2$
   $ = 10x^4 - 12x^2$

3. $f(x) = (x - 3)(2x - 1)$
   $f'(x) = (x - 3)(2x - 1)' + (2x - 1)(x - 3)'$ (using $\underline{1}$)
   $ = (x - 3)(2) + (2x - 1)(1)$
   $ = 2x - 6 + 2x - 1$
   $ = 4x - 7$

5. $f(x) = \dfrac{x}{x - 3}$
   $f'(x) = \dfrac{(x - 3)(x)' - x(x - 3)'}{(x - 3)^2}$ [using $\underline{2}$ with $T(x) = x$, $B(x) = x - 3$]
   $ = \dfrac{(x - 3)(1) - x(1)}{(x - 3)^2} = \dfrac{-3}{(x - 3)^2}$

7. $f(x) = \dfrac{2x + 3}{x - 2}$
   $f'(x) = \dfrac{(x - 2)(2x + 3)' - (2x + 3)(x - 2)'}{(x - 2)^2}$ (using $\underline{2}$)
   $ = \dfrac{(x - 2)(2) - (2x + 3)(1)}{(x - 2)^2} = \dfrac{2x - 4 - 2x - 3}{(x - 2)^2} = \dfrac{-7}{(x - 2)^2}$

9. $f(x) = (x^2 + 1)(2x - 3)$
   $f'(x) = (x^2 + 1)(2x - 3)' + (2x - 3)(x^2 + 1)'$ (using $\underline{1}$)
   $ = (x^2 + 1)(2) + (2x - 3)(2x)$
   $ = 2x^2 + 2 + 4x^2 - 6x$
   $ = 6x^2 - 6x + 2$

11. $f(x) = \dfrac{x^2 + 1}{2x - 3}$
    $f'(x) = \dfrac{(2x - 3)(x^2 + 1)' - (x^2 + 1)(2x - 3)'}{(2x - 3)^2}$ (using $\underline{2}$)
    $ = \dfrac{(2x - 3)(2x) - (x^2 + 1)(2)}{(2x - 3)^2}$
    $ = \dfrac{4x^2 - 6x - 2x^2 - 2}{(2x - 3)^2} = \dfrac{2x^2 - 6x - 2}{(2x - 3)^2}$

13. $f(x) = (x^2 + 2)(x^2 - 3)$
    $f'(x) = (x^2 + 2)(x^2 - 3)' + (x^2 - 3)(x^2 + 2)'$
    $ = (x^2 + 2)(2x) + (x^2 - 3)(2x)$
    $ = 2x^3 + 4x + 2x^3 - 6x$
    $ = 4x^3 - 2x$

15. $f(x) = \dfrac{x^2 + 2}{x^2 - 3}$

$f'(x) = \dfrac{(x^2 - 3)(x^2 + 2)' - (x^2 + 2)(x^2 - 3)'}{(x^2 - 3)^2}$

$= \dfrac{(x^2 - 3)(2x) - (x^2 + 2)(2x)}{(x^2 - 3)^2} = \dfrac{2x^3 - 6x - 2x^3 - 4x}{(x^2 - 3)^2} = \dfrac{-10x}{(x^2 - 3)^2}$

17. $h(x) = f(x)g(x);\ h'(1) = f(1)g'(1) + g(1)f'(1)$

$= 4(3) + 2(-2) = 12 - 4 = 8$

19. $h(x) = \dfrac{g(x)}{f(x)};\ h'(1) = \dfrac{f(1)g'(1) - g(1)f'(1)}{[f(1)]^2}$

$= \dfrac{4(3) - 2(-2)}{4^2} = \dfrac{16}{16} = 1$

21. $h(x) = \dfrac{1}{f(x)};\ h'(1) = \dfrac{f(1)(0) - f'(1)}{[f(1)]^2} = \dfrac{-(-2)}{4^2} = \dfrac{1}{8}$

23. $f(x) = (2x + 1)(x^2 - 3x)$

$f'(x) = (2x + 1)(x^2 - 3x)' + (x^2 - 3x)(2x + 1)'$

$= (2x + 1)(2x - 3) + (x^2 - 3x)(2)$

$= 6x^2 - 10x - 3$

25. $y = (2x - x^2)(5x + 2)$

$\dfrac{dy}{dx} = (2x - x^2)\dfrac{d}{dx}(5x + 2) + (5x + 2)\dfrac{d}{dx}(2x - x^2)$

$= (2x - x^2)(5) + (5x + 2)(2 - 2x)$

$= -15x^2 + 16x + 4$

27. $y = \dfrac{5x - 3}{x^2 + 2x}$

$y' = \dfrac{(x^2 + 2x)(5x - 3)' - (5x - 3)(x^2 + 2x)'}{(x^2 + 2x)^2}$

$= \dfrac{(x^2 + 2x)(5) - (5x - 3)(2x + 2)}{(x^2 + 2x)^2} = \dfrac{-5x^2 + 6x + 6}{(x^2 + 2x)^2}$

29. $\dfrac{d}{dx}\left[\dfrac{x^2 - 3x + 1}{x^2 - 1}\right] = \dfrac{(x^2 - 1)\dfrac{d}{dx}(x^2 - 3x + 1) - (x^2 - 3x + 1)\dfrac{d}{dx}(x^2 - 1)}{(x^2 - 1)^2}$

$= \dfrac{(x^2 - 1)(2x - 3) - (x^2 - 3x + 1)(2x)}{(x^2 - 1)^2}$

$= \dfrac{3x^2 - 4x + 3}{(x^2 - 1)^2}$

31. $f(x) = (1 + 3x)(5 - 2x)$

First find $f'(x)$:

$$\begin{aligned} f'(x) &= (1 + 3x)(5 - 2x)' + (5 - 2x)(1 + 3x)' \\ &= (1 + 3x)(-2) + (5 - 2x)(3) \\ &= -2 - 6x + 15 - 6x \\ &= 13 - 12x \end{aligned}$$

An equation for the tangent line at $x = 2$ is:

$$y - y_1 = m(x - x_1)$$

where $x_1 = 2$, $y_1 = f(x_1) = f(2) = 7$, and $m = f'(x_1) = f'(2) = -11$.
Thus, we have:

$$y - 7 = -11(x - 2) \quad \text{or} \quad y = -11x + 29$$

33. $f(x) = \dfrac{x - 8}{3x - 4}$

First find $f'(x)$:

$$\begin{aligned} f'(x) &= \frac{(3x - 4)(x - 8)' - (x - 8)(3x - 4)'}{(3x - 4)^2} \\ &= \frac{(3x - 4)(1) - (x - 8)(3)}{(3x - 4)^2} = \frac{20}{(3x - 4)^2} \end{aligned}$$

An equation for the tangent line at $x = 2$ is:

$$y - y_1 = m(x - x_1)$$

where $x_1 = 2$, $y_1 = f(x_1) = f(2) = -3$, and $m = f'(x_1) = f'(2) = 5$.
Thus, we have: $y - (-3) = 5(x - 2) \quad \text{or} \quad y = 5x - 13$

35. $f(x) = (2x - 15)(x^2 + 18)$

$$\begin{aligned} f'(x) &= (2x - 15)(x^2 + 18)' + (x^2 + 18)(2x - 15)' \\ &= (2x - 15)(2x) + (x^2 + 18)(2) \\ &= 6x^2 - 30x + 36 \end{aligned}$$

To find the values of x where $f'(x) = 0$, set: $f'(x) = 6x^2 - 30x + 36 = 0$

$$\text{or} \qquad x^2 - 5x + 6 = 0$$
$$(x - 2)(x - 3) = 0$$

Thus, $x = 2$, $x = 3$.

37. $f(x) = \dfrac{x}{x^2 + 1}$

$$f'(x) = \frac{(x^2 + 1)(x)' - x(x^2 + 1)'}{(x^2 + 1)^2} = \frac{(x^2 + 1)(1) - x(2x)}{(x^2 + 1)^2} = \frac{1 - x^2}{(x^2 + 1)^2}$$

Now, set $f'(x) = \dfrac{1 - x^2}{(x^2 + 1)^2} = 0$

$$\text{or} \qquad 1 - x^2 = 0$$
$$(1 - x)(1 + x) = 0$$

Thus, $x = 1$, $x = -1$.

39. $f(x) = x^3(x^4 - 1)$

First, we use the product rule:

$$\begin{aligned} f'(x) &= x^3(x^4 - 1)' + (x^4 - 1)(x^3)' \\ &= x^3(4x^3) + (x^4 - 1)(3x^2) \\ &= 7x^6 - 3x^2 \end{aligned}$$

Next, simplifying $f(x)$, we have $f(x) = x^7 - x^3$. Thus, $f'(x) = 7x^6 - 3x^2$.

41. $f(x) = \dfrac{x^3 + 9}{x^3}$

First, we use the quotient rule:

$$f'(x) = \frac{x^3(x^3 + 9)' - (x^3 + 9)(x^3)'}{(x^3)^2} = \frac{x^3(3x^2) - (x^3 + 9)(3x^2)}{x^6}$$

$$= \frac{-27x^2}{x^6} = \frac{-27}{x^4}$$

Next, simplifying $f(x)$, we have $f(x) = \dfrac{x^3 + 9}{x^3} = 1 + \dfrac{9}{x^3} = 1 + 9x^{-3}$

Thus, $f'(x) = -27x^{-4} = -\dfrac{27}{x^4}$.

43. $f(x) = (2x^4 - 3x^3 + x)(x^2 - x + 5)$

$f'(x) = (2x^4 - 3x^3 + x)(x^2 - x + 5)' + (x^2 - x + 5)(2x^4 - 3x^3 + x)'$

$\quad = (2x^4 - 3x^3 + x)(2x - 1) + (x^2 - x + 5)(8x^3 - 9x^2 + 1)$

$\quad = 4x^5 - 6x^4 + 2x^2 - 2x^4 + 3x^3 - x + 8x^5 - 8x^4 + 40x^3 - 9x^4 + 9x^3$
$\qquad\qquad\qquad\qquad\qquad\qquad\qquad\qquad - 45x^2 + x^2 - x + 5$

$\quad = 12x^5 - 25x^4 + 52x^3 - 42x^2 - 2x + 5$

45. $\dfrac{d}{dx} \dfrac{3x^2 - 2x + 3}{4x^2 + 5x - 1}$

$$= \frac{(4x^2 + 5x - 1)\dfrac{d}{dx}(3x^2 - 2x + 3) - (3x^2 - 2x + 3)\dfrac{d}{dx}(4x^2 + 5x - 1)}{(4x^2 + 5x - 1)^2}$$

$$= \frac{(4x^2 + 5x - 1)(6x - 2) - (3x^2 - 2x + 3)(8x + 5)}{(4x^2 + 5x - 1)^2}$$

$$= \frac{24x^3 + 30x^2 - 6x - 8x^2 - 10x + 2 - 24x^3 + 16x^2 - 24x - 15x^2 + 10x - 15}{(4x^2 + 5x - 1)^2}$$

$$= \frac{23x^2 - 30x - 13}{(4x^2 + 5x - 1)^2}$$

47. $y = 9x^{1/3}(x^3 + 5)$

$\dfrac{dy}{dx} = 9x^{1/3}\dfrac{d}{dx}(x^3 + 5) + (x^3 + 5)\dfrac{d}{dx}(9x^{1/3})$

$\quad = 9x^{1/3}(3x^2) + (x^3 + 5)\left(9 \cdot \dfrac{1}{3}x^{-2/3}\right) = 27x^{7/3} + (x^3 + 5)(3x^{-2/3})$

$\quad = 27x^{7/3} + \dfrac{3x^3 + 15}{x^{2/3}} = \dfrac{30x^3 + 15}{x^{2/3}}$

49. $f(x) = \dfrac{6\sqrt[3]{x}}{x^2 - 3} = \dfrac{6x^{1/3}}{x^2 - 3}$

$\quad f'(x) = \dfrac{(x^2 - 3)(6x^{1/3})' - 6x^{1/3}(x^2 - 3)'}{(x^2 - 3)^2}$

$\qquad = \dfrac{(x^2 - 3)\left(6 \cdot \frac{1}{3}x^{-2/3}\right) - 6x^{1/3}(2x)}{(x^2 - 3)^2} = \dfrac{(x^2 - 3)(2x^{-2/3}) - 12x^{4/3}}{(x^2 - 3)^2}$

$\qquad = \dfrac{\dfrac{2(x^2 - 3)}{x^{2/3}} - 12x^{4/3}}{(x^2 - 3)^2} = \dfrac{2x^2 - 6 - 12x^2}{(x^2 - 3)^2\, x^{2/3}} = \dfrac{-10x^2 - 6}{(x^2 - 3)^2\, x^{2/3}}$

51. $\dfrac{d}{dx}\, \dfrac{x^3 - 2x^2}{\sqrt[3]{x^2}} = \dfrac{d}{dx}\, \dfrac{x^3 - 2x^2}{x^{2/3}}$

$\qquad = \dfrac{x^{2/3}\dfrac{d}{dx}(x^3 - 2x^2) - (x^3 - 2x^2)\dfrac{d}{dx}(x^{2/3})}{(x^{2/3})^2}$

$\qquad = \dfrac{x^{2/3}(3x^2 - 4x) - (x^3 - 2x^2)\left(\frac{2}{3}x^{-1/3}\right)}{x^{4/3}}$

$\qquad = x^{-2/3}(3x^2 - 4x) - \frac{2}{3}x^{-5/3}(x^3 - 2x^2)$

$\qquad = 3x^{4/3} - 4x^{1/3} - \frac{2}{3}x^{4/3} + \frac{4}{3}x^{1/3}$

$\qquad = -\frac{8}{3}x^{1/3} + \frac{7}{3}x^{4/3}$

53. $f(x) = \dfrac{(2x^2 - 1)(x^2 + 3)}{x^2 + 1}$

$\quad f'(x) = \dfrac{(x^2 + 1)[(2x^2 - 1)(x^2 + 3)]' - (2x^2 - 1)(x^2 + 3)(x^2 + 1)'}{(x^2 + 1)^2}$

$\qquad = \dfrac{(x^2 + 1)[(2x^2 - 1)(x^2 + 3)' + (x^2 + 3)(2x^2 - 1)'] - (2x^2 - 1)(x^2 + 3)(2x)}{(x^2 + 1)^2}$

$\qquad = \dfrac{(x^2 + 1)[(2x^2 - 1)(2x) + (x^2 + 3)(4x)] - (2x^2 - 1)(x^2 + 3)(2x)}{(x^2 + 1)^2}$

$\qquad = \dfrac{(x^2 + 1)[4x^3 - 2x + 4x^3 + 12x] - [2x^4 + 5x^2 - 3](2x)}{(x^2 + 1)^2}$

$\qquad = \dfrac{(x^2 + 1)(8x^3 + 10x) - 4x^5 - 10x^3 + 6x}{(x^2 + 1)^2}$

$\qquad = \dfrac{8x^5 + 10x^3 + 8x^3 + 10x - 4x^5 - 10x^3 + 6x}{(x^2 + 1)^2}$

$\qquad = \dfrac{4x^5 + 8x^3 + 16x}{(x^2 + 1)^2}$

55. $f(x) = [u(x)]^2 = u(x) \cdot u(x)$

$f'(x) = u(x) \cdot u'(x) + u(x)u'(x) = 2u(x)u'(x)$

57. $f(x) = [u(x)]^n$

$f'(x) = n[u(x)]^{n-1} u'(x)$

59. $f(x) = \dfrac{5x - x^4}{x^2 + 1}$

Horizontal tangent line at $x \approx 0.75$.

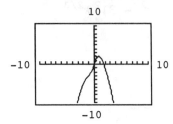

61. $f(x) = \dfrac{8 - x^3}{2 + x^2}$

Horizontal tangent lines at $x \approx -1.76$, $x = 0$.

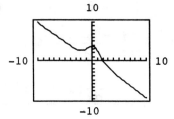

63. $f(x) = \dfrac{10x^2 + 9x}{x^4 + 2}$

Horizontal tangent lines at $x \approx -1.67$, $x \approx -0.43$, $x \approx 1.06$.

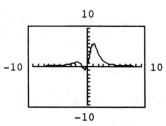

65. $S(t) = \dfrac{90t^2}{t^2 + 50}$

(A) $S'(t) = \dfrac{(t^2 + 50)(180t) - 90t^2(2t)}{(t^2 + 50)^2} = \dfrac{9000t}{(t^2 + 50)^2}$

(B) $S(10) = \dfrac{90(10)^2}{(10)^2 + 50} = \dfrac{9000}{150} = 60$;

$S'(10) = \dfrac{9000(10)}{[(10)^2 + 50]^2} = \dfrac{90,000}{22,500} = 4$

After 10 months, the total sales are 60,000 albums and the sales are INCREASING at the rate of 4,000 albums per month.

(C) The total sales after 11 months will be approximately 64,000 albums.

67. $x = \dfrac{4,000}{0.1p + 1}$, $10 \le p \le 70$

(A) $\dfrac{dx}{dp} = \dfrac{(0.1p + 1)(0) - 4,000(0.1)}{(0.1p + 1)^2} = \dfrac{-400}{(0.1p + 1)^2}$

(B) $x(40) = \dfrac{4{,}000}{0.1(40) + 1} = \dfrac{4{,}000}{5} = 800;$

$\dfrac{dx}{dp} = \dfrac{-400}{[0.1(40) + 1]^2} = \dfrac{-400}{25} = -16$

At a price level of \$40, the demand is 800 CD players and the demand is DECREASING at the rate of 16 CD players per dollar.

(C) At a price of \$41, the demand will be approximately 784 CD players.

69. $C(t) = \dfrac{0.14t}{t^2 + 1}$

(A) $C'(t) = \dfrac{(t^2 + 1)(0.14t)' - (0.14t)(t^2 + 1)'}{(t^2 + 1)^2}$

$= \dfrac{(t^2 + 1)(0.14) - (0.14t)(2t)}{(t^2 + 1)^2} = \dfrac{0.14 - 0.14t^2}{(t^2 + 1)^2} = \dfrac{0.14(1 - t^2)}{(t^2 + 1)^2}$

(B) $C'(0.5) = \dfrac{0.14(1 - [0.5]^2)}{([0.5]^2 + 1)^2} = \dfrac{0.14(1 - 0.25)}{(1.25)^2} = 0.0672$

Interpretation: At $t = 0.5$ hours, the concentration is increasing at the rate of 0.0672 units per hour.

$C'(3) = \dfrac{0.14(1 - 3^2)}{(3^2 + 1)^2} = \dfrac{0.14(-8)}{100} = -0.0112$

Interpretation: At $t = 3$ hours, the concentration is decreasing at the rate of 0.0112 units per hour.

71. $N(x) = \dfrac{100x + 200}{x + 32}$

(A) $N'(x) = \dfrac{(x + 32)(100x + 200)' - (100x + 200)(x + 32)'}{(x + 32)^2}$

$= \dfrac{(x + 32)(100) - (100x + 200)(1)}{(x + 32)^2}$

$= \dfrac{100x + 3200 - 100x - 200}{(x + 32)^2} = \dfrac{3000}{(x + 32)^2}$

(B) $N'(4) = \dfrac{3000}{(36)^2} = \dfrac{3000}{1296} \approx 2.31;$ $N'(68) = \dfrac{3000}{(100)^2} = \dfrac{3000}{10{,}000} = \dfrac{3}{10} = 0.30$

Things to remember:

<u>1</u>. GENERAL POWER RULE

If $u(x)$ is a differentiable function, n is any real number, and

$$y = f(x) = [u(x)]^n$$

then

$$f'(x) = n[u(x)]^{n-1}u'(x)$$

This rule is often written more compactly as

$$y' = nu^{n-1}u' \quad \text{or} \quad \frac{d}{dx}u^n = nu^{n-1}\frac{du}{dx}, \quad u = u(x)$$

---

1. $3; \dfrac{d}{dx}(3x + 4)^4 = 4(3x + 4)^3(3) = 12(3x + 4)^3$

3. $-4x; \dfrac{d}{dx}(4 - 2x^2)^3 = 3(4 - 2x^2)^2(-4x) = -12x(4 - 2x^2)^2$

5. $2 + 6x; \dfrac{d}{dx}(1 + 2x + 3x^2)^7 = 7(1 + 2x + 3x^2)^6(2 + 6x)$

$$= 7(2 + 6x)(1 + 2x + 3x^2)^6$$

7. $f(x) = (2x + 5)^3$
$f'(x) = 3(2x + 5)^2(2x + 5)'$
$\quad\ = 3(2x + 5)^2(2)$
$\quad\ = 6(2x + 5)^2$

9. $f(x) = (5 - 2x)^4$
$f'(x) = 4(5 - 2x)^3(5 - 2x)'$
$\quad\ = 4(5 - 2x)^3(-2)$
$\quad\ = -8(5 - 2x)^3$

11. $f(x) = (3x^2 + 5)^5$
$f'(x) = 5(3x^2 + 5)^4(3x^2 + 5)'$
$\quad\ = 5(3x^2 + 5)^4(6x)$
$\quad\ = 30x(3x^2 + 5)^4$

13. $f(x) = (x^3 - 2x^2 + 2)^8$
$f'(x) = 8(x^3 - 2x^2 + 2)^7(x^3 - 2x^2 + 2)'$
$\quad\ = 8(x^3 - 2x^2 + 2)^7(3x^2 - 4x)$

15. $f(x) = (2x - 5)^{1/2}$
$f'(x) = \dfrac{1}{2}(2x - 5)^{-1/2}(2x - 5)'$

$\quad\ = \dfrac{1}{2}(2x - 5)^{-1/2}(2) = \dfrac{1}{(2x - 5)^{1/2}}$

17. $f(x) = (x^4 + 1)^{-2}$
$f'(x) = -2(x^4 + 1)^{-3}(x^4 + 1)'$
$\quad\ = -2(x^4 + 1)^{-3}(4x^3)$

$\quad\ = -8x^3(x^4 + 1)^{-3} = \dfrac{-8x^3}{(x^4 + 1)^3}$

19. $f(x) = (2x - 1)^3$

$f'(x) = 3(2x - 1)^2(2) = 6(2x - 1)^2$

Tangent line at $x = 1$: $y - y_1 = m(x - x_1)$ where $x_1 = 1$, $y_1 = f(1) = (2(1) - 1)^3 = 1$, $m = f'(1) = 6[2(1) - 1]^2 = 6$. Thus, $y - 1 = 6(x - 1)$ or $y = 6x - 5$.

The tangent line is horizontal at the value(s) of x such that $f'(x) = 0$:

$6(2x - 1)^2 = 0$

$\quad 2x - 1 = 0$

$\qquad x = \dfrac{1}{2}$

21. $f(x) = (4x - 3)^{1/2}$

$f'(x) = \dfrac{1}{2}(4x - 3)^{-1/2}(4) = \dfrac{2}{(4x - 3)^{1/2}}$

Tangent line at $x = 3$: $y - y_1 = m(x - x_1)$ where $x_1 = 3$, $y_1 = f(3) = (4 \cdot 3 - 3)^{1/2} = 3$, $f'(3) = \dfrac{2}{(4 \cdot 3 - 3)^{1/2}} = \dfrac{2}{3}$. Thus, $y - 3 = \dfrac{2}{3}(x - 3)$ or

$y = \dfrac{2}{3}x + 1$.

The tangent line is horizontal at the value(s) of x such that $f'(x) = 0$.

Since $\dfrac{2}{(4x - 3)^{1/2}} \neq 0$ for all x $\left(x \neq \dfrac{3}{4} \right)$, there are no values of x where the tangent line is horizontal.

23. $y = 3(x^2 - 2)^4$

$\dfrac{dy}{dx} = 3 \cdot 4(x^2 - 2)^3(2x) = 24x(x^2 - 2)^3$

25. $y = 2(x^2 + 3x)^{-3}$

$\dfrac{dy}{dx} = 2 \cdot (-3)(x^2 + 3x)^{-4}(2x + 3) = -6(x^2 + 3x)^{-4}(2x + 3) = \dfrac{-6(2x + 3)}{(x^2 + 3x)^4}$

27. $y = \sqrt{x^2 + 8} = (x^2 + 8)^{1/2}$

$\dfrac{dy}{dx} = \dfrac{1}{2}(x^2 + 8)^{-1/2}(2x) = \dfrac{x}{(x^2 + 8)^{1/2}} = \dfrac{x}{\sqrt{x^2 + 8}}$

29. $y = \sqrt[3]{3x + 4} = (3x + 4)^{1/3}$

$\dfrac{dy}{dx} = \dfrac{1}{3}(3x + 4)^{-2/3}(3) = (3x + 4)^{-2/3} = \dfrac{1}{(3x + 4)^{2/3}} = \dfrac{1}{\sqrt[3]{(3x + 4)^2}}$

31. $y = (x^2 - 4x + 2)^{1/2}$

$\dfrac{dy}{dx} = \dfrac{1}{2}(x^2 - 4x + 2)^{-1/2}(2x - 4)$

$\quad = \dfrac{2(x - 2)}{2(x^2 - 4x + 2)^{1/2}} = \dfrac{x - 2}{(x^2 - 4x + 2)^{1/2}}$

33. $y = \dfrac{1}{2x + 4} = (2x + 4)^{-1}$

$\dfrac{dy}{dx} = -1(2x + 4)^{-2}(2) = \dfrac{-2}{(2x + 4)^2}$

35. $y = \dfrac{1}{(x^3 + 4)^5} = (x^3 + 4)^{-5}$

$\dfrac{dy}{dx} = -5(x^3 + 4)^{-6}(3x^2) = -15x^2(x^3 + 4)^{-6} = \dfrac{-15x^2}{(x^3 + 4)^6}$

37. $y = \dfrac{1}{4x^2 - 4x + 1} = (4x^2 - 4x + 1)^{-1}$

$\dfrac{dy}{dx} = -1(4x^2 - 4x + 1)^{-2}(8x - 4) = \dfrac{-4(2x - 1)}{(4x^2 - 4x + 1)^2}$

$= \dfrac{-4(2x - 1)}{(2x - 1)^4} = \dfrac{-4}{(2x - 1)^3}$

39. $y = \dfrac{4}{\sqrt{x^2 - 3x}} = \dfrac{4}{(x^2 - 3x)^{1/2}} = 4(x^2 - 3x)^{-1/2}$

$\dfrac{dy}{dx} = 4\left[-\dfrac{1}{2}(x^2 - 3x)^{-3/2}\right](2x - 3) = \dfrac{-2(2x - 3)}{(x^2 - 3x)^{3/2}}$

41. $f(x) = x(4 - x)^3$

$f'(x) = x[(4 - x)^3]' + (4 - x)^3(x)'$

$= x(3)(4 - x)^2(-1) + (4 - x)^3(1)$

$= (4 - x)^3 - 3x(4 - x)^2 = (4 - x)^2[4 - x - 3x] = 4(4 - x)^2(1 - x)$

An equation for the tangent line to the graph of f at $x = 2$ is:

$y - y_1 = m(x - x_1)$ where $x_1 = 2$, $y_1 = f(x_1) = f(2) = 16$, and

$m = f'(x_1) = f'(2) = -16$. Thus, $y - 16 = -16(x - 2)$ or $y = -16x + 48$.

43. $f(x) = \dfrac{x}{(2x - 5)^3}$

$f'(x) = \dfrac{(2x - 5)^3(1) - x(3)(2x - 5)^2(2)}{[(2x - 5)^3]^2}$

$= \dfrac{(2x - 5)^3 - 6x(2x - 5)^2}{(2x - 5)^6} = \dfrac{(2x - 5) - 6x}{(2x - 5)^4} = \dfrac{-4x - 5}{(2x - 5)^4}$

An equation for the tangent line to the graph of f at $x = 3$ is:

$y - y_1 = m(x - x_1)$ where $x_1 = 3$, $y_1 = f(x_1) = f(3) = 3$, and

$m = f'(x_1) = f'(3) = -17$. Thus, $y - 3 = -17(x - 3)$ or $y = -17x + 54$.

45. $f(x) = x\sqrt{2x + 2} = x(2x + 2)^{1/2}$

$f'(x) = x[(2x + 2)^{1/2}]' + (2x + 2)^{1/2}(x)'$

$= x\left(\dfrac{1}{2}\right)(2x + 2)^{-1/2}(2) + (2x + 2)^{1/2}(1) = \dfrac{x}{(2x + 2)^{1/2}} + (2x + 2)^{1/2}$

$= \dfrac{3x + 2}{(2x + 2)^{1/2}}$

An equation for the tangent line to the graph of f at $x = 1$ is:

$y - y_1 = m(x - x_1)$ where $x_1 = 1$, $y_1 = f(x_1) = f(1) = 2$, and

$m = f'(x_1) = f'(1) = \dfrac{5}{2}$. Thus, $y - 2 = \dfrac{5}{2}(x - 1)$ or $y = \dfrac{5}{2}x - \dfrac{1}{2}$.

47. $f(x) = x^2(x - 5)^3$

$f'(x) = x^2[(x - 5)^3]' + (x - 5)^3(x^2)'$

$\quad = x^2(3)(x - 5)^2(1) + (x - 5)^3(2x)$

$\quad = 3x^2(x - 5)^2 + 2x(x - 5)^3 = 5x(x - 5)^2(x - 2)$

The tangent line to the graph of f is horizontal at the values of x such that $f'(x) = 0$. Thus, we set $5x(x - 5)^2(x - 2) = 0$ and $x = 0$, $x = 2$, $x = 5$.

49. $f(x) = \dfrac{x}{(2x + 5)^2}$

$f'(x) = \dfrac{(2x + 5)^2(x)' - x[(2x + 5)^2]'}{[(2x + 5)^2]^2}$

$\quad = \dfrac{(2x + 5)^2(1) - x(2)(2x + 5)(2)}{(2x + 5)^4} = \dfrac{2x + 5 - 4x}{(2x + 5)^3} = \dfrac{5 - 2x}{(2x + 5)^3}$

The tangent line to the graph of f is horizontal at the values of x such that $f'(x) = 0$. Thus, we set

$\dfrac{5 - 2x}{(2x + 5)^3} = 0$

$5 - 2x = 0$

and $x = \dfrac{5}{2}$.

51. $f(x) = \sqrt{x^2 - 8x + 20} = (x^2 - 8x + 20)^{1/2}$

$f'(x) = \dfrac{1}{2}(x^2 - 8x + 20)^{-1/2}(2x - 8)$

$\quad = \dfrac{x - 4}{(x^2 - 8x + 20)^{1/2}}$

The tangent line to the graph of f is horizontal at the values of x such that $f'(x) = 0$. Thus, we set

$\dfrac{x - 4}{(x^2 - 8x + 20)^{1/2}} = 0$

$x - 4 = 0$

and $x = 4$.

53. $f(x) = 4x - \sqrt{x^4 + 10}$

Horizontal tangent
at $x \approx 2.32$.

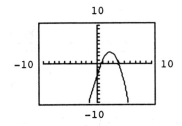

55. $f(x) = 5\sqrt{x^2 + 1} - \sqrt{x^4 + 1}$

Horizontal tangents
at $x \approx -2.34$, 0, 2.34.

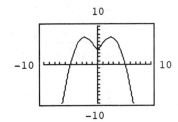

57. $f(x) = \sqrt{x^4 - 4x^3 + 6x + 10}$

Horizontal tangents
at $x \approx -0.64,\ 0.83,\ 2.81$.

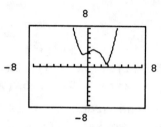

59. $\dfrac{d}{dx}[3x(x^2 + 1)^3] = 3x\dfrac{d}{dx}(x^2 + 1)^3 + (x^2 + 1)^3 \dfrac{d}{dx}3x$

$$= 3x \cdot 3(x^2 + 1)^2(2x) + (x^2 + 1)^3(3)$$
$$= 18x^2(x^2 + 1)^2 + 3(x^2 + 1)^3$$
$$= (x^2 + 1)^2[18x^2 + 3(x^2 + 1)]$$
$$= (x^2 + 1)^2(21x^2 + 3)$$
$$= 3(x^2 + 1)^2(7x^2 + 1)$$

61. $\dfrac{d}{dx}\dfrac{(x^3 - 7)^4}{2x^3} = \dfrac{2x^3 \dfrac{d}{dx}(x^3 - 7)^4 - (x^3 - 7)^4 \dfrac{d}{dx}2x^3}{(2x^3)^2}$

$$= \dfrac{2x^3 \cdot 4(x^3 - 7)^3(3x^2) - (x^3 - 7)^4 6x^2}{4x^6}$$
$$= \dfrac{3(x^3 - 7)^3 x^2[8x^3 - 2(x^3 - 7)]}{4x^6}$$
$$= \dfrac{3(x^3 - 7)^3(6x^3 + 14)}{4x^4} = \dfrac{3(x^3 - 7)^3(3x^3 + 7)}{2x^4}$$

63. $\dfrac{d}{dx}[(2x - 3)^2(2x^2 + 1)^3] = (2x - 3)^2 \dfrac{d}{dx}(2x^2 + 1)^3 + (2x^2 + 1)^3 \dfrac{d}{dx}(2x - 3)^2$

$$= (2x - 3)^2 3(2x^2 + 1)^2(4x) + (2x^2 + 1)^3 2(2x - 3)(2)$$
$$= 12x(2x - 3)^2(2x^2 + 1)^2 + 4(2x^2 + 1)^3(2x - 3)$$
$$= 4(2x - 3)(2x^2 + 1)^2[3x(2x - 3) + (2x^2 + 1)]$$
$$= 4(2x - 3)(2x^2 + 1)^2(6x^2 - 9x + 2x^2 + 1)$$
$$= 4(2x - 3)(2x^2 + 1)^2(8x^2 - 9x + 1)$$

65. $\dfrac{d}{dx}[4x^2\sqrt{x^2 - 1}] = \dfrac{d}{dx}[\sqrt{16x^4(x^2 - 1)}]$

$$= \dfrac{d}{dx}[(16x^6 - 16x^4)^{1/2}]$$
$$= \dfrac{1}{2}(16x^6 - 16x^4)^{-1/2}(96x^5 - 64x^3)$$
$$= \dfrac{96x^5 - 64x^3}{2(16x^6 - 16x^4)^{1/2}} = \dfrac{8x^2(12x^3 - 8x)}{2 \cdot 4x^2(x^2 - 1)^{1/2}} = \dfrac{12x^3 - 8x}{(x^2 - 1)^{1/2}}$$

or
$\dfrac{d}{dx}[4x^2\sqrt{x^2 - 1}] = \dfrac{d}{dx}[4x^2(x^2 - 1)^{1/2}]$

$$= 4x^2 \cdot \dfrac{1}{2}(x^2 - 1)^{-1/2}(2x) + (x^2 - 1)^{1/2}(8x)$$

$$= \frac{4x^3}{(x^2 - 1)^{1/2}} + 8x(x^2 - 1)^{1/2}$$

$$= \frac{4x^3 + 8x(x^2 - 1)}{(x^2 - 1)^{1/2}} = \frac{4x^3 + 8x^3 - 8x}{(x^2 - 1)^{1/2}} = \frac{12x^3 - 8x}{(x^2 - 1)^{1/2}}$$

67. $\dfrac{d}{dx} \dfrac{2x}{\sqrt{x - 3}} = \dfrac{(x - 3)^{1/2}(2) - 2x \cdot \frac{1}{2}(x - 3)^{-1/2}}{(x - 3)}$

$$= \frac{2(x - 3)^{1/2} - \dfrac{x}{(x - 3)^{1/2}}}{(x - 3)} = \frac{2(x - 3) - x}{(x - 3)(x - 3)^{1/2}}$$

$$= \frac{2x - 6 - x}{(x - 3)^{3/2}} = \frac{x - 6}{(x - 3)^{3/2}}$$

69. $\dfrac{d}{dx}\sqrt{(2x - 1)^3(x^2 + 3)^4} = \dfrac{d}{dx}[(2x - 1)^3(x^2 + 3)^4]^{1/2}$

$$= \frac{d}{dx}(2x - 1)^{3/2}(x^2 + 3)^2$$

$$= (2x - 1)^{3/2}\frac{d}{dx}(x^2 + 3)^2 + (x^2 + 3)^2\frac{d}{dx}(2x - 1)^{3/2}$$

$$= (2x - 1)^{3/2}2(x^2 + 3)(2x) + (x^2 + 3)^2 \cdot \frac{3}{2}(2x - 1)^{1/2}(2)$$

$$= (2x - 1)^{1/2}(x^2 + 3)[4x(2x - 1) + 3(x^2 + 3)]$$

$$= (2x - 1)^{1/2}(x^2 + 3)(8x^2 - 4x + 3x^2 + 9)$$

$$= (2x - 1)^{1/2}(x^2 + 3)(11x^2 - 4x + 9)$$

71. $C(x) = 10 + \sqrt{2x + 16} = 10 + (2x + 16)^{1/2}, \ 0 \le x \le 50$

(A) $C'(x) = \dfrac{1}{2}(2x + 16)^{-1/2}(2) = \dfrac{1}{(2x + 16)^{1/2}}$

(B) $C'(24) = \dfrac{1}{[2(24) + 16]^{1/2}} = \dfrac{1}{(64)^{1/2}} = \dfrac{1}{8}$ or $12.50; at a production
level of 24 calculators, total costs are INCREASING at the rate of
$12.50 per calculator; also, the cost of producing the 25th
calculator is approximately $12.50.

$C'(42) = \dfrac{1}{[2(42) + 16]^{1/2}} = \dfrac{1}{(100)^{1/2}} = \dfrac{1}{10}$ or $10.00; at a production
level of 42 calculators, total costs are INCREASING at the rate of
$10.00 per calculator; also the cost of producing the 43rd
calculator is approximatley $10.00.

73. $x = 80\sqrt{p + 25} - 400 = 80(p + 25)^{1/2} - 400,\ 20 \le p \le 100$

(A) $\dfrac{dx}{dp} = 80\left(\dfrac{1}{2}\right)(p + 25)^{-1/2}(1) = \dfrac{40}{(p + 25)^{1/2}}$

(B) At $p = 75$, $x = 80\sqrt{75 + 25} - 400 = 400$ and
$\dfrac{dx}{dp} = \dfrac{40}{(75 + 25)^{1/2}} = \dfrac{40}{(100)^{1/2}} = 4.$

At a price of \$75, the supply is 400 speakers, and the supply is INCREASING at a rate of 4 speakers per dollar.

75. $A = 1000\left(1 + \dfrac{1}{12}r\right)^{48}$

$\dfrac{dA}{dr} = 1000(48)\left(1 + \dfrac{1}{12}r\right)^{47}\left(\dfrac{1}{12}\right) = 4000\left(1 + \dfrac{1}{12}r\right)^{47}$

77. $y = (3 \times 10^6)\left[1 - \dfrac{1}{\sqrt[3]{(x^2 - 1)^2}}\right] = (3 \times 10^6)[1 - (x^2 - 1)^{-2/3}]$

$\dfrac{dy}{dx} = -(3 \times 10^6)\left(-\dfrac{2}{3}\right)(x^2 - 1)^{-5/3}(2x) = \dfrac{(4 \times 10^6)x}{(x^2 - 1)^{5/3}}$

79. $T = f(n) = 2n\sqrt{n - 2} = 2n(n - 2)^{1/2}$

(A) $f'(n) = 2n[(n - 2)^{1/2}]' + (n - 2)^{1/2}(2n)'$

$\qquad = 2n\left(\dfrac{1}{2}\right)(n - 2)^{-1/2}(1) + (n - 2)^{1/2}(2)$

$\qquad = \dfrac{n}{(n - 2)^{1/2}} + 2(n - 2)^{1/2}$

$\qquad = \dfrac{n + 2(n - 2)}{(n - 2)^{1/2}} = \dfrac{3n - 4}{(n - 2)^{1/2}}$

(B) $f'(11) = \dfrac{29}{3} = 9.67$; when the list contains 11 items, the learning time is increasing at the rate of 9.67 minutes per item;

$f'(27) = \dfrac{77}{5} = 15.4$; when the list contains 27 items, the learning time is increasing at the rate of 15.4 minutes per item.

Things to remember:

1. **MARGINAL COST, REVENUE, AND PROFIT**

   If x is the number of units of a product produced in some time interval, then:

   Total Cost $= C(x)$
   Marginal Cost $= C'(x)$
   Total Revenue $= R(x)$
   Marginal Revenue $= R'(x)$
   Total Profit $= P(x) = R(x) - C(x)$
   Marginal Profit $= P'(x) = R'(x) - C'(x)$
   $\qquad\qquad\qquad = $ (Marginal Revenue) $-$ (Marginal Cost)

   Marginal cost (or revenue or profit) is the instantaneous rate of change of cost (or revenue or profit) relative to production at a given production level.

2. **MARGINAL COST AND EXACT COST**

   If $C(x)$ is the cost of producing x items, then the marginal cost function approximates the exact cost of producing the $(x + 1)$st item:

   $\qquad$ Marginal Cost $\qquad\qquad$ Exact Cost
   $\qquad\qquad C'(x) \qquad\qquad \approx \qquad C(x + 1) - C(x)$

   Similar interpretations can be made for total revenue and total profit functions.

3. **BREAK-EVEN POINTS**

   The BREAK-EVEN POINTS are the points where total revenue equals total cost.

4. **MARGINAL AVERAGE COST, REVENUE, AND PROFIT**

   If x is the number of units of a product produced in some time interval, then:

   Average Cost $= \overline{C}(x) = \dfrac{C(x)}{x}$ $\qquad$ Cost per unit

   Marginal Average Cost $= \overline{C}'(x)$

   Average Revenue $= \overline{R}(x) = \dfrac{R(x)}{x}$ $\qquad$ Revenue per unit

   Marginal Average Revenue $= \overline{R}'(x)$

   Average Profit $= \overline{P}(x) = \dfrac{P(x)}{x}$ $\qquad$ Profit per unit

   Marginal Average Profit $= \overline{P}'(x)$

1. $C(x) = 2000 + 50x - 0.5x^2$

    (A) The exact cost of producing the 21st food processor is:

$$C(21) - C(20) = 2000 + 50(21) - \frac{(21)^2}{2} - \left[2000 + 50(20) - \frac{(20)^2}{2}\right]$$
$$= 2829.50 - 2800$$
$$= 29.50 \text{ or } \$29.50$$

    (B) $C'(x) = 50 - x$
$$C'(20) = 50 - 20 = 30 \text{ or } \$30$$

3. $C(x) = 60,000 + 300x$

    (A) $\overline{C}(x) = \dfrac{60,000 + 300x}{x} = \dfrac{60,000}{x} + 300 = 60,000x^{-1} + 300$

$$\overline{C}(500) = \frac{60,000 + 300(500)}{500} = \frac{210,000}{500} = 420 \text{ or } \$420$$

    (B) $\overline{C}'(x) = -60,000x^{-2} = \dfrac{-60,000}{x^2}$

$$\overline{C}'(500) = \frac{-60,000}{(500)^2} = -0.24 \text{ or } \$0.24$$

    Interpretation: At a production level of 500 frames, average cost is decreasing at the rate of 24¢ per frame.

    (C) The average cost per frame if 501 frames are produced is approximately $420 - $0.24 = $419.76.

5. $R(x) = 100x - 0.025x^2 = 100x - \dfrac{x^2}{40}$

$$R'(x) = 100 - \frac{x}{20}$$

    (A) $R'(1600) = 100 - \dfrac{1600}{20} = 100 - 80 = 20 \text{ or } \$20$

    Interpretation: At a production level of 1600 radios, revenue is increasing at the rate of $20 per radio.

    (B) $R'(2500) = 100 - \dfrac{2500}{20} = -25 \text{ or } -\$25$

    Interpretation: At a production level of 2500 radios, revenue is decreasing at the rate of $25 per radio.

7. $P(x) = 30x - 0.5x^2 - 250$

    (A) The exact profit from the sale of the 26th skateboard is:

$$P(26) - P(25) = 30(26) - \frac{(26)^2}{2} - 250 - \left[30(25) - \frac{(25)^2}{2} - 250\right]$$
$$= 192 - 187.50$$
$$= 4.50 \text{ or } \$4.50$$

    (B) $P'(x) = 30 - x$
$$P'(25) = 30 - 25 = 5 \text{ or } \$5.00$$

9. $P(x) = 5x - \dfrac{x^2}{200} - 450$

$P'(x) = 5 - \dfrac{x}{100}$

(A) $P'(450) = 5 - \dfrac{450}{100} = 0.5$ or \$0.50

Interpretation: At a production level of 450 cassettes, profit is increasing at the rate of 50¢ per cassette.

(B) $P'(750) = 5 - \dfrac{750}{100} = -2.5$ or -\$2.50

Interpretation: At a production level of 750 cassettes, profit is decreasing at the rate of \$2.50 per cassette.

11. $P(x) = 30x - 0.03x^2 - 750$

Average profit: $\bar{P}(x) = \dfrac{P(x)}{x} = 30 - 0.03x - \dfrac{750}{x} = 30 - 0.03x - 750x^{-1}$

(A) At $x = 50$, $\bar{P}(50) = 30 - (0.03)50 - \dfrac{750}{50} = 13.50$ or \$13.50.

(B) $\bar{P}'(x) = -0.03 + 750x^{-2} = -0.03 + \dfrac{750}{x^2}$

$\bar{P}'(50) = -0.03 + \dfrac{750}{(50)^2} = -0.03 + 0.3 = 0.27$ or \$0.27; at a production level of 50 mowers, the average profit per mower is INCREASING at the rate of \$0.27 per mower.

(C) The average profit per mower if 51 mowers are produced is approximately \$13.50 + \$0.27 = \$13.77.

13. $p = 200 - \dfrac{x}{30}$, $C(x) = 72,000 + 60x$

(A) $C'(x) = 60$ or \$60

(B) Revenue: $R(x) = xp = 200x - \dfrac{x^2}{30}$

(C) $R'(x) = 200 - \dfrac{x}{15}$

(D) $R'(1,500) = 200 - \dfrac{1,500}{15} = 100$; at a production level of 1,500 saws, revenue is INCREASING at the rate of \$100 per saw.

$R'(4,500) = 200 - \dfrac{4,500}{15} = -100$; at a production level of 4,500 saws, revenue is DECREASING at the rate of \$100 per saw.

(E) The graphs of $C(x)$ and $R(x)$ are shown below.

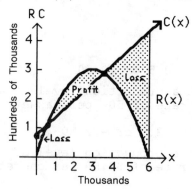

To find the break-even points, set $C(x) = R(x)$:

$$72,000 + 60x = 200x - \frac{x^2}{30}$$

$$x^2 - 4,200x + 2,160,000 = 0$$

$$(x - 600)(x - 3,600) = 0$$

$$x = 600 \quad \text{or} \quad x = 3,600$$

Now, $\quad C(600) = 72,000 + 60(600) = 108,000;$
$\quad\quad C(3,600) = 72,000 + 60(3,600) = 288,000.$
Thus, the break-even points are: $(600, 108,000)$, $(3,600, 288,000)$.

(F) $P(x) = R(x) - C(x) = 200x - \frac{x^2}{30} - [72,000 + 60x]$

$$= -\frac{x^2}{30} + 140x - 72,000$$

(G) $P'(x) = -\frac{x}{15} + 140$

(H) $P'(1,500) = -\frac{1,500}{15} + 140 = 40$; at a production level of 1,500 saws, the profit is INCREASING at the rate of $40 per saw.

$P'(3,000) = -\frac{3,000}{15} + 140 = -60$; at a production level of 3,000 saws, the profit is DECREASING at the rate of $60 per saw.

15. (A) We are given $p = 16$ when $x = 200$ and $p = 14$ when $x = 300$. Thus, we have the pair of equations:
$16 = 200m + b$
$14 = 300m + b$
Subtracting the second equation from the first, we get $-100m = 2$. Thus,
$$m = -\frac{1}{50}.$$
Substituting this into either equation yields $b = 20$. Therefore,
$$p = -\frac{x}{50} + 20 \quad \text{or} \quad p = 20 - \frac{x}{50} = 20 - 0.02x.$$

(B) $R(x) = x \cdot p(x) = 20x - \dfrac{x^2}{50} = 20x - 0.02x^2$.

(C) From the financial department's estimates, $m = 4$ and $b = 1400$. Thus, $C(x) = 4x + 1400$.

(D) The graphs of $R(x)$ and $C(x)$ are shown below:

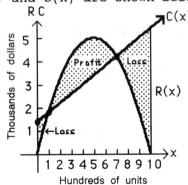

To find the break-even points, set $C(x) = R(x)$.

$$4x + 1400 = 20x - \dfrac{x^2}{50}$$

$$\dfrac{x^2}{50} - 16x + 1400 = 0$$

$$x^2 - 800x + 70,000 = 0$$

$$(x - 100)(x - 700) = 0$$

$$x = 100 \text{ or } x = 700$$

Now, $C(100) = 1800$ and $C(700) = 4200$. Thus, the break-even points are $(100, 1800)$ and $(700, 4200)$.

(E) $P(x) = R(x) - C(x) = 20x - 0.02x^2 - (4x + 1400) = 16x - 0.02x^2 - 1400$

(F) $P(x) = 16x - 0.02x^2 - 1400$

$P'(x) = 16 - 0.04x$

$P'(250) = 16 - \dfrac{250}{25} = 6$ or \$6

Interpretation: At a production level of 250 toasters, profit is increasing at the rate of \$6 per toaster.

$P'(475) = 16 - \dfrac{475}{25} = -3$ or $-$3$

Interpretation: At a production level of 475 toasters, profit is decreasing at the rate of \$3 per toaster.

17. Total cost: $C(x) = 24x + 21,900$

Total revenue: $R(x) = 200x - 0.2x^2$, $0 \le x \le 1,000$

(A) $R'(x) = 200 - 0.4x$

The graph of R has a horizontal tangent line at the value(s) of x where $R'(x) = 0$, i.e.,

$$200 - 0.4x = 0$$

$$\text{or } x = 500$$

(B) $P(x) = R(x) - C(x) = 200x - 0.2x^2 - (24x + 21,900)$
$$= 176x - 0.2x^2 - 21,900$$

(C) $P'(x) = 176 - 0.4x$. Setting $P'(x) = 0$, we have
$$176x - 0.4x = 0$$
$$\text{or } x = 440$$

(D) The graphs of C, R and P are shown below.

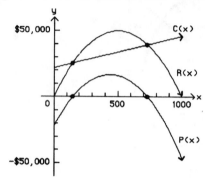

Break-even points: $R(x) = C(x)$
$$200x - 0.2x^2 = 24x + 21,900$$
$$0.2x^2 - 176x + 21,900 = 0$$
$$x = \frac{176 \pm \sqrt{(176)^2 - (4)(0.2)(21,900)}}{2(0.2)} \quad \text{(quadratic formula)}$$
$$= \frac{176 \pm \sqrt{30,976 - 17,520}}{0.4}$$
$$= \frac{176 \pm \sqrt{13,456}}{0.4} = \frac{176 \pm 116}{0.4} = 730, \ 150$$

Thus, the break-even points are: (730, 39,420) and (150, 25,500).

x-intercepts for P: $-0.2x^2 + 17.6x - 21,900 = 0$
$$\text{or } 0.2x^2 - 176x + 21,900 = 0$$
which is the same as the equation above.

Thus, $x = 150$ and $x = 730$.

19. Demand equation: $p = 20 - \sqrt{x} = 20 - x^{1/2}$
Cost equation: $C(x) = 500 + 2x$

   (A) Revenue $R(x) = xp = x(20 - x^{1/2})$
   $$\text{or } R(x) = 20x - x^{3/2}$$

   (B) The graphs for R and C for $0 \le x \le 400$
   are shown at the right.

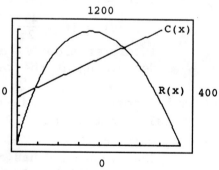

Break-even points (44, 588)
and (258, 1,016).

1. (A) $f(3) - f(1) = 2(3)^2 + 5 - [2(1)^2 + 5] = 16$

   (B) Average rate of change: $\dfrac{f(3) - f(1)}{3 - 1} = \dfrac{16}{2} = 8$

   (C) Slope of secant line: $\dfrac{f(3) - f(1)}{3 - 1} = \dfrac{16}{2} = 8$

   (D) Instantaneous rate of change at $x = 1$:

   Step 1. $\dfrac{f(1 + h) - f(1)}{h} = \dfrac{2(1 + h)^2 + 5 - [2(1)^2 + 5]}{h}$

   $= \dfrac{2(1 + 2h + h^2) + 5 - 7}{h} = \dfrac{4h + 2h^2}{h} = 4 + 2h$

   Step 2. $\lim\limits_{h \to 0} \dfrac{f(1 + h) - f(1)}{h} = \lim\limits_{h \to 0} (4 + 2h) = 4$

   (E) Slope of the tangent line at $x = 1$: 4

   (F) $f'(1) = 4$ $\hspace{6cm}$ (3-1, 3-3, 3-4)

2. $f'(-1) = -2;\ f'(1) = 1$ $\hspace{6cm}$ (3-1, (3-3)

3. (A) $\lim\limits_{x \to 1} (5f(x) + 3g(x)) = 5 \lim\limits_{x \to 1} f(x) + 3 \lim\limits_{x \to 1} g(x) = 5 \cdot 2 + 3 \cdot 4 = 22$

   (B) $\lim\limits_{x \to 1} [f(x)g(x)] = [\lim\limits_{x \to 1} f(x)][\lim\limits_{x \to 1} g(x)] = 2 \cdot 4 = 8$

   (C) $\lim\limits_{x \to 1} \dfrac{g(x)}{f(x)} = \dfrac{\lim\limits_{x \to 1} g(x)}{\lim\limits_{x \to 1} f(x)} = \dfrac{4}{2} = 2$ $\hspace{4cm}$ (3-2)

4. (A) $\lim\limits_{x \to 1^-} f(x) = 1$ $\quad$ (B) $\lim\limits_{x \to 1^+} f(x) = 1$ $\quad$ (C) $\lim\limits_{x \to 1} f(x) = 1$

   (D) $f(1) = 1$ $\hspace{8cm}$ (3-2)

5. (A) $\lim\limits_{x \to 2^-} f(x) = 2$ $\quad$ (B) $\lim\limits_{x \to 2^+} f(x) = 3$ $\quad$ (C) $\lim\limits_{x \to 2} f(x)$ does not exist

   (D) $f(2) = 3$ $\hspace{8cm}$ (3-2)

6. (A) $\lim\limits_{x \to 3^-} f(x) = 4$ $\quad$ (B) $\lim\limits_{x \to 3^+} f(x) = 4$ $\quad$ (C) $\lim\limits_{x \to 3} f(x) = 4$

   (D) $f(3)$ does not exist $\hspace{7cm}$ (3-2)

7. $f'(x) = \lim\limits_{h \to 0} \dfrac{f(x + h) - f(x)}{h} = \lim\limits_{h \to 0} \dfrac{3x^2 h + 3xh^2 + h^3 + 2xh + h^2}{h}$

   $= \lim\limits_{h \to 0} \dfrac{h(3x^2 + 3xh + h^2 + 2x + h)}{h}$

   $= \lim\limits_{h \to 0} (3x^2 + 3xh + h^2 + 2x + h)$

   $= 3x^2 + 2x$ $\hspace{7cm}$ (3-3)

8. (A) $h(x) = 2f(x) + 3g(x)$; $h(5) = 2f'(5) + 3g'(5) = 2(-1) + 3(-3) = -11$

(B) $h(x) = f(x)g(x)$; $h'(5) = f(5)g'(5) + g(5)f'(5) = 4(-3) + 2(-1) = -14$

(C) $h(x) = \dfrac{f(x)}{g(x)}$; $h'(5) = \dfrac{g(5)f'(5) - f(5)g'(5)}{[g(5)]^2} = \dfrac{2(-1) - 4(-3)}{2^2} = \dfrac{10}{4} = \dfrac{5}{2}$

(D) $h(x) = [f(x)]^2$; $h'(5) = 2f(5)f'(5) = 2(4)(-1) = -8$ (3-4, 3,5, 3-6)

9. $6x + 4$; $\dfrac{d}{dx}(3x^2 + 4x + 1)^5 = 5(3x^2 + 4x + 1)^4(6x + 4)$ (3-6)

10. $f(x) = 3x^4 - 2x^2 + 1$

$f'(x) = 12x^3 - 4x$ (3-4)

11. $f(x) = 2x^{1/2} - 3x$

$f'(x) = 2 \cdot \dfrac{1}{2}x^{-1/2} - 3 = \dfrac{1}{x^{1/2}} - 3$ (3-4)

12. $f(x) = 5$

$f'(x) = 0$

(3-4)

13. $f(x) = \dfrac{1}{2x^2} + \dfrac{x^2}{2}$

$\quad = \dfrac{1}{2}x^{-2} + \dfrac{x^2}{2}$

$f'(x) = -x^{-3} + x$

$\quad = -\dfrac{1}{x^3} + x$

(3-4)

14. $f(x) = (2x - 1)(3x + 2)$

$f'(x) = (2x - 1)(3) + (3x + 2)(2)$

$\quad = 6x - 3 + 6x + 4$

$\quad = 12x + 1$ (3-5)

15. $f(x) = (x^2 - 1)(x^3 - 3)$

$f'(x) = (x^2 - 1)(3x^2) + (x^3 - 3)(2x) = 3x^4 - 3x^2 + 2x^4 - 6x = 5x^4 - 3x^2 - 6x$

(3-5)

16. $f(x) = \dfrac{2x}{x^2 + 2}$

$f'(x) = \dfrac{(x^2 + 2)(2) - 2x(2x)}{(x^2 + 2)^2} = \dfrac{2x^2 + 4 - 4x^2}{(x^2 + 2)^2} = \dfrac{4 - 2x^2}{(x^2 + 2)^2}$ (3-5)

17. $f(x) = \dfrac{1}{3x + 2} = (3x + 2)^{-1}$

$f'(x) = -1(3x + 2)^{-2}(3) = \dfrac{-3}{(3x + 2)^2}$ (3-6)

18. $f(x) = (2x - 3)^3$

$f'(x) = 3(2x - 3)^2(2) = 6(2x - 3)^2$ (3-6)

19. $f(x) = (x^2 + 2)^{-2}$

$f'(x) = -2(x^2 + 2)^{-3}(2x) = \dfrac{-4x}{(x^2 + 2)^3}$ (3-6)

20. $f(x) = 0.5x^2 - 5$

(A) $\dfrac{f(4) - f(2)}{4 - 2} = \dfrac{0.5(4)^2 - 5 - [0.5(2)^2 - 5]}{2} = \dfrac{8 - 2}{2} = 3$

(B) $\dfrac{f(2 + h) - f(2)}{h} = \dfrac{0.5(2 + h)^2 - 5 - [0.5(2)^2 - 5]}{h}$

$\qquad\qquad = \dfrac{0.5(4 + 4h + h^2) - 5 + 3}{h}$

$\qquad\qquad = \dfrac{2h + 0.5h^2}{h} = \dfrac{h(2 + 0.5h)}{h} = 2 + 0.5h$

(C) $\displaystyle\lim_{h \to 0} \dfrac{f(2 + h) - f(2)}{h} = \lim_{h \to 0} (2 + 0.5h) = 2$ $\qquad$ (3–1, 3–3)

21.

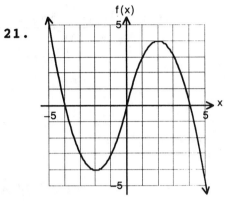

(3–1)

22. $y = 3x^4 - 2x^{-3} + 5$

$\qquad \dfrac{dy}{dx} = 12x^3 + 6x^{-4}$ $\qquad$ (3–4)

23. $y = (2x^2 - 3x + 2)(x^2 + 2x - 1)$

$\qquad y' = (2x^2 - 3x + 2)(2x + 2) + (x^2 + 2x - 1)(4x - 3)$

$\qquad\quad = 4x^3 - 6x^2 + 4x + 4x^2 - 6x + 4 + 4x^3 + 8x^2 - 4x - 3x^2 - 6x + 3$

$\qquad\quad = 8x^3 + 3x^2 - 12x + 7$ $\qquad\qquad\qquad\qquad\qquad\qquad$ (3–5,

24. $f(x) = \dfrac{2x - 3}{(x - 1)^2}$

$\qquad f'(x) = \dfrac{(x - 1)^2 2 - (2x - 3)2(x - 1)}{(x - 1)^4} = \dfrac{(x - 1)[2(x - 1) - 4x + 6]}{(x - 1)^4}$

$\qquad\quad = \dfrac{(2x - 2 - 4x + 6)}{(x - 1)^3} = \dfrac{4 - 2x}{(x - 1)^3}$ $\qquad\qquad\qquad\qquad$ (3–5)

25. $y = 2\sqrt{x} + \dfrac{4}{\sqrt{x}} = 2x^{1/2} + 4x^{-1/2}$

$\qquad y' = 2 \cdot \dfrac{1}{2} x^{-1/2} + 4\left(-\dfrac{1}{2}\right)x^{-3/2} = \dfrac{1}{x^{1/2}} - \dfrac{2}{x^{3/2}}$ or $\dfrac{1}{\sqrt{x}} - \dfrac{2}{\sqrt{x^3}}$ $\quad$ (3–4)

26. $\dfrac{d}{dx}[(x^2 - 1)(2x + 1)^2] = (x^2 - 1)\dfrac{d}{dx}(2x + 1)^2 + (2x + 1)^2 \dfrac{d}{dx}(x^2 - 1)$

$\qquad\qquad\qquad\qquad\quad = (x^2 - 1)[2(2x + 1)(2)] + (2x + 1)^2(2x)$

$\qquad\qquad\qquad\qquad\quad = 2(2x + 1)[2(x^2 - 1) + x(2x + 1)]$

$\qquad\qquad\qquad\qquad\quad = 2(2x + 1)(2x^2 - 2 + 2x^2 + x)$

$\qquad\qquad\qquad\qquad\quad = 2(2x + 1)(4x^2 + x - 2)$ $\qquad\qquad$ (3–5, 3–6)

27. $\dfrac{d}{dx}(x^3 - 5)^{1/3} = \dfrac{1}{3}(x^3 - 5)^{-2/3}(3x^2) = \dfrac{x^2}{(x^3 - 5)^{2/3}}$ $\qquad$ (3–6)

28. $y = \dfrac{3x^2 + 4}{x^2}$

$\dfrac{dy}{dx} = \dfrac{x^2(6x) - (3x^2 + 4)2x}{[x^2]^2} = \dfrac{-8x}{x^4} = \dfrac{-8}{x^3}$ (3-4)

29. $\dfrac{d}{dx} \dfrac{(x^2 + 2)^4}{2x - 3} = \dfrac{(2x - 3)4(x^2 + 2)^3(2x) - (x^2 + 2)^4(2)}{(2x - 3)^2}$

$\qquad = \dfrac{2(x^2 + 2)^3[4x(2x - 3) - (x^2 + 2)]}{(2x - 3)^2}$

$\qquad = \dfrac{2(x^2 + 2)^3(8x^2 - 12x - x^2 - 2)}{(2x - 3)^2} = \dfrac{2(x^2 + 2)^3(7x^2 - 12x - 2)}{(2x - 3)^2}$

(3-5, 3-6)

30. $f(x) = x^2 + 4$
$f'(x) = 2x$

(A) The slope of the graph at $x = 1$ is $m = f'(1) = 2$.

(B) $f(1) = 1^2 + 4 = 5$
The tangent line at $(1, 5)$, where the slope $m = 2$, is:
$(y - 5) = 2(x - 1)$ [Note: $(y - y_1) = m(x - x_1)$.]
$\qquad\qquad y = 5 + 2x - 2$
$\qquad\qquad y = 2x + 3$ (3-3, 3-4)

31. $f(x) = x^3(x + 1)^2$
$f'(x) = x^3(2)(x + 1)(1) + (x + 1)^2(3x^2)$
$\qquad = 2x^3(x + 1) + 3x^2(x + 1)^2$

(A) The slope of the graph of f at $x = 1$ is:
$f'(1) = 2 \cdot 1^3(1 + 1) + 3 \cdot 1^2(1 + 1)^2 = 16$

(B) $f(1) = 1^3(1 + 1)^2 = 4$
An equation for the tangent line to the graph of f at $x = 1$ is
$y - 4 = 16(x - 1)$ or $y = 16x - 12$. (3-3, 3-5)

32. $f(x) = 10x - x^2$
$f'(x) = 10 - 2x$
The tangent line is horizontal at the values of x such that $f'(x) = 0$:
$10 - 2x = 0$
$\qquad x = 5$ (3-4)

33. $f(x) = (x + 3)(x^2 - 45)$
$f'(x) = (x + 3)(2x) + (x^2 - 45)(1) = 3x^2 + 6x - 45$
Set $f'(x) = 0$:
$3x^2 + 6x - 45 = 0$
$x^2 + 2x - 15 = 0$
$(x - 3)(x + 5) = 0$
$\qquad\qquad x = 3, \; x = -5$ (3-5)

34. $f(x) = \dfrac{x}{x^2 + 4}$

$f'(x) = \dfrac{(x^2 + 4)(1) - x(2x)}{(x^2 + 4)^2} = \dfrac{4 - x^2}{(x^2 + 4)^2}$

Set $f'(x) = 0$:

$\dfrac{4 - x^2}{(x^2 + 4)^2} = 0$

$4 - x^2 = 0$

$(2 - x)(2 + x) = 0$

$x = 2, \ x = -2$ (3-5)

35. $f(x) = x^2(2x - 15)^3$

$f'(x) = x^2(3)(2x - 15)^2(2) + (2x - 15)^3(2x)$

$\quad = (2x - 15)^2[6x^2 + 4x^2 - 30x]$

$\quad = (2x - 15)^2 10x(x - 3)$

Set $f'(x) = 0$:

$10x(x - 3)(2x - 15)^2 = 0$

$x = 0, \ x = 3, \ x = \dfrac{15}{2}$ (3-5)

36.

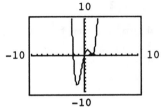

sHorizontal tangent lines at
$x \approx -1.37, \ 0.60, \ 1.52$ (3-4)

37.

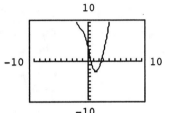

Horizontal tangent line
at $x \approx 1.43$ (3-5)

38.

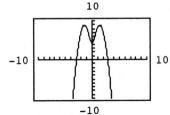

Horizontal tangent line at $x \approx -1.41, \ 1.41$ (3-6)

39. $y = f(x) = 16x^2 - 4x$

(A) Velocity function: $v(x) = f'(x) = 32x - 4$

(B) $v(3) = 32(3) - 4 = 92$ ft/sec   (3-4)

40. $y = f(x) = 96x - 16x^2$

(A) Velocity function: $v(x) = f'(x) = 96 - 32x$

(B) $v(x) = 0$ when $96 - 32x = 0$

$32x = 96$

$x = 3$ sec   (3-4)

41.

h	-0.1	-0.01	-0.001	→ 0	← 0.001	0.01	0.1
$\dfrac{f(h) - f(0)}{h}$	1.49	1.60	1.61	→ 1.61 ←	1.61	1.62	1.75

$f'(0) \approx 1.61$ (3-1, 3-3)

42. $f'(0) \approx -1.39$ (3-1, 3-3)

43. (A) $f(x) = x^3$, $g(x) = (x - 4)^3$, $h(x) = (x + 3)^3$

The graph of g is the graph of f shifted 4 units to the right;
the graph of h is the graph of f shifted 3 units to the left.

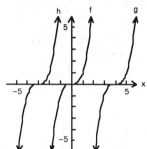

(B) $f'(x) = 3x^2$, $g'(x) = 3(x - 4)^2$, $h'(x) = 3(x + 3)^2$

The graph of g' is the graph of f' shifted 4 units to the right;
the graph of h' is the graph of f' shifted 3 units to the left.

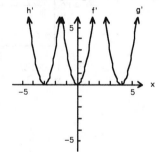

(1-2, 3-6)

44. (A) $g(x) = f(x + k)$; $g'(x) = f'(x + k)(1) = f'(x + k)$

The graph of g is a horizontal translation of the graph of f.
The graph of g' is a horizontal translation of the graph of f'.

(B) $g(x) = f(x) + k$, $g'(x) = f'(x)$

The graph of g is a vertical translation of the graph of f
(up k units if $k > 0$, down k units if $k < 0$). The graph of g' is
the same as the graph of f'. (1-2, 3-6)

45. $\lim\limits_{x \to 3} \dfrac{2x - 3}{x + 5} = \dfrac{2(3) - 3}{3 + 5} = \dfrac{6 - 3}{8} = \dfrac{3}{8}$ (3-2)

46. $\lim\limits_{x \to 3} (2x^2 - x + 1) = 2 \cdot 3^2 - 3 + 1 = 16$ (3-2)

47. $\lim\limits_{x \to 0} \dfrac{2x}{3x^2 - 2x} = \lim\limits_{x \to 0} \dfrac{2x}{x(3x - 2)} = \lim\limits_{x \to 0} \dfrac{2}{3x - 2} = \dfrac{2}{-2} = -1$ (3-2)

48. $\lim\limits_{x \to 3} \dfrac{x - 3}{x^2 - 9} = \lim\limits_{x \to 3} \dfrac{\cancel{x - 3}}{(x + 3)\cancel{(x - 3)}}$

$\qquad\qquad = \dfrac{1}{3 + 3} = \dfrac{1}{6}$ (3-2)

49. For $x < 4$, $|x - 4| = -(x - 4) = 4 - x$

$\quad \lim\limits_{x \to 4^-} \dfrac{|x - 4|}{x - 4} = \lim\limits_{x \to 4^-} \dfrac{-(x - 4)}{x - 4} = \lim\limits_{x \to 4^-} (-1) = -1$ (3-2)

50. For $x \geq 4$, $|x - 4| = x - 4$

$\quad \lim\limits_{x \to 4^+} \dfrac{|x - 4|}{x - 4} = \lim\limits_{x \to 4^+} \dfrac{x - 4}{x - 4} = \lim\limits_{x \to 4^+} 1 = 1$ (3-2)

51. It follows from Exercises 49 and 50 that

$\quad \lim\limits_{x \to 4} \dfrac{|x - 4|}{x - 4}$ does not exist.   (3-2)

52. $\lim\limits_{h \to 0} \dfrac{[(2 + h)^2 - 1] - [2^2 - 1]}{h} = \lim\limits_{h \to 0} \dfrac{4 + 4h + h^2 - 1 - 3}{h}$

$\qquad\qquad\qquad = \lim\limits_{h \to 0} \dfrac{4h + h^2}{h} = \lim\limits_{h \to 0} (4 + h) = 4$ (3-2)

53. $f(x) = x^2 + 4$

$\quad \lim\limits_{h \to 0} \dfrac{f(2 + h) - f(2)}{h} = \lim\limits_{h \to 0} \dfrac{[(2 + h)^2 + 4] - [2^2 + 4]}{h}$

$\qquad\qquad\qquad = \lim\limits_{h \to 0} \dfrac{4 + 4h + h^2 + 4 - 8}{h} = \lim\limits_{h \to 0} \dfrac{4h + h^2}{h}$

$\qquad\qquad\qquad = \lim\limits_{h \to 0} (4 + h) = 4$ (3-2)

54. Let $f(x) = \dfrac{1}{x + 2}$

$\quad \lim\limits_{h \to 0} \dfrac{f(x + h) - f(x)}{h} = \lim\limits_{h \to 0} \dfrac{\dfrac{1}{(x + h) + 2} - \dfrac{1}{x + 2}}{h}$

$\qquad\qquad\qquad = \lim\limits_{h \to 0} \dfrac{x + 2 - (x + h + 2)}{h(x + h + 2)(x + 2)}$

$\qquad\qquad\qquad = \lim\limits_{h \to 0} \dfrac{-h}{h(x + h + 2)(x + 2)}$

$\qquad\qquad\qquad = \lim\limits_{h \to 0} \dfrac{-1}{(x + h + 2)(x - 2)} = \dfrac{-1}{(x + 2)^2}$ (3-2)

55. (A) $\lim\limits_{x \to -2^-} f(x) = -6$, $\lim\limits_{x \to -2^+} f(x) = 6$; $\lim\limits_{x \to -2} f(x)$ does not exist

    (B) $\lim\limits_{x \to 0} f(x) = 4$

(C) $\lim\limits_{x \to 2^-} f(x) = 2$, $\lim\limits_{x \to 2^+} f(x) = -2$; $\lim\limits_{x \to 2} f(x)$ does not exist

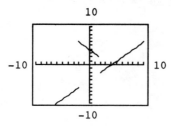

(3-2)

56. $f(x) = x^2 - x$

Step 1. Simplify $\dfrac{f(x + h) - f(x)}{h}$.

$$\frac{f(x + h) - f(x)}{h} = \frac{[(x + h)^2 - (x + h)] - (x^2 - x)}{h}$$

$$= \frac{x^2 + 2xh + h^2 - x - h - x^2 + x}{h}$$

$$= \frac{2xh + h^2 - h}{h} = 2x + h - 1$$

Step 2. Evaluate $\lim\limits_{h \to 0} \dfrac{f(x + h) - f(x)}{h}$.

$$\lim\limits_{h \to 0} \frac{f(x + h) - f(x)}{h} = \lim\limits_{h \to 0} (2x + h - 1) = 2x - 1$$

Thus, $f'(x) = 2x - 1$. (3-3)

57. $f(x) = \sqrt{x} - 3$

Step 1. Simplify $\dfrac{f(x + h) - f(x)}{h}$.

$$\frac{f(x + h) - f(x)}{h} = \frac{[\sqrt{x + h} - 3] - (\sqrt{x} - 3)}{h}$$

$$= \frac{\sqrt{x + h} - \sqrt{x}}{h} \cdot \frac{\sqrt{x + h} + \sqrt{x}}{\sqrt{x + h} + \sqrt{x}} = \frac{x + h - x}{h[\sqrt{x + h} + \sqrt{x}]}$$

$$= \frac{1}{\sqrt{x + h} + \sqrt{x}}$$

Step 2. Evaluate $\lim\limits_{h \to 0} \dfrac{f(x + h) - f(x)}{h}$.

$$\lim\limits_{h \to 0} \frac{f(x + h) - f(x)}{h} = \lim\limits_{h \to 0} \frac{1}{\sqrt{x + h} + \sqrt{x}} = \frac{1}{2\sqrt{x}}$$ (3-3)

58. f is not differentiable at $x = 0$, since f is not continuous at 0. (3-3)

59. f is not differentiable at $x = 1$; the curve has a vertical tangent line at this point. (3-3)

60. f is not differentiable at $x = 2$; the curve has a "corner" at this point. (3-3)

61. f is differentiable at $x = 3$. In fact, $f'(3) = 0$. (3-3)

62. $f(x) = (x - 4)^4(x + 3)^3$

$$f'(x) = (x - 4)^4(3)(x + 3)^2(1) + (x + 3)^3(4)(x - 4)^3(1)$$
$$= (x - 4)^3(x + 3)^2[3(x - 4) + 4(x + 3)]$$
$$= 7x(x - 4)^3(x + 3)^2 \qquad (3\text{-}5, \ 3\text{-}6)$$

63. $f(x) = \dfrac{x^5}{(2x + 1)^4}$

$$f'(x) = \frac{(2x + 1)^4(5x^4) - x^5(4)(2x + 1)^3(2)}{[(2x + 1)^4]^2}$$
$$= \frac{(2x + 1)(5x^4) - 8x^5}{(2x + 1)^5} = \frac{2x^5 + 5x^4}{(2x + 1)^5} = \frac{x^4(2x + 5)}{(2x + 1)^5} \qquad (3\text{-}5, \ 3\text{-}6)$$

64. $f(x) = \dfrac{\sqrt{x^2 - 1}}{x} = \dfrac{(x^2 - 1)^{1/2}}{x}$

$$f'(x) = \frac{x\left(\frac{1}{2}\right)(x^2 - 1)^{-1/2}(2x) - (x^2 - 1)^{1/2}(1)}{x^2} = \frac{\dfrac{x^2}{(x^2 - 1)^{1/2}} - (x^2 - 1)^{1/2}}{x^2}$$
$$= \frac{1}{x^2(x^2 - 1)^{1/2}} = \frac{1}{x^2\sqrt{x^2 - 1}} \qquad (3\text{-}5, \ 3\text{-}6)$$

65. $f(x) = \dfrac{x}{\sqrt{x^2 + 4}} = \dfrac{x}{(x^2 + 4)^{1/2}}$

$$f'(x) = \frac{(x^2 + 4)^{1/2}(1) - x\left(\frac{1}{2}\right)(x^2 + 4)^{-1/2}(2x)}{[(x^2 + 4)^{1/2}]^2}$$
$$= \frac{(x^2 + 4)^{1/2} - \dfrac{x^2}{(x^2 + 4)^{1/2}}}{(x^2 + 4)} = \frac{4}{(x^2 + 4)^{3/2}} \qquad (3\text{-}5, \ 3\text{-}6)$$

66. $f(x) = x^{1/5}$; $f'(x) = \dfrac{1}{5}x^{-4/5} = \dfrac{1}{5x^{4/5}}$

The domain of f' is all real numbers except $x = 0$. At $x = 0$, the graph of f is smooth, but the tangent line to the graph at $(0, 0)$ is vertical. (3-3)

67. $f(x) = \begin{cases} x^2 - m & \text{if } x \le 1 \\ -x^2 + m & \text{if } x > 1 \end{cases}$

(A)

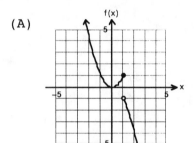

$$\lim_{x \to 1^-} f(x) = 1, \quad \lim_{x \to 1^+} f(x) = -1$$

(B)

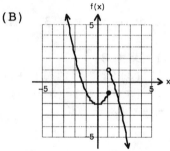

$$\lim_{x \to 1^-} f(x) = -1, \quad \lim_{x \to 1^+} f(x) = 1$$

(C) $\lim\limits_{x \to 1^-} f(x) = 1 - m, \quad \lim\limits_{x \to 1^+} f(x) = -1 + m$

We want $1 - m = -1 + m$ which implies $m = 1$.

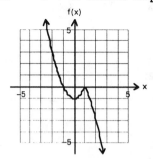

(D) The graphs in (A) and (B) have jumps at $x = 1$; the graph in (C) does not. (3-2)

68. $f(x) = 1 - |x - 1|, \ 0 \le x \le 2$

(A) $\lim\limits_{h \to 0^-} \dfrac{f(1 + h) - f(1)}{h} = \lim\limits_{h \to 0^-} \dfrac{1 - |1 + h - 1| - 1}{h} = \lim\limits_{h \to 0^-} \dfrac{-|h|}{h}$

$\qquad = \lim\limits_{h \to 0^-} \dfrac{h}{h} = 1 \quad (|h| = -h \text{ if } h < 0)$

(B) $\lim\limits_{h \to 0^+} \dfrac{f(1 + h) - f(1)}{h} = \lim\limits_{h \to 0^+} \dfrac{1 - |1 + h - 1| - 1}{h} = \lim\limits_{h \to 0^+} \dfrac{-|h|}{h}$

$\qquad = \lim\limits_{h \to 0^+} \dfrac{-h}{h} = -1 \quad (|h| = h \text{ if } h > 0)$

(C) $\lim\limits_{h \to 0} \dfrac{f(1 + h) - f(1)}{h}$ does not exist, since the left limit and the right limit are not equal.

(D) $f'(1)$ does not exist. (3-3)

69. $C(x) = 10,000 + 200x - 0.1x^2$

(A) $C(101) - C(100) = 10,000 + 200(101) - 0.1(101)^2$
$$- [10,000 + 200(100) - 0.1(100)^2]$$
$$= 29,179.90 - 29,000$$
$$= \$179.90$$

(B) $C'(x) = 200 - 0.2x$
$C'(100) = 200 - 0.2(100)$
$$= 200 - 20$$
$$= \$180 \tag{3-7}$$

70. $C(x) = 5,000 + 40x + 0.05x^2$

(A) Cost of producing 100 bicycles:
$C(100) = 5,000 + 40(100) + 0.05(100)^2$
$$= 9000 + 500 = 9500$$
Marginal cost:
$C'(x) = 40 + 0.1x$
$C'(100) = 40 + 0.1(100) = 40 + 10 = 50$

Interpretation: At a production level of 100 bicycles, the total cost is \$9,500 and is increasing at the rate of \$50 per additional bicycle.

(B) Average cost: $\overline{C}(x) = \dfrac{C(x)}{x} = \dfrac{5000}{x} + 40 + 0.05x$

$$\overline{C}(100) = \frac{5000}{100} + 40 + 0.05(100) = 50 + 40 + 5 = 95$$

Marginal average cost: $\overline{C}'(x) = -\dfrac{5000}{x^2} + 0.05$

$$\text{and } \overline{C}'(100) = -\frac{5000}{(100)^2} + 0.05 = -0.5 + 0.05 = -0.45$$

Interpretation: At a production level of 100 bicycles, the average cost is \$95 and the marginal average cost is decreasing at a rate of \$0.45 per additional bicycle. (3-7)

71. The approximate cost of producing the 201st printer is greater than that of producing the 601st printer (the slope of the tangent line at $x = 200$ is greater than the slope of the tangent line at $x = 600$). Since the marginal costs are decreasing, the manufacturing process is becoming more efficient. (3-7)

72. $p = 25 - 0.1x$, $C(x) = 2x + 9,000$

(A) Marginal cost: $C'(x) = 2$

Average cost: $\overline{C}(x) = \dfrac{C(x)}{x} = 2 + \dfrac{9,000}{x}$

Marginal cost: $\overline{C}'(x) = -\dfrac{9,000}{x^2}$

(B) Revenue: $R(x) = xp = 25x - 0.01x^2$
Marginal revenue: $R'(x) = 25 - 0.02x$

Average revenue: $\bar{R}(x) = \dfrac{R(x)}{x} = 25 - 0.01x$

Marginal average revenue: $\bar{R}'(x) = -0.01$

(C) Profit: $P(x) = R(x) - C(x) = 25x - 0.01x^2 - (2x + 9,000)$
$$= 23x - 0.01x^2 - 9,000$$
Marginal profit: $P'(x) = 23 - 0.02x$

Average profit: $\bar{P}(x) = \dfrac{P(x)}{x} = 23 - 0.01x - \dfrac{9,000}{x}$

Marginal average profit: $\bar{P}'(x) = -0.01 + \dfrac{9,000}{x^2}$

(D) Break-even points: $R(x) = C(x)$
$$25x - 0.01x^2 = 2x + 9,000$$
$$0.01x^2 - 23x + 9,000 = 0$$
$$x^2 - 2,300x + 900,000 = 0$$
$$(x - 500)(x - 1,800) = 0$$

Thus, the break-even points are: $x = 500$, $x = 1,800$.

(E) $P'(1,000) = 23 - 0.02(1000) = 3$; profit is increasing at the rate of \$3 per umbrella.

$P'(1,150) = 23 - 0.02(1,150) = 0$; profit is flat.

$P'(1,400) = 23 - 0.02(1,400) = -5$; profit is decreasing at the rate of \$5 per umbrella.

(F)

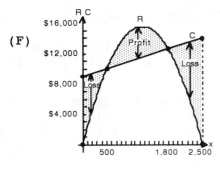

(3-7)

73. $N(t) = \dfrac{40t}{t + 2}$

(A) Average rate of change from $t = 3$ to $t = 6$:

$$\dfrac{N(6) - N(3)}{6 - 3} = \dfrac{\dfrac{40 \cdot 6}{6 + 2} - \dfrac{40 \cdot 3}{3 + 2}}{3} = \dfrac{30 - 24}{3} = 2 \text{ components per day}$$

(B) $N'(t) = \dfrac{(t + 2)(40) - 40t(1)}{(t + 2)^2} = \dfrac{80}{(t + 2)^2}$

$N'(3) = \dfrac{80}{25} = 3.2$ components per day

(3-5)

74. $N(t) = t\sqrt{4 + t} = t(4 + t)^{1/2}$

$N'(t) = t\left(\dfrac{1}{2}\right)(4 + t)^{-1/2} + (4 + t)^{1/2} = \dfrac{t}{2(4 + t)^{1/2}} + \dfrac{(4 + t)^{1/2}}{1}$

$\qquad\qquad = \dfrac{t + 2(4 + t)}{2(4 + t)^{1/2}} = \dfrac{8 + 3t}{2(4 + t)^{1/2}}$

$N(5) = 5\sqrt{4 + 5} = 15$; $N'(t) = \dfrac{8 + 3(5)}{2(4 + 5)^{1/2}} = \dfrac{23}{6} = 3.833$;

After 5 months, the total sales are 15,000 pools and sales are
INCREASING at the rate of 3,833 pools per month. $\qquad$ (3-6)

75. (A) $A(t) = 5{,}000(1.07)^t$; $A(10) = 5{,}000(1.07)^{10} \approx \$9836$

h	-0.1	-0.01	-0.001 →	0 ←	0.001	0.01	0.1
$\dfrac{A(10 + h) - A(10)}{h}$	663.23	665.25	665.45 →	665 ←	665.50	665.70	667.73

$A'(10) \approx 665$

(B) After 10 years, the amount in the account is \$9,836 and is GROWING
at the rate of \$665 per year. $\qquad$ (3-3)

76. $C(x) = 500(x + 1)^{-2}$

The instantaneous rate of change of concentration at x meters is:
$C'(x) = 500(-2)(x + 1)^{-3}$

$\qquad = -1000(x + 1)^{-3} = \dfrac{-1000}{(x + 1)^3}$

The rate of change of concentration at 9 meters is:
$C'(9) = \dfrac{-1000}{(9 + 1)^3} = \dfrac{-1000}{10^3} = -1$ part per million per meter

The rate of change of concentration at 99 meters is:
$C'(99) = \dfrac{-1000}{(99 + 1)^3} = \dfrac{-1000}{100^3} = \dfrac{-10^3}{10^6} = -10^{-3}$ or $= -\dfrac{1}{1000}$

$\qquad\qquad\qquad = -0.001$ parts per million
per meter $\qquad$ (3-4)

77. $F(t) = 98 + \dfrac{4}{\sqrt{t + 1}} = 98 + 4(t + 1)^{-1/2}$,

$F'(t) = 4\left(-\dfrac{1}{2}\right)(t + 1)^{-3/2} = \dfrac{-2}{(t + 1)^{3/2}}$

$F(3) = 98 + \dfrac{4}{\sqrt{3 + 1}} = 100$; $F'(3) = \dfrac{-2}{(3 + 1)^{3/2}} = -\dfrac{1}{4} = -0.25$

After 3 hours, the body temperature of the patient is 100° and is
DECREASING at the rate of 0.25° per hour. $\qquad$ (3-6)

78. $N(t) = 20\sqrt{t} = 20t^{1/2}$

The rate of learning is $N'(t) = 20\left(\dfrac{1}{2}\right)t^{-1/2} = 10t^{-1/2} = \dfrac{10}{\sqrt{t}}$.

(A) The rate of learning after one hour is $N'(1) = \dfrac{10}{\sqrt{1}}$

$\qquad\qquad\qquad\qquad\qquad = 10$ items per hour.

(B) The rate of learning after four hours is $N'(4) = \dfrac{10}{\sqrt{4}} = \dfrac{10}{2}$

$$= 5 \text{ items per hour.}$$

(3-4)

79. $M(t) = 40.7(1.01)^t$

(A) $M(50) = 40.7(1.01)^{50} \approx 66.9$

h	-0.1	-0.01	-0.001 $\to$ 0	$\gets$ 0.001	0.01	0.1
$\dfrac{M(50 + h) - M(50)}{h}$	0.666	0.666	0.666 $\to$ 0.666 $\gets$ 0.666	0.666	0.666	0.666

$M'(50) \approx 0.7$

(B) In 2010, there will be 66.9 million married couples and this number will be GROWING at the rate of 0.7 million = 700,000 couples per year.

(C) In 2011, there will be approximately 66.9 + 0.7 = 67.6 million married couples.

In 2012, there will be approximately 67.6 + 0.7 = 68.3 million married couples.

(3-3)

4 GRAPHING AND OPTIMIZATION

Things to remember:

1. CONTINUITY

    A function f is CONTINUOUS AT THE POINT $x = c$ if:

    (a) $\lim\limits_{x \to c} f(x)$ exists;

    (b) $f(c)$ exists;

    (c) $\lim\limits_{x \to c} f(x) = f(c)$

    If one or more of the three conditions fails, then f is DISCONTINUOUS at $x = c$.

    A function is CONTINUOUS ON THE OPEN INTERVAL (a, b) if it is continuous at each point on the interval.

2. ONE-SIDED CONTINUITY

    A function f is CONTINUOUS ON THE LEFT AT $x = c$ if $\lim\limits_{x \to c^-} f(x) = f(c)$; f is CONTINUOUS ON THE RIGHT AT $x = c$ if $\lim\limits_{x \to c^+} f(x) = f(c)$.

    The function f is continuous on the closed interval $[a, b]$ if it is continuous on the open interval (a, b), and is continuous on the right at a and continuous on the left at b.

3. CONTINUITY PROPERTIES OF SOME SPECIFIC FUNCTIONS

    (a) A constant function, $f(x) = k$, is continuous for all x.

    (b) For n a positive integer, $f(x) = x^n$ is continuous for all x.

    (c) A polynomial function
    $$P(x) = a_n x^n + a_{n-1} x^{n-1} + \ldots + a_1 x + a_0$$
    is continuous for all x.

    (d) A rational function
    $$R(x) = \frac{P(x)}{Q(x)},$$
    P and Q polynomial functions, is continuous for all x except those numbers $x = c$ such that $Q(c) = 0$.

    (e) For n an odd positive integer, $n > 1$, $\sqrt[n]{f(x)}$ is continuous wherever f is continuous.

(f) For n an even positive integer, $\sqrt[n]{f(x)}$ is continuous wherever f is continuous and non-negative.

<u>**4.**</u> **VERTICAL ASYMPTOTES**

Suppose that the limit of a function f fails to exist as x approaches c from the left because the values of $f(x)$ are becoming very large positive numbers (or very large negative numbers). This is denoted

$$\lim_{x \to c^-} f(x) = \infty \quad (\text{or } -\infty)$$

If this happens as x approaches c from the right, then

$$\lim_{x \to c^+} f(x) = \infty \quad (\text{or } -\infty)$$

If both one-sided limits exhibit the same behavior, then

$$\lim_{x \to c} f(x) = \infty \quad (\text{or } -\infty)$$

If any of the above hold, the line $x = c$ is a VERTICAL ASYMPTOTE for the graph of $y = f(x)$.

<u>**5.**</u> **SIGN PROPERTIES ON AN INTERVAL (a, b)**

If f is continuous or (a, b) and $f(x) \neq 0$ for all x in (a, b), then either $f(x) > 0$ for all x in (a, b) or $f(x) < 0$ for all x in (a, b).

<u>**6.**</u> **CONSTRUCTING SIGN CHARTS**

Given a function f:

Step 1. Find all partition numbers. That is:

(A) Find all numbers where f is discontinuous. (Rational functions are discontinuous for values of x that make a denominator 0.)

(B) Find all numbers where $f(x) = 0$. (For a rational function, this occurs where the numerator is 0 and the denominator is not 0.)

Step 2. Plot the numbers found in step 1 on a real number line, dividing the number line into intervals.

Step 3. Select a test number in each open interval determined in step 2, and evaluate $f(x)$ at each test number to determine whether $f(x)$ is positive (+) or negative (-) in each interval.

Step 4. Construct a sign chart using the real number line in step 2. This will show the sign of $f(x)$ on each open interval.

[*Note*: From the sign chart, it is easy to find the solution for the inequality $f(x) < 0$ or $f(x) > 0$.]

1. f is continuous at $x = 1$, since $\lim\limits_{x \to 1} f(x) = f(1) = 2$

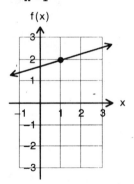

3. f is discontinuous at $x = 1$, since $\lim\limits_{x \to 1} f(x) \neq f(1)$

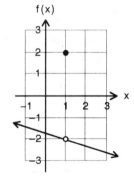

5. $\lim\limits_{x \to 1^-} f(x) = -2 = \lim\limits_{x \to 1^+} f(x)$
implies $\lim\limits_{x \to 1} f(x) = -2$;

f is continuous at $x = 1$,
since $\lim\limits_{x \to 1} f(x) = f(1) = -2$

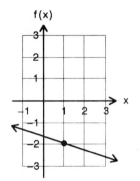

7. $\lim\limits_{x \to 1^-} f(x) = 2$, $\lim\limits_{x \to 1^+} f(x) = -2$
implies $\lim\limits_{x \to 1} f(x)$ does not exist;

f is discontinuous at $x = 1$,
since $\lim\limits_{x \to 1} f(x)$ does not exist

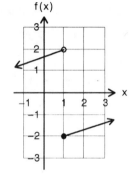

9.

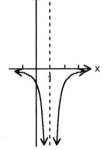

11.

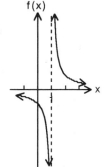

13. $f(x) = 2x - 3$ is a polynomial function. Therefore, f is continuous for all x [$\underline{3}$(c)].

15. $h(x) = \dfrac{2}{x - 5}$ is a rational function and the denominator $x - 5$ is 0 when $x = 5$. Thus, h is continuous for all x except $x = 5$ [$\underline{3}$(d)].

17. $g(x) = \dfrac{x - 5}{(x - 3)(x + 2)}$ is a rational function and the denominator $(x - 3)(x + 2)$ is 0 when $x = 3$ or $x = -2$. Thus, g is continuous for all x except $x = 3$, $x = -2$.

19. (A) $\lim\limits_{x \to 0^-} f(x) = 1$, $\lim\limits_{x \to 0^+} f(x) = 1$, $\lim\limits_{x \to 0} f(x) = 1$, $f(1) = 1$.

(B) f **is** continuous at $x = 0$ since $\lim\limits_{x \to 0} f(x)$ exists, $f(0)$ exists and $\lim\limits_{x \to 0} f(x) = f(0)$.

21. (A) $\lim\limits_{x \to 1^-} f(x) = 2$, $\lim\limits_{x \to 1^+} f(x) = 1$, $\lim\limits_{x \to 1} f(x)$ does not exist, $f(1) = 1$.

(B) f **is not** continuous at $x = 1$ since $\lim\limits_{x \to 0} f(x)$ does not exist.

23. (A) $\lim\limits_{x \to -2^-} f(x) = 1$, $\lim\limits_{x \to -2^+} f(x) = 1$, $\lim\limits_{x \to -2} f(x) = 1$, $f(-2) = 3$.

(B) f **is not** continuous at $x = -2$ since $\lim\limits_{x \to -2} f(x) \neq f(-2)$.

25. $f(x) = \begin{cases} 2 \text{ if } x \text{ is an integer} \\ 1 \text{ if } x \text{ is not an integer} \end{cases}$

(A) The graph of f is:

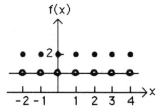

(B) $\lim\limits_{x \to 2} f(x) = 1$ (C) $f(2) = 2$

(D) f is not continuous at $x = 2$ since $\lim\limits_{x \to 2} f(x) \neq f(2)$.

(E) f is discontinuous at $x = n$ for all integers n.

27. $f(x) = \dfrac{1}{x + 3}$

f is discontinuous at $x = -3$; $\lim\limits_{x \to -3^-} f(x) = -\infty$, $\lim\limits_{x \to -3^+} f(x) = \infty$; the line $x = -3$ is a vertical asymptote.

29. $h(x) = \dfrac{x^2 + 4}{x^2 - 4} = \dfrac{x^2 + 4}{(x - 2)(x + 2)}$

h is discontinuous at $x = -2$; $\lim\limits_{x \to -2^-} h(x) = \infty$ and $\lim\limits_{x \to -2^+} h(x) = -\infty$; the line $x = -2$ is a vertical asymptote; h is discontinuous at $x = 2$; $\lim\limits_{x \to 2^-} h(x) = -\infty$ and $\lim\limits_{x \to 2^+} h(x) = \infty$; the line $x = 2$ is a vertical asymptote.

31. $F(x) = \dfrac{x^2 - 4}{x^2 + 4}$

$x^2 + 4 \neq 0$ for all x ($x^2 + 4 \geq 4$). Therefore, F is continuous for all real numbers x; there are no vertical asymptotes.

33. $H(x) = \dfrac{x^2 - 2x - 3}{x^2 - 4x + 3} = \dfrac{(x - 3)(x + 1)}{(x - 3)(x - 1)} = \dfrac{x + 1}{x - 1}$, $x \neq 3$

H is discontinuous at $x = 1$; $\lim\limits_{x \to 1^-} H(x) = -\infty$ and $\lim\limits_{x \to 1^+} H(x) = \infty$; the line $x = 1$ is a vertical asymptote; H is discontinuous at $x = 3$ (H is not defined at $x = 3$); $\lim\limits_{x \to 3} H(x) = 2$; there is not a vertical asymptote at $x = 3$.

35. $T(x) = \dfrac{8x - 16}{x^4 - 8x^3 + 16x^2} = \dfrac{8(x - 2)}{x^2(x - 4)^2}$

T is discontinuous at $x = 0$; $\lim\limits_{x \to 0^-} T(x) = -\infty$ and $\lim\limits_{x \to 0^+} T(x) = -\infty$. Therefore $\lim\limits_{x \to 0} T(x) = -\infty$; the line $x = 0$ (the y axis) is a vertical asymptote. T is discontinuous at $x = 4$; $\lim\limits_{x \to 4^-} T(x) = \infty$ and $\lim\limits_{x \to 4^+} T(x) = \infty$. Therefore, $\lim\limits_{x \to 4} T(x) = \infty$; the line $x = 4$ is a vertical asymptote.

37. $x^2 - x - 12 < 0$

Let $f(x) = x^2 - x - 12 = (x - 4)(x + 3)$. Then f is continuous for all x and $f(-3) = f(4) = 0$. Thus, $x = -3$ and $x = 4$ are partition numbers.

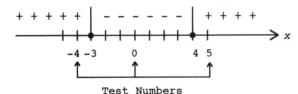

Test Numbers	
x	$f(x)$
-4	8 (+)
0	-12 (−)
5	8 (+)

Thus, $x^2 - x - 12 < 0$ for:

$-3 < x < 4$ (inequality notation)

$(-3, 4)$ (interval notation)

39. $x^2 + 21 > 10x$ or $x^2 - 10x + 21 > 0$

Let $f(x) = x^2 - 10x + 21 = (x - 7)(x - 3)$. Then f is continuous for all x and $f(3) = f(7) = 0$. Thus, $x = 3$ and $x = 7$ are partition numbers.

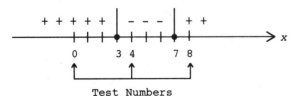

Test Numbers	
x	$f(x)$
0	21 (+)
4	-3 (−)
8	5 (+)

Thus, $x^2 - 10x + 21 > 0$ for:

$x < 3$ or $x > 7$ (inequality notation)

$(-\infty, 3) \cup (7, \infty)$ (interval notation)

41. $\dfrac{x^2 + 5x}{x - 3} > 0$

Let $f(x) = \dfrac{x^2 + 5x}{x - 3} = \dfrac{x(x + 5)}{x - 3}$. Then f is discontinuous at $x = 3$ and

$f(0) = f(-5) = 0$. Thus, $x = -5$, $x = 0$, and $x = 3$ are partition numbers.

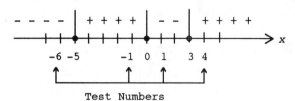

Test Numbers

x	$f(x)$
-6	$-\frac{2}{3}$ $(-)$
-1	1 $(+)$
1	-3 $(-)$
4	36 $(+)$

Thus, $\dfrac{x^2 + 5x}{x - 3} > 0$ for: $-5 < x < 0$ or $x > 3$ (inequality notation)

$(-5, 0) \cup (3, \infty)$ (interval notation)

43. $f(x) = x^3 - 3x^2 - 2x + 5$.
Partition numbers: $x_1 \approx -1.33$, $x_2 \approx 1.20$, $x_3 \approx 3.13$

(A) $f(x) > 0$ on $(-1.33, 1.20) \cup (3.13, \infty)$

(B) $f(x) < 0$ on $(-\infty, -1.33) \cup (1.20, 3.13)$

45. $f(x) = x^4 - 6x^2 + 3x + 5$.
Partition numbers: $x_1 \approx -2.53$, $x_2 \approx -0.72$

(A) $f(x) > 0$ on $(-\infty, -2.53) \cup (-0.72, \infty)$

(B) $f(x) < 0$ on $(-2.53, -0.72)$

47. $f(x) = \dfrac{x^3 + x + 6}{-x^3 - 2x + 5}$
Partition numbers: $x_1 \approx -1.63$, $x_2 \approx 1.33$

(A) $f(x) > 0$ on $(-1.63, 1.33)$

(B) $f(x) < 0$ on $(-\infty, -1.63) \cup (1.33, \infty)$

49. $F(x) = 2x^8 - 3x^4 + 5$ is a polynomial function. Thus F is continuous for all x, i.e., F is continuous on $(-\infty, \infty)$ [see 3(c)].

51. Since $f(x) = x - 5$ is a polynomial function, it is continuous for all x. Thus, $g(x) = \sqrt{f(x)} = \sqrt{x - 5}$ is continuous for all x such that $x - 5 \geq 0$, i.e., g is continuous on $[5, \infty)$ [see 3(f)].

53. Since $f(x) = x - 5$ is continuous for all x, $K(x) = \sqrt[3]{x - 5}$ is continuous for all x, i.e., K is continuous on $(-\infty, \infty)$ [see 3(e)].

55. $f(x) = \dfrac{x^2 - 1}{x^2 - 3x + 2} = \dfrac{x^2 - 1}{(x - 1)(x - 2)}$

Since f is a rational function, it is continuous for all x except the numbers $x = c$ at which the denominator is 0. Thus, f is continuous for all x except $x = 1$ and $x = 2$, i.e., f is continuous on $(-\infty, 1)$, $(1, 2)$, and $(2, \infty)$.

57. The graph of f is shown at the right. This function is discontinuous at $x = 1$. [$\lim\limits_{x \to 1} f(x)$ does not exist.]

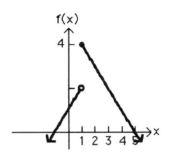

59. The graph of f is:

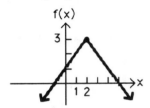

This function is continuous for all x. [Note: $\lim\limits_{x \to 2} f(x) = f(2) = 3$.]

61. The graph of f is:

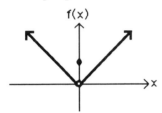

This function is discontinuous at $x = 0$. [Note: $\lim\limits_{x \to 0} f(x) = 0 \neq f(0) = 1$.]

63. (A) Yes; g is continuous on $(-1, 2)$.

(B) Since $\lim\limits_{x \to -1^+} g(x) = -1 = g(-1)$, g is continuous from the right at $x = -1$.

(C) Since $\lim\limits_{x \to 2^-} g(x) = 2 = g(2)$, g is continuous from the left at $x = 2$.

(D) Yes; g is continuous on the closed interval $[-1, 2]$.

65. (A) Since $\lim\limits_{x \to 0^+} f(x) = f(0) = 0$, f is continuous from the right at $x = 0$.

(B) Since $\lim\limits_{x \to 0^-} f(x) = -1 \neq f(0) = 0$, f is not continuous from the left at $x = 0$.

(C) f is continuous on the open interval $(0, 1)$.

(D) *f* is *not* continuous on the closed interval [0, 1] since
$\lim\limits_{x \to 1^-} f(x) = 0 \neq f(1) = 1$, i.e., *f* is not continuous from the left at
x = 1.

(E) *f* is continuous on the half-closed interval [0, 1).

67. *x* intercepts: *x* = -5, 2

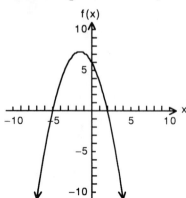

69. *x* intercepts: *x* = -6, -1, 4

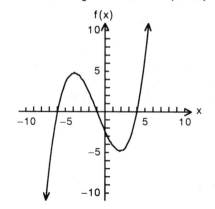

71. $f(x) = \dfrac{2}{1 - x} \neq 0$ for all *x*. This does not contradict Theorem 2 because *f*
is not continuous on (-1, 3); *f* is discontinuous at *x* = 1.

73. The following sketches illustrate that either condition is possible.
Theorem 2 implies that one of these two conditions must occur.

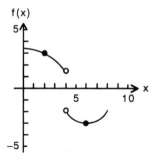

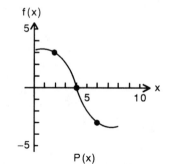

75. (A) $P(x) = \begin{cases} 0.32 & 0 < x \le 1 \\ 0.55 & 1 < x \le 2 \\ 0.78 & 2 < x \le 3 \\ 1.01 & 3 < x \le 4 \\ 1.24 & 4 < x \le 5 \end{cases}$

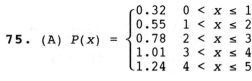

The graph of *P* is:

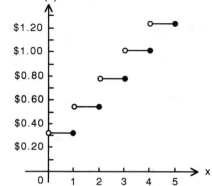

(B) $\lim\limits_{x \to 4.5} P(x) = \$1.24; \quad P(4.5) = \$1.24$

(C) $\lim\limits_{x \to 4} P(x)$ does not exist; $P(4)$ = \$1.01

(D) P is continuous at x = 4.5;
P is not continuous at x = 4.

77. (A) $p(x) = \begin{cases} 0.49 & 150 \le x < 250 \\ 0.39 & 250 \le x < 500 \\ 0.29 & 500 \le x < 1000 \\ 0.24 & 1000 \le x \le 1500 \end{cases}$

The graph of p is:

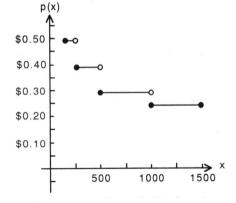

(B) p is discontinuous at x = 250, x = 500, x = 1,000. In each case, the limit from the left is greater than the limit from the right reflecting the corresponding drop in price at these order quantities.

(C) $C(x) = \begin{cases} 0.49x & 150 \le x < 250 \\ 0.39x & 250 \le x < 500 \\ 0.29x & 500 \le x < 1000 \\ 0.24x & 1000 \le x \le 1500 \end{cases}$

The graph of C is shown at the right.

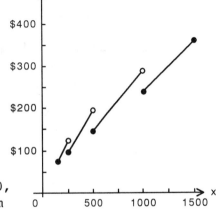

(D) C is discontinuous at x = 150, 250, 500, and 1,000. In each case, the limit from the left is greater than the limit from the right, reflecting savings to the customer due to the corresponding drop in price at these order quantities.

79. (A) $E(s) = \begin{cases} 1000, & 0 \le s \le 10,000 \\ 1000 + 0.05(s - 10,000), & 10,000 < s < 20,000 \\ 1500 + 0.05(s - 10,000), & s \ge 20,000 \end{cases}$

The graph of E is:

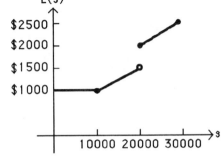

(B) From the graph, $\lim\limits_{s \to 10,000} E(s) = \$1000$ and $E(10,000) = \$1000$.

(C) From the graph, $\lim\limits_{s \to 20,000} E(s)$ does not exist. $E(20,000) = \$2000$.

(D) E is continuous at 10,000; E is not continuous at 20,000.

81. (A) From the graph, N is discontinuous at $t = t_2$, $t = t_3$, $t = t_4$, $t = t_6$, and $t = t_7$.

(B) From the graph, $\lim\limits_{t \to t_5} N(t) = 7$ and $N(t_5) = 7$.

(C) From the graph, $\lim\limits_{t \to t_3} N(t)$ does not exist; $N(t_3) = 4$.

EXERCISE 4-2

Things to remember:

1. INCREASING AND DECREASING FUNCTIONS

   For the interval (a, b):

$f'(x)$	$f(x)$	Graph of f	Examples
+	Increases ↗	Rises ↗	
−	Decreases ↘	Falls ↘	

2. CRITICAL VALUES

   The values of x in the domain of f where $f'(x) = 0$ or $f'(x)$ does not exist are called the CRITICAL VALUES of f. The critical values of f are always partition numbers for f', but f' may have partition numbers that are not critical values.

3. LOCAL EXTREMA

   Given a function f. The value $f(c)$ is a LOCAL MAXIMUM of f if there is an interval (m, n) containing c such that $f(x) \le f(c)$ for all x in (m, n). The value $f(e)$ is a LOCAL MINIMUM of f if there is an interval (p, q) containing e such that $f(x) \ge f(e)$ for all x in (p, q). Local maxima and local minima are called LOCAL EXTREMA.

   A point on the graph where a local extremum occurs is also called a TURNING POINT.

4. EXISTENCE OF LOCAL EXTREMA: CRITICAL VALUES

If f is continuous on the interval (a, b) and $f(c)$ is a local extremum, then either $f'(c) = 0$ or $f'(c)$ does not exist (is not defined).

5. FIRST DERIVATIVE TEST FOR LOCAL EXTREMA

Let c be a critical value of f [$f(c)$ is defined and either $f'(c) = 0$ or $f'(c)$ is not defined.]

Construct a sign chart for $f'(x)$ close to and on either side of c.

Sign Chart	$f(c)$
$f'(x)$ $---\mid+++$ $\longrightarrow x$ m c n $f(x)$ Decreasing $\mid$ Increasing	$f(c)$ is a local minimum. If $f'(x)$ changes from negative to positive at c, then $f(c)$ is a local minimum.
$f'(x)$ $+++\mid---$ $\longrightarrow x$ m c n $f(x)$ Increasing $\mid$ Decreasing	$f(c)$ is a local maximum. If $f'(x)$ changes from positive to negative at c, then $f(c)$ is a local maximum.
$f'(x)$ $---\mid---$ $\longrightarrow x$ m c n $f(x)$ Decreasing $\mid$ Decreasing	$f(c)$ is not a local extremum. If $f'(x)$ does not change sign at c, then $f(c)$ is neither a local maximum nor a local minimum.
$f'(x)$ $+++\mid+++$ $\longrightarrow x$ m c n $f(x)$ Increasing $\mid$ Increasing	$f(c)$ is not a local extremum. If $f'(x)$ does not change sign at c, then $f(c)$ is neither a local maximum nor a local minimum.

6. INTERCEPTS AND LOCAL EXTREMA FOR POLYNOMIAL FUNCTIONS

If $f(x) = a_n x^n + a_{n-1} x^{n-1} + \ldots + a_1 x + a_0$, $a_n \neq 0$ is an nth degree polynomial then f has at most n x intercepts and at most $n-1$ local extrema.

---

1. (a, b), (d, f), (g, h) **3.** (b, c), (c, d), (f, g)

5. $x = c, d, f$ **7.** $x = b, f$

9. f has a local maximum at $x = a$, and a local minimum at $x = c$; f does not have a local extremum at $x = b$ or at $x = d$.

11.

x	$f'(x)$	$f(x)$	GRAPH OF f
$(-\infty, -1)$	+	Increasing	Rising
$x = -1$	0	Neither local maximum nor local minimum	Horizontal tangent
$(-1, 1)$	+	Increasing	Rising
$x = 1$	0	Local maximum	Horizontal tangent
$(1, \infty)$	–	Decreasing	Falling

Using this information together with the points $(-2, -1)$, $(-1, 1)$, $(0, 2)$, $(1, 3)$, $(2, 1)$ on the graph, we have

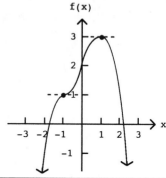

13.

x	$f'(x)$	$f(x)$	GRAPH OF $f(x)$
$(-\infty, -1)$	–	Decreasing	Falling
$x = -1$	0	Local minimum	Horizontal tangent
$(-1, 0)$	+	Increasing	Rising
$x = 0$	Not defined	Local maximum	Vertical tangent line
$(0, 2)$	–	Decreasing	Falling
$x = 2$	0	Neither local maximum nor local minimum	Horizontal tangent
$(2, \infty)$	–	Decreasing	Falling

Using this information together with the points $(-2, 2)$, $(-1, 1)$, $(0, 2)$, $(2, 1)$, $(4, 0)$ on the graph, we have

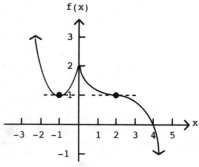

15.

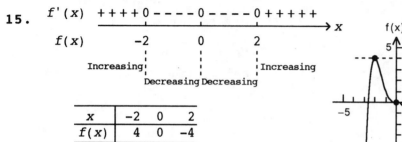

x	-2	0	2
$f(x)$	4	0	-4

17.

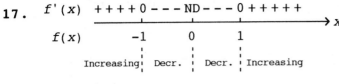

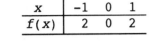

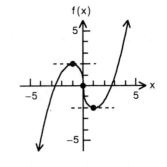

x	-1	0	1
$f(x)$	2	0	2

19. $f_1' = g_4$ **21.** $f_3' = g_6$ **23.** $f_5' = g_2$

25. $f(x) = x^2 - 16x + 12$
$f'(x) = 2x - 16$
f' is continuous for all x and
$f'(x) = 2x - 16 = 0$
$\qquad\qquad x = 8$
Thus, $x = 8$ is a partition number for f'.
Next, we construct a sign chart for f' (partition number is 8).

$f'(x)$ $\quad - - - - - - - - \mid + + + +$ $\quad\quad$ Test Numbers

$\qquad\qquad\qquad\qquad 0 \qquad\qquad 8\ 9 \qquad\longrightarrow x$

$f(x)$ $\qquad$ Decreasing $\mid$ Increasing

x	$f'(x)$
0	-16 (−)
9	2 (+)

Therefore, f is decreasing on $(-\infty, 8)$ and increasing on $(8, \infty)$; f has a local minimum at $x = 8$.

27. $f(x) = 4 + 10x - x^2$
$f'(x) = 10 - 2x$
f' is continuous for all x and
$f'(x) = 10 - 2x = 0$
$\qquad\qquad x = 5$
Thus, $x = 5$ is a partition number for f'.
Next, we construct a sign chart for f' (partition number is 5).

$f'(x)$ $\quad + + + + + \mid - - - -$ $\quad\quad$ Test Numbers

$\qquad\qquad\qquad\qquad 0 \qquad\qquad 5\ 6 \qquad\longrightarrow x$

$f(x)$ $\qquad$ Increasing $\mid$ Decreasing

x	$f'(x)$
0	10 (+)
6	-2 (−)

Therefore, f is increasing on $(-\infty, 5)$ and decreasing on $(5, \infty)$; f has a local maximum at $x = 5$.

29. $f(x) = 2x^3 + 4$
$f'(x) = 6x^2$
f' is continuous for all x and
$f'(x) = 6x^2 = 0$
$\qquad\qquad x = 0$
Thus, $x = 0$ is a partition number for f'.

Next, we construct a sign chart for f' (partition number is 0).

$f'(x)$ $+ + + +\ \vdots\ + + + +$

$$\text{-1}\ 0\ 1$$

$f(x)$ Increasing $\vdots$ Increasing

Test Numbers	
x	$f'(x)$
-1	6 (+)
1	6 (+)

Therefore, f is increasing for all x; i.e., on $(-\infty, \infty)$; f has no local extrema.

31. $f(x) = 2 - 6x - 2x^3$
$f'(x) = -6 - 6x^2$
f' is continuous for all x and
$f'(x) = -6 - 6x^2 = 0$
$\qquad -6(1 + x^2) = 0$
There are no real numbers that satisfy this equation.
The sign chart for f' is:

$f'(x)$ $- - - - - - - -$

$$0$$

$f(x)$ Decreasing

Test Numbers	
x	$f'(x)$
0	-6 (−)

Thus, f is decreasing for all x; i.e., on $(-\infty, \infty)$; f has no local extrema.

33. $f(x) = x^3 - 12x + 8$
$f'(x) = 3x^2 - 12$
f' is continuous for all x and
$f'(x) = 3x^2 - 12 = 0$
$\qquad 3(x^2 - 4) = 0$
$3(x - 2)(x + 2) = 0$
Thus, $x = -2$ and $x = 2$ are partition numbers for f'.
Next, we construct the sign chart for f' (partition numbers are -2 and 2).

$f'(x)$ $+ + + + 0 - - - - - - - - 0 + + + +$

$$\text{-3 -2}\qquad 0 \qquad 2\ 3$$

$f(x)$ Increasing $\vdots$ Decreasing $\vdots$ Increasing

Test Numbers	
x	$f'(x)$
-3	15 (+)
0	-12 (−)
3	15 (+)

Therefore, f is increasing on $(-\infty, -2)$ and on $(2, \infty)$, f is decreasing on $(-2, 2)$; f has a local maximum at $x = -2$ and a local minimum at $x = 2$.

35. $f(x) = x^3 - 3x^2 - 24x + 7$
$f'(x) = 3x^2 - 6x - 24$
f' is continuous for all x and
$f'(x) = 3x^2 - 6x - 24 = 0$
$\qquad 3(x^2 - 2x - 8) = 0$
$\qquad 3(x + 2)(x - 4) = 0$
Thus, $x = -2$ and $x = 4$ are partition numbers for f'.
The sign chart for f' is:

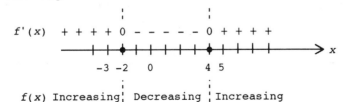

Test Numbers

x	$f'(x)$
-3	21 (+)
0	-24 (−)
5	21 (+)

Therefore, f is increasing on $(-\infty, -2)$ and on $(4, \infty)$, f is decreasing on $(-2, 4)$; f has a local maximum at $x = -2$ and a local minimum at $x = 4$.

37. $f(x) = 2x^2 - x^4$
$f'(x) = 4x - 4x^3$
f' is continuous for all x and
$f'(x) = 4x - 4x^3 = 0$
$\qquad 4x(1 - x^2) = 0$
$4x(1 - x)(1 + x) = 0$
Thus, $x = -1$, $x = 0$, and $x = 1$ are partition numbers for f'.

The sign chart for f' is:

Test Numbers

x	$f'(x)$
-2	24 (+)
$-\frac{1}{2}$	$-\frac{3}{2}$ (−)
$\frac{1}{2}$	$\frac{3}{2}$ (+)
2	-24 (−)

Therefore, f is increasing on $(-\infty, -1)$ and on $(0, 1)$, f is decreasing on $(-1, 0)$ and on $(1, \infty)$; f has local maxima at $x = -1$ and $x = 1$ and a local minimum at $x = 0$.

39. $f(x) = 4 + 8x - x^2$
$f'(x) = 8 - 2x$
f' is continuous for all x and
$f'(x) = 8 - 2x = 0$
$\qquad\qquad x = 4$
Thus, $x = 4$ is a partition number for f'.

The sign chart for f' is:

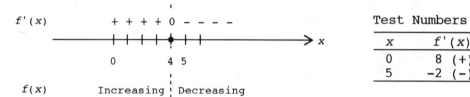

	Test Numbers	
x	$f'(x)$	
0	8	(+)
5	−2	(−)

Therefore, f is increasing on $(-\infty, 4)$ and decreasing on $(4, \infty)$; f has a local maximum at $x = 4$.

x	$f'(x)$	f	GRAPH OF f
$(-\infty, 4)$	+	Increasing	Rising
$x = 4$	0	Local maximum	Horizontal tangent
$(4, \infty)$	−	Decreasing	Falling

x	$f(x)$
0	4
4	20

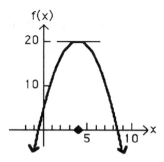

41. $f(x) = x^3 - 3x + 1$
    $f'(x) = 3x^2 - 3$
    f' is continuous for all x and
    $f'(x) = 3x^2 - 3 = 0$
    $\qquad 3(x^2 - 1) = 0$
    $3(x + 1)(x - 1) = 0$
    Thus, $x = -1$ and $x = 1$ are partition numbers for f'.
    The sign chart for f' is:

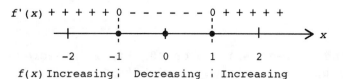

	Test Numbers	
x	$f'(x)$	
−2	9	(+)
0	−3	(−)
2	9	(+)

Therefore, f is increasing on $(-\infty, -1)$ and on $(1, \infty)$, f is decreasing on $(-1, 1)$; f has a local maximum at $x = -1$ and a local minimum at $x = 1$.

x	$f'(x)$	f	GRAPH OF f
$(-\infty, -1)$	+	Increasing	Rising
$x = -1$	0	Local maximum	Horizontal tangent
$(-1, 1)$	−	Decreasing	Falling
$x = 1$	0	Local minimum	Horizontal tangent
$(1, \infty)$	+	Increasing	Rising

x	$f(x)$
-1	3
0	1
1	-1

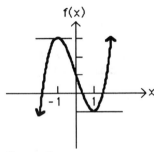

43. $f(x) = 10 - 12x + 6x^2 - x^3$
$f'(x) = -12 + 12x - 3x^2$
f' is continuous for all x and
$$f'(x) = -12 + 12x - 3x^2 = 0$$
$$-3(x^2 - 4x + 4) = 0$$
$$-3(x - 2)^2 = 0$$
Thus, $x = 2$ is a partition number for f'.
The sign chart for f' is:

$f'(x)$ - - - - - 0 - - - - -

     0   1   2   3

$f(x)$ Decreasing ¦ Decreasing

Test Numbers

x	$f'(x)$
0	-12 (−)
3	-3 (−)

Therefore, f is decreasing for all x, i.e., on $(-\infty, \infty)$, and there is a horizontal tangent line at $x = 2$.

x	$f'(x)$	f	GRAPH OF f
$(-\infty, 2)$	−	Decreasing	Falling
$x = 2$	0		Horizontal tangent
$x > 2$	−	Decreasing	Falling

x	$f(x)$
0	10
2	2

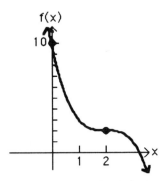

45. Critical values: $x = -1.26$; increasing on $(-1.26, \infty)$; decreasing on $(-\infty, -1.26)$; local minimum at $x = -1.26$

47. Critical values: $x = -0.43$, $x = 0.54$, $x = 2.14$; increasing on $(-0.43, 0.54)$ and $(2.14, \infty)$; decreasing on $(-\infty, -0.43)$ and $(0.54, 2.14)$; local maximum at $x = 0.54$, local minima at $x = -0.43$ and $x = 2.14$

49. Increasing on $(-1, 2)$ $[f'(x) > 0]$;
decreasing on $(-\infty, -1)$ and on $(2, \infty)$ $[f'(x) < 0]$;
local minimum at $x = -1$; local maximum at $x = 2$.

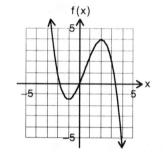

51. Increasing on $(-1, 2)$ and on $(2, \infty)$ $[f'(x) > 0]$;
decreasing on $(-\infty, -1)$ $[f'(x) < 0]$;
local minimum at $x = -1$.

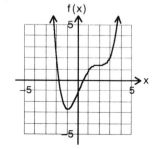

53. $f'(x) > 0$ on $(-\infty, -1)$ and on $(3, \infty)$;
$f'(x) < 0$ on $(-1, 3)$; $f'(x) = 0$ at $x = -1$ and $x = 3$.

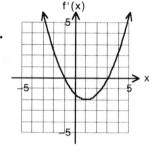

55. $f'(x) > 0$ on $(-2, 1)$ and on $(3, \infty)$;
$f'(x) < 0$ on $(-\infty, -2)$ and on $(1, 3)$:
$f'(x) = 0$ at $x = -2$, $x = 1$, and $x = 3$.

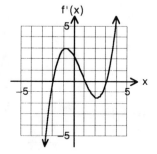

57. $f(x) = \dfrac{x - 1}{x + 2}$ [<u>Note</u>: f is not defined at $x = -2$.]

$f'(x) = \dfrac{(x + 2)(1) - (x - 1)(1)}{(x + 2)^2} = \dfrac{3}{(x + 2)^2}$

Critical values: $f'(x) \neq 0$ for all x, and f' is defined at all points *in
the domain of f* (i.e., -2 is not a critical value since -2 is not in the
domain of f). Thus, f does not have any critical values; $x = -2$ is a
partition number for f'.

The sign chart for f' is:

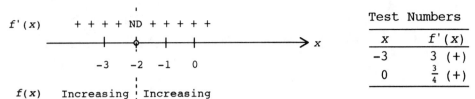

Test Numbers

x	$f'(x)$
-3	3 (+)
0	$\frac{3}{4}$ (+)

Therefore, f is increasing on $(-\infty, -2)$ and on $(-2, \infty)$; f has no local extrema.

59. $f(x) = x + \dfrac{4}{x}$ [<u>Note</u>: f is not defined at $x = 0$.]

$f'(x) = 1 - \dfrac{4}{x^2}$

Critical values: $x = 0$ is *not* a critical value of f since 0 is not in the domain of f, but $x = 0$ is a partition number for f'.

$f'(x) = 1 - \dfrac{4}{x^2} = 0$

$x^2 - 4 = 0$

$(x + 2)(x - 2) = 0$

Thus, the critical values are $x = -2$ and $x = 2$; $x = -2$ and $x = 2$ are also partition numbers for f'.

The sign chart for f' is:

$f'(x)$ $+ + + + 0 - - \text{ND} - - 0 + + + +$

$f(x)$ Increasing : Decreasing : Increasing

Test Numbers

x	$f'(x)$
-3	$\frac{5}{9}$ (+)
-1	-3 (−)
1	-3 (−)
3	$\frac{5}{9}$ (+)

Therefore, f is increasing on $(-\infty, -2)$ and on $(2, \infty)$, f is decreasing on $(-2, 0)$ and on $(0, 2)$; f has a local maximum at $x = -2$ and a local minimum at $x = 2$.

61. $f(x) = 1 + \dfrac{1}{x} + \dfrac{1}{x^2}$ [<u>Note</u>: f is not defined at $x = 0$.]

$f'(x) = -\dfrac{1}{x^2} - \dfrac{2}{x^3}$

Critical values: $x = 0$ is not a critical value of f since 0 is not in the domain of f; $x = 0$ is a partition number for f'.

$f'(x) = -\dfrac{1}{x^2} - \dfrac{2}{x^3} = 0$

$-x - 2 = 0$

$x = -2$

Thus, the critical value is $x = -2$; -2 is also a partition number for f'.

The sign chart for f' is:

$f'(x)$ - - - - - 0 + + + + + ND - - - - -

$f(x)$ Decreasing | Increasing | Decreasing

positions: -3 -2 -1 0 1

Test Numbers

x	$f'(x)$
-3	$-\frac{1}{27}$ (−)
-1	1 (+)
1	-3 (−)

Therefore, f is increasing on $(-2, 0)$ and f is decreasing on $(-\infty, -2)$ and on $(0, \infty)$; f has a local minimum at $x = -2$.

63. $f(x) = \dfrac{x^2}{x - 2}$ [<u>Note</u>: f is not defined at $x = 2$.]

$f'(x) = \dfrac{(x - 2)(2x) - x^2(1)}{(x - 2)^2} = \dfrac{x^2 - 4x}{(x - 2)^2}$

Critical values: $x = 2$ is *not* a critical value of f since 2 is not in the domain of f; $x = 2$ is a partition number for f'.

$f'(x) = \dfrac{x^2 - 4x}{(x - 2)^2} = 0$

$x^2 - 4x = 0$

$x(x - 4) = 0$

Thus, the critical values are $x = 0$ and $x = 4$; 0 and 4 are also partition numbers for f'.

The sign chart for f' is:

$f'(x)$ + + + + 0 - - ND - - 0 + + + +

$f(x)$ Increasing | Decreasing | Increasing

positions: -1 0 1 2 3 4 5

Test Numbers

x	$f'(x)$
-1	$\frac{5}{9}$ (+)
1	-3 (−)
3	-3 (−)
5	$\frac{5}{9}$ (+)

Therefore, f is increasing on $(-\infty, 0)$ and on $(4, \infty)$, f is decreasing on $(0, 2)$ and on $(2, 4)$; f has a local maximum at $x = 0$ and a local minimum at $x = 4$.

65. $f(x) = x^4(x - 6)^2$

$f'(x) = x^4(2)(x - 6)(1) + (x - 6)^2(4x^3)$

$\quad\;\; = 2x^3(x - 6)[x + 2(x - 6)]$

$\quad\;\; = 2x^3(x - 6)(3x - 12)$

$\quad\;\; = 6x^3(x - 4)(x - 6)$

Thus, the critical values of f are $x = 0$, $x = 4$, and $x = 6$.

Now we construct the sign chart for f' ($x = 0$, $x = 4$, $x = 6$ are partition numbers).

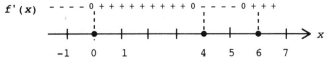

Test Numbers	
x	$f'(x)$
-1	-210 (-)
1	90 (+)
5	-750 (-)
7	+

Therefore, f is increasing on $(0, 4)$ and on $(6, \infty)$, f is decreasing on $(-\infty, 0)$ and on $(4, 6)$; f has a local maximum at $x = 4$ and local minima at $x = 0$ and $x = 6$.

67. $f(x) = 3(x - 2)^{2/3} + 4$

$f'(x) = 3\left(\dfrac{2}{3}\right)(x - 2)^{-1/3} = \dfrac{2}{(x - 2)^{1/3}}$

Critical values: f' is not defined at $x = 2$. [Note: $f(2)$ is defined, $f(2) = 4$.] $f'(x) \neq 0$ for all x. Thus, the critical value for f is $x = 2$; $x = 2$ is also a partition number for f'.

Test Numbers	
x	$f'(x)$
1	-2 (-)
3	2 (+)

Therefore, f is increasing on $(2, \infty)$ and decreasing on $(-\infty, 2)$; f has a local minimum at $x = 2$.

69. $f(x) = 2\sqrt{x} - x = 2x^{1/2} - x, \quad x > 0$

$f'(x) = x^{-1/2} - 1 = \dfrac{1}{x^{1/2}} - 1 = \dfrac{1 - \sqrt{x}}{\sqrt{x}}, \quad x > 0$

Critical values: $f'(x) = \dfrac{1 - \sqrt{x}}{\sqrt{x}} = 0, \quad x > 0$

$1 - \sqrt{x} = 0$

$\sqrt{x} = 1$

$x = 1$

Thus, the critical value for f is $x = 1$; $x = 1$ is also a partition number for f'.

The sign chart for f' is:

Test Numbers	
x	$f'(x)$
$\frac{1}{4}$	1 (+)
4	$-\frac{1}{2}$ (-)

Therefore, f is increasing on $(0, 1)$ and decreasing on $(1, \infty)$; f has a local maximum at $x = 1$.

71. Let $f(x) = x^3 + kx$

(A) $k > 0$
$f'(x) = 3x^2 + k > 0$ for all x.
There are no critical values and no local extrema; f is increasing on $(-\infty, \infty)$.

(B) $k < 0$
$f'(x) = 3x^2 + k$; $3x^2 + k = 0$
$$x^2 = -\frac{k}{3}$$
$$x = \pm\sqrt{-\frac{k}{3}}$$
Critical vlaues: $x = -\sqrt{-\frac{k}{3}}$, $x = \sqrt{-\frac{k}{3}}$;

$f'(x)$ $+++++0--------0+++++$

$-\sqrt{-\frac{k}{3}}$ 0 $\sqrt{-\frac{k}{3}}$

f is increasing on $\left(-\infty, -\sqrt{-\frac{k}{3}}\right)$ and on $\left(\sqrt{-\frac{k}{3}}, \infty\right)$; f is decreasing on $\left(-\sqrt{-\frac{k}{3}}, \sqrt{-\frac{k}{3}}\right)$; f has a local maximum at $x = -\sqrt{-\frac{k}{3}}$ and a local minimum at $x = \sqrt{-\frac{k}{3}}$.

(C) The only critical value is $x = 0$. There are no extrema, the function is increasing for all x.

73. (A) The marginal profit function, P', is positive on $(0, 600)$, zero at $x = 600$, and negative on $(600, 1,000)$.

(B)

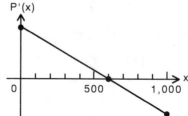

75. (A) The price function, $B(t)$, decreases for the first 15 months to a local minimum, increases for the next 40 months to a local maximum, and then decreases for the remaining 15 months.

(B)

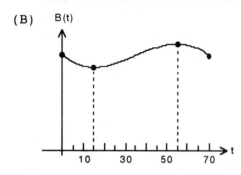

77. $C(x) = \frac{x^2}{20} + 20x + 320$

(A) $\overline{C}(x) = \frac{C(x)}{x} = \frac{x}{20} + 20 + \frac{320}{x}$

(B) Critical values:

$$\overline{C}'(x) = \frac{1}{20} - \frac{320}{x^2} = 0$$

$$x^2 - 320(20) = 0$$

$$x^2 - 6400 = 0$$

$$(x - 80)(x + 80) = 0$$

Thus, the critical value of $\overline{C}$ on the interval $(0, 150)$ is $x = 80$. Next, construct the sign chart for $\overline{C}'$ ($x = 80$ is a partition number for $\overline{C}'$).

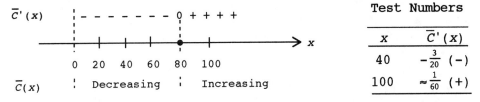

Test Numbers

x	$\overline{C}'(x)$
40	$-\frac{3}{20}$ (−)
100	$\approx \frac{1}{60}$ (+)

Therefore, $\overline{C}$ is increasing for $80 < x < 150$ and decreasing for $0 < x < 80$; $\overline{C}$ has a local minimum at $x = 80$.

79. $P(x) = R(x) - C(x)$
$P'(x) = R'(x) - C'(x)$
Thus, if $R'(x) > C'(x)$ on the interval (a, b), then $P'(x) = R'(x) - C'(x) > 0$ on this interval and P is increasing.

81. $C(t) = \frac{0.14t}{t^2 + 1}$, $0 < t < 24$

$$C'(t) = \frac{(t^2 + 1)(0.14) - 0.14t(2t)}{(t^2 + 1)^2} = \frac{0.14(1 - t^2)}{(t^2 + 1)^2}$$

Critical values: C' is continuous for all t on the interval $(0, 24)$:

$$C'(t) = \frac{0.14(1 - t^2)}{(t^2 + 1)^2} = 0$$

$$1 - t^2 = 0$$

$$(1 - t)(1 + t) = 0$$

Thus, the critical value of C on the interval $(0, 24)$ is $t = 1$. The sign chart for C' ($t = 1$ is a partition number for C') is:

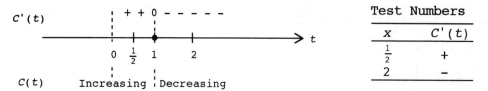

Test Numbers	
x	$C'(t)$
$\frac{1}{2}$	+
2	−

Therefore, C is increasing for $0 < t < 1$ and decreasing for $1 < t < 24$; C has a local maximum at $t = 1$.

83. $P(t) = \dfrac{8.4t}{t^2 + 49} + 0.1, \quad 0 < t < 24$

$P'(t) = \dfrac{(t^2 + 49)(8.4) - 8.4t(2t)}{(t^2 + 49)^2} = \dfrac{8.4(49 - t^2)}{(t^2 + 49)^2}$

Critical values: P is continuous for all t on the interval $(0, 24)$:

$P'(t) = \dfrac{8.4(49 - t^2)}{(t^2 + 49)^2} = 0$

$49 - t^2 = 0$

$(7 - t)(7 + t) = 0$

Thus, the critical value of P on $(0, 24)$ is $t = 7$.

The sign chart for P' ($t = 7$ is a partition number for P') is:

Test Numbers	
x	$P'(t)$
6	+
8	−

Therefore, P is increasing for $0 < t < 7$ and decreasing for $7 < t < 24$; P has a local maximum at $t = 7$.

EXERCISE 4-3

Things to remember:

<u>1</u>. SECOND DERIVATIVE

For $y = f(x)$, the SECOND DERIVATIVE of f, provided it exists, is:

$$f''(x) = \frac{d}{dx} f'(x)$$

Other notations for $f''(x)$ are:

$$\frac{d^2y}{dx^2} \quad \text{and} \quad y''.$$

<u>2</u>. CONCAVITY

For the interval (a, b):

$f''(x)$	$f'(x)$	Graph of $y = f(x)$	Example
+	Increasing	Concave upward	$\smile$
–	Decreasing	Concave downward	$\frown$

<u>3</u>. INFLECTION POINT

An INFLECTION POINT is a point on the graph of a function where the concavity changes (from upward to downward, or from downward to upward).

If $y = f(x)$ is continuous on (a, b) and has an inflection point at $x = c$, then either $f''(c) = 0$ or $f''(c)$ does not exist.

<u>4</u>. SECOND-DERIVATIVE TEST FOR LOCAL MAXIMA AND MINIMA
Let c be a critical value for $f(x)$.

$f'(c)$	$f''(c)$	Graph of f is	$f(c)$	Example
0	+	Concave upward	Local minimum	$\smile$
0	–	Concave downward	Local maximum	$\frown$
0	0	?	Test fails	

The first-derivative test must be used whenever $f''(c) = 0$ [or $f''(c)$ does not exist].

1. From the graph, f is concave upward on (a, c), (c, d), and (e, g).

3. $f''(x) < 0$ on (d, e) and (g, h)

5. $f'(x)$ is increasing on (a, c), (c, d), and (e, g).

7. From the graph, f has inflection points at $x = d, e,$ and g.

9. $f'(x) > 0$, $f''(x) > 0$; (c)

11. $f'(x) < 0$, $f''(x) > 0$; (d)

13. f has a local minimum at $x = 2$.

15. Unable to determine from the given information ($f'(-3) = f''(-3) = 0$).

17. Neither a local maximum nor a local minimum at $x = 6$; $x = 6$ is not a critical value of f.

19.

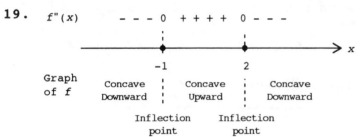

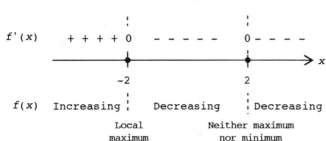

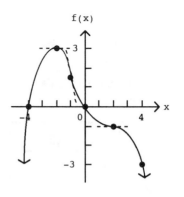

Using this information together with the points $(-4, 0)$, $(-2, 3)$, $(-1, 1.5)$, $(0, 0)$, $(2, -1)$, $(4, -3)$ on the graph, we have

21.

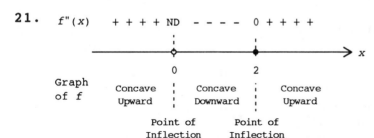

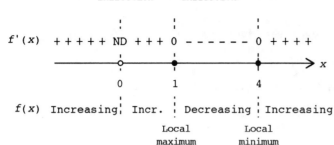

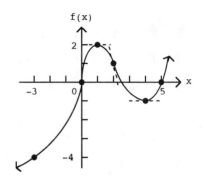

Using this information together with the points $(-3, -4)$, $(0, 0)$, $(1, 2)$, $(2, 1)$, $(4, -1)$, $(5, 0)$ on the graph, we have

23.

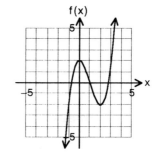

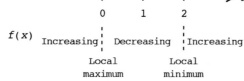

x	0	1	2
$f(x)$	2	0	-2

25.

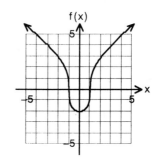

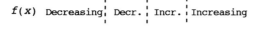

x	-1	0	1
$f(x)$	0	-2	0

27. $f(x) = x^3 - 2x^2 - 1$

$f'(x) = 3x^2 - 4x$

$f''(x) = 6x - 4$

29. $y = 2x^5 - 3$

$\dfrac{dy}{dx} = 10x^4$

$\dfrac{d^2y}{dx^2} = 40x^3$

31. $y = 1 - 2x + x^3$

$y' = -2 + 3x^2$

$y'' = 6x$

33. $y = (x^2 - 1)^3$

$y' = 3(x^2 - 1)^2(2x) = 6x(x^2 - 1)^2$

$y'' = 6x(2)(x^2 - 1)(2x) + (x^2 - 1)^2(6)$

$\quad = 24x^2(x^2 - 1) + 6(x^2 - 1)^2$

$\quad = 6(x^2 - 1)[4x^2 + x^2 - 1]$

$\quad = 6(x^2 - 1)(5x^2 - 1)$

35. $f(x) = 3x^{-1} + 2x^{-2} + 5$

$f'(x) = -3x^{-2} - 4x^{-3}$

$f''(x) = 6x^{-3} + 12x^{-4}$

37. $f(x) = 2x^2 - 8x + 6$

$f'(x) = 4x - 8 = 4(x - 2)$

$f''(x) = 4$

Critical value: $x = 2$

Now, $f''(2) = 4 > 0$. Therefore, $f(2) = 2 \cdot 2^2 - 8 \cdot 2 + 6 = -2$ is a local minimum.

39. $f(x) = 2x^3 - 3x^2 - 12x - 5$

$f'(x) = 6x^2 - 6x - 12 = 6(x^2 - x - 2) = 6(x - 2)(x + 1)$

$f''(x) = 12x - 6 = 6(2x - 1)$

Critical values: $x = 2, -1$.

Now, $f''(2) = 6(2 \cdot 2 - 1) = 18 > 0$.

Therefore, $f(2) = 2 \cdot 2^3 - 3 \cdot 2^2 - 12 \cdot 2 - 5 = -25$ is a local minimum.

$f''(-1) = 6[2(-1) - 1] = -18 < 0$.

Therefore, $f(-1) = 2(-1)^3 - 3(-1)^2 - 12(-1) - 5 = 2$ is a local maximum.

41. $f(x) = 3 - x^3 + 3x^2 - 3x$

$f'(x) = -3x^2 + 6x - 3 = -3(x^2 - 2x + 1) = -3(x - 1)^2$

$f''(x) = -6(x - 1)$

Critical value: $x = 1$

Now, $f''(1) = -6(1 - 1) = 0$.

Thus, the second-derivative test fails. Since $f'(x) = -3(x - 1)^2 < 0$ for all $x \neq 1$, $f(x)$ is decreasing on $(-\infty, \infty)$. Therefore, $f(x)$ has no local extrema.

43. $f(x) = x^4 - 8x^2 + 10$

$f'(x) = 4x^3 - 16x = 4x(x^2 - 4) = 4x(x + 2)(x - 2)$

$f''(x) = 12x^2 - 16$

Critical values: $x = 0, -2, 2$

Now $f''(0) = 0 - 16 = -16 < 0$. Therefore, $f(0) = 10$ is a local maximum.

$f''(-2) = 12(-2)^2 - 16 = 32 > 0$. Therefore, $f(-2) = -6$ is a local minimum. $f''(2) = 12 \cdot 2^2 - 16 = 32 > 0$. Therefore, $f(2) = -6$ is a local minimum.

45. $f(x) = x^6 + 3x^4 + 2$

$f'(x) = 6x^5 + 12x^3 = 6x^3(x^2 + 2)$

$f''(x) = 30x^4 + 36x^2 = 6x^2(5x^2 + 6)$

Critical value: $x = 0$ [<u>Note</u>: $x^2 + 2 \neq 0$ for all x.]

Now, $f''(0) = 0$. Thus, the second-derivative test fails, so the first-derivative test must be used.

The sign chart for f' (partition number is 0) is:

	Test Numbers

f'(x)   - - - - 0 + + + +

$$\xrightarrow{\quad\quad\quad\quad\quad\quad\quad\quad} x$$

    -1   0   1

f(x)   Decreasing | Increasing

x	f'(x)
-1	-18 (−)
1	18 (+)

Therefore, $f(0) = 2$ is a local minimum of f.

47. $f(x) = x + \dfrac{16}{x}$

$f'(x) = 1 - \dfrac{16}{x^2} = \dfrac{x^2 - 16}{x^2} = \dfrac{(x + 4)(x - 4)}{x^2}$

$f''(x) = \dfrac{32}{x^3}$

Critical values: $x = -4, 4$ [Note: $x = 0$ is not a critical value, since $f(0)$ is not defined.]

$f''(-4) = \dfrac{32}{(-4)^3} = -\dfrac{1}{2} < 0$

Therefore, $f(-4) = -8$ is a local maximum.

$f''(4) = \dfrac{32}{(4)^3} = \dfrac{1}{2} > 0$

Therefore, $f(4) = 8$ is a local minimum.

49. $f(x) = x^2 - 4x + 5$
$f'(x) = 2x - 4$
$f''(x) = 2 > 0$ Thus, the graph of f is concave upward for all x; there are no inflection points.

51. $f(x) = x^3 - 18x^2 + 10x - 11$
$f'(x) = 3x^2 - 36x + 10$
$f''(x) = 6x - 36$
Now, $f''(x) = 6x - 36 = 0$
$x - 6 = 0$
$x = 6$

The sign chart for f'' (partition number is 6) is:

f''(x)   - - - - - - 0 + + + +

$$\xrightarrow{\quad\quad\quad\quad\quad\quad\quad\quad} x$$

    0   5 6 7

Graph   Concave | Concave
of f   Downward | Upward

x	f''(x)
5	-6 (−)
7	6 (+)

Therefore, the graph of f is concave upward on $(6, \infty)$ and concave downward on $(-\infty, 6)$; there is an inflection point at $x = 6$.

53. $f(x) = x^4 - 24x^2 + 10x - 5$
$f'(x) = 4x^3 - 48x + 10$
$f''(x) = 12x^2 - 48$
Now, $f''(x) = 12x^2 - 48 = 0$
$12(x^2 - 4) = 0$
$12(x + 2)(x - 2) = 0$
$x = -2, 2$
The sign chart for f'' (partition numbers are -2 and 2) is:

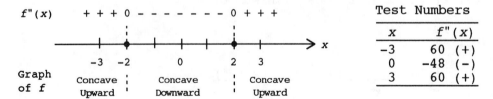

Test Numbers

x	$f''(x)$
-3	60 (+)
0	-48 (−)
3	60 (+)

Thus, the graph of f is concave upward on $(-\infty, -2)$ and on $(2, \infty)$, the graph is concave downward on $(-2, 2)$; the graph has inflection points at $x = -2$ and at $x = 2$.

55. $f(x) = -x^4 + 4x^3 + 3x + 7$
$f'(x) = -4x^3 + 12x^2 + 3$
$f''(x) = -12x^2 + 24x$
Now, $f''(x) = -12x^2 + 24x = 0$
$12x(-x + 2) = 0$
or $-12x(x - 2) = 0$
$x = 0, 2$
The sign chart for f'' (partition numbers are 0 and 2) is:

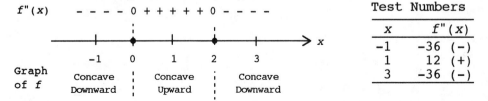

Test Numbers

x	$f''(x)$
-1	-36 (−)
1	12 (+)
3	-36 (−)

Thus, the graph of f is concave downward on $(-\infty, 0)$ and on $(2, \infty)$. The graph is concave upward on $(0, 2)$; the graph has inflection points at $x = 0$ and at $x = 2$.

57. $f(x) = x^3 - 6x^2 + 16$
$f'(x) = 3x^2 - 12x = 3x(x - 4)$
$f''(x) = 6x - 12 = 6(x - 2)$
Critical values: $x = 0, 4$
$f''(0) = -12 < 0$. Therefore, f has a local maximum at $x = 0$. $f''(4) = 6(4 - 2) = 12 > 0$. Therefore, f has a local minimum at $x = 4$.

The sign chart for $f''(x)$ (partition number is 2) is:

$f''(x)$ $- - - - 0 + + + +$

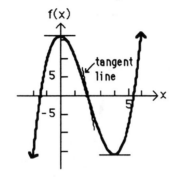

Test Numbers	
x	$f''(x)$
0	-12 $(-)$
3	6 $(+)$

Graph of f Concave Downward   Concave Upward

The graph of f has an inflection point at $x = 2$. The graph of f is:

x	$f(x)$
0	16
2	0
4	-16

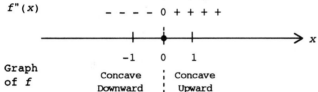

59. $f(x) = x^3 + x + 2$
$f'(x) = 3x^2 + 1$
$f''(x) = 6x$
Since $f'(x) = 3x^2 + 1 > 0$ for all x, f does not have any critical values. Now, $f''(x) = 6x = 0$
$$x = 0$$
The sign chart for f'' (partition number is 0) is:

$f''(x)$ $- - - - 0 + + + +$

Test Numbers	
x	$f''(x)$
-1	-6 $(-)$
1	6 $(+)$

Graph of f Concave Downward   Concave Upward

The graph of f has an inflection point at $x = 0$.
The graph of f is:

x	$f(x)$
-1	0
0	2
1	4

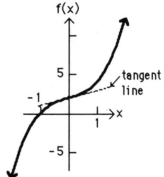

61. $f(x) = (2 - x)^3 + 1$
$f'(x) = 3(2 - x)^2(-1) = -3(2 - x)^2$
$f''(x) = -6(2 - x)(-1) = 6(2 - x)$
Critical value: $x = 2$
$f''(2) = 0$. Thus, the second derivative test fails. Note that $f'(x) = -3(2 - x)^2 < 0$ for all $x \neq 2$. Therefore, f is decreasing on $(-\infty, 2)$ and on $(2, \infty)$, and f does not have any local extrema. The sign chart for f'' (partition number is 2) is:

$f''(x)$ $+ + + + 0 - - - -$

Test Numbers	
x	$f''(x)$
0	12 (+)
3	-6 (-)

Graph
of f Concave Upward | Concave Downward

The graph of f has an inflection point at $x = 2$.
The graph of f is:

x	$f(x)$
0	9
2	1
3	0

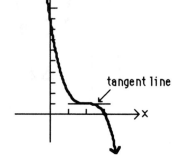

tangent line

63. $f(x) = x^3 - 12x$
$f'(x) = 3x^2 - 12 = 3(x^2 - 4) = 3(x + 2)(x - 2)$
$f''(x) = 6x$
Critical values: $x = -2$, $x = 2$
$f''(-2) = 6(-2) = -12 < 0$. Therefore, f has a local maximum at $x = -2$.
$f''(2) = 6(2) = 12 > 0$. Therefore, f has a local minimum at $x = 2$. The sign chart for $f''(x) = 6x$ (partition number is 0) is:

$f''(x)$ $- - - - 0 + + + +$

Test Numbers	
x	$f''(x)$
-1	-6 (-)
1	6 (+)

Graph
of f Concave Downward | Concave Upward

The graph of f has an inflection point at $x = 0$. The graph of f is:

x	$f(x)$
-2	+16
0	0
2	-16

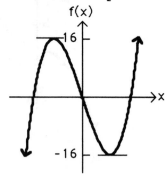

65.

x	f'(x)	f(x)
-∞ < x < -1	Positive and decreasing	Increasing and concave downward
x = -1	x intercept	Local maximum
-1 < x < 0	Negative and decreasing	Decreasing and concave downward
x = 0	Local minimum	Inflection point
0 < x < 2	Negative and increasing	Decreasing and concave upward
x = 2	Local maximum	Inflection point
2 < x < ∞	Negative and decreasing	Decreasing and concave downward

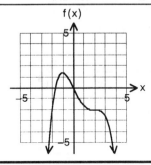

67.

x	f'(x)	f(x)
-∞ < x < -2	Negative and increasing	Decreasing and concave upward
x = -2	Local maximum	Inflection point
-2 < x < 0	Negative and decreasing	Decreasing and concave downward
x = 0	Local minimum	Inflection point
0 < x < 2	Negative and increasing	Decreasing and concave upward
x = 2	Local maximum	Inflection point
2 < x < ∞	Negative and decreasing	Decreasing and concave downward

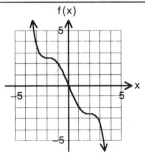

69. Inflection point at $x = -1.40$; concave upward on $(-1.40, \infty)$; concave downward on $(-\infty, -1.40)$

71. Inflection points at $x = -0.61$, $x = 0.66$, and $x = 1.74$; concave upward on $(-0.61, 0.66)$ and $(1.74, \infty)$; concave downward on $(-\infty, -0.61)$ and $(0.66, 1.74)$

73. If $f'(x)$ has a local extremum at $x = c$, then $f'(x)$ must change from increasing to decreasing, or from decreasing to increasing at $x = c$. It follows from this that the graph of f must change its concavity at $x = c$ and so there must be an inflection point at $x = c$.

75. If there is an inflection point on the graph of f at $x = c$, then the graph changes its concavity at $x = c$. Consequently, f' must change from increasing to decreasing, or from decreasing to increasing at $x = c$ and so $x = c$ is a local extremum of f'.

77. $f(x) = \dfrac{1}{x^2 + 12}$

$f'(x) = \dfrac{(x^2 + 12)(0) - 1(2x)}{(x^2 + 12)^2} = \dfrac{-2x}{(x^2 + 12)^2}$

$f''(x) = \dfrac{(x^2 + 12)^2(-2) - (-2x)(2)(x^2 + 12)(2x)}{(x^2 + 12)^4}$

$\quad = \dfrac{(x^2 + 12)(-2) + 8x^2}{(x^2 + 12)^3} = \dfrac{6x^2 - 24}{(x^2 + 12)^3}$

Now $f''(x) = \dfrac{6x^2 - 24}{(x^2 + 12)^3} = 0$

$\qquad\qquad 6x^2 - 24 = 0$

$\qquad\qquad 6(x^2 - 4) = 0$

$\qquad 6(x + 2)(x - 2) = 0$

$\qquad\qquad\qquad x = -2,\ 2$

The sign chart for f'' (partition numbers are -2 and 2) is:

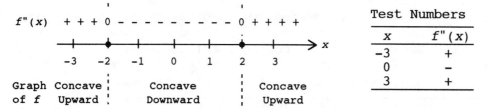

Test Numbers	
x	$f''(x)$
-3	$+$
0	$-$
3	$+$

Thus, the graph of f has inflection points at $x = -2$ and $x = 2$.

79. $f(x) = \dfrac{x}{x^2 + 12}$

$f'(x) = \dfrac{(x^2 + 12)(1) - x(2x)}{(x^2 + 12)^2} = \dfrac{12 - x^2}{(x^2 + 12)^2}$

$f''(x) = \dfrac{(x^2 + 12)^2(-2x) - (12 - x^2)(2)(x^2 + 12)(2x)}{(x^2 + 12)^4}$

$\quad = \dfrac{(x^2 + 12)(-2x) - 4x(12 - x^2)}{(x^2 + 12)^3} = \dfrac{2x^3 - 72x}{(x^2 + 12)^3} = \dfrac{2x(x^2 - 36)}{(x^2 + 12)^3}$

Now $f''(x) = \dfrac{2x(x^2 - 36)}{(x^2 + 12)^3} = 0$

$\qquad 2x(x + 6)(x - 6) = 0$

$\qquad\qquad\qquad x = 0,\ -6,\ 6$

The sign chart for f'' (partition numbers are 0, –6, 6) is:

$f''(x)$ – – – 0 + + + + + 0 – – – – – 0 + + +

	Test Numbers	
	x	$f''(x)$
	–7	–
	–1	+
	1	–
	7	+

Graph of f: Concave Downward | Concave Upward | Concave Downward | Concave Upward

Thus, the graph of f has inflection points at $x = -6$, $x = 0$, and $x = 6$.

81. The graph of the CPI is concave up.

83. The graph of C is increasing and concave down. Therefore, the graph of C' is positive and decreasing. Since the marginal costs are decreasing, the production process is becoming more efficient.

85. $R(x) = xp = 1296x - 0.12x^3$, $0 < x < 80$
$R'(x) = 1296 - 0.36x^2$
Critical values: $R'(x) = 1296 - 0.36x^2 = 0$
$$x^2 = \frac{1296}{0.36} = 3600$$
$$x = \pm 60$$
Thus, $x = 60$ is the only critical value on the interval $(0, 80)$.
$R''(x) = -0.72x$
$R''(60) = -43.2 < 0$
(A) R has a local maximum at $x = 60$.
(B) Since $R''(x) = -0.72x < 0$ for $0 < x < 80$, R is concave downward on this interval.

87. $N(x) = -3x^3 + 225x^2 - 3600x + 17{,}000$, $10 \le x \le 40$
$N'(x) = -9x^2 + 450x - 3600$
$N''(x) = -18x + 450$
(A) To determine when N' is increasing or decreasing, we must solve the inequalities $N''(x) > 0$ and $N''(x) < 0$, respectively. Now
$N''(x) = -18x + 450 = 0$
$x = 25$
The sign chart for N'' (partition number is 25) is:

$N''(x)$ + + + + + 0 – – – – –

	Test Numbers	
	x	$N''(x)$
	10	270 (+)
	30	–90 (–)

$N'(x)$: Increasing | Decreasing

Thus N' is increasing on (10, 25) and decreasing on (25, 40).

(B) Using the results in (A), the graph of N has an inflection point at $x = 25$.

(C)

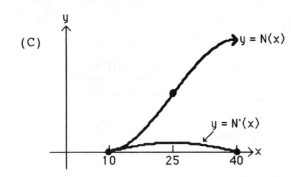

(D) Using the results in (A), N' has a local maximum at $x = 25$:
$N'(25) = 2025$

89. $N(t) = 1000 + 30t^2 - t^3$, $0 \le t \le 20$
$N'(t) = 60t - 3t^2$
$N''(t) = 60 - 6t$

(A) To determine when N' is increasing or decreasing, we must solve the inequalities $N''(t) > 0$ and $N''(t) < 0$, respectively. Now
$N''(t) = 60 - 6t = 0$
$t = 10$

The sign chart for N'' (partition number is 10) is:

$N''(t)$ $\quad$ $+ + + + 0 - - - -$

Test Numbers	
t	$N''(t)$
0	60 (+)
20	−60 (−)

$N'(t)$ $\quad$ Increasing ┊ Decreasing

Thus, N' is increasing on $(0, 10)$ and decreasing on $(10, 20)$.

(B) From the results in (A), the graph of N has an inflection point at $t = 10$.

(C)

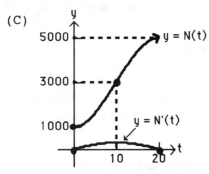

(D) Using the results in (A), N' has a local maximum at $t = 10$:
$N'(10) = 300$

91. $T(n) = 0.08n^3 - 1.2n^2 + 6n$, $n \ge 0$
$T'(n) = 0.24n^2 - 2.4n + 6$, $n \ge 0$
$T''(n) = 0.48n - 2.4$

(A) To determine when the rate of change of T, i.e., T', is increasing or decreasing, we must solve the inequalities $T''(n) > 0$ and $T''(n) < 0$, respectively. Now
$T''(n) = 0.48n - 2.4 = 0$
$n = 5$

The sign chart for T'' (partition number is 5) is:

$T''(n)$ $- - - - \ 0 + + + +$

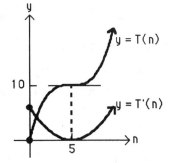

$T'(n)$ Decreasing $\vdots$ Increasing

Test Numbers	
t	$N''(t)$
1	-1.92 (−)
10	2.4 (+)

Thus, T' is increasing on $(5, \infty)$ and decreasing on $(0, 5)$.

(B) Using the results in (A), the graph of T has an inflection point at $n = 5$. The graphs of T and T' are shown at the right.

(C) Using the results in (A), T' has a local minimum at $n = 5$:

$$T'(5) = 0.24(5)^2 - 2.4(5) + 6 = 0$$

EXERCISE 4-4

Things to remember:

1. HORIZONTAL ASYMPTOTES

A line $y = b$ is a HORIZONTAL ASYMPTOTE for the graph of $y = f(x)$ if $f(x)$ approaches b as either x increases without bound or decreases without bound. That is,

$$\lim_{x \to \infty} f(x) = b \quad \text{or} \quad \lim_{x \to -\infty} f(x) = b$$

2. LIMITS AT INFINITY FOR POWER FUNCTIONS

If p is a positive real number and k is any real constant, then

a. $\displaystyle \lim_{x \to -\infty} \frac{k}{x^p} = 0$ b. $\displaystyle \lim_{x \to \infty} \frac{k}{x^p} = 0$

c. $\displaystyle \lim_{x \to -\infty} kx^p = \pm\infty$ d. $\displaystyle \lim_{x \to \infty} kx^p = \pm\infty$

provided that x^p names a real number for negative values of x. The limits in 3 and 4 will be either $-\infty$ or ∞, depending on k and p.

3. LIMITS AT INFINITY FOR POLYNOMIAL FUNCTIONS

If

$$p(x) = a_n x^n + a_{n-1} x^{n-1} + \dots + a_1 x + a_0, \ a_n \neq 0, \ n \geq 1$$

then

$$\lim_{x \to \infty} p(x) = \lim_{x \to \infty} a_n x^n = \pm\infty$$

and

$$\lim_{x \to -\infty} p(x) = \lim_{x \to -\infty} a_n x^n = \pm\infty$$

Each limit will be either $-\infty$ or ∞, depending on a_n and n.

4. **LIMITS AT INFINITY AND HORIZONTAL ASYMPTOTES FOR RATIONAL FUNCTIONS**

If

$$f(x) = \frac{a_m x^m + a_{m-1} x^{m-1} + \ldots + a_1 x + a_0}{b_n x^n + b_{n-1} x^{n-1} + \ldots + b_1 x + b_0}, \quad a_m \neq 0, \ b_n \neq 0$$

then

$$\lim_{x \to \infty} f(x) = \lim_{x \to \infty} \frac{a_m x^m}{b_n x^n} \quad \text{and} \quad \lim_{x \to -\infty} f(x) = \lim_{x \to -\infty} \frac{a_m x^m}{b_n x^n}$$

There are three possible cases for these limits:

a. If $m < n$, then $\lim\limits_{x \to \infty} f(x) = \lim\limits_{x \to -\infty} f(x) = 0$ and the line $y = 0$ (the x axis) is a horizontal asymptote for $f(x)$.

b. If $m = n$, then $\lim\limits_{x \to \infty} f(x) = \lim\limits_{x \to -\infty} f(x) = \frac{a_m}{b_n}$ and the line $y = \frac{a_m}{b_n}$ is a horizontal asymptote for $f(x)$.

c. If $m > n$, then each limit will be ∞ or $-\infty$, depending on m, n, a_m, and b_n, and $f(x)$ does not have a horizontal asymptote.

5. **LOCATING VERTICAL ASYMPTOTES**

Let $f(x) = \frac{n(x)}{d(x)}$, where both n and d are continuous at $x = c$.

If at $x = c$ the denominator $d(x)$ is 0 and the numerator $n(x)$ is not 0, then the line $x = c$ is a vertical asymptote for the graph of f.

[*Note*: Since a rational function is a ratio of two polynomial functions and polynomial functions are continuous for all real numbers, this theorem includes rational functions as a special case.]

6. **A GRAPHING STRATEGY FOR $y = f(x)$**

Omit any of the following steps if procedures involved appear to be too difficult or impossible (what may seem too difficult now, will become less so with a little practice).

Step 1. Analyze $f(x)$:

(A) Find the domain of f. [The domain of f is the set of all real numbers x that produce real values for $f(x)$.]

(B) Find intercepts. [The y intercept is $f(0)$, if it exists; the x intercepts are the solutions to $f(x) = 0$, if they exist.]

(C) Find asymptotes. [Use Theorems 3 and 4, if they apply, otherwise calculate limits at points of discontinuity and as x increases and decreases without bound.]

Step 2. Analyze $f'(x)$: Find any critical values for $f(x)$ and any partition numbers for $f'(x)$. [Remember, every critical value for $f(x)$ is also a partition number for $f'(x)$, but some partition numbers for $f'(x)$ may not be critical values for $f(x)$.] Construct a sign chart for $f'(x)$, determine the intervals where $f(x)$ is increasing and decreasing, and find local maxima and minima.

Step 3. Analyze $f''(x)$: Construct a sign chart for $f''(x)$, determine where the graph of f is concave upward and concave downward, and find any inflection points.

Step 4. Sketch the graph of f: Draw asymptotes and locate intercepts, local maxima and minima, and inflection points. Sketch in what you know from steps 1—3. In regions of uncertainty, use point-by-point plotting to complete the graph.

1. From the graph, $f'(x) < 0$ on $(-\infty, b)$, $(0, e)$, (e, g).

3. From the graph, $f(x)$ is increasing on (b, d), $(d, 0)$, (g, ∞).

5. From the graph, $f(x)$ has a local maximum at $x = 0$.

7. From the graph, $f''(x) < 0$ on $(-\infty, a)$, (d, e), (h, ∞).

9. The graph of f is concave upward on (a, d), (e, h).

11. From the graph, f has inflection points at $x = a$ and $x = h$.

13. From the graph, the lines $x = d$ and $x = e$ are vertical asymptotes.

15. Step 1. Analyze $f(x)$:

(A) Domain: All real numbers
(B) Intercepts: y-intercept: 0
               x-intercepts: -4, 0, 4

Step 2. Analyze $f'(x)$:

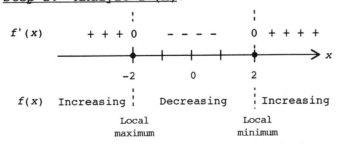

Step 3. Analyze $f''(x)$:

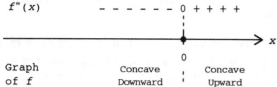

f''(x)   – – – – – – 0 + + + +

→ x

                         0

Graph   Concave   Concave
of f Downward   Upward

                       Point of
                      Inflection

Step 4. Sketch the graph of f:

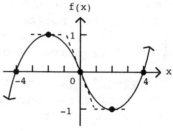

17. Step 1. Analyze $f(x)$:

(A) Domain: All real numbers
(B) Intercepts: y-intercept: 0
                 x-intercepts: -4, 0, 4
(C) Asymptotes: Horizontal asymptote: $y = 2$

Step 2. Analyze $f'(x)$:

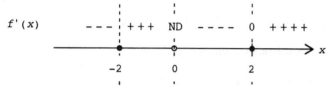

f'(x)   – – – ┊ + + + ND – – – – 0 + + + +

→ x

              -2   0   2

$f(x)$ Decreasing┊ Incr. ┊Decreasing┊ Increasing

          Local   Local   Local
        minimum maximum   minimum

Step 3. Analyze $f''(x)$:

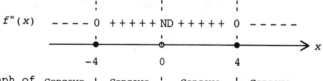

f''(x)   – – – – 0 + + + + + ND + + + + + 0 – – – – –

→ x

              -4   0   4

Graph of Concave ┊ Concave ┊ Concave ┊ Concave
   $f(x)$ Downward ┊ Upward ┊ Upward ┊ Downward

Step 4. Sketch the graph of f:

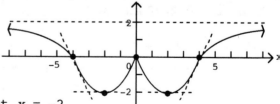

19. Step 1. Analyze $f(x)$:
(A) Domain: All real numbers except $x = -2$
(B) Intercepts: y-intercept: 0
                 x-intercepts: -4, 0
(C) Asymptotes: Horizontal asymptote: $y = 1$
                Vertical asymptote: $x = -2$

Step 2. Analyze $f'(x)$:

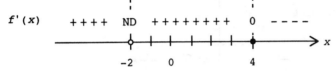

f'(x)   + + + +   ND   + + + + + + + +   0   - - - -

        ○————┼—┼—┼—┼—┼—┼————●————————→ x
        -2   0   4

f(x)   Increasing ┊ Increasing ┊ Decreasing
                                      Local
                                      maximum

Step 3. Analyze $f''(x)$:

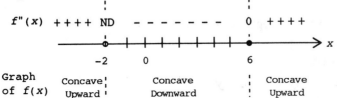

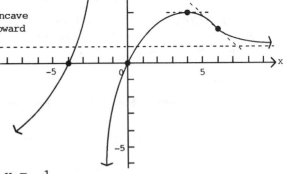

f"(x) + + + + ND - - - - - - - -   0   + + + +

       ○———┼—┼—┼—┼—┼—┼—┼—┼—————●————→ x
       -2   0   6

Graph   Concave ┊   Concave   ┊   Concave
of $f(x)$ Upward ┊   Downward   ┊   Upward

Step 4. Sketch the graph of f:

21. Step 1. Analyze $f(x)$:

(A) Domain: All real numbers except $x = -1$
(B) Intercepts: y-intercept: -1
               x-intercept: 1
(C) Asymptotes: Horizontal asymptote: $y = 1$
               Vertical asymptote: $x = -1$

Step 2. Analyze $f'(x)$:

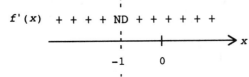

f'(x)   + + + + ND + + + + + +

        ———————————┼———┼————————→ x
                  -1   0

f(x)   Increasing ┊ Increasing

Step 3. Analyze $f''(x)$:

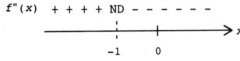

f"(x)   + + + + ND - - - - - -

        ———————————┼———┼————————→ x
                  -1   0

Graph   Concave ┊ Concave
of f Upward ┊ Downward

Step 4. Sketch the graph of f:

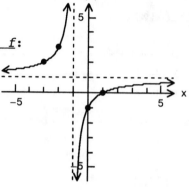

23. Step 1. Analyze $f(x)$:

(A) Domain: All real numbers except $x = -2$, $x = 2$

(B) Intercepts: y-intercept: 0
$\qquad\qquad\quad\;$ x-intercept: 0

(C) Asymptotes: Horizontal asymptote: $y = 0$
$\qquad\qquad\quad\;$ Vertical asymptotes: $x = -2$, $x = 2$

Step 2. Analyze $f'(x)$:

$f'(x) \quad - - - - \text{ND} + + + + + + \text{ND} - - - -$

$\qquad\qquad\qquad\qquad\qquad\qquad\qquad\qquad \to x$

$\qquad\qquad\quad -2 \qquad 0 \qquad 2$

$f(x)$ Decreasing ¦ Increasing ¦ Decreasing

Step 3. Analyze $f''(x)$:

$f''(x) \quad - - - - \text{ND} - - \; 0 + + \; \text{ND} + + + +$

$\qquad\qquad\qquad\qquad\qquad\qquad\qquad\qquad \to x$

$\qquad\qquad\quad -2 \qquad 0 \qquad 2$

Graph $\quad$ Concave ¦Concave¦Concave¦ Concave
of f $\quad\;$ Downward ¦Downward¦Upward ¦ Upward
$\qquad\qquad\qquad\qquad\qquad$ Inflection
$\qquad\qquad\qquad\qquad\qquad\quad$ Point

Step 4. Sketch the graph of f:

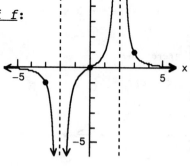

25. $\lim\limits_{x \to \infty} (4x^3 + 2x - 9) = \lim\limits_{x \to \infty} 4x^3 = \infty$ (by $\underline{3}$)

27. $\lim\limits_{x \to -\infty} (-3x^6 + 9x^5 + 4) = \lim\limits_{x \to -\infty} (-3x^6) = -\infty$ (by $\underline{3}$)

29. $\lim\limits_{x \to \infty} \dfrac{4x + 7}{5x - 9} = \dfrac{4}{5}$ (by 4(b))

31. $\lim\limits_{x \to \infty} \dfrac{5x^2 + 11}{7x - 2} = \infty$ (by 4(c))

33. $\lim\limits_{x \to -\infty} \dfrac{7x^4 - 11x^2}{6x^5 + 3} = 0$ (by 4(a))

35. $f(x) = \dfrac{2x}{x + 2}$; f is a rational function

<u>Horizontal asymptotes</u>: $\dfrac{a_n x^n}{b_m x^m} = \dfrac{2x}{x} = 2$ and the line $y = 2$ is a horizontal asymptote.

<u>Vertical asymptotes</u>: Using 5, $D(-2) = 0$ and $N(-2) = -4 \neq 0$. Thus, the line $x = -2$ is a vertical asymptote.

37. $f(x) = \dfrac{x^2 + 1}{x^2 - 1} = \dfrac{x^2 + 1}{(x - 1)(x + 1)}$; f is a rational function

<u>Horizontal asymptotes</u>: $\dfrac{a_n x^n}{b_m x^m} = \dfrac{x^2}{x^2} = 1$ and the line $y = 1$ is a horizontal asymptote.

<u>Vertical asymptotes</u>: Using 5, $D(-1) = 0$ and $N(-1) = 2$, so the line $x = -1$ is a vertical asymptote; $D(1) = 0$, $N(1) = 2$, so the line $x = 1$ is a vertical asymptote.

39. $f(x) = \dfrac{x^3}{x^2 + 6}$; f is a rational function

<u>Horizontal asymptotes</u>: $\dfrac{a_n x^n}{b_m x^m} = \dfrac{x^3}{x^2} = x$; f does not have a horizontal asymptote.

<u>Vertical asymptotes</u>: Using 5, $D(x) = x^2 + 6 \neq 0$ for all x; f has no vertical asymptotes.

41. $f(x) = \dfrac{x}{x^2 + 4}$; f is a rational function

<u>Horizontal asymptotes</u>: $\dfrac{a_n x^n}{b_m x^m} = \dfrac{x}{x^2} = \dfrac{1}{x}$; the line $y = 0$ (the x-axis) is a horizontal asymptote.

<u>Vertical asymptotes</u>: Using 5, $D(x) = x^2 + 4 \neq 0$ for all x; f has no vertical asymptotes.

43. $f(x) = \dfrac{x^2}{x - 3}$; f is a rational function

<u>Horizontal asymptotes</u>: $\dfrac{a_n x^n}{b_m x^m} = \dfrac{x^2}{x} = x$; f does not have a horizontal asymptote.

<u>Vertical asymptotes</u>: Using 5, $D(3) = 0$, $N(3) = 9$, so the line $x = 3$ is a vertical asymptote.

45. $f(x) = \dfrac{2x^2 + 3x - 2}{x^2 - x - 2} = \dfrac{(2x - 1)(x + 2)}{(x - 2)(x + 1)}$; f is a rational function

<u>Horizontal asymptotes</u>: $\dfrac{a_n x^n}{b_m x^m} = \dfrac{2x^2}{x^2} = 2$; the line $y = 2$ is a horizontal asymptote.

<u>Vertical asymptotes</u>: Using <u>5</u>, $D(2) = 0$, $N(2) = 12$, so the line $x = 2$ is a vertical asymptote; $D(-1) = 0$, $N(-1) = -3$, so the line $x = -1$ is a vertical asymptote.

47. $f(x) = \dfrac{2x^2 - 5x + 2}{x^2 - x - 2} = \dfrac{(2x - 1)(x - 2)}{(x - 2)(x + 1)} = \dfrac{2x - 1}{x + 1}$, $x \neq 2$

<u>Horizontal asymptotes</u>: $\dfrac{a_n x^n}{b_m x^m} = \dfrac{2x^2}{x^2} = 2$; the line $y = 2$ is a horizontal asymptote.

<u>Vertical asymptotes</u>: Using <u>5</u>, $D(-1) = 0$, $N(-1) = -3$, so the line $x = -1$ is a vertical asymptote. Since $\lim\limits_{x \to 2} f(x) = \lim\limits_{x \to 2} \dfrac{2x - 1}{x + 1} = 1$, f does not have a vertical asymptote at $x = 2$.

49. $f(x) = x^2 - 6x + 5$

<u>Step 1. Analyze $f(x)$</u>:
(A) Domain: All real numbers, $(-\infty, \infty)$.
(B) Intercepts: y intercept: $f(0) = 5$

$$x \text{ intercepts: } x^2 - 6x + 5 = 0$$
$$(x - 5)(x - 1) = 0$$
$$x = 1, 5$$

(C) Asymptotes: Since f is a polynomial, there are no horizontal or vertical asymptotes.

<u>Step 2. Analyze $f'(x)$</u>:
$f'(x) = 2x - 6 = 2(x - 3)$
Critical value: $x = 3$
Partition number: $x = 3$
Sign chart for f':

Test Numbers	
x	$f'(x)$
0	-6 $(-)$
4	2 $(+)$

Thus, f is decreasing on $(-\infty, 3)$ and increasing on $(3, \infty)$; f has a local minimum at $x = 3$.

Step 3. Analyze $f''(x)$:
$f''(x) = 2 > 0$ for all x.
Thus, the graph of f is concave upward on $(-\infty, \infty)$.
Step 4. Sketch the graph of f:

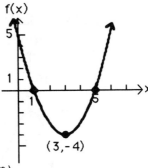

51. $f(x) = x^3 - 6x^2$
Step 1. Analyze $f(x)$:
(A) Domain: All real numbers, $(-\infty, \infty)$.

(B) Intercepts: y intercept: $f(0) = 0$
$\qquad\qquad\qquad x$ intercepts: $x^3 - 6x^2 = 0$
$\qquad\qquad\qquad\qquad\qquad x^2(x - 6) = 0$
$\qquad\qquad\qquad\qquad\qquad\qquad x = 0, 6$

(C) Asymptotes: Since f is a polynomial, there are no horizontal or vertical asymptotes.

Step 2. Analyze $f'(x)$:
$f'(x) = 3x^2 - 12x = 3x(x - 4)$
Critical values: $x = 0$, $x = 4$
Partition numbers: $x = 0$, $x = 4$
The sign chart for f' is:

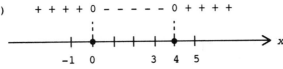

Test Numbers	
x	$f'(x)$
-1	15 (+)
3	-9 (-)
5	15 (+)

Thus, f is increasing on $(-\infty, 0)$ and on $(4, \infty)$; f is decreasing on $(0, 4)$; f has a local maximum at $x = 0$ and a local minimum at $x = 4$.

Step 3. Analyze $f''(x)$:
$f''(x) = 6x - 12 = 6(x - 2)$
Partition numbers for f'': $x = 2$

Sign chart for f'':

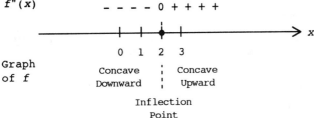

Test Numbers	
x	$f''(x)$
1	-6 (-)
3	6 (+)

Thus, the graph of f is concave downward on $(-\infty, 2)$ and concave upward on $(2, \infty)$; there is an inflection point at $x = 2$.

<u>Step 4.</u> <u>Sketch the graph of f:</u>

x	$f(x)$
0	0
2	-16
4	-32
6	0

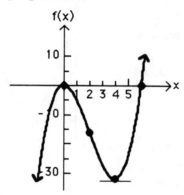

53. $f(x) = (x + 4)(x - 2)^2$

<u>Step 1.</u> <u>Analyze $f(x)$:</u>
(A) Domain: All real numbers, $(-\infty, \infty)$.
(B) Intercepts: y intercept: $f(0) = 4(-2)^2 = 16$
     x intercepts: $(x + 4)(x - 2)^2 = 0$
     $x = -4, 2$
(C) Asymptotes: Since f is a polynomial, there are no horizontal or vertical asymptotes.

<u>Step 2.</u> <u>Analyze $f'(x)$:</u>
$f'(x) = (x + 4)2(x - 2)(1) + (x - 2)^2(1)$
$\quad = (x - 2)[2(x + 4) + (x - 2)]$
$\quad = (x - 2)(3x + 6)$
$\quad = 3(x - 2)(x + 2)$
Critical values: $x = -2$, $x = 2$
Partition numbers: $x = -2$, $x = 2$
Sign chart for f':

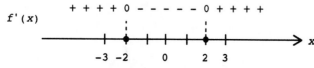

Test Numbers	
x	$f'(x)$
-3	15 (+)
0	-12 (-)
3	15 (+)

Thus, f is increasing on $(-\infty, -2)$ and on $(2, \infty)$; f is decreasing on $(-2, 2)$; f has a local maximum at $x = -2$ and a local minimum at $x = 2$.

<u>Step 3.</u> <u>Analyze $f''(x)$:</u>
$f''(x) = 3(x + 2)(1) + 3(x - 2)(1) = 6x$
Partition number for f'': $x = 0$

Sign chart for f'':

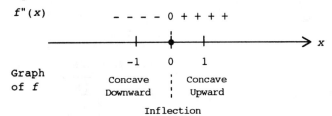

x	$f''(x)$
1	-6 (−)
1	6 (+)

Test Numbers

Thus, the graph of f is concave downward on $(-\infty, 0)$ and concave upward on $(0, \infty)$; there is an inflection point at $x = 0$.

Step 4. <u>Sketch the graph of f</u>:

x	$f(x)$
-2	32
0	16
2	0

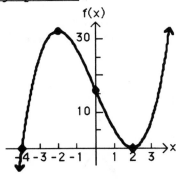

55. $f(x) = 8x^3 - 2x^4$

Step 1. <u>Analyze $f(x)$</u>:

(A) Domain: All real numbers, $(-\infty, \infty)$.

(B) Intercepts: y intercept: $f(0) = 0$

x intercepts: $8x^3 - 2x^4 = 0$

$2x^3(4 - x) = 0$

$x = 0, 4$

(C) Asymptotes: No horizontal or vertical asymptotes.

Step 2. <u>Analyze $f'(x)$</u>:

$f'(x) = 24x^2 - 8x^3 = 8x^2(3 - x)$

Critical values: $x = 0, x = 3$

Partition numbers: $x = 0, x = 3$

Sign chart for f':

x	$f'(x)$
-1	32 (+)
1	16 (+)
4	-128 (−)

Test Numbers

Thus, f is increasing on $(-\infty, 3)$ and decreasing on $(3, \infty)$; f has a local maximum at $x = 3$.

Step 3.   Analyze $f''(x)$:

$f''(x) = 48x - 24x^2 = 24x(2 - x)$
Partition numbers for f'': $x = 0$, $x = 2$
Sign chart for f'':

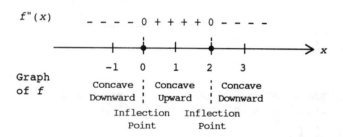

Test Numbers	
x	$f''(x)$
-1	-72 (-)
1	24 (+)
3	-72 (-)

Thus, the graph of f is concave downward on $(-\infty, 0)$ and on $(2, \infty)$;   the graph is concave upward on $(0, 2)$; there are inflection points at $x = 0$ and $x = 2$.

Step 4.   Sketch the graph of f:

x	$f(x)$
0	0
2	32
3	54

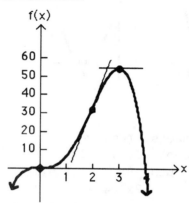

57. $f(x) = \dfrac{x + 3}{x - 3}$

Step 1.   Analyze $f(x)$:
(A) Domain: All real numbers except $x = 3$.

(B) Intercepts: y intercept: $f(0) = \dfrac{3}{-3} = -1$

$\qquad\qquad\qquad x$ intercepts: $\dfrac{x + 3}{x - 3} = 0$

$\qquad\qquad\qquad\qquad\qquad x + 3 = 0$

$\qquad\qquad\qquad\qquad\qquad\qquad x = -3$

(C) Asymptotes:

Horizontal asymptote: $\displaystyle\lim_{x \to \infty} \dfrac{x + 3}{x - 3} = \lim_{x \to \infty} \dfrac{x\left(1 + \dfrac{3}{x}\right)}{x\left(1 - \dfrac{3}{x}\right)} = 1.$

Thus, $y = 1$ is a horizontal asymptote.

Vertical asymptote: The denominator is 0 at $x = 3$ and the numerator is not 0 at $x = 3$.   Thus, $x = 3$ is a vertical asymptote.

Step 2. Analyze $f'(x)$:

$$f'(x) = \frac{(x - 3)(1) - (x + 3)(1)}{(x - 3)^2} = \frac{-6}{(x - 3)^2} = -6(x - 3)^{-2}$$

Critical values: None
Partition number: $x = 3$
Sign chart for f':

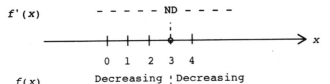

Test Numbers	
x	$f'(x)$
2	-6 $(-)$
4	-6 $(-)$

Thus, f is decreasing on $(-\infty, 3)$ and on $(3, \infty)$; there are no local extrema.

Step 3. Analyze $f''(x)$:

$$f''(x) = 12(x - 3)^{-3} = \frac{12}{(x - 3)^3}$$

Partition number for f'': $x = 3$
Sign chart for f'':

Test Numbers	
x	$f''(x)$
2	-12 $(-)$
4	12 $(+)$

Thus, the graph of f is concave downward on $(-\infty, 3)$ and concave upward on $(3, \infty)$.

Step 4. Sketch the graph of f:

x	$f(x)$
-3	0
0	-1
5	4

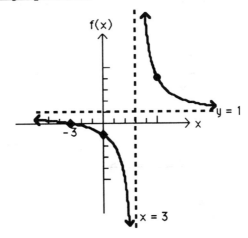

59. $f(x) = \dfrac{x}{x - 2}$

Step 1. Analyze $f(x)$:

(A) Domain: All real numbers except $x = 2$.

(B) Intercepts: y intercept: $f(0) = \dfrac{0}{-2} = 0$

$\quad\quad\quad\quad\quad\quad x$ intercepts: $\dfrac{x}{x - 2} = 0$

$\quad\quad\quad\quad\quad\quad\quad\quad\quad\quad\quad x = 0$

(C) Asymptotes:

<u>Horizontal asymptote</u>: $\displaystyle\lim_{x \to \infty} \dfrac{x}{x - 2} = \lim_{x \to \infty} \dfrac{x}{x\left(1 - \dfrac{2}{x}\right)} = 1.$

Thus, $y = 1$ is a horizontal asymptote.

<u>Vertical asymptote</u>: The denominator is 0 at $x = 2$ and the numerator is not 0 at $x = 2$. Thus, $x = 2$ is a vertical asymptote.

Step 2. Analyze $f'(x)$:

$f'(x) = \dfrac{(x - 2)(1) - x(1)}{(x - 2)^2} = \dfrac{-2}{(x - 2)^2} = -2(x - 2)^{-2}$

Critical values: None

Partition number: $x = 2$

Sign chart for f':

Test Numbers	
x	$f'(x)$
0	$-\frac{1}{2}$ (−)
3	-2 (−)

Thus, f is decreasing on $(-\infty, 2)$ and on $(2, \infty)$; there are no local extrema.

Step 3. Analyze $f''(x)$:

$f''(x) = 4(x - 2)^{-3} = \dfrac{4}{(x - 2)^3}$

Partition number for f'': $x = 2$

Sign chart for f'':

Test Numbers	
x	$f''(x)$
0	$-\frac{1}{2}$ (−)
3	4 (+)

Thus, the graph of f is concave downward on $(-\infty, 2)$ and concave upward on $(2, \infty)$.

Step 4. Sketch the graph of f:

x	$f(x)$
0	0
4	2

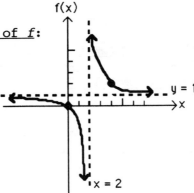

61. For any $n \geq 1$, $\lim\limits_{n \to \infty} (a_n x^n + a_{n-1} x^{n-1} + \ldots + a_1 x + a_0)$

$= \lim\limits_{x \to \infty} a_n x^n = \begin{cases} \infty & \text{if} \quad a_n > 0 \\ -\infty & \text{if} \quad a_n < 0 \end{cases}$

63. $p(x) = x^3 - 2x^2$; $p'(x) = 3x^2 - 4x$; $p''(x) = 6x - 4$

(A) $\lim\limits_{x \to \infty} p'(x) = \lim\limits_{x \to \infty} 3x^2 = \infty$

$\lim\limits_{x \to \infty} p''(x) = \lim\limits_{x \to \infty} 6x = \infty$

The graph of f is increasing and concave upward for large positive values of x.

(B) $\lim\limits_{x \to -\infty} p'(x) = \lim\limits_{x \to -\infty} 3x^2 = \infty$

$\lim\limits_{x \to -\infty} p'(x) = \lim\limits_{x \to -\infty} 6x = -\infty$

The graph of f is increasing and concave downward for large negative values of x.

65. For $|x|$ very large, $f(x) = x + \dfrac{1}{x} \approx x$. Thus, the line $y = x$ is an oblique asymptote.

Step 1. Analyze $f(x)$:
(A) Domain: All real numbers except $x = 0$.
(B) Intercepts: y intercept: There is no y intercept since f is not
   defined at $x = 0$.

   x intercepts: $x + \dfrac{1}{x} = 0$

   $x^2 + 1 = 0$

Thus, there are no x intercepts.
(C) Asymptotes:
   Oblique asymptote: $y = x$

   Vertical asymptote: $f(x) = x + \dfrac{1}{x} = \dfrac{x^2 + 1}{x}$

The denominator is 0 at $x = 0$ and the numerator is not 0 at $x = 0$.
Thus, $x = 0$ is a vertical asymptote.

<u>Step 2. Analyze $f'(x)$:</u>

$f'(x) = 1 - \dfrac{1}{x^2} = \dfrac{x^2 - 1}{x^2} = \dfrac{(x + 1)(x - 1)}{x^2}$

Critical values: $\dfrac{(x - 1)(x + 1)}{x^2} = 0$

$(x - 1)(x + 1) = 0$

$x = 1, -1$

Partition numbers: $x = 0$, $x = 1$, $x = -1$

Sign chart for f':

Test Numbers	
x	$f'(x)$
-2	$\frac{3}{4}$ (+)
$-\frac{1}{2}$	-3 (−)
$\frac{1}{2}$	-3 (−)
2	$\frac{3}{4}$ (+)

$f'(x)$ $\quad$ + + + + $\ 0$ − − − − ND − − − − 0 + + + +

$f(x)$ $\quad$ Increasing $\vdots$ Decreasing $\vdots$ Decreasing $\vdots$ Increasing

$\qquad\qquad$ Local $\qquad\qquad$ Local
$\qquad\qquad$ maximum $\qquad\qquad$ minimum

Thus, f is increasing on $(-\infty, -1)$ and on $(1, \infty)$; f is decreasing on $(-1, 0)$ and on $(0, 1)$; f has a local maximum at $x = -1$ and a local minimum at $x = 1$.

<u>Step 3. Analyze $f''(x)$:</u>

$f''(x) = 2x^{-3} = \dfrac{2}{x^3}$

Partition number for f'': $x = 0$
Sign chart for f'':

Test Numbers	
x	$f''(x)$
-1	-2 (−)
1	2 (+)

$f''(x)$ $\qquad$ − − − − − ND + + + +

Graph of f $\qquad$ Concave $\vdots$ Concave
$\qquad\qquad\qquad$ Downward $\vdots$ Upward

Thus, the graph of f is concave downward on $(-\infty, 0)$ and concave upward on $(0, \infty)$.

<u>Step 4. Sketch the graph of f:</u>

x	$f(x)$
-1	-2
1	2

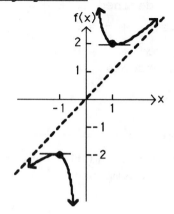

67. $f(x) = x^3 - x$

Step 1. Analyze $f(x)$:

(A) Domain: All real numbers, $(-\infty, \infty)$.

(B) Intercepts: y intercept: $f(0) = 0^3 - 0 = 0$

x intercepts: $x^3 - x = 0$

$$x(x^2 - 1) = 0$$
$$x(x - 1)(x + 1) = 0$$
$$x = 0,\ 1,\ -1$$

(C) Asymptotes: There are no asymptotes.

Step 2. Analyze $f'(x)$:

$f'(x) = 3x^2 - 1 = (\sqrt{3}x + 1)(\sqrt{3}x - 1)$

Critical values: $x = -\dfrac{\sqrt{3}}{3},\ x = \dfrac{\sqrt{3}}{3}$

Partition numbers: $x = -\dfrac{\sqrt{3}}{3},\ x = \dfrac{\sqrt{3}}{3}$

Sign chart for f':

Test Numbers

x	$f'(x)$
-1	2 (+)
0	-1 (−)
1	2 (+)

Thus, f is increasing on $\left(-\infty,\ -\dfrac{\sqrt{3}}{3}\right)$ and on $\left(\dfrac{\sqrt{3}}{3},\ \infty\right)$; f is decreasing on $\left(-\dfrac{\sqrt{3}}{3},\ \dfrac{\sqrt{3}}{3}\right)$; f has a local maximum at $x = -\dfrac{\sqrt{3}}{3}$ and a local minimum at $x = \dfrac{\sqrt{3}}{3}$.

Step 3. Analyze $f''(x)$:

$f''(x) = 6x$

Partition number for f'': $x = 0$

Sign chart for f'':

Thus, the graph of f is concave downward on $(-\infty, 0)$ and concave upward on $(0, \infty)$. There is an inflection point at $x = 0$.

Step 4. Sketch the graph of f:

x	$f(x)$
-1	0
$-\dfrac{\sqrt{3}}{3}$	≈ 0.4
0	0
$\dfrac{\sqrt{3}}{3}$	≈ -0.4

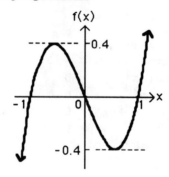

69. $f(x) = (x^2 + 3)(9 - x^2)$

Step 1. Analyze $f(x)$:

(A) Domain: All real numbers, $(-\infty, \infty)$.

(B) Intercepts: y intercept: $f(0) = 3(9) = 27$

x intercepts: $(x^2 + 3)(9 - x^2) = 0$

$(3 - x)(3 + x) = 0$

$x = 3, -3$

(C) Asymptotes: There are no asymptotes.

Step 2. Analyze $f'(x)$:

$f'(x) = (x^2 + 3)(-2x) + (9 - x^2)(2x)$

$= 2x[9 - x^2 - (x^2 + 3)]$

$= 2x(6 - 2x^2)$

$= 4x(\sqrt{3} + x)(\sqrt{3} - x)$

Critical values: $x = 0$, $x = -\sqrt{3}$, $x = \sqrt{3}$

Partition numbers: $x = 0$, $x = -\sqrt{3}$, $x = \sqrt{3}$

Sign chart for f':

```
f'(x)   + + + + 0 - - - - - 0 + + + + + 0 - - - -
      ──┼──●──┼──●──┼──●──┼──→ x
       -2  -√3  -1   0   1  √3   2

f(x)   Increasing ┊ Decreasing ┊ Increasing ┊ Decreasing
          Local        Local        Local
         maximum      minimum      maximum
```

Test Numbers	
x	$f'(x)$
-2	$8\ (+)$
-1	$-8\ (-)$
1	$8\ (+)$
2	$-8\ (-)$

Thus, f is increasing on $(-\infty, -\sqrt{3})$ and on $(0, \sqrt{3})$; f is decreasing on $(-\sqrt{3}, 0)$ and on $(\sqrt{3}, \infty)$; f has local maxima at $x = -\sqrt{3}$ and $x = \sqrt{3}$ and a local minimum at $x = 0$.

Step 3. Analyze $f''(x)$:

$f''(x) = 2x(-4x) + (6 - 2x^2)(2) = 12 - 12x^2 = -12(x - 1)(x + 1)$

Partition numbers for f'': $x = 1$, $x = -1$

Sign chart for f'':

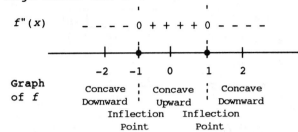

Test Numbers		
x	$f''(x)$	
-2	-36	(-)
0	12	(+)
2	-36	(-)

Thus, the graph of f is concave downward on $(-\infty, -1)$ and on $(1, \infty)$; the graph of f is concave upward on $(-1, 1)$; the graph has inflection points at $x = -1$ and $x = 1$.

Step 4. Sketch the graph of f:

x	$f(x)$
$-\sqrt{3}$	36
-1	32
0	27
1	32
$\sqrt{3}$	36

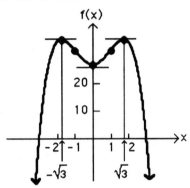

71. $f(x) = (x^2 - 4)^2$

Step 1. Analyze $f(x)$:
(A) Domain: All real numbers, $(-\infty, \infty)$.
(B) Intercepts: y intercept: $f(0) = (-4)^2 = 16$
x intercepts: $(x^2 - 4)^2 = 0$
$[(x - 2)(x + 2)]^2 = 0$
$(x - 2)^2 (x + 2)^2 = 0$
$x = 2, -2$
(C) Asymptotes: There are no asymptotes.

Step 2. Analyze $f'(x)$:
$f'(x) = 2(x^2 - 4)(2x) = 4x(x - 2)(x + 2)$
Critical values: $x = 0$, $x = 2$, $x = -2$
Partition numbers: Same as critical values.

Sign chart for f':

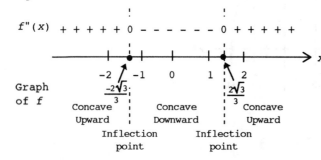

Test Numbers	
x	$f'(x)$
-3	-60 (-)
-1	12 (+)
1	-12 (-)
3	60 (+)

Thus, f is decreasing on $(-\infty, -2)$ and on $(0, 2)$; f is increasing on $(-2, 0)$ and on $(2, \infty)$; f has local minima at $x = -2$ and $x = 2$ and a local maximum at $x = 0$.

Step 3. Analyze $f''(x)$:

$$f''(x) = 4x(2x) + (x^2 - 4)(4) = 12x^2 - 16 = 12\left(x^2 - \frac{4}{3}\right)$$

$$= 12\left(x - \frac{2\sqrt{3}}{3}\right)\left(x + \frac{2\sqrt{3}}{3}\right)$$

Partition numbers for f'': $x = \frac{2\sqrt{3}}{3}$, $x = \frac{-2\sqrt{3}}{3}$

Sign chart for f'':

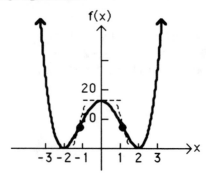

Test Numbers	
x	$f''(x)$
-2	32 (+)
0	-16 (-)
2	32 (+)

Thus, the graph of f is concave upward on $\left(-\infty, \frac{-2\sqrt{3}}{3}\right)$ and on $\left(\frac{2\sqrt{3}}{3}, \infty\right)$;

the graph of f is concave downward on $\left(\frac{-2\sqrt{3}}{3}, \frac{2\sqrt{3}}{3}\right)$; the graph has

inflection points at $x = \frac{-2\sqrt{3}}{3}$ and $x = \frac{2\sqrt{3}}{3}$.

Step 4. Sketch the graph of f:

x	$f(x)$
-2	0
$-\frac{2\sqrt{3}}{3}$	$\frac{64}{9}$
0	16
$\frac{2\sqrt{3}}{3}$	$\frac{64}{9}$
2	0

73. $f(x) = 2x^6 - 3x^5$

Step 1. Analyze $f(x)$:

(A) Domain: All real numbers, $(-\infty, \infty)$.

(B) Intercepts: y intercept: $f(0) = 2 \cdot 0^6 - 3 \cdot 0^5 = 0$

x intercepts: $2x^6 - 3x^5 = 0$

$x^5(2x - 3) = 0$

$x = 0, \dfrac{3}{2}$

(C) Asymptotes: There are no asymptotes.

Step 2. Analyze $f'(x)$:

$f'(x) = 12x^5 - 15x^4 = 12x^4\left(x - \dfrac{5}{4}\right)$

Critical values: $x = 0, \quad x = \dfrac{5}{4}$

Partition numbers: Same as critical values.

Sign chart for f':

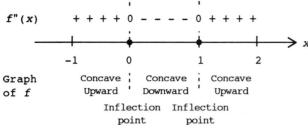

	Test Numbers	
x	$f'(x)$	
-1	-27	$(-)$
1	-3	$(-)$
2	144	$(+)$

Thus, f is decreasing on $(-\infty, 0)$ and $\left(0, \dfrac{5}{4}\right)$; f is increasing on $\left(\dfrac{5}{4}, \infty\right)$; f has a local minimum at $x = \dfrac{5}{4}$.

Step 3. Analyze $f''(x)$:

$f''(x) = 60x^4 - 60x^3 = 60x^3(x - 1)$

Partition numbers for f'': $x = 0, \quad x = 1$

Sign chart for f'':

$f''(x)$	$+ + + + 0 - - - - - 0 + + + +$	
	-1 $\quad$ 0 $\quad$ 1 $\quad$ 2 $\longrightarrow x$	
Graph of f	Concave Upward $\vdots$ Concave Downward $\vdots$ Concave Upward	
	Inflection point $\quad$ Inflection point	

	Test Numbers	
x	$f''(x)$	
-1	120	$(+)$
$\dfrac{1}{2}$	$-\dfrac{15}{4}$	$(-)$
2	480	$(+)$

Thus, the graph of f is concave upward on $(-\infty, 0)$ and on $(1, \infty)$; the graph of f is concave downward on $(0, 1)$; the graph has inflection points at $x = 0$ and $x = 1$.

Step 4. Sketch the graph of f:

x	$f(x)$
0	0
1	-1
$\frac{5}{4}$	≈ -1.5

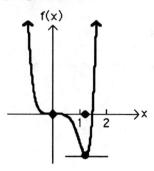

75. $f(x) = \dfrac{x}{x^2 - 4} = \dfrac{x}{(x-2)(x+2)}$

Step 1. Analyze $f(x)$:

(A) Domain: All real numbers except $x = 2$, $x = -2$.

(B) Intercepts: y intercept: $f(0) = \dfrac{0}{-4} = 0$

$\qquad\qquad\qquad x$ intercept: $\dfrac{x}{x^2 - 4} = 0$

$\qquad\qquad\qquad\qquad\qquad x = 0$

(C) Asymptotes:
   Horizontal asymptote:

$$\lim_{x \to \infty} \frac{x}{x^2 - 4} = \lim_{x \to \infty} \frac{x}{x^2\left(1 - \dfrac{4}{x^2}\right)} = \lim_{x \to \infty} \frac{1}{x}\left(\frac{1}{1 - \dfrac{4}{x^2}}\right) = 0$$

Thus, $y = 0$ (the x axis) is a horizontal asymptote.

Vertical asymptotes: The denominator is 0 at $x = 2$ and $x = -2$. The numerator is nonzero at each of these points. Thus, $x = 2$ and $x = -2$ are vertical asymptotes.

Step 2. Analyze $f'(x)$:

$$f'(x) = \frac{(x^2 - 4)(1) - x(2x)}{(x^2 - 4)^2} = \frac{-(x^2 + 4)}{(x^2 - 4)^2}$$

Critical values: None ($x^2 + 4 \neq 0$ for all x)
Partition numbers: $x = 2$, $x = -2$
Sign chart for f':

```
f'(x)   - - - - - ND - - - - - - - - ND - - - - -
       ─────────┼──◇──┼──┼──┼──┼──◇──┼──────────→ x
               -2    -1  0  1  2

f(x)  Decreasing ┊  Decreasing  ┊ Decreasing
```

Thus, f is decreasing on $(-\infty, -2)$, on $(-2, 2)$, and on $(2, \infty)$; f has no local extrema.

Step 3. Analyze $f''(x)$:

$$f''(x) = \frac{(x^2 - 4)^2(-2x) - [-(x^2 + 4)](2)(x^2 - 4)(2x)}{(x^2 - 4)^4}$$

$$= \frac{(x^2 - 4)(-2x) + 4x(x^2 + 4)}{(x^2 - 4)^3} = \frac{2x^3 + 24x}{(x^2 - 4)^3} = \frac{2x(x^2 + 12)}{(x^2 - 4)^3}$$

Partition numbers for f'': $x = 0$, $x = 2$, $x = -2$

Sign chart for f'':

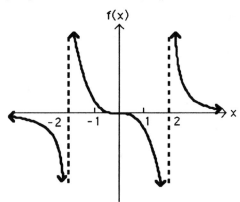

	Test Numbers	
x	$f''(x)$	
-3	$-\frac{126}{125}$	$(-)$
-1	$\frac{26}{27}$	$(+)$
1	$-\frac{26}{27}$	$(-)$
3	$\frac{126}{127}$	$(+)$

Thus, the graph of f is concave downward on $(-\infty, -2)$ and on $(0, 2)$; the graph of f is concave upward on $(-2, 0)$ and on $(2, \infty)$; the graph has an inflection point at $x = 0$.

Step 4. Sketch the graph of f:

x	$f(x)$
0	0
1	$-\frac{1}{3}$
-1	$\frac{1}{3}$
3	$\frac{3}{5}$
-3	$-\frac{3}{5}$

77. $f(x) = \dfrac{1}{1 + x^2}$

Step 1. Analyze $f(x)$:

(A) Domain: All real numbers ($1 + x^2 \neq 0$ for all x).

(B) Intercepts: y intercept: $f(0) = 1$

$\qquad\qquad\qquad x$ intercept: $\dfrac{1}{1 + x^2} \neq 0$ for all x; no x intercepts

(C) Asymptotes:

<u>Horizontal asymptote</u>: $\lim\limits_{x \to \infty} \dfrac{1}{1 + x^2} = 0$

Thus, $y = 0$ (the x axis) is a horizontal asymptote.

<u>Vertical asymptotes</u>: Since $1 + x^2 \neq 0$ for all x, there are no vertical asymptotes.

Step 2. Analyze $f'(x)$:

$$f'(x) = \frac{(1 + x^2)(0) - 1(2x)}{(1 + x^2)^2} = \frac{-2x}{(1 + x^2)^2}$$

Critical values: $x = 0$

Partition numbers: $x = 0$

Sign chart for f':

Test Numbers

x	$f'(x)$
-1	$\frac{1}{2}$ $(+)$
1	$-\frac{1}{2}$ $(-)$

Thus, f is increasing on $(-\infty, 0)$; f is decreasing on $(0, \infty)$; f has a local maximum at $x = 0$.

Step 3. Analyze $f''(x)$:

$$f''(x) = \frac{(1 + x^2)^2(-2) - (-2x)(2)(1 + x^2)2x}{(1 + x^2)^4} = \frac{(-2)(1 + x^2) + 8x^2}{(1 + x^2)^3}$$

$$= \frac{6x^2 - 2}{(1 + x^2)^3} = \frac{6\left(x + \frac{\sqrt{3}}{3}\right)\left(x - \frac{\sqrt{3}}{3}\right)}{(1 + x^2)^3}$$

Partition numbers for f'': $x = -\frac{\sqrt{3}}{3}$, $x = \frac{\sqrt{3}}{3}$

Sign chart for f'':

Test Numbers

x	$f''(x)$
-1	$\frac{1}{2}$ $(+)$
0	-2 $(-)$
1	$\frac{1}{2}$ $(+)$

Thus, the graph of f is concave upward on $\left(-\infty, \frac{-\sqrt{3}}{3}\right)$ and on $\left(\frac{\sqrt{3}}{3}, \infty\right)$; the graph of f is concave downward on $\left(\frac{-\sqrt{3}}{3}, \frac{\sqrt{3}}{3}\right)$; the graph has inflection points at $x = \frac{-\sqrt{3}}{3}$ and $x = \frac{\sqrt{3}}{3}$.

Step 4. Sketch the graph of f:

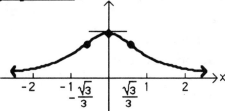

x	$f(x)$
$-\dfrac{\sqrt{3}}{3}$	$\dfrac{3}{4}$
0	1
$\dfrac{\sqrt{3}}{3}$	$\dfrac{3}{4}$

79. $f(x) = -x^4 - x^3 + 2x^2 - 2x + 3$

Step 1. Analyze $f(x)$:

(A) Domain: all real numbers

(B) Intercepts: y-intercept: $f(0) = 3$
x-intercepts: $x \approx -2.40, \; 1.16$

(C) Asymptotes: None

Step 2. Analyze $f'(x)$: $\quad f'(x) = -4x^3 - 3x^2 + 4x - 2$
Critical value: $x \approx -1.58$
f is increasing on $(-\infty, -1.58)$; f is decreasing on $(-1.58, \infty)$; f has a local maximum at $x = -1.58$.

Step 3. Analyze $f''(x)$: $\quad f''(x) = -12x^2 - 6x + 4$
The graph of f is concave downward on $(-\infty, -0.88)$ and $(0.38, \infty)$; the graph of f is concave upward on $(-0.88, 0.38)$; the graph has inflection points at $x = -0.88$ and $x = 0.38$.

81. $f(x) = x^4 - 5x^3 + 3x^2 + 8x - 5$

Step 1. Analyze $f(x)$:

(A) Domain: all real numbers

(B) Intercepts: y-intercept: $f(0) = -5$
x-intercepts: $x \approx -1.18, \; 0.61, \; 1.87, \; 3.71$

(C) Asymptotes: None

Step 2. Analyze $f'(x)$: $\quad f'(x) = 4x^3 - 15x^2 + 6x + 8$
Critical values: $x \approx -0.53, \; 1.24, \; 3.04$
f is decreasing on $(-\infty, -0.53)$ and $(1.24, 3.04)$; f is increasing on $(-0.53, 1.24)$ and $(3.04, \infty)$; f has local minima at $x = -0.53$ and 3.04; f has a local maximum at $x = 1.24$

Step 3. Analyze $f''(x)$: $\quad f''(x) = 12x^2 - 30x + 6$
The graph of f is concave upward on $(-\infty, 0.22)$ and $(2.28, \infty)$; the graph of f is concave downward on $(0.22, 2.28)$; the graph has inflection points at $x = 0.22$ and 2.28.

83. $f(x) = 0.01x^5 + 0.03x^4 - 0.4x^3 - 0.5x^2 + 4x + 3$

Step 1. Analyze $f(x)$:

(A) Domain: all real numbers

(B) Intercepts: y-intercept: $f(0) = 3$
x-intercepts: $x \approx -6.68, -3.64, -0.72$

(C) Asymptotes: None

Step 2. Analyze $f'(x)$: $f'(x) = 0.05x^4 + 0.12x^3 - 1.2x^2 - x + 4$
Critical values: $x \approx -5.59, -2.27, 1.65, 3.82$
f is increasing on $(-\infty, -5.59)$, $(-2.27, 1.65)$, and $(3.82, \infty)$; f is decreasing on $(-5.59, -2.27)$ and $(1.65, 3.82)$; f has local minima at $x = -2.27$ and 3.82; f has local maxima at $x = -5.59$ and 1.65

Step 3. Analyze $f''(x)$: $f''(x) = 0.2x^3 + 0.36x^2 - 2.4x - 1$
The graph of f is concave downward on $(-\infty, -4.31)$ and $(-0.40, 2.91)$; the graph of f is concave upward on $(-4.31, -0.40)$ and $(2.91, \infty)$; the graph has inflection points at $x = -4.31, -0.40$ and 2.91.

85. $R(x) = xp = 1296x - 0.12x^3$, $0 < x < 80$

Step 1. Analyze $R(x)$:
(A) Domain: $0 < x < 80$ or $(0, 80)$.
(B) Intercepts: There are no intercepts on $(0, 80)$.
(C) Asymptotes: Since R is a polynomial, there are no asymptotes.

Step 2. Analyze $R'(x)$:
$$R'(x) = 1296 - 0.36x^2$$
$$= -0.36(x^2 - 3600)$$
$$= -0.36(x - 60)(x + 60), \quad 0 < x < 80$$
Critical values: [on $(0, 80)$]: $x = 60$
Partition numbers: $x = 60$
Sign chart for R':

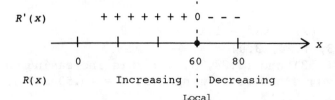

Test Numbers	
x	$R'(x)$
1	1295.64 (+)
61	−43.56 (−)

Thus, R is increasing on $(0, 60)$ and decreasing on $(60, 80)$, R has a local maximum at $x = 60$.

Step 3. Analyze $R''(x)$:
$R''(x) = -0.72x < 0$ for $0 < x < 80$
Thus, the graph of R is concave downward on $(0, 80)$.

Step 4. Sketch the graph of R:

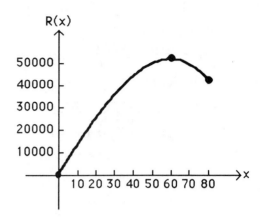

87. $P(x) = \dfrac{2x}{1 - x}$, $0 \le x < 1$

    (A) $P'(x) = \dfrac{(1 - x)(2) - 2x(-1)}{(1 - x)^2} = \dfrac{2}{(1 - x)^2}$

       $P'(x) > 0$ for $0 \le x < 1$. Thus, P is increasing on $(0, 1)$.

    (B) From (A), $P'(x) = 2(1 - x)^{-2}$. Thus,

       $P''(x) = -4(1 - x)^{-3}(-1) = \dfrac{4}{(1 - x)^3}.$

       $P''(x) > 0$ for $0 \le x < 1$, and the graph of P is concave upward on $(0, 1)$.

    (C) Since the domain of P is $[0, 1)$, there are no horizontal asymptotes. The denominator is 0 at $x = 1$ and the numerator is nonzero there. Thus, $x = 1$ is a vertical asymptote.

    (D) $P(0) = \dfrac{2 \cdot 0}{1 - 0} = 0.$

       Thus, the origin is both an x and a y intercept of the graph.

    (E) The graph of P is:

x	$P(x)$
0	0
$\frac{1}{2}$	2
$\frac{3}{4}$	6

89. $C(n) = 3200 + 250n + 50n^2$, $0 < n < \infty$

(A) Average cost per year:

$$\overline{C}(n) = \frac{C(n)}{n} = \frac{3200}{n} + 250 + 50n, \quad 0 < n < \infty$$

(B) Graph $\overline{C}(n)$:

Step 1. Analyze $\overline{C}(n)$:
Domain: $0 < n < \infty$
Intercepts: C intercept: None $(n > 0)$

n intercepts: $\dfrac{3200}{n} + 250 + 50n > 0$ on $(0, \infty)$;

there are no n intercepts.

Asymptotes: For large n, $C(n) = \dfrac{3200}{n} + 250 + 50n \approx 250 + 50n$.

Thus, $y = 250 + 50n$ is an oblique asymptote. As $n \to 0$, $\overline{C}(n) \to \infty$; thus, $n = 0$ is a vertical asymptote.

Step 2. Analyze $\overline{C}'(n)$:

$$\overline{C}'(n) = -\frac{3200}{n^2} + 50 = \frac{50n^2 - 3200}{n^2} = \frac{50(n^2 - 64)}{n^2}$$

$$= \frac{50(n - 8)(n + 8)}{n^2}, \quad 0 < n < \infty$$

Critical value: $n = 8$

Sign chart for $\overline{C}'$:

n	$\overline{C}'(n)$
7	$(-)$
9	$(+)$

Test Numbers

Thus, $\overline{C}$ is decreasing on $(0, 8)$ and increasing on $(8, \infty)$; $n = 8$ is a local minimum.

Step 3: Analyze $\overline{C}''(n)$:

$$\overline{C}''(n) = \frac{6400}{n^3}, \quad 0 < n < \infty$$

$\overline{C}''(n) > 0$ on $(0, \infty)$. Thus, the graph of $\overline{C}$ is concave upward on $(0, \infty)$.

Step 4. Sketch the graph of $\overline{C}$:

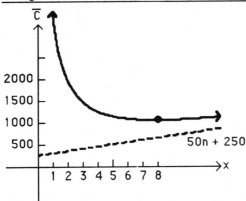

(C) The average cost per year is a minimum when $n = 8$ years.

91. $C(x) = 1000 + 5x + 0.1x^2$, $0 < x < \infty$.

(A) The average cost function is: $\overline{C}(x) = \dfrac{1000}{x} + 5 + 0.1x$.

Now, $\overline{C}'(x) = -\dfrac{1000}{x^2} + \dfrac{1}{10} = \dfrac{x^2 - 10,000}{10x^2} = \dfrac{(x + 100)(x - 100)}{10x^2}$

Sign chart for $\overline{C}'$:

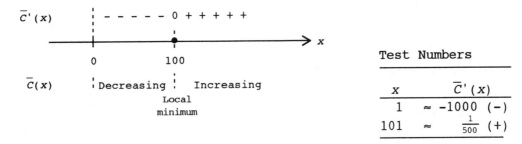

Test Numbers	
x	$\overline{C}'(x)$
1	≈ -1000 (−)
101	$\approx \dfrac{1}{500}$ (+)

Thus, $\overline{C}$ is decreasing on (0, 100) and increasing on (100, ∞); $\overline{C}$ has a minimum at $x = 100$.

Since $\overline{C}''(x) = \dfrac{2000}{x^3} > 0$ for $0 < x < \infty$, the graph of $\overline{C}$ is concave upward on (0, ∞). The line $x = 0$ is a vertical asymptote and the line

$y = 5 + 0.1x$ is an oblique asymptote for the graph of $\overline{C}$.
The marginal cost function is $C'(x) = 5 + 0.2x$.

The graphs of $\overline{C}$ and C' are:

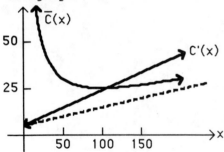

(B) The minimum average cost is:
$$\overline{C}(100) = \frac{1000}{100} + 5 + \frac{1}{10}(100) = 25$$

93. $C(t) = \dfrac{0.14t}{t^2 + 1}$

<u>Step 1. Analyze $C(t)$</u>:

Domain: $t \geq 0$, i.e., $[0, \infty)$

Intercepts: y intercept: $C(0) = 0$

$\qquad\qquad$ t intercepts: $\dfrac{0.14t}{t^2 + 1} = 0$

$\qquad\qquad\qquad\qquad\qquad\qquad\quad t = 0$

Asymptotes:

<u>Horizontal asymptote</u>: $\displaystyle\lim_{t\to\infty} \frac{0.14t}{t^2 + 1} = \lim_{t\to\infty} \frac{0.14t}{t^2\left(1 + \frac{1}{t^2}\right)} = \lim_{t\to\infty} \frac{0.14}{t\left(1 + \frac{1}{t^2}\right)} = 0$

Thus, $y = 0$ (the t axis) is a horizontal asymptote.

<u>Vertical asymptotes</u>: Since $t^2 + 1 > 0$ for all t, there are no vertical asymptotes.

<u>Step 2. Analyze $C'(t)$</u>:

$C'(t) = \dfrac{(t^2 + 1)(0.14) - 0.14t(2t)}{(t^2 + 1)^2} = \dfrac{0.14(1 - t^2)}{(t^2 + 1)^2} = \dfrac{0.14(1 - t)(1 + t)}{(t^2 + 1)^2}$

Critical values on $[0, \infty)$: $t = 1$

Sign chart for C':

$C'(t) \qquad\qquad + + + + + 0 - - - - - $

$\qquad\qquad\overline{\quad\quad\quad\quad\quad\bullet\quad\quad\quad\quad\quad} \to t$

$\qquad\qquad\qquad 0 \qquad\quad 1$

$C(t) \qquad\quad$ Increasing $\vdots$ Decreasing
$\qquad\qquad\qquad\qquad\quad$ Local
$\qquad\qquad\qquad\quad$ maximum

Test Numbers	
t	$C'(t)$
0	$(+)$
2	$(-)$

Thus, C is increasing on $(0, 1)$ and decreasing on $(1, \infty)$; C has a maximum value at $t = 1$.

Step 3. Analyze $C''(t)$:

$$C''(t) = \frac{(t^2 + 1)^2(-0.28t) - 0.14(1 - t^2)(2)(t^2 + 1)(2t)}{(t^2 + 1)^4}$$

$$= \frac{(t^2 + 1)(-0.28t) - 0.56t(1 - t^2)}{(t^2 + 1)^3} = \frac{0.28t^3 - 0.84t}{(t^2 + 1)^3}$$

$$= \frac{0.28t(t^2 - 3)}{(t^2 + 1)^3} = \frac{0.28t(t - \sqrt{3})(t + \sqrt{3})}{(t^2 + 1)^3}, \quad 0 \le t < \infty$$

Partition numbers for C'' on $[0, \infty)$: $t = \sqrt{3}$
Sign chart for C'':

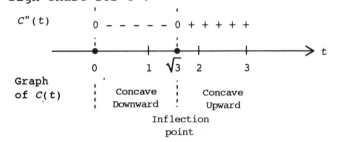

Test Numbers		
t	$C''(t)$	
1	-0.07	$(-)$
2	≈ 0.005	$(+)$

Thus, the graph of C is concave downward on $(0, \sqrt{3})$ and concave upward on $(\sqrt{3}, \infty)$; the graph has an inflection point at $t = \sqrt{3}$.

Step 4. Sketch the graph of $C(t)$:

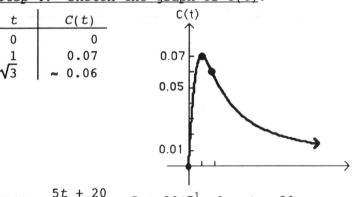

t	$C(t)$
0	0
1	0.07
$\sqrt{3}$	≈ 0.06

95. $N(t) = \dfrac{5t + 20}{t} = 5 + 20t^{-1}$, $1 \le t \le 30$

Step 1. Analyze $N(t)$:
Domain: $1 \le t \le 30$, or $[1, 30]$.
Intercepts: There are no t or N intercepts.
Asymptotes: Since N is defined only for $1 \le t \le 30$, there are no horizontal asymptotes. Also, since $t \ne 0$ on $[1, 30]$, there are no vertical asymptotes.

Step 2. Analyze $N'(t)$:
$$N'(t) = -20t^{-2} = \frac{-20}{t^2}, \quad 1 \le t \le 30$$

Since $N'(t) < 0$ for $1 \le t \le 30$, N is decreasing on $(1, 30)$; N has no local extrema.

Step 3. Analyze $N''(t)$:

$N''(t) = \dfrac{40}{t^3}$, $1 \le t \le 30$

Since $N''(t) > 0$ for $1 \le t \le 30$, the graph of N is concave upward on $(1, 30)$.

Step 4. Sketch the graph of N:

t	$N(t)$
1	25
5	9
10	7
30	5.67

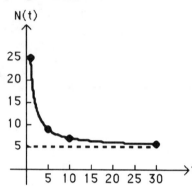

EXERCISE 4-5

Things to remember:

1. A function f continuous on a closed interval $[a, b]$ assumes both an absolute maximum and an absolute minimum on that interval. Absolute extrema (if they exist) must always occur at critical values or at endpoints.

2. STEPS FOR FINDING ABSOLUTE EXTREMA:

    To find the absolute maximum and absolute minimum of a function f on the closed interval $[a, b]$:

    (a) Verify that f is continuous on $[a, b]$.

    (b) Determine the critical values of f on the open interval (a, b).

    (c) Evaluate f at the endpoints a and b and at the critical values found in (b).

    (d) The absolute maximum $f(x)$ on $[a, b]$ is the largest of the values found in step (c).

    (e) The absolute minimum $f(x)$ on $[a, b]$ is the smallest of the values found in step (c).

3. SECOND-DERIVATIVE TEST FOR ABSOLUTE MAXIMUM AND MINIMUM

Suppose f is continuous on an interval I and has only one critical value C on I:

$f'(c)$	$f''(c)$	$f(c)$	Example
0	+	Absolute minimum	
0	−	Absolute maximum	
0	0	Test fails	

4. STRATEGY FOR SOLVING APPLIED OPTIMIZATION PROBLEMS

Step 1: Introduce variables and a function f, including the domain I of f, and then construct a mathematical model of the form:

Maximize (or minimize) f on the interval I

Step 2: Find the absolute maximum (or minimum) value of f on the interval I and the value(s) of x where this occurs.

Step 3: Use the solution to the mathematical model to answer the questions asked in the problem.

---

1. On $[0, 10]$; absolute minimum: $f(0) = 1$; absolute maximum: $f(10) = 9$

3. On $[0, 8]$; absolute minimum: $f(0) = 1$; absolute maximum: $f(3) = 8$

5. On $[4, 6]$; absolute minimum: $f(6) = 3$; absolute maximum: $f(4) = 7$

7. On $[1, 5]$; absolute minimum: $f(1) = f(5) = 5$; absolute maximum: $f(3) = 8$

9. $f(x) = x^2 - 4x + 5$, $I = (-\infty, \infty)$
$f'(x) = 2x - 4 = 2(x - 2)$
$x = 2$ is the *only critical value* on I, and $f(2) = 2^2 - 4 \cdot 2 + 5 = 1$.
$f''(x) = 2$
$f''(2) = 2 > 0$. Therefore, $f(2) = 1$ is the absolute minimum.
The function does not have an absolute maximum.

11. $f(x) = 10 + 8x - x^2$, $I = (-\infty, \infty)$
$f'(x) = 8 - 2x = 2(4 - x)$
$x = 4$ is the *only critical value* on I, and $f(4) = 10 + 32 - 16 = 26$.
$f''(x) = -2$
$f''(4) = -2 < 0$. Therefore, $f(4) = 26$ is the absolute maximum.
The function does not have an absolute minimum.

13. $f(x) = 1 - x^3$, $I = (-\infty, \infty)$

$f'(x) = -3x^2$

$x = 0$ is the *only critical value* on I, and $f(0) = 1 - 0^3 = 1$.

$f''(x) = -6x$

$f''(0) = 0$.   Therefore, the test fails.

Since $f'(x) = -3x^2 < 0$ on $(-\infty, 0)$ and on $(0, \infty)$, f is decreasing on I and f does not have any absolute extrema.

15. $f(x) = 24 - 2x - \dfrac{8}{x}$, $x > 0$

$f'(x) = -2 + \dfrac{8}{x^2} = \dfrac{-2x^2 + 8}{x^2} = \dfrac{-2(x^2 - 4)}{x^2} = \dfrac{-2(x + 2)(x - 2)}{x^2}$

Critical values: $x = 2$ [Note: $x = -2$ is not a critical value, since the domain of f is $x > 0$.]

$f''(x) = -\dfrac{16}{x^3}$ and $f''(2) = -\dfrac{16}{8} = -2 < 0$.

Thus, the absolute maximum value of f is $f(2) = 24 - 2 \cdot 2 - \dfrac{8}{2} = 16$.

17. $f(x) = 5 + 3x + \dfrac{12}{x^2}$, $x > 0$

$f'(x) = 3 - \dfrac{24}{x^3} = \dfrac{3x^3 - 24}{x^3} = \dfrac{3(x^3 - 8)}{x^3} = \dfrac{3(x - 2)(x^2 + 2x + 4)}{x^3}$

Critical value: $x = 2$

$f''(x) = \dfrac{72}{x^4}$ and $f''(2) = \dfrac{72}{2^4} = \dfrac{72}{16} > 0$.

Thus, the absolute minimum of f is: $f(2) = 5 + 3(2) + \dfrac{12}{2^2} = 14$.

19. $f(x) = x^3 - 6x^2 + 9x - 6$

$f'(x) = 3x^2 - 12x + 9 = 3(x^2 - 4x + 3) = 3(x - 3)(x - 1)$

Critical values: $x = 1, 3$

(A) On the interval $[-1, 5]$: $f(-1) = -1 - 6 - 9 - 6 = -22$

$\qquad\qquad\qquad\qquad\qquad\quad f(1) = 1 - 6 + 9 - 6 = -2$

$\qquad\qquad\qquad\qquad\qquad\quad f(3) = 27 - 54 + 27 - 6 = -6$

$\qquad\qquad\qquad\qquad\qquad\quad f(5) = 125 - 150 + 45 - 6 = 14$

Thus, the absolute maximum of f is $f(5) = 14$, and the absolute minimum of f is $f(-1) = -22$.

(B) On the interval $[-1, 3]$: $f(-1) = -22$

$\qquad\qquad\qquad\qquad\qquad\quad f(1) = -2$

$\qquad\qquad\qquad\qquad\qquad\quad f(3) = -6$

Absolute maximum of f: $f(1) = -2$

Absolute minimum of f: $f(-1) = -22$

(C) On the interval $[2, 5]$: $f(2) = 8 - 24 + 18 - 6 = -4$

$\qquad\qquad\qquad\qquad\qquad\quad f(3) = -6$

$\qquad\qquad\qquad\qquad\qquad\quad f(5) = 14$

Absolute maximum of f: $f(5) = 14$

Absolute minimum of f: $f(3) = -6$

21. $f(x) = (x - 1)(x - 5)^3 + 1$
$f'(x) = (x - 1)3(x - 5)^2 + (x - 5)^3$
$\quad = (x - 5)^2(3x - 3 + x - 5)$
$\quad = (x - 5)^2(4x - 8)$
Critical values: $x = 2, 5$

(A) Interval $[0, 3]$: $f(0) = (-1)(-5)^3 + 1 = 126$
$\qquad\qquad\qquad\quad f(2) = (2 - 1)(2 - 5)^3 + 1 = -26$
$\qquad\qquad\qquad\quad f(3) = (3 - 1)(3 - 5)^3 + 1 = -15$

　　Absolute maximum of f: $f(0) = 126$
　　Absolute minimum of f: $f(2) = -26$

(B) Interval $[1, 7]$: $f(1) = 1$
$\qquad\qquad\qquad\quad f(2) = -26$
$\qquad\qquad\qquad\quad f(5) = 1$
$\qquad\qquad\qquad\quad f(7) = (7 - 1)(7 - 5)^3 + 1 = 6 \cdot 8 + 1 = 49$

　　Absolute maximum of f: $f(7) = 49$
　　Absolute minimum of f: $f(2) = -26$

(C) Interval $[3, 6]$: $f(3) = (3 - 1)(3 - 5)^3 + 1 = -15$
$\qquad\qquad\qquad\quad f(5) = 1$
$\qquad\qquad\qquad\quad f(6) = (6 - 1)(6 - 5)^3 + 1 = 6$

　　Absolute maximum of f: $f(6) = 6$
　　Absolute minimum of f: $f(3) = -15$

23. Let one length $= x$ and the other $= 10 - x$.
Since neither length can be negative, we have $x \geq 0$ and $10 - x \geq 0$, or
$x \leq 10$. We want the maximum value of the product $x(10 - x)$, where
$0 \leq x \leq 10$.

Let $f(x) = x(10 - x) = 10x - x^2$; domain $I = [0, 10]$
$\quad f'(x) = 10 - 2x$; $x = 5$ is the only critical value
$\quad f''(x) = -2$
$\quad f''(5) = -2 < 0$
Thus, $f(5) = 25$ is the absolute maximum; divide the line in half.

25. Let one number $= x$. Then the other number $= x + 30$.
$f(x) = x(x + 30) = x^2 + 30x$; domain $I = (-\infty, \infty)$
$f'(x) = 2x + 30$; $x = -15$ is the only critical value
$f''(x) = 2$
$f''(-15) = 2 > 0$
Thus, the absolute minimum of f occurs at $x = -15$. The numbers, then,
are -15 and $-15 + 30 = 15$.

27. Let $x =$ the length of the rectangle
and $y =$ the width of the rectangle.
Then, $2x + 2y = 100$
$\qquad\quad x + y = 50$
$\qquad\qquad y = 50 - x$
We want to find the maximum of the area:
$A(x) = x \cdot y = x(50 - x) = 50x - x^2$.

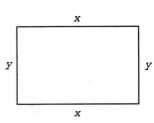

Since $x \geq 0$ and $y \geq 0$, we must have $0 \leq x \leq 50$. [<u>Note</u>: $A(0) = A(50) = 0$.]
$A'(x) = \dfrac{dA}{dx} = 50 - 2x$; $x = 25$ is the only critical value.
Now, $A'' = -2$ and $A''(25) = -2 < 0$. Thus, $A(25)$ is the absolute maximum.
The maximum area is $A(25) = 25(50 - 25) = 625$ cm^2, which means that the rectangle is actually a square with sides measuring 25 cm each.

29. Let the rectangle of fixed area A have dimensions x and y. Then $A = xy$ and $y = \dfrac{A}{x}$.

The cost of the fence is

$$C = 2Bx + 2By = 2Bx + \dfrac{2AB}{x}, \quad x > 0$$

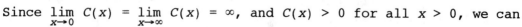

Thus, we want to find the absolute minimum of

$$C(x) = 2Bx + \dfrac{2AB}{x}, \quad x > 0$$

Since $\lim\limits_{x \to 0} C(x) = \lim\limits_{x \to \infty} C(x) = \infty$, and $C(x) > 0$ for all $x > 0$, we can conclude that C has an absolute minimum on $(0, \infty)$. This agrees with our intuition that there should be a cheapest way to build the fence.

31. Let x and y be the dimensions of the rectangle and let C be the fixed amount which can be spent. Then

$$C = 2Bx + 2By \quad \text{and} \quad y = \dfrac{C - 2Bx}{2B}$$

The area enclosed by the fence is:

$$A = xy = x\left[\dfrac{C - 2Bx}{2B}\right]$$

Thus, we want to find the absolute maximum value of

$$A(x) = \dfrac{C}{2B}x - x^2, \quad 0 \leq x \leq \dfrac{C}{2B}$$

Since $A(x)$ is a continuous function on the closed interval $\left[0, \dfrac{C}{2B}\right]$, it has an absolute maximum value. This agrees with our intuition that there should be a largest rectangular area that can be enclosed with a fixed amount of fencing.

33. (A) Revenue $R(x) = x \cdot p(x) = x\left(200 - \dfrac{x}{30}\right) = 200x - \dfrac{x^2}{30}$, $0 \leq x \leq 6{,}000$

$R'(x) = 200 - \dfrac{2x}{30} = 200 - \dfrac{x}{15}$

Now $R'(x) = 200 - \dfrac{x}{15} = 0$
implies $x = 3000$.
$R''(x) = -\dfrac{1}{15} < 0$.

Thus, $R''(3000) = -\dfrac{1}{15} < 0$ and we conclude that R has an absolute maximum at $x = 3000$. The maximum revenue is

$R(3000) = 200(3000) - \dfrac{(3000)^2}{30} = \$300{,}000$

(B) Profit $P(x) = R(x) - C(x) = 200x - \dfrac{x^2}{30} - (72,000 + 60x)$

$$= 140x - \dfrac{x^2}{30} - 72,000$$

$P'(x) = 140 - \dfrac{x}{15}$

Now $140 - \dfrac{x}{15} = 0$ implies $x = 2,100$.

$P''(x) = -\dfrac{1}{15}$ and $P''(2,100) = -\dfrac{1}{15} < 0$. Thus, the maximum profit occurs when 2,100 television sets are produced. The maximum profit is

$$P(2,100) = 140(2,100) - \dfrac{(2,100)^2}{30} - 72,000 = \$75,000$$

the price that the company should charge is

$$p(2,100) = 200 - \dfrac{2,100}{30} = \$130 \text{ for each set.}$$

(C) If the government taxes the company \$5 for each set, then the profit $P(x)$ is given by

$$P(x) = 200x - \dfrac{x^2}{30} - (72,000 + 60x) - 5x$$

$$= 135x - \dfrac{x^2}{30} - 72,000.$$

$P'(x) = 135 - \dfrac{x}{15}.$

Now $135 - \dfrac{x}{15} = 0$ implies $x = 2,025$.

$P''(x) = -\dfrac{1}{15}$ and $P''(2,025) = -\dfrac{1}{15} < 0$. Thus, the maximum profit in this case occurs when 2,025 television sets are produced. The maximum profit is

$$P(2,025) = 135(2,025) - \dfrac{(2,025)^2}{30} - 72,000$$

$$= \$64,687.50$$

and the company should charge $p(2,025) = 200 - \dfrac{2,025}{30}$

$$= \$132.50 \text{ per set.}$$

35. Let x = number of dollar increases in the rate per day. Then
$200 - 5x$ = total number of cars rented and $30 + x$ = rate per day.
Total income = (total number of cars rented)(rate)

$\quad y(x) = (200 - 5x)(30 + x), \quad 0 \le x \le 40$

$\quad y'(x) = (200 - 5x)(1) + (30 + x)(-5)$

$\qquad = 200 - 5x - 150 - 5x$

$\qquad = 50 - 10x$

$\qquad = 10(5 - x)$

Thus, $x = 5$ is the only critical value and $y(5) = (200 - 25)(30 + 5) = 6125$.

$y''(x) = -10$

$y''(5) = -10 < 0$

Therefore, the absolute maximum income is $y(5) = \$6125$ when the rate is \$35 per day.

37. Let x = number of additional trees planted per acre. Then
$30 + x$ = total number of trees per acre and $50 - x$ = yield per tree.
Yield per acre = (total number of trees per acre)(yield per tree)

$$y(x) = (30 + x)(50 - x), \quad 0 \le x \le 20$$
$$y'(x) = (30 + x)(-1) + (50 - x)$$
$$= 20 - 2x$$
$$= 2(10 - x)$$

The only critical value is $x = 10$, $y(10) = 40(40) = 1600$ pounds per acre.
$y''(x) = -2$
$y''(10) = -2 < 0$
Therefore, the absolute maximum yield is $y(10) = 1600$ pounds per acre
when the number of trees per acre is 40.

39. Volume =
$$V(x) = (12 - 2x)(8 - 2x)x, \quad 0 \le x \le 4$$
$$= 96x - 40x^2 + 4x^3$$
$$V'(x) = 96 - 80x + 12x^2$$
$$= 4(24 - 20x + 3x^2)$$

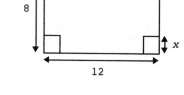

We solve $24 - 20x + 3x^2 = 0$ by using the
quadratic formula:
$$x = \frac{20 \pm \sqrt{400 - 4 \cdot 24 \cdot 3}}{6} = \frac{10 \pm 2\sqrt{7}}{3}$$

Thus, $x = \dfrac{10 - 2\sqrt{7}}{3} \approx 1.57$ is the only critical value on the interval
$[0, 4]$.
$V''(x) = -80 + 24x$
$V''(1.57) = -80 + 24(1.57) < 0$
Therefore, a square with a side of length $x = 1.57$ inches should be cut
from each corner to obtain the maximum volume.

41. Area = 800 square feet = xy (1)
Cost = $18x + 6(2y + x)$

From (1), we have $y = \dfrac{800}{x}$.

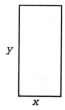

Hence, cost $C(x) = 18x + 6\left(\dfrac{1600}{x} + x\right)$, or

$$C(x) = 24x + \frac{9600}{x}, \quad x > 0,$$

$$C'(x) = 24 - \frac{9600}{x^2} = \frac{24(x^2 - 400)}{x^2} = \frac{24(x - 20)(x + 20)}{x^2}.$$

Therefore, $x = 20$ is the only critical value.

$$C''(x) = \frac{19,200}{x^3}$$

$C''(20) = \dfrac{19,200}{8000} > 0$. Therefore, $x = 20$ for the

minimum cost.
The dimensions of the fence are shown in the
diagram at the right.

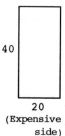

43. Let x = number of books produced each printing. Then, the number of printings = $\dfrac{50,000}{x}$.

Cost = $C(x)$ = cost of storage + cost of printing

$$= \frac{x}{2} + \frac{50,000}{x}(1000), \quad x > 0$$

[Note: $\dfrac{x}{2}$ is the average number in storage each day.]

$$C'(x) = \frac{1}{2} - \frac{50,000,000}{x^2} = \frac{x^2 - 100,000,000}{2x^2} = \frac{(x + 10,000)(x - 10,000)}{2x^2}$$

Critical value: $x = 10,000$

$$C''(x) = \frac{100,000,000}{x^3}$$

$$C''(10,000) = \frac{100,000,000}{(10,000)^3} > 0$$

Thus, the minimum cost occurs when $x = 10,000$ and the number of printings is $\dfrac{50,000}{10,000} = 5$.

45. (A) Let the cost to lay the pipe on the land be 1 unit; then the cost to lay the pipe in the lake is 1.4 units.

$$C(x) = \text{total cost} = (1.4)\sqrt{x^2 + 25} + (1)(10 - x), \quad 0 \le x \le 10$$
$$= (1.4)(x^2 + 25)^{1/2} + 10 - x$$

$$C'(x) = (1.4)\frac{1}{2}(x^2 + 25)^{-1/2}(2x) - 1$$
$$= (1.4)x(x^2 + 25)^{-1/2} - 1$$
$$= \frac{1.4x - \sqrt{x^2 + 25}}{\sqrt{x^2 + 25}}$$

$C'(x) = 0$ when $1.4x - \sqrt{x^2 + 25} = 0$ or $1.96x^2 = x^2 + 25$

$$.96x^2 = 25$$
$$x^2 = \frac{25}{.96} = 26.04$$
$$x = \pm5.1$$

Thus, the critical value is $x = 5.1$.

$$C''(x) = (1.4)(x^2 + 25)^{-1/2} + (1.4)x\left(-\frac{1}{2}\right)(x^2 + 25)^{-3/2}2x$$

$$= \frac{1.4}{(x^2 + 25)^{1/2}} - \frac{(1.4)x^2}{(x^2 + 25)^{3/2}} = \frac{35}{(x^2 + 25)^{3/2}}$$

$$C''(5.1) = \frac{35}{[(5.1)^2 + 25]^{3/2}} > 0$$

Thus, the cost will be a minimum when $x = 5.1$.

Note that: $C(0) = (1.4)\sqrt{25} + 10 = 17$

$\quad\quad\quad\quad C(5.1) = (1.4)\sqrt{51.01} + (10 - 5.1) = 14.9$

$\quad\quad\quad\quad C(10) = (1.4)\sqrt{125} = 15.65$

Thus, the absolute minimum occurs when $x = 5.1$ miles.

(B) $C(x) = (1.1)\sqrt{x^2 + 25} + (1)(10 - x)$, $0 \le x \le 10$

$$C'(x) = \frac{(1.1)x - \sqrt{x^2 + 25}}{\sqrt{x^2 + 25}}$$

$C'(x) = 0$ when $1.1x - \sqrt{x^2 + 25} = 0$ or $(1.21)x^2 = x^2 + 25$

$$.21x^2 = 25$$
$$x^2 = \frac{25}{.21} = 119.05$$
$$x = \pm10.91$$

Critical value: $x = 10.91 > 10$, i.e., there are no critical values on the interval $[0, 10]$. Now,

$C(0) = (1.1)\sqrt{25} + 10 = 15.5$,
$C(10) = (1.1)\sqrt{125} \approx 12.30$.

Therefore, the absolute minimum occurs when $x = 10$ miles.

47. $C(t) = 30t^2 - 240t + 500$, $0 \le t \le 8$
$C'(t) = 60t - 240$; $t = 4$ is the only critical value.
$C''(t) = 60$
$C''(4) = 60 > 0$
Now, $C(0) = 500$
$C(4) = 30(4)^2 - 240(4) + 500 = 20$,
$C(8) = 30(8)^2 - 240(8) + 500 = 500$.
Thus, 4 days after a treatment, the concentration will be minimum; the minimum concentration is 20 bacteria per cm^3.

49. Let $x = $ the number of mice ordered in each order. Then the number of orders $= \dfrac{500}{x}$.

$C(x) = $ Cost $= \dfrac{x}{2} \cdot 4 + \dfrac{500}{x}(10)$ [Note: Cost = cost of feeding + cost of order, $\dfrac{x}{2}$ is the average number of mice at any one time.]

$C(x) = 2x + \dfrac{5000}{x}$, $0 < x \le 500$

$C'(x) = 2 - \dfrac{5000}{x^2} = \dfrac{2x^2 - 5000}{x^2} = \dfrac{2(x^2 - 2500)}{x^2} = \dfrac{2(x + 50)(x - 50)}{x^2}$

Critical value: $x = 50$ (-50 is not a critical value, since the domain of C is $x > 0$.

$C''(x) = \dfrac{10,000}{x^3}$ and $C''(50) = \dfrac{10,000}{50^3} > 0$

Therefore, the minimum cost occurs when 50 mice are ordered each time. The total number of orders is $\dfrac{500}{50} = 10$.

51. $H(t) = 4t^{1/2} - 2t$, $0 \le t \le 2$
$H'(t) = 2t^{-1/2} - 2$
Thus, $t = 1$ is the only critical value.
Now, $H(0) = 4 \cdot 0^{1/2} - 2(0) = 0$,
$H(1) = 4 \cdot 1^{1/2} - 2(1) = 2$,
$H(2) = 4 \cdot 2^{1/2} - 4 \approx 1.66$.
Therefore, $H(1)$ is the absolute maximum, and after one month the maximum height will be 2 feet.

53. $N(t) = 30 + 12t^2 - t^3$, $0 \leq t \leq 8$
The rate of increase $= R(t) = N'(t) = 24t - 3t^2$, and
$R'(t) = N''(t) = 24 - 6t$.
Thus, $t = 4$ is the only critical value of $R(t)$.
Now, $R(0) = 0$,
$$R(4) = 24 \cdot 4 - 3 \cdot 4^2 = 48,$$
$$R(8) = 24 \cdot 8 - 3 \cdot 8^2 = 0.$$
Therefore, the absolute maximum value of R occurs when $t = 4$; the maximum rate of increase will occur four years from now.

CHAPTER 4 REVIEW

1. The function f is increasing on (a, c_1), (c_3, c_6).   (4-2, 4-3)

2. $f'(x) < 0$ on (c_1, c_3), (c_6, b).   (4-2, 4-3)

3. The graph of f is concave downward on (a, c_2), (c_4, c_5), (c_7, b).
    (4-2, 4-3)

4. A local minimum occurs at $x = c_3$.   (4-2)

5. The absolute maximum occurs at $x = c_6$.   (4-5)

6. $f'(x)$ appears to be zero at $x = c_1$, c_3, c_5.   (4-2)

7. $f'(x)$ does not exist at $x = c_6$.   (4-2)

8. $x = c_2$, c_4, c_5, c_7 are inflection points.   (4-3)

9. (A) From the graph, $\lim\limits_{x \to 1} f(x)$ does not exist since
    $\lim\limits_{x \to 1^-} f(x) = 2 \neq \lim\limits_{x \to 1^+} f(x) = 3$.

    (B) $f(1) = 3$
    (C) f is NOT continuous at $x = 1$, since $\lim\limits_{x \to 1} f(x)$ does not exist.   (4-1)

10. (A) $\lim\limits_{x \to 2} f(x) = 2$ (B) $f(2)$ is not defined

    (C) f is NOT continous at $x = 2$ since $f(2)$ is not defined.   (4-1)

11. (A) $\lim\limits_{x \to 3} f(x) = 1$ (B) $f(3) = 1$

    (C) f is continous at $x = 3$ since $\lim\limits_{x \to 3} f(x) = f(3)$.   (4-1)

12. $f''(x)$ ---- 0 +++ 0 ----- ND ----

                          -1   0   2

Graph of f Concave | Concave | Concave | Concave
              Downward | Upward | Downward | Downward

              Point of   Point of
              Inflection   Inflection

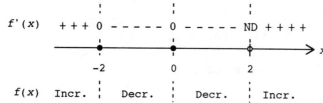

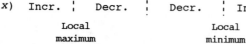

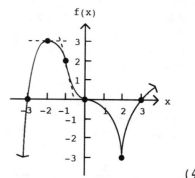

Using this information together with the points $(-3, 0)$, $(-2, 3)$, $(-1, 2)$, $(0, 0)$, $(2, -3)$, $(3, 0)$ on the graph, we have

(4-4)

13. Domain: all real numbers
Intercepts: y intercept: $f(0) = 0$
$\quad\quad\quad\quad\quad$ x-intercepts: $x = 0$
Asymptotes: Horizontal asymptote: $y = 2$
$\quad\quad\quad\quad\quad$ no vertical asymptotes
Critical values: $x = 0$

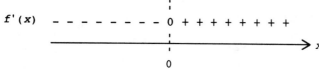

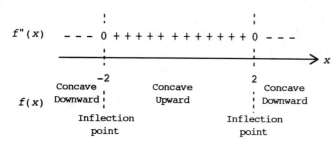

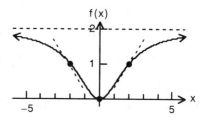

(4-4)

14. $f(x) = x^4 + 5x^3$

$f'(x) = 4x^3 + 15x^2$

$f''(x) = 12x^2 + 30x$ $\quad$ (4-3)

15. $y = 3x + \dfrac{4}{x}$

$y' = 3 - \dfrac{4}{x^2}$

$y'' = \dfrac{8}{x^3}$ $\quad$ (4-3)

16. From the graph:
(A) $\lim\limits_{x \to 2^-} f(x) = 4$
$\quad\quad$ (B) $\lim\limits_{x \to 2^+} f(x) = 6$

(C) $\lim\limits_{x \to 2} f(x)$ does not exist since $\lim\limits_{x \to 2^-} f(x) \neq \lim\limits_{x \to 2^+} f(x)$

(D) $f(2) = 6$
$\quad\quad\quad\quad$ (E) No, since $\lim\limits_{x \to 2} f(x)$ does not exist. $\quad$ (4-1)

17. From the graph:

(A) $\lim\limits_{x \to 5^-} f(x) = 3$ (B) $\lim\limits_{x \to 5^+} f(x) = 3$ (C) $\lim\limits_{x \to 5} f(x) = 3$ (D) $f(5) = 3$

(E) Yes, since $\lim\limits_{x \to 5} f(x) = f(5) = 3$.   (4-1)

18. $x^2 - x < 12$ or $x^2 - x - 12 < 0$

Let $f(x) = x^2 - x - 12 = (x + 3)(x - 4)$. Then f is continuous for all x and $f(-3) = f(4) = 0$. Thus, $x = -3$ and $x = 4$ are partition numbers.

$f(x)$ + + + + + | - - - - - - | + + + +

-4 -3   0   4 5

Test Numbers

x	$f(x)$
-4	8 (+)
0	-12 (−)
5	8 (+)

Thus, $x^2 - x < 12$ for: $-3 < x < 4$ or $(-3, 4)$.   (4-1)

19. $\dfrac{x - 5}{x^2 + 3x} > 0$ or $\dfrac{x - 5}{x(x + 3)} > 0$

Let $f(x) = \dfrac{x - 5}{x(x + 3)}$. Then f is discontinuous at $x = 0$ and $x = -3$, and $f(5) = 0$. Thus, $x = -3$, $x = 0$, and $x = 5$ are partition numbers.

$f(x)$ - - - | + + | - - - - | + + +

-4 -3 -1 0 1   5 6

Test Numbers

x	$f(x)$
-4	$-\frac{9}{4}$ (−)
-1	3 (+)
1	-1 (−)
6	$\frac{1}{54}$ (+)

Thus, $\dfrac{x - 5}{x^2 + 3x} > 0$ for $-3 < x < 0$ or $x > 5$

or $(-3, 0) \cup (5, \infty)$.   (4-1)

20. $x^3 + x^2 - 4x - 2 > 0$

Let $f(x) = x^3 + x^2 - 4x - 2$. The f is continuous for all x and $f(x) = 0$ at $x \approx -2.34$, -0.47 and 1.81.

$f(x)$ - - - 0 + + + 0 - - - - 0 + + +

-2.34 -0.47 0   1.81

Thus, $x^3 + x^2 - 4x - 2 > 0$ for $-2.34 < x < -0.47$ or $1.81 < x < \infty$,
or $(-2.34, -0.47) \cup (1.81, \infty)$.   (4-1)

21. $f(x) = x^3 - 18x^2 + 81x$

(A) Domain: f is defined for all real numbers.

(B) Intercepts: y intercept: $f(0) = 0^3 - 18(0)^2 + 81(0) = 0$

x intercepts: $x^3 - 18x^2 + 81x = 0$

$x(x^2 - 18x + 81) = 0$

$x(x - 9)^2 = 0$

$x = 0, 9$

(C) Since f is a polynomial, there are no horizontal or vertical asymptotes.

(4-4)

22. $f(x) = x^3 - 18x^2 + 81x$
$f'(x) = 3x^2 - 36x + 81 = 3(x^2 - 12x + 27) = 3(x - 3)(x - 9)$

(A) Critical values: $x = 3$, $x = 9$

(B) Partition numbers: $x = 3$, $x = 9$

(C) Sign chart for f':

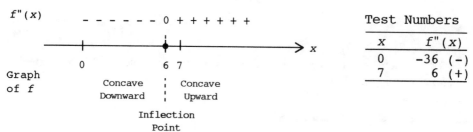

Test Numbers	
x	$f'(x)$
0	81 (+)
5	−24 (−)
10	21 (+)

Thus, f is increasing on $(-\infty, 3)$ and on $(9, \infty)$; f is decreasing on $(3, 9)$.

(D) There is a local maximum at $x = 3$ and a local minimum at $x = 9$. (4-4)

23. $f'(x) = 3x^2 - 36x + 81$
$f''(x) = 6x - 36 = 6(x - 6)$

Thus, $x = 6$ is a partition number for f''.

(A) Sign chart for f'':

Test Numbers	
x	$f''(x)$
0	−36 (−)
7	6 (+)

Thus, the graph of f is concave downward on $(-\infty, 6)$ and concave upward on $(6, \infty)$.

(B) The point $x = 6$ is an inflection point.

(4-4)

24. The graph of f:

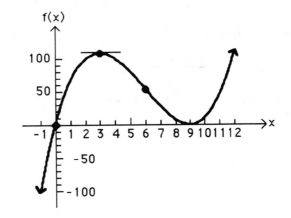

(4-4)

25. $f(x) = \dfrac{3x}{x + 2}$

(A) The domain of f is all real numbers except $x = -2$.

(B) Intercepts: y intercept: $f(0) = \dfrac{3(0)}{0 + 2} = 0$

$\qquad\qquad$ x intercepts: $\dfrac{3x}{x + 2} = 0$

$\qquad\qquad\qquad\qquad\qquad$ $3x = 0$

$\qquad\qquad\qquad\qquad\qquad$ $x = 0$

(C) Asymptotes:

Horizontal asymptotes: $\dfrac{a_n x^n}{b_m x^m} = \dfrac{3x}{x} = 3$

Thus, the line $y = 3$ is a horizontal asymptote.

Vertical asymptote(s): The denominator is 0 at $x = -2$ and the numerator is nonzero at $x = -2$. Thus, the line $x = -2$ is a vertical asymptote. (4-4)

26. $f(x) = \dfrac{3x}{x + 2}$

$f'(x) = \dfrac{(x + 2)(3) - 3x(1)}{(x + 2)^2} = \dfrac{6}{(x + 2)^2}$

(A) Critical values: $f'(x) = \dfrac{6}{(x + 2)^2} \neq 0$ for all x ($x \neq -2$).

Thus, f does not have any critical values.

(B) Partition numbers: $x = -2$ is a partition number for f'.

(C) Sign chart for f':

	Test Numbers	
	x	$f'(x)$
	-3	6 (+)
	0	$\frac{3}{2}$ (+)

$f'(x)$ $\quad$ + + + ND + + + + + +

$\qquad\qquad\qquad$ -3 -2 -1 0 1 $\qquad\qquad$ x

$f(x)$ $\quad$ Increasing $\vdots$ Increasing

Thus, f is increasing on $(-\infty, -2)$ and on $(-2, \infty)$.

(D) $f'(x) > 0$ for all x ($x \neq -2$). Thus, from (C), f does not have any local extrema. (4-4)

27. $f'(x) = \dfrac{6}{(x + 2)^2} = 6(x + 2)^{-2}$

$f''(x) = -12(x + 2)^{-3} = \dfrac{-12}{(x + 2)^3}$

(A) Partition numbers for f'': $x = -2$
Sign chart for f'':

$f''(x)$ + + + ND - - - -

Graph of f Concave Upward   Concave Downward

Test Numbers

x	$f''(x)$
-3	12 (+)
0	$-\frac{3}{2}$ (-)

The graph of f is concave upward on $(-\infty, -2)$ and concave downward on $(-2, \infty)$.

(B) The graph of f does not have any inflection points.   (4-4)

28. The graph of f is:

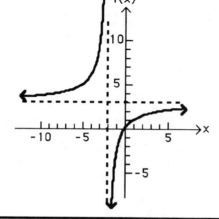

(4-4)

29.

x	$f'(x)$	$f(x)$
$-\infty < x < -2$	Negative and increasing	Decreasing and concave upward
$x = -2$	x intercept	Local minimum
$-2 < x < -1$	Positive and increasing	Increasing and concave upward
$x = -1$	Local maximum	Inflection point
$-1 < x < 1$	Positive and decreasing	Increasing and concave downward
$x = 1$	Local minimum	Inflection point
$1 < x < \infty$	Positive and increasing	Increasing and concave upward

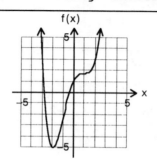

(4-3)

30. The graph in (C) could be the graph of $y = f''(x)$. (4-3)

31. $f(x) = x^3 - 6x^2 - 15x + 12$
$\quad f'(x) = 3x^2 - 12x - 15$
$\quad\; 3x^2 - 12x - 15 = 0$
$\quad\; 3(x^2 - 4x - 5) = 0$
$\quad\; 3(x - 5)(x + 1) = 0$
$\quad$ Thus, $x = -1$ and $x = 5$ are critical values of f.
$\qquad f''(x) = 6x - 12$
$\quad$ Now, $f''(-1) = 6(-1) - 12 = -18 < 0$. Thus, f has a local maximum at $x = -1$.
$\quad$ Also, $f''(5) = 6(5) - 12 = 18 > 0$ and f has a local minimum at $x = 5$.
(4-3)

32. $y = f(x) = x^3 - 12x + 12,\; -3 \le x \le 5$
$\quad f'(x) = 3x^2 - 12$
$\quad$ Critical values: f' is defined for all x:
$\quad f'(x) = 3x^2 - 12 = 0$
$\qquad\quad 3(x^2 - 4) = 0$
$\quad 3(x - 2)(x + 2) = 0$
$\quad$ Thus, the critical values of f are: $x = -2,\; x = 2$.
$\quad f(-3) = (-3)^3 - 12(-3) + 12 = 21$
$\quad f(-2) = (-2)^3 - 12(-2) + 12 = 28$
$\quad\; f(2) = 2^3 - 12(2) + 12 = -4$ Absolute minimum
$\quad\; f(5) = 5^3 - 12(5) + 12 = 77$ Absolute maximum
(4-5)

33. $y = f(x) = x^2 + \dfrac{16}{x^2},\; x > 0$

$\quad f'(x) = 2x - \dfrac{32}{x^3} = \dfrac{2x^4 - 32}{x^3} = \dfrac{2(x^4 - 16)}{x^3} = \dfrac{2(x - 2)(x + 2)(x^2 + 4)}{x^3}$

$\quad f''(x) = 2 + \dfrac{96}{x^4}$

$\quad$ The only critical value of f in the interval $(0, \infty)$ is $x = 2$.
$\quad$ Since

$\quad f''(2) = 2 + \dfrac{96}{2^4} = 8 > 0,$

$\quad f(2) = 8$ is the absolute minimum of f on $(0, \infty)$. (4-5)

34. $f(x) = 2x^2 - 3x + 1$. Since f is a polynomial function, $\;$ is continuous for all x, i.e., f is continuous on $(-\infty, \infty)$. (4-1)

35. $f(x) = \dfrac{1}{x + 5}$. Since f is a rational function, f is continuous for all x such that the denominator $x + 5 \ne 0$, i.e., for all x such that $x \ne -5$. Thus, f is continuous on $(-\infty, -5)$ and on $(-5, \infty)$. (4-1)

36. $f(x) = \dfrac{x - 3}{x^2 - x - 6}$

Since f is a rational function, f is continuous for all x such that the denominator $x^2 - x - 6 \neq 0$. Now $x^2 - x - 6 = (x - 3)(x + 2) = 0$ for $x = -2$ and $x = 3$. Thus, f is continuous on $(-\infty, -2)$, $(-2, 3)$, and $(3, \infty)$.

(4-1)

37. $f(x) = \sqrt{x - 3}$. f is continuous for all x such that $x - 3 \geq 0$, or $x \geq 3$. Thus, f is continuous on $[3, \infty)$.

(4-1)

38. $f(x) = \sqrt[3]{1 - x^2}$. f is continuous for all x such that $g(x) = 1 - x^2$ is continuous. Since g is a polynomial, g is continuous for all x. Thus, f is continuous on $(-\infty, \infty)$.

(4-1)

39. $\displaystyle\lim_{x \to 2^-} \dfrac{x}{2 - x} = \infty$ (4-1) **40.** $\displaystyle\lim_{x \to 2^+} \dfrac{x}{2 - x} = -\infty$ (4-1)

41. $\displaystyle\lim_{x \to 2} \dfrac{x}{2 - x}$ does not exist (4-1) **42.** $\displaystyle\lim_{x \to 2} \dfrac{x}{(2 - x)^2} = \infty$ (4-1)

43. $\displaystyle\lim_{x \to \infty} (-3x^5) = -\infty$ (4-4) **44.** $\displaystyle\lim_{x \to -\infty} (3x^2 - 4x^3) = \lim_{x \to -\infty} (-4x^3) = \infty$ (4-4)

45. $\displaystyle\lim_{x \to \infty} (4x^6 - 2x^5) = \lim_{x \to \infty} 4x^6 = \infty$

(4-4)

46. $\displaystyle\lim_{x \to \infty} \dfrac{6x^3 + 4x^2 + 5}{3x^3 + 2x + 7} = \lim_{x \to \infty} \dfrac{6x^3}{3x^3} = \lim_{x \to \infty} 2 = 2$

(4-4)

47. $\displaystyle\lim_{x \to \infty} \dfrac{6x^4 + 4x^2 + 5}{3x^3 + 2x + 7} = \lim_{x \to \infty} \dfrac{6x^4}{3x^3} = \lim_{x \to \infty} 2x = \infty$

(4-4)

48. $\displaystyle\lim_{x \to \infty} \dfrac{6x^3 + 4x^2 + 5}{3x^4 + 2x + 7} = \lim_{x \to \infty} \dfrac{6x^3}{3x^4} = \lim_{x \to \infty} \dfrac{2}{x} = 0$

(4-4)

49. $f(x) = \dfrac{x}{x^2 + 9}$

$\displaystyle\lim_{x \to \infty} f(x) = \lim_{x \to \infty} \dfrac{x}{x^2 + 9} = 0$

Thus, the line $y = 0$, or the x axis, is a horizontal asymptote.
Since $x^2 + 9 \neq 0$ for all x, there are no vertical asymptotes. (4-4)

50. $f(x) = \dfrac{x^3}{x^2 - 9}$

$\displaystyle\lim_{x \to \infty} f(x) = \lim_{x \to \infty} \dfrac{x^3}{x^2} = \lim_{x \to \infty} x = \infty$

Thus, there are no horizontal asymptotes.
Since the denominator $x^2 - 9 = (x + 3)(x - 3) = 0$ when $x = -3$ and when $x = 3$, and since the numerator $x^3 \neq 0$ at these values, the lines $x = -3$ and $x = 3$ are vertical asymptotes.

(4-4)

51. Yes. Consider f on the interval $[a, b]$. Since f is a polynomial, f is continuous on $[a, b]$. Therefore, f has an absolute maximum on $[a, b]$. Since f has a local minimum at $x = a$ and $x = b$, the absolute maximum of f on $[a, b]$ must occur at some point c in (a, b); f has a local maximum at $x = c$.
(4-5)

52. No, increasing/decreasing properties are stated in terms of intervals in the domain of f. A correct statement is: $f(x)$ is decreasing on $(-\infty, 0)$ and $(0, \infty)$.
(4-2)

53. A critical value for $f(x)$ is a partition number for $f'(x)$ that is also in the domain of f. However, $f'(x)$ may have partition numbers that are not in the domain of f and hence are not critical values for $f(x)$. For example, let $f(x) = \dfrac{1}{x}$. Then $f'(x) = -\dfrac{1}{x^2}$ and 0 is a partition number for $f'(x)$, but 0 is NOT a critical value for $f(x)$ since it is not in the domain of f.
(4-2)

54. $f(x) = 6x^2 - x^3 + 8,\ 0 \le x \le 4$
$f'(x) = 12x - 3x^2$
$f''(x) = 12 - 6x$

Now, $f''(x)$ is defined for all x and $f''(x) = 12 - 6x = 0$ implies $x = 2$. Thus, f' has a critical value at $x = 2$. Since this is the only critical value of f' and $(f'(x))'' = f'''(x) = -6$ so that $f'''(2) = -6 < 0$, it follows that $f'(2) = 12$ is the absolute maximum of f'. The graph is shown at the right.

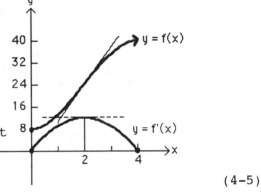

(4-5)

55. Let $x > 0$ be one of the numbers. Then $\dfrac{400}{x}$ is the other number. Now, we have:

$S(x) = x + \dfrac{400}{x},\ x > 0,$

$S'(x) = 1 - \dfrac{400}{x^2} = \dfrac{x^2 - 400}{x^2} = \dfrac{(x - 20)(x + 20)}{x^2}$

Thus, $x = 20$ is the only critical value of S on $(0, \infty)$.

$S''(x) = \dfrac{800}{x^3}$ and $S''(20) = \dfrac{800}{8000} = \dfrac{1}{10} > 0$

Therefore, $S(20) = 20 + \dfrac{400}{20} = 40$ is the absolute minimum sum, and this occurs when each number is 20.
(4-5)

56. $f(x) = (x - 1)^3(x + 3)$

<u>Step 1. Analyze $f(x)$</u>:
(A) Domain: All real numbers.
(B) Intercepts: y intercept: $f(0) = (-1)^3(3) = -3$
$\qquad\qquad\qquad x$ intercepts: $(x - 1)^3(x + 3) = 0$
$\qquad\qquad\qquad\qquad\qquad\qquad\qquad x = 1, -3$

(C) Asymptotes: Since f is a polynomial (of degree 4), the graph of f has no asymptotes.

Step 2. Analyze $f'(x)$:
$$f'(x) = (x - 1)^3(1) + (x + 3)(3)(x - 1)^2(1)$$
$$= (x - 1)^2[(x - 1) + 3(x + 3)]$$
$$= 4(x - 1)^2(x + 2)$$
Critical values: $x = -2$, $x = 1$
Partition numbers: $x = -2$, $x = 1$
Sign chart for f':

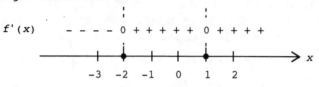

f'(x)   – – – – 0 + + + + + 0 + + + +

$f(x)$ Decreasing ¦ Increasing ¦ Increasing

Test Numbers

x	$f'(x)$
-3	-64 (−)
0	8 (+)
2	16 (+)

Thus, f is decreasing on $(-\infty, -2)$; f is increasing on $(-2, 1)$ and $(1, \infty)$; f has a local minimum at $x = -2$.

Step 3. Analyze $f''(x)$:
$$f''(x) = 4(x - 1)^2(1) + 4(x + 2)(2)(x - 1)(1)$$
$$= 4(x - 1)[(x - 1) + 2(x + 2)]$$
$$= 12(x - 1)(x + 1)$$
Partition numbers for f'': $x = -1$, $x = 1$.
Sign chart for f'':

f''(x)   + + + + 0 – – – – – 0 + + + +

Graph
of f Concave ¦ Concave ¦ Concave
           Upward ¦ Downward ¦ Upward

Test Numbers

x	$f''(x)$
-2	36 (+)
0	-12 (−)
2	36 (+)

Thus, the graph of f is concave upward on $(-\infty, -1)$ and on $(1, \infty)$; the graph of f is concave downward on $(-1, 1)$; the graph has inflection points at $x = -1$ and at $x = 1$.

Step 4. Sketch the graph of f:

x	$f(x)$
-2	-27
0	-3
1	0

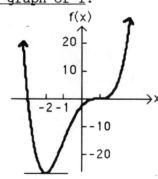

(4-4)

57. $f(x) = x^4 + x^3 + 4x^2 - 3x + 4$.
　　Step 1. Analyze $f(x)$:
　　(A) Domain: All real numbers (f is a polynomial function)
　　(B) Intercepts: y-intercept: $f(0) = 4$
　　　　　　　　　　x-intercepts: $x \approx 0.79, 1.64$
　　(C) Asymptotes: Since f is a polynomial function (of degree 4), the
　　　　graph of f has no asymptotes.

　　Step 2. Analyze $f'(x)$:
　　　　$f'(x) = 4x^3 + 3x^2 - 8x - 3$
　　Critical values: $x \approx -1.68, -0.35, 1.28$;
　　f is increasing on $(-1.68, -0.35)$ and $(1.28, \infty)$; f is decreasing on
　　$(-\infty, -1.68)$ and $(-0.35, 1.28)$. f has local minima at $x = -1.68$ and
　　$x = 1.28$. f has a local maximum at $x = -0.35$.

　　Step 3. Analyze $f''(x)$:
　　　　$f''(x) = 12x^2 + 6x - 8$
　　The graph of f is concave downward on $(-1.10, 0.60)$; the graph of f is
　　concave upward on $(-\infty, -1.10)$ and $(0.60, \infty)$; the graph has inflection
　　points at $x \approx -1.10$ and 0.60.　　　　　　　　　　　　　　　(4-4)

58. (A)

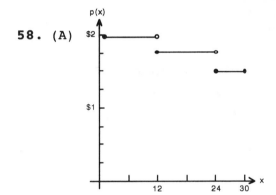

(B) p is discontinuous at $x = 12$ and
　　$x = 24$. In each case, the limit from
　　the left is greater then the limit
　　from the right, reflecting the
　　corresponding drop in price at these
　　order quantities.

(C)

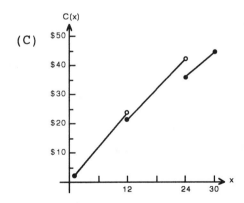

(D) C is discontinuous at $x = 12$ and
　　$x = 24$. In each case, the limit from
　　the left is greater than the limit
　　from the right, reflecting savings to
　　the customer due to the corresponding
　　drop in price at these order
　　quantities.　　　　　　　　　　　　(4-1)

59. (A) For the first 15 months, the price is increasing and concave down,
　　with a local maximum at $t = 15$. For the next 15 months, the price is
　　decreasing and concave down, with an inflection point at $t = 30$.
　　For the next 15 months, the price is decreasing and concave up, with

a local minimum at $t = 45$. For the remaining 15 months, the price is increasing and concave up.

(B)

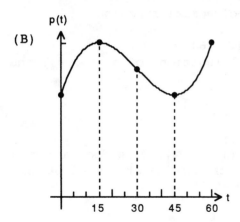

60. (A) $R(x) = xp(x) = 500x - 0.025x^2$, $0 \le x \le 20,000$

$R'(x) = 500 - 0.05x$; $500 - 0.05x = 0$

$$x = 10,000$$

Thus, $x = 10,000$ is a critical value.

Now, $R(0) = 0$

$R(10,000) = 2,500,000$

$R(20,000) = 0$

Thus, $R(10,000) = \$2,500,000$ is the absolute maximum of R.

(B) $P(x) = R(x) - C(x) = 500x - 0.025x^2 - (350x + 50,000)$

$= 150x - 0.025x^2 - 50,000$, $0 \le x \le 20,000$

$P'(x) = 150 - 0.05x$; $150 - 0.05x = 0$

$$x = 3,000$$

Now, $P(0) = -50,000$

$P(3,000) = 175,000$

$P(20,000) = -7,050,000$

Thus, the maximum profit is $175,000 when 3000 stoves are manufactured and sold at $p(3,000) = \$425$ each.

(C) If the government taxes the company $20 per stove, then the cost equation is:

$C(x) = 370x + 50,000$

and

$P(x) = 500x - 0.025x^2 - (370x + 50,000)$

$= 130x - 0.025x^2 - 50,000$, $0 \le x \le 20,000$

$P'(x) = 130 - 0.05x$; $130 - 0.05x = 0$

$$x = 2,600$$

The maximum profit is $P(2,600) = \$119,000$ when 2,600 stoves are produced and sold for $p(2,600) = \$435$ each. (4-5)

61.

$5/ft (top)
$5/ft (left)
$15/ft (right)
$5/ft (bottom)

Let x be the length and y the width of the rectangle.

(A) $C(x, y) = 5x + 5x + 5y + 15y = 10x + 20y$

Also, Area $A = xy = 5000$, so $y = \dfrac{5000}{x}$

and $C(x) = 10x + \dfrac{100,000}{x}$, $x \geq 0$

Now, $C'(x) = 10 - \dfrac{100,000}{x^2}$ and

$10 - \dfrac{100,000}{x^2} = 0$ implies $10x^2 = 100,000$

$$x^2 = 10,000$$
$$x = \pm 100$$

Thus, $x = 100$ is the critical value.

Now, $C''(x) = \dfrac{200,000}{x^3}$ and $C''(100) = \dfrac{200,000}{1,000,000} = 0.2 > 0$

and the most economical (i.e. least cost) fence will have

dimensions: length $x = 100$ feet and width $y = \dfrac{5000}{100} = 50$ feet.

(B) We want to maximize $A = xy$ subject to

$C(x, y) = 10x + 20y = 3000$ or $x = 300 - 2y$

Thus, $A = y(300 - 2y) = 300y - 2y^2$, $0 \leq y \leq 150$.

Now, $A'(y) = 300 - 4y$ and

$300 - 4y = 0$ implies $y = 75$.

Therefore, $y = 75$ is the critical value.

Now, $A''(y) = -4$ and $A''(75) = -4 < 0$. Thus, A has an absolute maximum when $y = 75$. Therefore the dimensions of the rectangle that will enclose maximum area are:

length $x = 300 - 2(75) = 150$ feet and width $y = 75$ feet.   (4-5)

62. $C(x) = 4000 + 10x + 0.1x^2$, $x > 0$

Average cost $= \overline{C}(x) = \dfrac{4000}{x} + 10 + 0.1x$

Marginal cost $= C'(x) = 10 + \dfrac{2}{10}x = 10 + 0.2x$

The graph of $C'(x)$ is a straight line with slope $\dfrac{1}{5}$ and y intercept 10.

$\overline{C}'(x) = \dfrac{-4000}{x^2} + \dfrac{1}{10} = \dfrac{-40,000 + x^2}{10x^2} = \dfrac{(x + 200)(x - 200)}{10x^2}$

Thus, $\overline{C}'(x) < 0$ on $(0, 200)$ and $\overline{C}'(x) > 0$ on $(200, \infty)$. Therefore, $\overline{C}(x)$ is decreasing on $(0, 200)$, increasing on $(200, \infty)$, and a minimum occurs at $x = 200$.

$$\text{Min } \overline{C}(x) = \overline{C}(200) = \frac{4000}{200} + 10 + \frac{1}{10}(200) = 50$$

$$\overline{C}''(x) = \frac{8000}{x^3} > 0 \text{ on } (0, \infty).$$

Therefore, the graph of $\overline{C}(x)$ is concave upward on $(0, \infty)$.

Using this information and point-by-point plotting (use a calculator), the graphs of $C'(x)$ and $\overline{C}(x)$ are as shown in the diagram at the right.
The line $y = 0.1x + 10$ is an oblique asymptote for $y = \overline{C}(x)$.

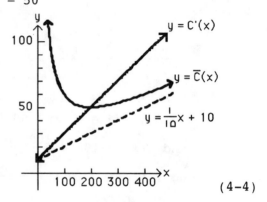

(4-4)

63. Let x = the number of dollars increase in the nightly rate, $x \geq 0$. Then $200 - 4x$ rooms will be rented at $(40 + x)$ dollars per room. [Note: Since $200 - 4x \geq 0$, $x \leq 50$.] The cost of service for $200 - 4x$ rooms at $8 per room is $8(200 - 4x)$. Thus:

$$\begin{aligned}\text{Gross profit: } P(x) &= (200 - 4x)(40 + x) - 8(200 - 4x) \\ &= (200 - 4x)(32 + x) \\ &= 6400 + 72x - 4x^2, \ 0 \leq x \leq 50\end{aligned}$$

$P'(x) = 72 - 8x$
Critical value: $72 - 8x = 0$
$$x = 9$$

Now, $P(0) = 6400$
$\quad\quad P(9) = 6724$ Absolute maximum
$\quad\quad P(50) = 0$

Thus, the maximum gross profit is $6724 and this occurs at $x = 9$, i.e., the rooms should be rented at $49 per night. (4-5)

64. Let x = number of times the company should order. Then, the number of discs per order = $\frac{7200}{x}$. The average number of unsold discs is given by:

$$\frac{7200}{2x} = \frac{3600}{x}$$

Total cost: $C(x) = 5x + 0.2\left(\frac{3600}{x}\right)$, $x > 0$

$$C(x) = 5x + \frac{720}{x}$$

$$C'(x) = 5 - \frac{720}{x^2} = \frac{5x^2 - 720}{x^2} = \frac{5(x^2 - 144)}{x^2}$$
$$= \frac{5(x + 12)(x - 12)}{x^2}$$

Critical value: $x = 12$ [Note: $x > 0$, so $x = -12$ is not a critical value.]

$$C''(x) = \frac{1440}{x^3} \text{ and } C''(12) = \frac{1440}{12^3} > 0$$

Therefore, $C(x)$ is a minimum when $x = 12$. (4-5)

65. $C(t) = 20t^2 - 120t + 800,\ 0 \le t \le 9$
$C'(t) = 40t - 120 = 40(t - 3)$
Critical value: $t = 3$
$C''(t) = 40$ and $C''(3) = 40 > 0$
Therefore, a local minimum occurs at $t = 3$.
$C(3) = 20(3^2) - 120(3) + 800 = 620$ Absolute minimum
$C(0) = 800$
$C(9) = 20(81) - 120(9) + 800 = 1340$
Therefore, the bacteria count will be at a minimum three days after a treatment. $\hfill$ (4-2)

66. $N = 10 + 6t^2 - t^3,\ 0 \le t \le 5$
$\dfrac{dN}{dt} = 12t - 3t^2$
Now, find the critical values of the rate function $R(t)$:
$R(t) = \dfrac{dN}{dt} = 12t - 3t^2$
$R'(t) = \dfrac{dR}{dt} = \dfrac{d^2N}{dt^2} = 12 - 6t$
Critical value: $t = 2$
$R''(t) = -6$ and $R''(2) = -6 < 0$
$R(0) = 0$
$R(2) = 12$ Absolute maximum
$R(5) = -15$
Therefore, $R(t)$ has an absolute maximum at $t = 2$. The rate of increase will be a maximum after two years. $\hfill$ (4-2)

5 ADDITIONAL DERIVATIVE TOPICS

Things to remember:

1. THE NUMBER e

    The irrational number e is defined by
    $$e = \lim_{n \to \infty} \left(1 + \frac{1}{n}\right)^n$$
    or alternatively,
    $$e = \lim_{s \to 0} (1 + s)^{1/s}$$
    $$e = 2.7182818\ldots$$

2. CONTINUOUS COMPOUND INTEREST

    $$A = Pe^{rt}$$
    where P = Principal
    $\quad\quad\quad\ r$ = Annual nominal interest rate compounded continuously
    $\quad\quad\quad\ t$ = Time in years
    $\quad\quad\quad\ A$ = Amount at time t

1. $A = \$1000e^{0.1t}$
    When $t = 2$, $A = \$1000e^{(0.1)2} = \$1000e^{0.2} = \$1221.40$.
    When $t = 5$, $A = \$1000e^{(0.1)5} = \$1000e^{0.5} = \$1628.72$.
    When $t = 8$, $A = \$1000e^{(0.1)8} = \$1000e^{0.8} = \$2225.54$

3. $2 = e^{0.06t}$
    Take the natural log of both sides of this equation
    $\ln(e^{0.06t}) = \ln 2$
    $0.06t \ln e = \ln 2$
    $\quad\ 0.06t = \ln 2 \quad\quad (\ln e = 1)$
    $\quad\quad\quad t = \dfrac{\ln 2}{0.06} \approx 11.55$

5. $3 = e^{0.1t}$
    $\ln(e^{0.1t}) = \ln 3$
    $\quad 0.1t = \ln 3$
    $\quad\quad t = \dfrac{\ln 3}{0.1} \approx 10.99$

7. $2 = e^{5r}$
    $\ln(e^{5r}) = \ln 2$
    $\quad 5r = \ln 2$
    $\quad\ r = \dfrac{\ln 2}{5} \approx 0.14$

9.

n	$\left(1 + \frac{1}{n}\right)^n$
10	2.59374
100	2.70481
1000	2.71692
10,000	2.71815
100,000	2.71827
1,000,000	2.71828
10,000,000	2.71828
$\downarrow$	$\downarrow$
∞	$e = 2.7182818\ldots$

11. The graphs of $y_1 = \left(1 + \frac{1}{n}\right)^n$,

$y_2 = 2.718281828 \approx e$, and

$y_3 = \left(1 + \frac{1}{n}\right)^{n+1}$ for $0 \le n \le 20$ are

given below:

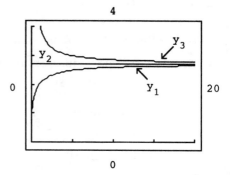

13. $A = Pe^{rt} = \$20,000e^{0.12(8.5)}$
$= \$20,000e^{1.02}$
$\approx \$55,463.90$

15. $A = Pe^{rt}$
$\$20,000 = Pe^{0.07(10)} = Pe^{0.7}$
Therefore,
$P = \dfrac{\$20,000}{e^{0.7}} = \$20,000e^{-0.7}$

$\approx \$9931.71$

17. $30,000 = 20,000e^{4r}$
$e^{4r} = 1.5$
$4r = \ln 1.5$
$r = \dfrac{\ln 1.5}{4} \approx 0.1014$ or 10.14%

19. $P = 10,000e^{-0.08t}$, $0 \le t \le 50$

(A)

t	0	10	20	30	40	50
P	10,000	4493.30	2019	907.18	407.62	183.16

The graph of P is shown below.

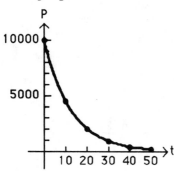

(B) $\lim\limits_{t\to\infty} 10,000e^{-0.08t} = 0$

21.
$$2P = Pe^{0.25t}$$
$$e^{0.25t} = 2$$
$$\ln(e^{0.25t}) = \ln 2$$
$$0.25t = \ln 2$$
$$t = \frac{\ln 2}{0.25} \approx 2.77 \text{ years}$$

23.
$$2P = Pe^{r(5)}$$
$$\ln(e^{5r}) = \ln 2$$
$$5r = \ln 2$$
$$r = \frac{\ln 2}{5} \approx .1386 \text{ or} = 13.86\%$$

25. The total investment in the two accounts is given by
$$A = 10{,}000e^{0.072t} + 10{,}000(1 + 0.084)^t$$
$$= 10{,}000[e^{0.072t} + (1.084)^t]$$

On a graphing utility, locate the intersection point of
$$y_1 = 10{,}000[e^{0.072x} + (1.084)^x]$$
and $y_2 = 35{,}000$.

The result is: $x = t \approx 7.3$ years.

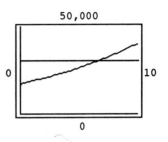

27. (A) $A = Pe^{rt}$; set $A = 2P$
$$2P = Pe^{rt}$$
$$e^{rt} = 2$$
$$rt = \ln 2$$
$$t = \frac{\ln 2}{r}$$

(B)

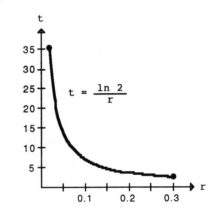

$$t = \frac{\ln 2}{r}$$

In theory, r could be any positive number. However, the restrictions on r are reasonable in the sense that most investments would be expected to earn between 2% and 30%.

(C) $r = 5\%$; $t = \dfrac{\ln 2}{0.05} \approx 13.86$ years

$\quad\ r = 10\%$; $t = \dfrac{\ln 2}{0.10} \approx 6.93$ years

$\quad\ r = 15\%$; $t = \dfrac{\ln 2}{0.15} \approx 4.62$ years

$\quad\ r = 20\%$; $t = \dfrac{\ln 2}{0.20} \approx 3.47$ years

$\quad\ r = 25\%$; $t = \dfrac{\ln 2}{0.25} \approx 2.77$ years

$\quad\ r = 30\%$; $t = \dfrac{\ln 2}{0.30} \approx 2.31$ years

29.
$$Q = Q_0 e^{-0.0004332t}$$
$$\tfrac{1}{2}Q_0 = Q_0 e^{-0.0004332t}$$
$$e^{-0.0004332t} = \tfrac{1}{2}$$
$$\ln(e^{-0.0004332t}) = \ln\!\left(\tfrac{1}{2}\right) = \ln 1 - \ln 2$$
$$-0.0004332t = -\ln 2 \quad (\ln 1 = 0)$$
$$t = \frac{\ln 2}{0.0004332}$$
$$\approx \frac{0.6931}{0.0004332} \approx 1599.95$$

Thus, the half-life of radium is approximately 1600 years.

31.
$$Q = Q_0 e^{rt} \quad (r < 0)$$
$$\tfrac{1}{2}Q_0 = Q_0 e^{r(30)}$$
$$e^{30r} = \tfrac{1}{2}$$
$$\ln(e^{30r}) = \ln\!\left(\tfrac{1}{2}\right) = \ln 1 - \ln 2$$
$$30r = -\ln 2 \quad (\ln 1 = 0)$$
$$r = \frac{-\ln 2}{30} \approx \frac{-0.6931}{30}$$
$$\approx -0.0231$$

Thus, the continuous compound rate of decay of the cesium isotope is approximately -0.0231.

33. $2P_0 = P_0 e^{0.02t}$ or $e^{0.02t} = 2$

Thus, $\ln(e^{0.02t}) = \ln 2$
and $\qquad 0.02t = \ln 2$.

Therefore, $t = \dfrac{\ln 2}{0.02} \approx 34.66$ years.

35.
$$2P_0 = P_0 e^{r(20)}$$
$$e^{20r} = 2$$
$$\ln(e^{20r}) = \ln 2$$
$$20r = \ln 2$$
$$r = \frac{\ln 2}{20} \approx 0.0347$$
$$\text{or } 3.47\%$$

37.
$$A = Pe^{rt}$$
$$1.68 \times 10^{14} = 5 \times 10^9 e^{0.02t}$$
$$e^{0.02t} = \frac{1.68 \times 10^{14}}{5 \times 10^9} = 33{,}600$$
$$\ln(e^{0.02t}) = \ln(33{,}600)$$
$$0.02t = \ln 33{,}600$$
$$t = \frac{\ln 33{,}600}{0.02} \approx 521.11$$

Thus, there will be one square yard of land per person in approximately 521 years.

Things to remember:

1. DERIVATIVES OF THE NATURAL LOGARITHMIC AND EXPONENTIAL FUNCTIONS

$$\frac{d}{dx} \ln\ x = \frac{1}{x} \qquad \frac{d}{dx} e^x = e^x$$

---

1. $f(x) = 6e^x - 7\ \ln\ x$

$f'(x) = 6\dfrac{d}{dx}e^x - 7\dfrac{d}{dx}\ \ln\ x = 6e^x - 7\!\left(\dfrac{1}{x}\right) = 6e^x - \dfrac{7}{x}$

3. $f(x) = 2x^e + 3e^x$

$f'(x) = 2\dfrac{d}{dx}x^e + 3\dfrac{d}{dx}e^x = 2ex^{e-1} + 3e^x$

[Note: $e \approx 2.71828$ is a constant and so we use the power rule on the first term.]

5. $f(x) = \ln\ x^5 = 5\ \ln\ x$ (Property of logarithms)

$f'(x) = 5\dfrac{d}{dx}\ \ln\ x = 5\!\left(\dfrac{1}{x}\right) = \dfrac{5}{x}$

7. $f(x) = (\ln\ x)^2$

$f'(x) = 2(\ln\ x)\dfrac{d}{dx}\ \ln\ x$ (Power rule for functions)

$\qquad = 2(\ln\ x)\dfrac{1}{x} = \dfrac{2\ \ln\ x}{x}$

9. $f(x) = x^4\ \ln\ x$

$f'(x) = x^4\dfrac{d}{dx}\ \ln\ x + \ln\ x\ \dfrac{d}{dx}x^4$ (Product rule)

$\qquad = x^4\!\left(\dfrac{1}{x}\right) + (\ln\ x)4x^3 = x^3 + 4x^3\ \ln\ x = x^3(1 + 4\ \ln\ x)$

11. $f(x) = x^3 e^x$

$f'(x) = x^3\ \dfrac{d}{dx}e^x + e^x\ \dfrac{d}{dx}x^3$ (Product rule)

$\qquad = x^3 e^x + e^x 3x^2 = x^2 e^x(x + 3)$

13. $f(x) = \dfrac{e^x}{x^2 + 9}$

$f'(x) = \dfrac{(x^2 + 9)\dfrac{d}{dx}e^x - e^x\dfrac{d}{dx}(x^2 + 9)}{(x^2 + 9)^2}$ (Quotient rule)

$\qquad = \dfrac{(x^2 + 9)e^x - e^x(2x)}{(x^2 + 9)^2} = \dfrac{e^x(x^2 - 2x + 9)}{(x^2 + 9)^2}$

15. $f(x) = \dfrac{\ln x}{x^4}$

$$f'(x) = \frac{x^4 \dfrac{d}{dx} \ln x - \ln x \dfrac{d}{dx} x^4}{(x^4)^2} \quad \text{(Quotient rule)}$$

$$= \frac{x^4\left(\dfrac{1}{x}\right) - (\ln x) 4x^3}{x^8} = \frac{x^3 - 4x^3 \ln x}{x^8} = \frac{1 - 4 \ln x}{x^5}$$

17. $f(x) = (x + 2)^3 \ln x$

$$f'(x) = (x + 2)^3 \frac{d}{dx} \ln x + (\ln x) \frac{d}{dx}(x + 2)^3$$

$$= (x + 2)^3\left(\frac{1}{x}\right) + (\ln x)[3(x + 2)^2(1)]$$

$$= 3(x + 2)^2 \ln x + \frac{(x + 2)^3}{x} = (x + 2)^2\left[3 \ln x + \frac{x + 2}{x}\right]$$

19. $f(x) = (x + 1)^3 e^x$

$$f'(x) = (x + 1)^3 \frac{d}{dx} e^x + e^x \frac{d}{dx}(x + 1)^3$$

$$= (x + 1)^3 e^x + e^x(3)(x + 1)^2(1)$$

$$= (x + 1)^2 e^x[x + 1 + 3] = (x + 1)^2(x + 4) e^x$$

21. $f(x) = \dfrac{x^2 + 1}{e^x}$

$$f'(x) = \frac{e^x \dfrac{d}{dx}(x^2 + 1) - (x^2 + 1)\dfrac{d}{dx} e^x}{(e^x)^2} = \frac{e^x(2x) - (x^2 + 1)e^x}{e^{2x}} = \frac{2x - x^2 - 1}{e^x}$$

23. $f(x) = x(\ln x)^3$

$$f'(x) = x \frac{d}{dx}(\ln x)^3 + (\ln x)^3 \frac{d}{dx} x$$

$$= x(3)(\ln x)^2\left(\frac{1}{x}\right) + (\ln x)^3(1) = (\ln x)^2[3 + \ln x]$$

25. $f(x) = (4 - 5e^x)^3$

$$f'(x) = 3(4 - 5e^x)^2(-5e^x) = -15e^x(4 - 5e^x)^2$$

27. $f(x) = \sqrt{1 + \ln x} = (1 + \ln x)^{1/2}$

$$f'(x) = \frac{1}{2}(1 + \ln x)^{-1/2}\left(\frac{1}{x}\right)$$

$$= \frac{1}{2x(1 + \ln x)^{1/2}} = \frac{1}{2x\sqrt{1 + \ln x}}$$

29. $f(x) = xe^x - e^x$

$$f'(x) = x\frac{d}{dx} e^x + e^x \frac{d}{dx} x - \frac{d}{dx} e^x = xe^x + e^x - e^x = xe^x$$

31. $f(x) = 2x^2 \ln x - x^2$

$f'(x) = 2x^2 \dfrac{d}{dx} \ln x + \ln x \dfrac{d}{dx} 2x^2 - \dfrac{d}{dx} x^2 = 2x^2\left(\dfrac{1}{x}\right) + 4x \ln x - 2x$

$$= 4x \ln x$$

33. $f(x) = e^x$

$f'(x) = \dfrac{d}{dx} e^x = e^x$

The tangent line at $x = 1$ has an equation of the form
$y - y_1 = m(x - x_1)$
where $x_1 = 1$, $y_1 = f(1) = e$, and $m = f'(1) = e$. Thus, we have:
$y - e = e(x - 1)$ or $y = ex$

35. $f(x) = \ln x$

$f'(x) = \dfrac{d}{dx} (\ln x) = \dfrac{1}{x}$

The tangent line at $x = e$ has an equation of the form
$y - y_1 = m(x - x_1)$

where $x_1 = e$, $y_1 = f(e) = \ln e = 1$, and $m = f'(e) = \dfrac{1}{e}$. Thus, we have:

$y - 1 = \dfrac{1}{e} (x - e)$ or $y = \dfrac{1}{e} x$

37. An equation for the tangent line to the graph of $f(x) = e^x$ at the point
$(3, f(3)) = (3, e^3)$ is:

$$y - e^3 = e^3 (x - 3)$$
or $\qquad y = xe^3 - 2e^3 = e^3 (x - 2)$

Clearly, $y = 0$ when $x = 2$, that is the tangent line passes through the
point $(2, 0)$.

In general, an equation for the tangent line to the graph of $f(x) = e^x$
at the point $(c, f(c)) = (c, e^c)$ is:

$$y - e^c = e^c (x - c)$$
or $\qquad y = e^c (x - [c - 1])$

Thus, the tangent line at the point (c, e^c) passes through $(c - 1, 0)$;
then tangent line at the point $(4, e^4)$ passes through $(3, 0)$.

39. $f(x) = 4x - x \ln x,\ x > 0$

$f'(x) = 4 - x\dfrac{d}{dx} \ln x - (\ln x)\dfrac{d}{dx} x$

$$= 4 - x\left(\dfrac{1}{x}\right) - (\ln x)(1) = 3 - \ln x,\ x > 0$$

Critical values: $f'(x) = 3 - \ln x = 0$
$$\ln x = 3$$
$$x = e^3$$

Thus, $x = e^3$ is the only critical value of f on $(0, \infty)$.

Now, $f''(x) = \dfrac{d}{dx}(3 - \ln x) = -\dfrac{1}{x}$

and $f''(e^3) = -\dfrac{1}{e^3} < 0$.

Therefore, f has a maximum value at $x = e^3$, and $f(e^3) = 4e^3 - e^3 \ln e^3$
$= 4e^3 - 3e^3 = e^3 \approx 20.086$ is the absolute maximum of f.

41. $f(x) = \dfrac{e^x}{x}$, $x > 0$

$$f'(x) = \dfrac{x\dfrac{d}{dx}e^x - e^x\dfrac{d}{dx}x}{x^2} = \dfrac{xe^x - e^x(1)}{x^2} = \dfrac{e^x(x-1)}{x^2}, \quad x > 0$$

Critical values: $f'(x) = \dfrac{e^x(x-1)}{x^2} = 0$

$$e^x(x-1) = 0$$
$$x = 1 \quad [\underline{\text{Note}}\colon e^x \neq 0 \text{ for all } x.]$$

Thus, $x = 1$ is the only critical value of f on $(0, \infty)$.
Sign chart for f' [$\underline{\text{Note}}$: This approach is a litle easier than calculating $f''(x)$]:

x	$f'(x)$
$\frac{1}{2}$	$-2\sqrt{e}$ (−)
2	$\frac{e^2}{4}$ (+)

By the first derivative test, f has a minimum value at $x = 1$;
$f(1) = \dfrac{e}{1} = e \approx 2.718$ is the absolute minimum of f.

43. $f(x) = \dfrac{1 + 2 \ln x}{x}$, $x > 0$

$$f'(x) = \dfrac{x\dfrac{d}{dx}(1 + 2\ln x) - (1 + 2\ln x)\dfrac{d}{dx}x}{x^2} = \dfrac{x\left(\dfrac{2}{x}\right) - (1 + 2\ln x)(1)}{x^2}$$

$$= \dfrac{1 - 2\ln x}{x^2}, \quad x > 0$$

Critical values: $f'(x) = \dfrac{1 - 2\ln x}{x^2} = 0$

$$1 - 2\ln x = 0$$
$$\ln x = \dfrac{1}{2}$$
$$x = e^{1/2} = \sqrt{e} \approx 1.65$$

Sign chart for f':

$f'(x)$ + + + + + 0 - - - - - →x

0   1 $\sqrt{e}$ 2   3

$f(x)$ Increasing Decreasing

Test Numbers	
x	$f'(x)$
1	1 (+)
e	$-\frac{1}{e^2}$ (−)

By the first derivative test, f has a maximum value at $x = \sqrt{e}$;

$$f(\sqrt{e}) = \frac{1 + 2 \ln \sqrt{e}}{\sqrt{e}} = \frac{1 + 2\left(\frac{1}{2}\right)}{\sqrt{e}} = \frac{2}{\sqrt{e}} \approx 1.213 \text{ is the absolute maximum of } f.$$

45. $f(x) = 1 - e^x$

Step 1. Analyze $f(x)$:
(A) Domain: All real numbers, $(-\infty, \infty)$.

(B) Intercepts: y intercept: $f(0) = 1 - e^0 = 0$
$\qquad\qquad\quad$ x intercept: $1 - e^x = 0$
$\qquad\qquad\qquad\qquad\qquad\quad e^x = 1$
$\qquad\qquad\qquad\qquad\qquad\quad\; x = 0$

(C) Asymptotes:
$\qquad$ <u>Horizontal asymptote</u>: $\lim\limits_{x \to -\infty} (1 - e^x) = 1 - \lim\limits_{x \to -\infty} e^x = 1 - 0 = 1$

$\qquad\qquad\qquad\qquad\qquad\qquad\qquad\qquad [\lim\limits_{x \to \infty}(1 - e^x) = -\infty]$

$\qquad\qquad\qquad\qquad$ Thus, $y = 1$ is a horizontal asymptote.
$\qquad$ <u>Vertical asymptotes</u>: There are no vertical asymptotes.

Step 2. Analyze $f'(x)$:
$f'(x) = -e^x$
Critical values: $f'(x) = -e^x$ is continuous and nonzero (negative) for
$\qquad\qquad\qquad\;$ all x; there are no critical values.
Partition numbers: There are no partition numbers. Since $f'(x) < 0$ for
$\qquad\qquad\qquad\qquad$ all x, f is decreasing on $(-\infty, \infty)$; f has no local
$\qquad\qquad\qquad\qquad$ extrema.

Step 3. Analyze $f''(x)$:
$f''(x) = -e^x < 0$ for all x
Thus, the graph of f is concave downward on $(-\infty, \infty)$.

Step 4. Sketch the graph of f:

x	$f(x)$
0	0
1	$1 - e \approx -1.718$
−1	$1 - \frac{1}{e}$

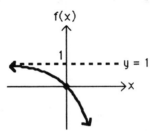

47. $f(x) = x - \ln x$

Step 1. Analyze $f(x)$:

(A) Domain: All positive real numbers, $(0, \infty)$.
[<u>Note</u>: $\ln x$ is defined only for positive numbers.]

(B) Intercepts: y intercept: There is no y intercept; $f(0) = 0 - \ln(0)$ is not defined.

x intercept: $x - \ln x = 0$

$$\ln x = x$$

Since the graph of $y = \ln x$ is below the graph of $y = x$, there are no solutions to this equation; there are no x intercepts.

(C) **Asymptotes:**
<u>Horizontal asymptote</u>: None
<u>Vertical asymptotes</u>: Since $\lim\limits_{x \to 0^+} \ln x = -\infty$, $\lim\limits_{x \to 0^+} (x - \ln x) = \infty$.

Thus, $x = 0$ is a vertical asymptote for $f(x) = x - \ln x$.

Step 2. Analyze $f'(x)$:

$$f'(x) = 1 - \frac{1}{x} = \frac{x - 1}{x}, \quad x > 0$$

Critical values: $\dfrac{x - 1}{x} = 0$

$$x = 1$$

Partition numbers: $x = 1$

Sign chart for $f'(x) = \dfrac{x - 1}{x}$:

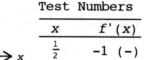

Test Numbers	
x	$f'(x)$
$\frac{1}{2}$	-1 $(-)$
2	$\frac{1}{2}$ $(+)$

Thus, f is decreasing on $(0, 1)$ and increasing on $(1, \infty)$; f has a local minimum at $x = 1$.

Step 3. Analyze $f''(x)$:

$$f''(x) = \frac{1}{x^2}, \quad x > 0$$

Thus, $f''(x) > 0$ and the graph of f is concave upward on $(0, \infty)$.

Step 4. Sketch the graph of f:

x	$f(x)$
0.1	≈ 2.4
1	1
10	≈ 7.7

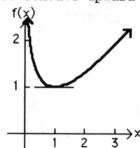

49. $f(x) = (3 - x)e^x$

Step 1. Analyze $f(x)$:

(A) Domain: All real numbers, $(-\infty, \infty)$.

(B) Intercepts: y intercept: $f(0) = (3 - 0)e^0 = 3$
x intercept: $(3 - x)e^x = 0$
$3 - x = 0$
$x = 3$

(C) Asymptotes:

Horizontal asymptote: Consider the behavior of f as $x \to \infty$ and as $x \to -\infty$.

Using the following tables,

x	-1	-10	-20
$f(x)$	1.47	0.00059	0.000000047

x	5	10
$f(x)$	-296.83	-154,185.26

we conclude that $\lim\limits_{x \to -\infty} f(x) = 0$ and $\lim\limits_{x \to \infty} f(x)$ does not exist. Because of the first limit, $y = 0$ is a horizontal asymptote.

Vertical asymptotes: There are no vertical asymptotes.

Step 2. Analyze $f'(x)$:

$f'(x) = (3 - x)e^x + e^x(-1) = (2 - x)e^x$
Critical values: $(2 - x)e^x = 0$
$x = 2$ [Note: $e^x > 0$]
Partition numbers: $x = 2$
Sign chart for f':

```
f'(x)    + + + + + 0 - - - - -
         ├──┼──●──┼──────────→ x
         0  1  2  3
f(x)     Increasing ┊ Decreasing
```

Test Numbers	
x	$f'(x)$
0	2 (+)
3	$-e^3$ (-)

Thus, f is increasing on $(-\infty, 2)$ and decreasing on $(2, \infty)$; f has a local maximum at $x = 2$.

Step 3. Analyze $f''(x)$:

$f''(x) = (2 - x)e^x + e^x(-1) = (1 - x)e^x$
Partition number for f'': $x = 1$
Sign chart for f'':

```
f"(x)      + + + + 0 - - - -
         ──┼──●──┼──────────→ x
           0  1  2
Graph     Concave ┊ Concave
of f      Upward  ┊ Downward
```

Test Numbers	
x	$f''(x)$
0	1 (+)
2	$-e^2$ (-)

Thus, the graph of f is concave upward on $(-\infty, 1)$ and concave downward on $(1, \infty)$; the graph has an inflection point at $x = 1$.

Sketch the graph of f:

x	$f(x)$
0	3
2	$e^2 \approx 7.4$
3	0

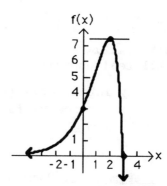

51. $f(x) = x^2 \ln x$.

Step 1. Analyze $f(x)$:

(A) Domain: All positive numbers, $(0, \infty)$.

(B) Intercepts: y intercept: There is no y intercept.

x intercept: $x^2 \ln x = 0$

$\ln x = 0$

$x = 1$

(C) Asymptotes: Consider the behavior of f as $x \to \infty$ and as $x \to 0$. It is clear that $\lim\limits_{x \to \infty} f(x)$ does not exist; f is unbounded as x approaches ∞. The following table indicates that f approaches 0 as x approaches 0.

x	1	0.1	0.01	0.001
$f(x)$	0	-0.023	-0.00046	-0.000007

Thus, there are no vertical or horizontal asymptotes.

Step 2. Analyze $f'(x)$:

$f'(x) = x^2\left(\dfrac{1}{x}\right) + (\ln x)(2x) = x(1 + 2 \ln x)$

Critical values: $x(1 + 2 \ln x) = 0$

$1 + 2 \ln x = 0$ [Note: $x > 0$]

$\ln x = -\dfrac{1}{2}$

$x = e^{-1/2} = \dfrac{1}{\sqrt{e}} \approx 0.6065$

Partition number: $x = \dfrac{1}{\sqrt{e}} \approx 0.6065$

Sign chart for f':

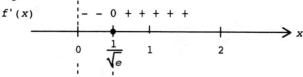

Test Numbers	
x	$f'(x)$
$\frac{1}{2}$	≈ -0.19 $(-)$
1	1 $(+)$

Thus, f is decreasing on $(0, e^{-1/2})$ and increasing on $(e^{-1/2}, \infty)$; f has a local minimum at $x = e^{-1/2}$.

Step 3. Analyze $f''(x)$:

$f''(x) = x\left(\dfrac{2}{x}\right) + (1 + 2 \ln x) = 3 + 2 \ln x$

Partition number for f'': $3 + 2 \ln x = 0$

$$\ln x = -\frac{3}{2}$$

$$x = e^{-3/2} \approx 0.2231$$

Sign chart for f'':

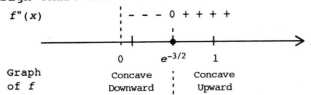

x	$f''(x)$
$\frac{1}{10}$	≈ -1.61 (−)
1	3 (+)

Test Numbers

Thus, the graph of f is concave downward on $(0, e^{-3/2})$ and concave upward on $(e^{-3/2}, \infty)$; the graph has an inflection point at $x = e^{-3/2}$.

Step 4. Sketch the graph of f:

x	$f(x)$
$e^{-3/2}$	≈ -0.075
$e^{-1/2}$	≈ -0.18
1	0

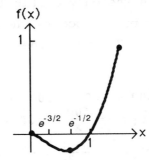

53. $f(x) = e^x - 2x^2 \qquad -\infty < x < \infty$

$f'(x) = e^x - 4x$

Critical values:

Solve $f'(x) = e^x - 4x = 0$

To two decimal places, $x = 0.36$ and $x = 2.15$

Increasing/Decreasing: $f(x)$ is increasing on $(-\infty, 0.36)$ and on $(2.15, \infty)$; $f(x)$ is decreasing on $(0.36, 2.15)$

Local extrema: $f(x)$ has a local maximum at $x = 0.36$ and a local minimum at $x = 2.15$

55. $f(x) = 20 \ln x - e^x \qquad 0 < x < \infty$

$f'(x) = \dfrac{20}{x} - e^x$

Critical values:

Solve $f'(x) = \dfrac{20}{x} - e^x = 0$

To two decimal places, $x = 2.21$

Increasing/Decreasing: $f(x)$ is increasing on $(0, 2.21)$ and decreasing on $(2.21, \infty)$

Local extrema: $f(x)$ has a local maximum at $x = 2.21$

57. On a graphing utility, graph $y_1 = e^x$ and $y_2 = x^4$. Rounded off to two decimal places, the points of intersection are: $(-0.82, 0.44)$, $(1.43, 4.18)$, $(8.61, 5503.66)$.

59. Demand: $p = 5 - \ln x$, $5 \le x \le 50$

Revenue: $R = xp = x(5 - \ln x) = 5x - x \ln x$

Cost: $C = x(1) = x$

Profit = Revenue - Cost: $P = 5x - x \ln x - x$

$$\text{or} \quad P(x) = 4x - x \ln x$$

$$P'(x) = 4 - x\left(\frac{1}{x}\right) - \ln x$$

$$= 3 - \ln x$$

Critical value(s): $P'(x) = 3 - \ln x = 0$

$$\ln x = 3$$

$$x = e^3$$

$P''(x) = -\dfrac{1}{x}$ and $P''(e^3) = -\dfrac{1}{e^3} < 0$.

Since $x = e^3$ is the only critical value and $P''(e^3) < 0$, the maximum weekly profit occurs when $x = e^3 \approx 20.09$ and the price $p = 5 - \ln(e^3) = 2$. Thus, the hot dogs should be sold at $2.

61. Cost: $C(x) = 600 + 100x - 100 \ln x$, $x \ge 1$

Average cost: $\overline{C}(x) = \dfrac{600}{x} + 100 - \dfrac{100}{x} \ln x$

$$\overline{C}'(x) = \frac{-600}{x^2} - \frac{100}{x^2} + \frac{100 \ln x}{x^2} = \frac{-700 + 100 \ln x}{x^2}, \quad x \ge 1$$

Critical value(s): $\overline{C}'(x) = \dfrac{-700 + 100 \ln x}{x^2} = 0$

$$-700 + 100 \ln x = 0$$

$$\ln x = 7$$

$$x = e^7$$

$$\overline{C}''(x) = \frac{x^2 \dfrac{100}{x} - (-700 + 100 \ln x)(2x)}{x^4}$$

$$= \frac{100x + 1400x - 200x \ln x}{x^4} = \frac{1500 - 200 \ln x}{x^3}$$

$$\overline{C}''(e^7) = \frac{1500 - 200 \ln(e^7)}{e^{21}} = \frac{100}{e^{21}} > 0$$

Since $x = e^7$ is the only critical value and $\overline{C}''(e^7) > 0$, the minimum average cost is

$$\overline{C}(e^7) = \frac{600}{e^7} + 100 - \frac{100}{e^7}\ln(e^7) = \frac{600}{e^7} + 100 - \frac{700}{e^7} = 100 - \frac{100}{e^7} \approx 99.91$$

Thus, the minimal average cost is approximately $99.91.

63. Demand: $p = 10e^{-x} = \dfrac{10}{e^x}$, $0 \le x \le 2$

Revenue: $R(x) = xp = 10xe^{-x} = \dfrac{10x}{e^x}$

(A) $R'(x) = \dfrac{e^x(10) - 10xe^x}{e^{2x}} = \dfrac{10 - 10x}{e^x}$, $0 \le x \le 2$

Critical value(s): $\dfrac{10 - 10x}{e^x} = 0$

$$10 - 10x = 0$$
$$x = 1$$

$R''(x) = \dfrac{e^x(-10) - (10 - 10x)e^x}{e^{2x}} = \dfrac{-20 + 10x}{e^x}$

$R''(1) = -\dfrac{10}{e} < 0$

Now, $R(0) = 0$

$R(1) = \dfrac{10}{e} \approx 3.68$ Absolute maximum

$R(2) = \dfrac{20}{e^2} \approx 2.71$

Thus, the maximum weekly revenue occurs at price $p = \dfrac{10}{e} \approx \$3.68$.
The maximum weekly revenue is $R(1) = 3.68$ thousand dollars, or $\$3680$.

(B) The sign chart for R' is:

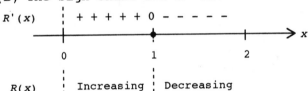

Test Numbers	
x	$f'(x)$
0	10 $(+)$
2	$-\frac{10}{e^2}$ $(-)$

Thus, R is increasing on $(0, 1)$ and decreasing on $(1, 2)$; the maximum value of R occurs at $x = 1$, as noted in (A).

$R''(x) = 10e^{-x}(x - 2) < 0$ on $(0, 2)$
Thus, the graph of R is concave downward on $(0, 2)$. The graph is shown at the right.

x	$R(x)$
0	0
1	3.68
2	2.71

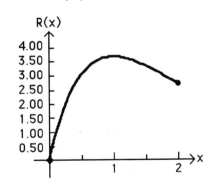

65. $P(x) = 17.5(1 + \ln x)$, $10 \leq x \leq 100$

$P'(x) = \dfrac{17.5}{x}$

$P'(40) = \dfrac{17.5}{40} \approx 0.44$

$P'(90) = \dfrac{17.5}{90} \approx 0.19$

Thus, at the 40 pound weight level, blood pressure would increase at the rate of 0.44 mm of mercury per pound of weight gain; at the 90 pound weight level, blood pressure would increase at the rate of 0.19 mm of mercury per pound of weight gain.

67. $C(t) = 4.35e^{-t} = \dfrac{4.35}{e^t}$, $0 \leq t \leq 5$

(A) $C'(t) = \dfrac{-4.35e^t}{e^{2t}} = \dfrac{-4.35}{e^t} = -4.35e^{-t}$

$C'(1) = -4.35e^{-1} \approx -1.60$

$C'(4) = -4.35e^{-4} \approx -0.08$

Thus, after one hour, the concentration is decreasing at the rate of 1.60 mg/ml per hour; after four hours, the concentration is decreasing at the rate of 0.08 mg/ml per hour.

(B) $C'(t) = -4.35e^{-t} < 0$ on $(0, 5)$

Thus, C is decreasing on $(0, 5)$; there are no local extrema.

$C''(t) = \dfrac{4.35e^t}{e^{2t}} = \dfrac{4.35}{e^t} = 4.35e^{-t} > 0$ on $(0, 5)$

Thus, the graph of C is concave upward on $(0, 5)$. The graph of C is shown at the right.

t	$C(t)$
0	4.35
1	1.60
4	0.08
5	0.03

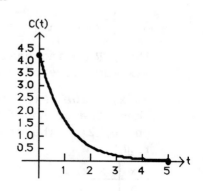

69. $R = k \ln(S/S_0)$

$\quad = k[\ln S - \ln S_0]$

$\dfrac{dR}{dS} = \dfrac{k}{S}$

Things to remember:

1. COMPOSITE FUNCTIONS

    A function m is the composite of functions f and g (in this order) if

    $m(x) = f[g(x)]$.

    The domain of m is the set of all numbers x in the domain of g such that $g(x)$ is in the domain of f.

2. CHAIN RULE

    If $y = f(u)$ and $u = g(x)$, define the composite function
    $y = m(x) = f[g(x)]$.
    Then
    $$\frac{dy}{dx} = \frac{dy}{du} \cdot \frac{du}{dx}$$
    provided $\frac{dy}{du}$ and $\frac{du}{dx}$ exist.
    Or, equivalently, $m'(x) = f'[g(x)]g'(x)$ provided $f'[g(x)]$ and $g'(x)$ exist.

3. GENERAL DERIVATIVE RULES

    (a) $\frac{d}{dx}[f(x)]^n = n[f(x)]^{n-1}f'(x)$

    (b) $\frac{d}{dx}\ln[f(x)] = \frac{1}{f(x)} \cdot f'(x)$

    (c) $\frac{d}{dx}e^{f(x)} = e^{f(x)} \cdot f'(x)$

4. OTHER LOGARITHMIC AND EXPONENTIAL FUNCTIONS

    $\frac{d}{dx}[\log_b x] = \frac{1}{\ln b} \cdot \frac{1}{x},\ b \neq e$

    $\frac{d}{dx}[b^x] = b^x \ln b,\ b \neq e$

---

1. Let $u = g(x) = 2x + 5$ and $f(u) = u^3$. Then $y = f(u) = u^3$.

3. Let $u = g(x) = 2x^2 + 7$ and $f(u) = \ln u$. Then $y = f(u) = \ln u$.

5. Let $u = g(x) = x^2 - 2$ and $f(u) = e^u$. Then $y = f(u) = e^u$.

7. $y = u^2$, $u = 2 + e^x$. Thus, $y = (2 + e^x)^2$ and $\dfrac{dy}{dx} = \dfrac{dy}{du} \cdot \dfrac{du}{dx} = 2u(e^x)$

$$= 2e^x(2 + e^x).$$

9. $y = e^u$, $u = 2 - x^4$. Thus, $y = e^{2-x^4}$ and $\dfrac{dy}{dx} = \dfrac{dy}{du} \cdot \dfrac{du}{dx} = e^u(-4x^3)$

$$= -4x^3 e^{2-x^4}.$$

11. $y = \ln u$, $u = 4x^5 - 7$, so $y = \ln(4x^5 - 7)$ and

$$\dfrac{dy}{dx} = \dfrac{dy}{du} \cdot \dfrac{du}{dx} = \dfrac{1}{u}(20x^4) = \dfrac{1}{4x^5 - 7}(20x^4) = \dfrac{20x^4}{4x^5 - 7}$$

13. $\dfrac{d}{dx}\ln(x - 3) = \dfrac{1}{x - 3}(1)$ (using $\underline{3b}$)

$$= \dfrac{1}{x - 3}$$

15. $\dfrac{d}{dt}\ln(3 - 2t) = \dfrac{1}{3 - 2t}(-2)$ (using $\underline{3b}$)

$$= \dfrac{-2}{3 - 2t}$$

17. $\dfrac{d}{dx}3e^{2x} = 3\dfrac{d}{dx}e^{2x} = 3e^{2x}(2)$ (using $\underline{3c}$)

$$= 6e^{2x}$$

19. $\dfrac{d}{dt}2e^{-4t} = 3\dfrac{d}{dt}e^{-4t} = 2e^{-4t}(-4) = -8e^{-4t}$

21. $\dfrac{d}{dx}100e^{-0.03x} = 100\dfrac{d}{dx}e^{-0.03x} = 100e^{-0.03x}(-0.03) = -3e^{-0.03x}$

23. $\dfrac{d}{dx}\ln(x + 1)^4 = \dfrac{d}{dx}4\ln(x + 1) = 4\dfrac{d}{dx}\ln(x + 1) = 4\dfrac{1}{x + 1}(1) = \dfrac{4}{x + 1}$

25. $\dfrac{d}{dx}(2e^{2x} - 3e^x + 5) = 2\dfrac{d}{dx}e^{2x} - 3\dfrac{d}{dx}e^x + \dfrac{d}{dx}5 = 2e^{2x}(2) - 3e^x = 4e^{2x} - 3e^x$

27. $\dfrac{d}{dx}e^{3x^2-2x} = e^{3x^2-2x}(6x - 2) = (6x - 2)e^{3x^2-2x}$

29. $\dfrac{d}{dt}\ln(t^2 + 3t) = \dfrac{1}{t^2 + 3t}(2t + 3) = \dfrac{2t + 3}{t^2 + 3t}$

31. $\dfrac{d}{dx}\ln(x^2 + 1)^{1/2} = \dfrac{d}{dx}\dfrac{1}{2}\ln(x^2 + 1) = \dfrac{1}{2}\dfrac{d}{dx}\ln(x^2 + 1)$

$$= \dfrac{1}{2}\left(\dfrac{1}{x^2 + 1}\right)(2x) = \dfrac{x}{x^2 + 1}$$

33. $\dfrac{d}{dt}[\ln(t^2 + 1)]^4 = 4[\ln(t^2 + 1)]^3\dfrac{1}{t^2 + 1}(2t) = \dfrac{8t}{t^2 + 1}[\ln(t^2 + 1)]^3$

35. $\dfrac{d}{dx}(e^{2x} - 1)^4 = 4(e^{2x} - 1)^3[e^{2x}(2)] = 8e^{2x}(e^{2x} - 1)^3$

37. $\dfrac{d}{dx}\dfrac{e^{2x}}{x^2 + 1} = \dfrac{(x^2 + 1)\dfrac{d}{dx}e^{2x} - e^{2x}\dfrac{d}{dx}(x^2 + 1)}{(x^2 + 1)^2} = \dfrac{(x^2 + 1)e^{2x}(2) - e^{2x}(2x)}{(x^2 + 1)^2}$

$$= \dfrac{2e^{2x}(x^2 - x + 1)}{(x^2 + 1)^2}$$

39. $\dfrac{d}{dx}(x^2 + 1)e^{-x} = (x^2 + 1)\dfrac{d}{dx}e^{-x} + e^{-x}\dfrac{d}{dx}(x^2 + 1)$

$$= (x^2 + 1)e^{-x}(-1) + e^{-x}(2x) = e^{-x}(2x - x^2 - 1)$$

41. $\dfrac{d}{dx}e^{-x}\ln x = e^{-x}\dfrac{d}{dx}\ln x + \ln x\dfrac{d}{dx}e^{-x} = e^{-x}\left(\dfrac{1}{x}\right) + (\ln x)(e^{-x})(-1)$

$$= \dfrac{e^{-x}}{x} - e^{-x}\ln x = \dfrac{e^{-x}[1 - x\ln x]}{x}$$

43. $\dfrac{d}{dx}\dfrac{1}{\ln(1 + x^2)} = \dfrac{d}{dx}[\ln(1 + x^2)]^{-1} = -1[\ln(1 + x^2)]^{-2}\dfrac{d}{dx}\ln(1 + x^2)$

$$= -[\ln(1 + x^2)]^{-2}\dfrac{1}{1 + x^2}(2x) = \dfrac{-2x}{(1 + x^2)[\ln(1 + x^2)]^2}$$

45. $\dfrac{d}{dx}\sqrt[3]{\ln(1 - x^2)} = \dfrac{d}{dx}[\ln(1 - x^2)]^{1/3} = \dfrac{1}{3}[\ln(1 - x^2)]^{-2/3}\dfrac{d}{dx}\ln(1 - x^2)$

$$= \dfrac{1}{3}[\ln(1 - x^2)]^{-2/3}\dfrac{1}{1 - x^2}(-2x) = \dfrac{-2x}{3(1 - x^2)[\ln(1 - x^2)]^{2/3}}$$

47. $f(x) = 1 - e^{-x}$

<u>Step 1. Analyze $f(x)$</u>:

(A) Domain: All real numbers, $(-\infty, \infty)$.

(B) Intercepts: y intercept: $f(0) = 1 - e^{-0} = 0$
$\qquad\qquad\qquad\quad\; x$ intercept: $1 - e^{-x} = 0$
$\qquad\qquad\qquad\qquad\qquad\qquad e^{-x} = 1$
$\qquad\qquad\qquad\qquad\qquad\qquad\;\; x = 0$

(C) Asymptotes:

$\qquad$<u>Horizontal asymptote</u>: $\displaystyle\lim_{x \to \infty}(1 - e^{-x}) = \lim_{x \to \infty}\left(1 - \dfrac{1}{e^x}\right) = 1$

$\qquad\qquad\qquad\qquad\qquad\quad$ $\displaystyle\lim_{x \to -\infty}(1 - e^{-x})$ does not exist.

$\qquad\qquad\qquad\qquad\qquad\quad$ $y = 1$ is a horizontal asymptote.
$\qquad$<u>Vertical asymptotes</u>: There are no vertical asymptotes.

<u>Step 2. Analyze $f'(x)$</u>:
$f'(x) = -e^{-x}(-1) = e^{-x}$
Since $e^{-x} > 0$ for all x, f is increasing on $(-\infty, \infty)$; there are no local extrema.

<u>Step 3. Analyze $f''(x)$</u>:
$f''(x) = e^{-x}(-1) = -e^{-x}$
Since $-e^{-x} < 0$ for all x, the graph of f is concave downward on $(-\infty, \infty)$.

Step 4. Sketch the graph of f:

x	f(x)
0	0
-1	≈ -1.72
1	≈ 0.63

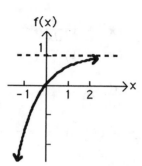

49. $f(x) = \ln(1 - x)$

Step 1. Analyze $f(x)$:

(A) Domain: All real numbers x such that $1 - x > 0$, i.e., $x < 1$
 or $(-\infty, 1)$.

(B) Intercepts: y intercept: $f(0) = \ln(1 - 0) = \ln 1 = 0$
 x intercepts: $\ln(1 - x) = 0$
 $1 - x = 1$
 $x = 0$

(C) Asymptotes:
 Horizontal asymptote: $\lim\limits_{x \to -\infty} f(x) = \lim\limits_{x \to -\infty} \ln(1 - x)$ does not exist.
 Thus, there are no horizontal asymptotes.
 Vertical asymptote: From the table,

x	0.9	0.99	0.99999	0.9999999	→ 1
f(x)	-2.30	-4.61	-11.51	-16.12	→ -∞

 We conclude that $x = 1$ is a vertical asymptote.

Step 2. Analyze $f'(x)$:
$$f'(x) = \frac{1}{1 - x}(-1), \; x < 1$$
$$= \frac{1}{x - 1}$$
Now, $f'(x) = \dfrac{1}{x - 1} < 0$ on $(-\infty, 1)$.
Thus, f is decreasing on $(-\infty, 1)$; there are no critical values and no local extrema.

Step 3. Analyze $f''(x)$:
$$f'(x) = (x - 1)^{-1}$$
$$f''(x) = -1(x - 1)^{-2} = \frac{-1}{(x - 1)^2}$$
Since $f''(x) = \dfrac{-1}{(1 - x)^2} < 0$ on $(-\infty, 1)$, the graph of f is concave
downward on $(-\infty, 1)$; there are no inflection points.

Step 4. Sketch the graph of f:

x	$f(x)$
0	0
-2	≈ 1.10
.9	≈ -2.30

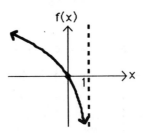

51. $f(x) = e^{-(1/2)x^2}$

Step 1. Analyze $f(x)$:

(A) Domain: All real numbers, $(-\infty, \infty)$.

(B) Intercepts: y intercept: $f(0) = e^{-(1/2)0} = e^0 = 1$

$\qquad$ x intercepts: Since $e^{-(1/2)x^2} \neq 0$ for all x, there are no x intercepts.

(C) Asymptotes: $\displaystyle\lim_{x \to \infty} f(x) = \lim_{x \to \infty} e^{-(1/2)x^2} = \lim_{x \to \infty} \frac{1}{e^{(1/2)x^2}} = 0$

$\qquad\qquad\qquad$ $\displaystyle\lim_{x \to -\infty} f(x) = \lim_{x \to -\infty} e^{-(1/2)x^2} = \lim_{x \to -\infty} \frac{1}{e^{(1/2)x^2}} = 0$

Thus, $y = 0$ is a horizontal asymptote.

Since

$f(x) = e^{-(1/2)x^2} = \dfrac{1}{e^{(1/2)x^2}}$ and $g(x) = e^{(1/2)x^2} \neq 0$ for all x, there

are no vertical asymptotes.

Step 2. Analyze $f'(x)$:

$f'(x) = e^{-(1/2)x^2}(-x) = -xe^{-(1/2)x^2}$

Critical values: $-xe^{-(1/2)x^2} = 0$

$\qquad\qquad\qquad\qquad x = 0$

Partition numbers: $x = 0$

Sign chart for f':

$f'(x)$	$+ + + + + 0 - - - - -$	
	$\xrightarrow{\qquad\qquad}$ x	
	$-1 \qquad 0 \qquad 1$	
$f(x)$	Increasing $\vdots$ Decreasing	

Test Numbers

x	$f'(x)$
-1	$\frac{1}{e^{1/2}}$ (+)
1	$-\frac{1}{e^{1/2}}$ (−)

Thus, f is increasing on $(-\infty, 0)$ and decreasing on $(0, \infty)$; f has a local maximum at $x = 0$.

Step 3. Analyze $f''(x)$:

$f''(x) = -xe^{-(1/2)x^2}(-x) - e^{-(1/2)x^2}$

$\qquad = e^{-(1/2)x^2}(x^2 - 1) = e^{-(1/2)x^2}(x - 1)(x + 1)$

Partition numbers for f'': $e^{-(1/2)x^2}(x - 1)(x + 1) = 0$

$\qquad\qquad\qquad\qquad\qquad\qquad (x - 1)(x + 1) = 0$

$\qquad\qquad\qquad\qquad\qquad\qquad\qquad\qquad x = -1, 1$

Sign chart for f'':

$f''(x)$ $+ + + + 0 - - - - 0 + + + +$

$$\xrightarrow{\quad\quad\quad\quad\quad\quad\quad\quad\quad\quad} x$$

$-2 \quad -1 \quad 0 \quad 1 \quad 2$

Graph
of f

Concave : Concave : Concave
Upward : Downward : Upward

Test Numbers

x	$f''(x)$
-2	$\frac{3}{e^2}$ (+)
0	-1 (−)
2	$\frac{3}{e^2}$ (+)

Thus, the graph of f is concave upward on $(-\infty, -1)$ and on $(1, \infty)$; the graph of f is concave downward on $(-1, 1)$; the graph has inflection points at $x = -1$ and at $x = 1$.

Step 4.   Sketch the graph of f:

x	$f(x)$
0	1
-1	≈ 0.61
1	≈ 0.61

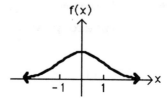

53. $y = 1 + w^2$, $w = \ln u$, $u = 2 + e^x$
Thus, $y = 1 + [\ln(2 + e^x)]^2$ and
$$\frac{dy}{dx} = 2[\ln(2 + e^x)]\left(\frac{1}{2 + e^x}\right)(e^x) = \frac{2e^x \ln(2 + e^x)}{2 + e^x}$$

55. $\frac{d}{dx} \log_2(3x^2 - 1) = \frac{1}{\ln 2} \cdot \frac{1}{3x^2 - 1} \cdot 6x = \frac{1}{\ln 2} \cdot \frac{6x}{3x^2 - 1}$

57. $\frac{d}{dx} 10^{x^2+x} = 10^{x^2+x}(\ln 10)(2x + 1) = (2x + 1)10^{x^2+x} \ln 10$

59. $\frac{d}{dx} \log_3(4x^3 + 5x + 7) = \frac{1}{\ln 3} \cdot \frac{1}{4x^3 + 5x + 7}(12x^2 + 5)$
$$= \frac{12x^2 + 5}{\ln 3(4x^3 + 5x + 7)}$$

61. $\frac{d}{dx} 2^{x^3-x^2+4x+1} = 2^{x^3-x^2+4x+1} \ln 2(3x^2 - 2x + 4)$
$$= \ln 2(3x^2 - 2x + 4)2^{x^3-x^2+4x+1}$$

63. Let $u = f(x)$ and $y = \ln u$.   Then, by the chain rule,
$$\frac{dy}{dx} = \frac{dy}{du} \cdot \frac{du}{dx} = \frac{1}{u} \cdot f'(x) = \frac{1}{f(x)} \cdot f'(x).$$

65. (A) f is decreasing on $(-\infty, 0)$ but $g(x)$ is not negative on this interval.

(B) $f'(x) = \dfrac{1}{x^2 + 1}(2x) = \dfrac{2x}{x^2 + 1}$

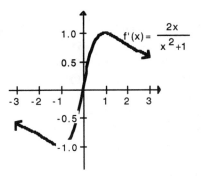

67. $f'(x) = \dfrac{1}{5(x^2 + 3)^4}[20(x^2 + 3)^3](2x) = \dfrac{8x}{x^2 + 3}$

$g'(x) = 4 \cdot \dfrac{1}{x^2 + 3}(2x) = \dfrac{8x}{x^2 + 3}$

For another way to see this, recall the properties of logarithms discussed in Section 2.3:

$f(x) = \ln[5(x^2 + 3)^4] = \ln 5 + \ln(x^2 + 3)^4 = \ln 5 + 4 \ln(x^2 + 3)$
$$= \ln 5 + g(x)$$

Now $\dfrac{d}{dx} f(x) = \dfrac{d}{dx} \ln 5 + \dfrac{d}{dx} g(x) = 0 + \dfrac{d}{dx} g(x) = \dfrac{d}{dx} g(x)$

Conclusion: $f'(x)$ and $g'(x)$ ARE the same function.

69. Price: $p = 100e^{-0.05x}$, $x \ge 0$
Revenue: $R(x) = xp = 100xe^{-0.05x}$
$\quad\quad R'(x) = 100xe^{-0.05x}(-0.05) + 100e^{-0.05x}$
$\quad\quad\quad\quad = 100e^{-0.05x}(1 - 0.05x)$
Critical value(s): $R'(x) = 100e^{-0.05x}(1 - 0.05x) = 0$
$\quad\quad\quad\quad\quad\quad\quad\quad\quad\quad\quad 1 - 0.05x = 0$
$\quad\quad\quad\quad\quad\quad\quad\quad\quad\quad\quad\quad\quad\quad x = 20$
$R''(x) = 100e^{-0.05x}(-0.05) + (1 - 0.05x)100e^{-0.05x}(-0.05)$
$\quad\quad\quad = 100e^{-0.05x}(0.0025x - 0.1)$
$R''(20) = -100e^{-1}(0.05) = \dfrac{-5}{e} < 0$

Since $x = 20$ is the only critical value and $R''(20) < 0$, the production level that maximizes the revenue is 20 units. The maximum revenue is $R(20) = 20(36.79) = 735.80$ or \$735.80, and the price is $p(20) = 36.79$ or \$36.79 each.

71. The cost function $C(x)$ is given by
$$C(x) = 400 + 6x$$
and the revenue function $R(x)$ is
$$R(x) = xp = 100xe^{-0.05x}$$
The profit function $P(x)$ is
$$P(x) = R(x) - C(x)$$
$$= 100xe^{-0.05x} - 400 - 6x$$
and $P'(x) = 100e^{-0.05x} - 5xe^{-0.05x} - 6$
We graph $y = P(x)$ and $y = P'(x)$ in the
viewing rectangle $0 \le x \le 50$,
$-400 \le y \le 300$

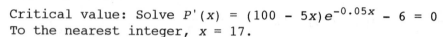

Critical value: Solve $P'(x) = (100 - 5x)e^{-0.05x} - 6 = 0$
To the nearest integer, $x = 17$.
$P(x)$ is increasing on $(0, 17)$ and decreasing on $(17, \infty)$; $P(x)$ has a
maximum at $x = 17$. Thus, the maximum profit $P(17) = \$224.61$ is realized
at a production level of 17 units at a price of $\$42.74$ per unit.

73. $S(t) = 300,000e^{-0.1t}$, $t \ge 0$
$S'(t) = 300,000e^{-0.1t}(-0.1) = -30,000e^{-0.1t}$
The rate of depreciation after one year is:
$S'(1) = -30,000e^{-0.1} \approx -\$27,145.12$ per year.
The rate of depreciation after five years is:
$S'(5) = -30,000e^{-0.5} \approx -\$18,195.92$ per year.
The rate of depreciation after ten years is:
$S'(10) = -30,000e^{-1} \approx -\$11,036.38$ per year.

75. Revenue: $R(t) = 200,000(1 - e^{-0.03t})$, $t \ge 0$
Cost: $C(t) = 4000 + 3000t$, $t \ge 0$
Profit: $P(t) = R(t) - C(t) = 200,000(1 - e^{-0.03t}) - (4000 + 3000t)$
$$= 200,000(1 - e^{-0.03t}) - 3000t - 4000$$
(A) $P'(t) = -200,000e^{-0.03t}(-0.03) - 3000 = 6000e^{-0.03t} - 3000$
Critical value(s): $P'(t) = 6000e^{-0.03t} - 3000 = 0$
$$e^{-0.03t} = \frac{1}{2}$$
$$-0.03t = \ln\left(\frac{1}{2}\right) = -\ln 2$$
$$t = \frac{\ln 2}{0.03} \approx 23$$
$$P''(t) = 6000e^{-0.03t}(-0.03) = -180e^{-0.03t}$$
$$P''(23) = -180e^{-0.69} < 0$$
Since $t = 23$ is the only critical value and $P''(23) < 0$, 23 days of
TV promotion should be used to maximize profits. The maximum profit
is: $P(23) = 200,000(1 - e^{-0.03(23)}) - 3000(23) - 4000 \approx \$26,685$
The proportion of people buying the disk after t days is:
$p(t) = 1 - e^{-0.03t}$
Thus, $p(23) = 1 - e^{-0.03(23)} \approx 0.50$ or approximately 50%.

(B) From A, the sign chart for P' is:

$$P'(t) \quad + + + + + 0 - - - - -$$

0   23   50 $\to t$

$P(t)$ Increasing   Decreasing

Test Numbers	
t	$P'(t)$
0	3000 (+)
50	−1661.22 (−)

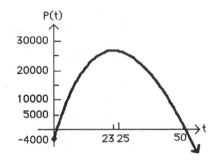

Thus, P is increasing on $(0, 23)$ and decreasing on $(23, \infty)$; P has a maximum at $t = 23$.
Since $P''(t) = -180e^{-0.03t} < 0$ on $(0, \infty)$, the graph of P is concave downward on $(0, \infty)$; $P(0) = -4000$ and $P(50) \approx 0$.
The graph of P is shown at the right.

77. $P(x) = 40 + 25 \ln(x + 1) \quad 0 \le x \le 65$

$$P'(x) = 25\left(\frac{1}{x + 1}\right)(1) = \frac{25}{x + 1}$$

$$P'(10) = \frac{25}{11} \approx 2.27$$

$$P'(30) = \frac{25}{31} \approx 0.81$$

$$P'(60) = \frac{25}{61} \approx 0.41$$

Thus, the rate of change of pressure at the end of 10 years is 2.27 millimeters of mercury per year; at the end of 30 years the rate of change is 0.81 millimeters of mercury per year; at the end of 60 years the rate of change is 0.41 millimeters of mercury per year.

79. $A(t) = 5000 \cdot 2^{2t}$

$A'(t) = 5000 \cdot 2^{2t}(2)(\ln 2) = 10{,}000 \cdot 2^{2t}(\ln 2)$

$A'(1) = 10{,}000 \cdot 2^2(\ln 2) = 40{,}000 \ln 2$

      $\approx 27{,}726$ rate of change of bacteria at the end of the first hour.

$A'(5) = 10{,}000 \cdot 2^{2 \cdot 5}(\ln 2) = 10{,}000 \cdot 2^{10}(\ln 2)$

      $\approx 7{,}097{,}827$ rate of change of bacteria at the end of the fifth hour.

81. $N(n) = 1{,}000{,}000e^{-0.09(n-1)}, \quad 1 \le n \le 20$

There are no asymptotes and no intercepts.
Using the first derivative:

$N'(n) = 1{,}000{,}000e^{-0.09(n-1)}(-0.09)$

      $= -90{,}000e^{-0.09(n-1)} < 0, \quad 1 \le n \le 20$

Thus, N is decreasing on $(0, 20)$.
Using the second derivative:

$N''(n) = -90{,}000e^{-0.09(n-1)}(-0.09)$

      $= 8100e^{-0.09(n-1)} > 0, \quad 1 \le n \le 20$

Thus, the graph of N is concave upward on $(0, 20)$.

The graph of N is:

n	N
1	1,000,000
20	$\approx$ 180,866

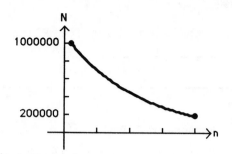

EXERCISE 5-4

Things to remember:

1. Let $y = y(x)$. Then

(a) $\dfrac{d}{dx} y^n = n y^{n-1} y'$ (General Power Rule)

(b) $\dfrac{d}{dx} \ln y = \dfrac{1}{y} \cdot y' = \dfrac{y'}{y}$

(c) $\dfrac{d}{dx} e^y = e^y \cdot y' = y' e^y$

---

1. $y - 3x^2 + 5 = 0$
Using implicit differentiation:
$$y' - \frac{d}{dx}(3x^2) + \frac{d}{dx}(5) = \frac{d}{dx}(0)$$
$$y' - 6x + 0 = 0$$
$$y' = 6x$$
$$y'\Big|_{(1, -2)} = 6 \cdot 1 = 6$$

3. $y^2 - 3x^2 + 8 = 0$
$$\frac{d}{dx}(y^2) - \frac{d}{dx}(3x^2) + \frac{d}{dx}(8) = \frac{d}{dx}(0)$$
$$2yy' - 6x + 0 = 0 \text{ (using } \underline{1})$$
$$2yy' = 6x$$
$$y' = \frac{3x}{y}$$
$$y'\Big|_{(2, 2)} = \frac{3 \cdot 2}{2} = 3$$

5. $y^2 + y - x = 0$
$$\frac{d}{dx} y^2 + \frac{d}{dx} y - \frac{d}{dx} x = \frac{d}{dx}(0)$$
$$2yy' + y' - 1 = 0$$
$$y'(2y + 1) = 1$$
$$y' = \frac{1}{2y + 1}$$
$$y' \text{ at } (2, 1) = \frac{1}{2 \cdot 1 + 1} = \frac{1}{3}$$

7. $xy - 6 = 0$
$$\frac{d}{dx} xy - \frac{d}{dx} 6 = \frac{d}{dx}(0)$$
$$xy' + y - 0 = 0$$
$$xy' = -y$$
$$y' = -\frac{y}{x}$$
$$y' \text{ at } (2, 3) = -\frac{3}{2}$$

9. $2xy + y + 2 = 0$
$$2\frac{d}{dx} xy + \frac{d}{dx} y + \frac{d}{dx} 2 = \frac{d}{dx}(0)$$
$$2xy' + 2y + y' + 0 = 0$$
$$y'(2x + 1) = -2y$$
$$y' = \frac{-2y}{2x + 1}$$
$$y' \text{ at } (-1, 2) = \frac{-2(2)}{2(-1) + 1} = 4$$

11. $x^2y - 3x^2 - 4 = 0$

$$\frac{d}{dx}x^2y - \frac{d}{dx}3x^2 - \frac{d}{dx}4 = \frac{d}{dx}(0)$$

$$x^2y' + y\frac{d}{dx}(x^2) - 6x - 0 = 0$$

$$x^2y' + y2x - 6x = 0$$

$$x^2y' = 6x - 2yx$$

$$y' = \frac{6x - 2yx}{x^2} \text{ or } \frac{6 - 2y}{x}$$

$$y'\Big|_{(2, 4)} = \frac{6\cdot2 - 2\cdot4\cdot2}{2^2} = \frac{12 - 16}{4} = -1$$

13. $e^y = x^2 + y^2$

$$\frac{d}{dx}e^y = \frac{d}{dx}x^2 + \frac{d}{dx}y^2$$

$$e^yy' = 2x + 2yy'$$

$$y'(e^y - 2y) = 2x$$

$$y' = \frac{2x}{e^y - 2y}$$

$$y'\Big|_{(1, 0)} = \frac{2\cdot1}{e^0 - 2\cdot0} = \frac{2}{1} = 2$$

15. $x^3 - y = \ln y$

$$\frac{d}{dx}x^3 - \frac{d}{dx}y = \frac{d}{dx}\ln y$$

$$3x^2 - y' = \frac{y'}{y}$$

$$3x^2 = \left(1 + \frac{1}{y}\right)y'$$

$$3x^2 = \frac{y + 1}{y}y'$$

$$y' = \frac{3x^2y}{y + 1}$$

$$y'\Big|_{(1, 1)} = \frac{3\cdot1^2\cdot1}{1 + 1} = \frac{3}{2}$$

17. $x \ln y + 2y = 2x^3$

$$\frac{d}{dx}[x \ln y] + \frac{d}{dx}2y = \frac{d}{dx}2x^3$$

$$\ln y\cdot\frac{d}{dx}x + x\frac{d}{dx}\ln y + 2y' = 6x^2$$

$$\ln y\cdot1 + x\cdot\frac{y'}{y} + 2y' = 6x^2$$

$$y'\left(\frac{x}{y} + 2\right) = 6x^2 - \ln y$$

$$y' = \frac{6x^2y - y \ln y}{x + 2y}$$

$$y'\Big|_{(1, 1)} = \frac{6\cdot1^2\cdot1 - 1\cdot\ln 1}{1 + 2\cdot1} = \frac{6}{3} = 2$$

19. $x^2 - t^2x + t^3 + 11 = 0$

$$\frac{d}{dt}x^2 - \frac{d}{dt}(t^2x) + \frac{d}{dt}t^3 + \frac{d}{dt}11 = \frac{d}{dt}0$$

$$2xx' - [t^2x' + x(2t)] + 3t^2 + 0 = 0$$

$$2xx' - t^2x' - 2tx + 3t^2 = 0$$

$$x'(2x - t^2) = 2tx - 3t^2$$

$$x' = \frac{2tx - 3t^2}{2x - t^2}$$

$$x'\Big|_{(-2, 1)} = \frac{2(-2)(1) - 3(-2)^2}{2(1) - (-2)^2}) = \frac{-4 - 12}{2 - 4} = \frac{-16}{-2} = 8$$

21. $xy - x - 4 = 0$

When $x = 2$, $2y - 2 - 4 = 0$, so $y = 3$. Thus, we want to find the equation of the tangent line at $(2, 3)$.

First, find y'.

$$\frac{d}{dx}xy - \frac{d}{dx}x - \frac{d}{dx}4 = \frac{d}{dx}0$$

$$xy' + y - 1 - 0 = 0$$

$$xy' = 1 - y$$

$$y' = \frac{1 - y}{x}$$

$$y'\Big|_{(2,\ 3)} = \frac{1 - 3}{2} = -1$$

Thus, the slope of the tangent line at $(2, 3)$ is $m = -1$. The equation of the line through $(2, 3)$ with slope $m = -1$ is:

$$(y - 3) = -1(x - 2)$$

$$y - 3 = -x + 2$$

$$y = -x + 5$$

23. $y^2 - xy - 6 = 0$

When $x = 1$,

$$y^2 - y - 6 = 0$$

$$(y - 3)(y + 2) = 0$$

$$y = 3 \text{ or } -2.$$

Thus, we want to find the equations of the tangent lines at $(1, 3)$ and $(1, -2)$. First, find y'.

$$\frac{d}{dx}y^2 - \frac{d}{dx}xy - \frac{d}{dx}6 = \frac{d}{dx}0$$

$$2yy' - xy' - y - 0 = 0$$

$$y'(2y - x) = y$$

$$y' = \frac{y}{2y - x}$$

$$y'\Big|_{(1,\ 3)} = \frac{3}{2(3) - 1} = \frac{3}{5} \qquad \text{[Slope at } (1, 3)\text{]}$$

The equation of the tangent line at $(1, 3)$ with $m = \frac{3}{5}$ is:

$$(y - 3) = \frac{3}{5}(x - 1)$$

$$y - 3 = \frac{3}{5}x - \frac{3}{5}$$

$$y = \frac{3}{5}x + \frac{12}{5}$$

$$y'\Big|_{(1,\ -2)} = \frac{-2}{2(-2) - 1} = \frac{2}{5} \qquad \text{[Slope at } (1, -2)\text{]}$$

Thus, the equation of the tangent line at $(1, -2)$ with $m = \frac{2}{5}$ is:

$$(y + 2) = \frac{2}{5}(x - 1)$$

$$y + 2 = \frac{2}{5}x - \frac{2}{5}$$

$$y = \frac{2}{5}x - \frac{12}{5}$$

25. $xe^y = 1$

Implicit differentiation: $x \cdot \dfrac{d}{dx} e^y + e^y \dfrac{d}{dx} x = \dfrac{d}{dx} 1$

$$xe^y y' + e^y = 0$$

$$y' = -\frac{e^y}{xe^y} = -\frac{1}{x}$$

Solve for y: $e^y = \dfrac{1}{x}$

$$y = \ln\left(\frac{1}{x}\right) = -\ln x \quad \text{(see Section 2.3)}$$

$$y' = -\frac{1}{x}$$

In this case, solving for y first and then differentiating is a little easier than differentiating implicity.

27. $(1 + y)^3 + y = x + 7$

$$\frac{d}{dx}(1 + y)^3 + \frac{d}{dx} y = \frac{d}{dx} x + \frac{d}{dx} 7$$

$$3(1 + y)^2 y' + y' = 1$$

$$y'[3(1 + y)^2 + 1] = 1$$

$$y' = \frac{1}{3(1 + y)^2 + 1}$$

$$y'\Big|_{(2,\ 1)} = \frac{1}{3(1 + 1)^2 + 1} = \frac{1}{13}$$

29. $(x - 2y)^3 = 2y^2 - 3$

$$\frac{d}{dx}(x - 2y)^3 = \frac{d}{dx}(2y^2) - \frac{d}{dx}(3)$$

$$3(x - 2y)^2 (1 - 2y') = 4yy' - 0 \quad \begin{array}{l}[\underline{\text{Note}}\text{: The chain rule is applied to the}\\ \text{left-hand side.}]\end{array}$$

$$3(x - 2y)^2 - 6(x - 2y)^2 y' = 4yy'$$

$$-6(x - 2y)^2 y' - 4yy' = -3(x - 2y)^2$$

$$-y'[6(x - 2y)^2 + 4y] = -3(x - 2y)^2$$

$$y' = \frac{3(x - 2y)^2}{6(x - 2y)^2 + 4y}$$

$$y'\Big|_{(1,\ 1)} = \frac{3(1 - 2 \cdot 1)^2}{6(1 - 2)^2 + 4} = \frac{3}{10}$$

31. $\sqrt{7 + y^2} - x^3 + 4 = 0 \quad \text{or} \quad (7 + y^2)^{1/2} - x^3 + 4 = 0$

$$\frac{d}{dx}(7 + y^2)^{1/2} - \frac{d}{dx} x^3 + \frac{d}{dx} 4 = \frac{d}{dx} 0$$

$$\frac{1}{2}(7 + y^2)^{-1/2} \frac{d}{dx}(7 + y^2) - 3x^2 + 0 = 0$$

$$\frac{1}{2}(7 + y^2)^{-1/2} 2yy' - 3x^2 = 0$$

$$\frac{yy'}{(7 + y^2)^{1/2}} = 3x^2$$

$$y' = \frac{3x^2 (7 + y^2)^{1/2}}{y}$$

$$y'\Big|_{(2,\ 3)} = \frac{3 \cdot 2^2 (7 + 3^2)^{1/2}}{3} = \frac{12(16)^{1/2}}{3} = 16$$

33. $\ln(xy) = y^2 - 1$

$$\frac{d}{dx}[\ln(xy)] = \frac{d}{dx}y^2 - \frac{d}{dx}1$$

$$\frac{1}{xy} \cdot \frac{d}{dx}(xy) = 2yy'$$

$$\frac{1}{xy}(x \cdot y' + y) = 2yy'$$

$$\frac{1}{y} \cdot y' - 2yy' + \frac{1}{x} = 0$$

$$xy' - 2xy^2y' + y = 0$$

$$y'(x - 2xy^2) = -y$$

$$y' = \frac{-y}{x - 2xy^2} = \frac{y}{2xy^2 - x}$$

$$y'\Big|_{(1,\ 1)} = \frac{1}{2 \cdot 1 \cdot 1^2 - 1} = 1$$

35. First find point(s) on the graph of the equation with abscissa $x = 1$:
Setting $x = 1$, we have

$$y^3 - y - 1 = 2 \quad \text{or} \quad y^3 - y - 3 = 0$$

Graphing this equation on a graphing utility, we get $y \approx 1.67$.

Now, differentiate implicitly to find the slope of the tangent line at
the point $(1,\ 1.67)$: $\dfrac{d}{dx}y^3 + x\dfrac{d}{dx}y + y\dfrac{d}{dx}x - \dfrac{d}{dx}x^3 = \dfrac{d}{dx}2$

$$3y^2y' - xy' - y - 3x^2 = 0$$

$$(3y^2 - x)y' = 3x^2 + y$$

$$y' = \frac{3x^2 + y}{3y^2 - x};$$

$$y'\Big|_{(1,\ 1.67)} = \frac{3 + 1.67}{3(1.67)^2 - 1} = \frac{4.67}{7.37} \approx 0.63$$

Tangent line: $y - 1.67 = 0.63(x - 1)$ or $y = 0.63x + 1.04$

37. $x = p^2 - 2p + 1000$

$$\frac{d(x)}{dx} = \frac{d(p^2)}{dx} - \frac{d(2p)}{dx} + \frac{d(1000)}{dx}$$

$$1 = 2p\frac{dp}{dx} - 2\frac{dp}{dx} + 0$$

$$1 = (2p - 2)\frac{dp}{dx}$$

Thus, $\dfrac{dp}{dx} = p' = \dfrac{1}{2p - 2}.$

39. $x = \sqrt{10,000 - p^2} = (10,000 - p^2)^{1/2}$

$\dfrac{d}{dx} x = \dfrac{d}{dx} (10,000 - p^2)^{1/2}$

$1 = \dfrac{1}{2}(10,000 - p^2)^{-1/2} \dfrac{d}{dx}[10,000 - p^2]$

$1 = \dfrac{1}{2(10,000 - p^2)^{1/2}} \cdot (-2pp')$

$1 = \dfrac{-pp'}{\sqrt{10,000 - p^2}}$

$p' = \dfrac{-\sqrt{10,000 - p^2}}{p}$

41. $(L + m)(V + n) = k$

$$\dfrac{d(L + m)(V + n)}{dV} = \dfrac{d(k)}{dV}$$

$$(V + n)\dfrac{d(L + m)}{dV} + (L + m)\dfrac{d(V + n)}{dV} = 0$$

$$(V + n)\dfrac{dL}{dV} + (L + m) = 0$$

$$\dfrac{dL}{dV} = \dfrac{-(L + m)}{V + n}$$

EXERCISE 5-5

Things to remember:

1. SUGGESTIONS FOR SOLVING RELATED RATE PROBLEMS

   (a) Sketch a figure.

   (b) Identify all relevant variables, including those whose rates are given and those whose rates are to be found.

   (c) Express all given rates and rates to be found as derivatives.

   (d) Find an equation connecting the variables in (b).

   (e) Differentiate the equation implicitly, using the chain rule where appropriate, and substitute in all given values.

   (f) Solve for the derivative that will give the unknown rate.

---

1. $y = 2x^2 - 1$

Differentiating with respect to t:

$\dfrac{dy}{dt} = 4x\dfrac{dx}{dt}$

$\dfrac{dy}{dt} = 4 \cdot 30 \cdot 2 \quad \left(x = 30, \ \dfrac{dx}{dt} = 2\right)$

$\quad\ = 240$

3. $x^2 + y^2 = 25$

Differentiating with respect to t:

$$2x\frac{dx}{dt} + 2y\frac{dy}{dt} = 0$$

$$2y\frac{dy}{dt} = -2x\frac{dx}{dt}$$

$$\frac{dy}{dt} = -\frac{x}{y}\frac{dx}{dt} \quad \left(\text{at } x = 3, \ y = 4, \text{ and } \frac{dx}{dt} = -3\right)$$

$$= -\frac{3}{4}(-3) = \frac{9}{4}$$

5. $x^2 + xy + 2 = 0$

Differentiating with respect to t:

$$2x\frac{dx}{dt} + x\frac{dy}{dt} + y\frac{dx}{dt} = 0$$

$$x\frac{dy}{dt} = -2x\frac{dx}{dt} - y\frac{dx}{dt}$$

$$\frac{dy}{dt} = -2\frac{dx}{dt} - \frac{y}{x}\frac{dx}{dt}$$

Given: $\frac{dx}{dt} = -1$ when $x = 2$ and $y = -3$. Thus, at $(2, -3)$ and $\frac{dx}{dt} = -1$,

$$\frac{dy}{dt} = -2(-1) - \frac{-3}{2}(-1) = 2 - \frac{3}{2} = \frac{1}{2}.$$

7. $xy = 36$

Differentiate with respect to t:

$$\frac{d(xy)}{dt} = \frac{d(36)}{dt}$$

$$x\frac{dy}{dt} + y\frac{dx}{dt} = 0$$

Given: $\frac{dx}{dt} = 4$ when $x = 4$ and $y = 9$. Therefore,

$$4\frac{dy}{dt} + 9(4) = 0$$

$$4\frac{dy}{dt} = -36$$

and $\quad \dfrac{dy}{dt} = -9.$

The y coordinate is decreasing at 9 units per second.

9.

From the triangle,

$$x^2 + y^2 = z^2$$

or $\quad x^2 + 16 = z^2$, since $y = 4$.

Differentiate with respect to t:

$$2x\frac{dx}{dt} = 2z\frac{dz}{dt}$$

or $\quad x\dfrac{dx}{dt} = z\dfrac{dz}{dt}$

Given: $\frac{dz}{dt} = -3$. Also, when $x = 30$, $900 + 16 = z^2$ or $z = \sqrt{916}$.

Therefore,

$$30 \frac{dx}{dt} = \sqrt{916}(-3) \qquad \text{and} \qquad \frac{dx}{dt} = \frac{-3\sqrt{916}}{30} = \frac{-\sqrt{916}}{10} \approx \frac{-30.27}{10}$$

$$\approx -3.03 \text{ feet per second.}$$

[Note: The negative sign indicates that the distance between the boat and the dock is decreasing.]

11. Area: $A = \pi R^2$

$$\frac{dA}{dt} = \frac{d\pi R^2}{dt} = \pi \cdot 2R \frac{dR}{dt}$$

Given: $\frac{dR}{dt} = 2$ ft/sec

$$\frac{dA}{dt} = 2\pi R \cdot 2 = 4\pi R$$

$$\left. \frac{dA}{dt} \right|_{R = 10 \text{ ft}} = 4\pi(10) = 40\pi \text{ ft}^2/\text{sec}$$
$$\approx 126 \text{ ft}^2/\text{sec}$$

13. $V = \frac{4}{3}\pi R^3$

$$\frac{dV}{dt} = \frac{4}{3}\pi 3R^2 \frac{dR}{dt} = 4\pi R^2 \frac{dR}{dt}$$

Given: $\frac{dR}{dt} = 3$ cm/min

$$\frac{dV}{dt} = 4\pi R^2 3 = 12\pi R^2$$

$$\left. \frac{dV}{dt} \right|_{R = 10 \text{ cm}} = 12\pi(10)^2 = 1200\pi$$
$$\approx 3768 \text{ cm}^3/\text{min}$$

15. $\frac{P}{T} = k \qquad (1)$

$P = kT$

Differentiate with respect to t:

$$\frac{dP}{dt} = k \frac{dT}{dt}$$

Given: $\frac{dT}{dt} = 3$ degrees per hour, $T = 250°$, $P = 500$ pounds per square inch.

From (1), for $T = 250$ and $P = 500$,

$$k = \frac{500}{250} = 2.$$

Thus, we have

$$\frac{dP}{dt} = 2 \frac{dT}{dt}$$

$$\frac{dP}{dt} = 2(3) = 6$$

Pressure increases at 6 pounds per square inch per hour.

17. By the Pythagorean theorem,

$$x^2 + y^2 = 10^2$$

or $x^2 + y^2 = 100 \qquad (1)$

Differentiate with respect to t:

$$2x \frac{dx}{dt} + 2y \frac{dy}{dt} = 0$$

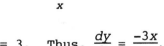

Therefore, $\frac{dy}{dt} = -\frac{x}{y}\frac{dx}{dt}$. Given: $\frac{dx}{dt} = 3$. Thus, $\frac{dy}{dt} = \frac{-3x}{y}$.

From (1), $y^2 = 100 - x^2$ and, when $x = 6$,

$$y^2 = 100 - 6^2$$
$$= 100 - 36 = 64.$$

Thus, $y = 8$ when $x = 6$, and

$$\left. \frac{dy}{dt} \right|_{(6,\ 8)} = \frac{-3(6)}{8} = \frac{-18}{8} = \frac{-9}{4} \text{ ft/sec.}$$

19. y = length of shadow
x = distance of man from light
z = distance of tip of shadow from light

We want to compute $\dfrac{dz}{dt}$. Triangles ABE and CDE are similar triangles; thus, the ratios of corresponding sides are equal.

Therefore, $\dfrac{z}{20} = \dfrac{y}{5} = \dfrac{z-x}{5}$ [Note: $y = z - x$.]

or $\qquad \dfrac{z}{20} = \dfrac{z-x}{5}$

$$z = 4(z-x)$$
$$z = 4z - 4x$$
$$4x = 3z$$

Differentiate with respect to t:

$$4\frac{dx}{dt} = 3\frac{dz}{dt}$$
$$\frac{dz}{dt} = \frac{4}{3}\frac{dx}{dt}$$

Given: $\dfrac{dx}{dt} = 5$. Thus, $\dfrac{dz}{dt} = \dfrac{4}{3}(5) = \dfrac{20}{3}$ ft/sec.

21. $V = \dfrac{4}{3}\pi r^3$ (1)

Differentiate with respect to t:

$$\frac{dV}{dt} = 4\pi r^2 \frac{dr}{dt} \quad \text{and} \quad \frac{dr}{dt} = \frac{1}{4\pi r^2} \cdot \frac{dV}{dt}$$

Since $\dfrac{dV}{dt} = 4$ cu ft/sec,

$$\frac{dr}{dt} = \frac{1}{4\pi r^2}(4) = \frac{1}{\pi r^2} \text{ ft/sec} \quad (2)$$

At $t = 1$ minute $= 60$ seconds,
$\quad V = 4(60) = 240$ cu ft and, from (1),

$$r^3 = \frac{3V}{4\pi} = \frac{3(240)}{4\pi} = \frac{180}{\pi}; \quad r = \left(\frac{180}{\pi}\right)^{1/3} \approx 3.855.$$

From (2)

$$\frac{dr}{dt} = \frac{1}{\pi(3.855)^2} \approx 0.0214 \text{ ft/sec}$$

At $t = 2$ minutes $= 120$ seconds,
$\quad V = 4(120) = 480$ cu ft and

$$r^3 = \frac{3V}{4\pi} = \frac{3(480)}{4\pi} = \frac{360}{\pi}; \quad r = \left(\frac{360}{\pi}\right)^{1/3} \approx 4.857$$

From (2),

$$\frac{dr}{dt} = \frac{1}{\pi(4.857)^2} \approx 0.0135 \text{ ft/sec}$$

To find the time at which $\frac{dr}{dt} = 100$ ft/sec, solve

$$\frac{1}{\pi r^2} = 100$$

$$r^2 = \frac{1}{100\pi}$$

$$r = \frac{1}{\sqrt{100\pi}} = \frac{1}{10\sqrt{\pi}}$$

Now, when $r = \frac{1}{10\sqrt{\pi}}$,

$$V = \frac{4}{3}\pi\left(\frac{1}{10\sqrt{\pi}}\right)^3$$

$$= \frac{4}{3} \cdot \frac{1}{1000\sqrt{\pi}}$$

$$= \frac{1}{750\sqrt{\pi}}$$

Since the volume at time t is $4t$, we have

$$4t = \frac{1}{750\sqrt{\pi}} \quad \text{and}$$

$$t = \frac{1}{3000\sqrt{\pi}} \approx 0.00019 \text{ secs.}$$

23. $y = e^x + x + 1;\ \frac{dx}{dt} = 3.$

Differentiate with respect to t:
$$\frac{dy}{dt} = e^x \frac{dx}{dt} + \frac{dx}{dt} = e^x(3) + 3 = 3(e^x + 1)$$

To find where the point crosses the x axis, use a graphing utility to solve
$$e^x + x + 1 = 0$$
The result is $x \approx -1.278$.

Now, at $x = -1.278$,
$$\frac{dy}{dt} = 3(e^{-1.278} + 1) \approx 3.835 \text{ units/sec.}$$

25. $C = 90,000 + 30x$ (1)

$R = 300x - \dfrac{x^2}{30}$ (2)

$P = R - C$ (3)

(A) Differentiating (1) with respect to t:
$$\frac{dC}{dt} = \frac{d(90,000)}{dt} + \frac{d(30x)}{dt}$$
$$\frac{dC}{dt} = 30\frac{dx}{dt}$$

Thus, $\dfrac{dC}{dt} = 30(500)$ $\left(\dfrac{dx}{dt} = 500\right)$

$\phantom{Thus, \frac{dC}{dt}} = \$15,000$ per week.

Costs are increasing at \$15,000 per week at this production level.

(B) Differentiating (2) with respect to t:

$$\frac{dR}{dt} = \frac{d(300x)}{dt} - \frac{d\frac{x^2}{30}}{dt}$$

$$= 300\frac{dx}{dt} - \frac{2x}{30}\frac{dx}{dt}$$

$$= \left(300 - \frac{x}{15}\right)\frac{dx}{dt}$$

Thus, $\dfrac{dR}{dt} = \left(300 - \dfrac{6000}{15}\right)(500) \qquad \left(x = 6000, \dfrac{dx}{dt} = 500\right)$

$$= (-100)500$$

$$= -\$50{,}000 \text{ per week.}$$

Revenue is decreasing at \$50,000 per week at this production level.

(C) Differentiating (3) with respect to t:

$$\frac{dP}{dt} = \frac{dR}{dt} - \frac{dC}{dt}$$

Thus, from parts (A) and (B), we have:

$$\frac{dP}{dt} = -50{,}000 - 15{,}000 = -\$65{,}000$$

Profits are decreasing at \$65,000 per week at this production level.

27. $S = 60{,}000 - 40{,}000e^{-0.0005x}$

Differentiating implicitly with respect to t, we have

$$\frac{ds}{dt} = -40{,}000(-0.0005)e^{-0.0005x}\frac{dx}{dt} \text{ and } \frac{ds}{dt} = 20e^{-0.0005x}\frac{dx}{dt}$$

Now, for $x = 2000$ and $\dfrac{dx}{dt} = 300$, we have

$$\frac{ds}{dt} = 20(300)e^{-0.0005(2000)}$$

$$= 6000e^{-1} = 2{,}207$$

Thus, sales are increasing at the rate of \$2,207 per week.

29. Price p and demand x are related by the equation

$$2x^2 + 5xp + 50p^2 = 80{,}000 \qquad (1)$$

Differentiating implicitly with respect to t, we have

$$4x\frac{dx}{dt} + 5x\frac{dp}{dt} + 5p\frac{dx}{dt} + 100p\frac{dp}{dt} = 0 \qquad (2)$$

(A) From (2), $\dfrac{dx}{dt} = \dfrac{-(5x + 100p)\frac{dp}{dt}}{4x + 5p}$

Setting $p = 30$ in (1), we get

$$2x^2 + 150x + 45{,}000 = 80{,}000$$

or $\quad x^2 + 75x - 17{,}500 = 0$

Thus, $x = \dfrac{-75 \pm \sqrt{(75)^2 + 70{,}000}}{2}$

$$= \frac{-75 \pm 275}{2} = 100, -175$$

Since $x \geq 0$, $x = 100$

Now, for $x = 100$, $p = 30$ and $\frac{dp}{dt} = 2$, we have

$$\frac{dx}{dt} = \frac{-[5(100) + 100(30)] \cdot 2}{4(100) + 5(30)} = -\frac{7000}{550} \text{ and } \frac{dx}{dt} = -12.73$$

The demand is decreasing at the rate of -12.73 units/month.

(B) From (2), $\dfrac{dp}{dt} = \dfrac{-(4x + 5p)\frac{dx}{dt}}{(5x + 100p)}$

Setting $x = 150$ in (1), we get

$$45{,}000 + 750p + 50p^2 = 80{,}000$$

or $\qquad p^2 + 15p - 700 = 0$

and $p = \dfrac{-15 \pm \sqrt{225 + 2800}}{2} = \dfrac{-15 \pm 55}{2} = -35,\ 20$

Since $p \geq 0$, $p = 20$.

Now, for $x = 150$, $p = 20$ and $\frac{dx}{dt} = -6$, we have

$$\frac{dp}{dt} = -\frac{[4(150) + 5(20)](-6)}{5(150) + 100(20)} = \frac{4200}{2750} \approx 1.53$$

Thus, the price is increasing at the rate of \$1.53 per month.

31. Volume $V = \pi R^2 h$, where h = thickness of the circular oil slick.

Since $h = 0.1 = \dfrac{1}{10}$, we have:

$$V = \frac{\pi}{10} R^2$$

Differentiating with respect to t:

$$\frac{dV}{dt} = \frac{d\left(\frac{\pi}{10}R^2\right)}{dt} = \frac{\pi}{10} 2R \frac{dR}{dt} = \frac{\pi}{5} R \frac{dR}{dt}$$

Given: $\dfrac{dR}{dt} = 0.32$ when $R = 500$. Therefore,

$$\frac{dV}{dt} = \frac{\pi}{5}(500)(0.32) = 100\pi(0.32) \approx 100.48 \text{ cubic feet per minute.}$$

CHAPTER 5 REVIEW

1. $A(t) = 2000e^{0.09t}$

$A(5) = 2000e^{0.09(5)} = 2000e^{0.45} \approx 3136.62 \quad \text{or} \quad \$3136.62$

$A(10) = 2000e^{0.09(10)} = 2000e^{0.9} \approx 4919.21 \quad \text{or} \quad \$4919.21$

$A(20) = 2000e^{0.09(20)} = 2000e^{1.8} \approx 12{,}099.29 \quad \text{or} \quad \$12{,}099.29 \qquad (5\text{-}1)$

2. $\dfrac{d}{dx}(2 \ln x + 3e^x) = 2\dfrac{d}{dx} \ln x + 3\dfrac{d}{dx}e^x = \dfrac{2}{x} + 3e^x \qquad (5\text{-}2)$

3. $\dfrac{d}{dx}e^{2x-3} = e^{2x-3}\dfrac{d}{dx}(2x - 3) \qquad$ (by the chain rule)

$\qquad = 2e^{2x-3} \qquad\qquad\qquad\qquad\qquad\qquad\qquad\qquad (5\text{-}2)$

4. $y = \ln(2x + 7)$

$y' = \dfrac{1}{2x + 7}(2)$ (by the chain rule)

$\quad = \dfrac{2}{2x + 7}$ (5-2)

5. $y = \ln u$, where $u = 3 + e^x$.

(A) $y = \ln[3 + e^x]$

(B) $\dfrac{dy}{dx} = \dfrac{dy}{du} \cdot \dfrac{du}{dx} = \dfrac{1}{u}(e^x) = \dfrac{1}{3 + e^x}(e^x) = \dfrac{e^x}{3 + e^x}$ (5-3)

6. $\dfrac{d}{dx}2y^2 - \dfrac{d}{dx}3x^3 - \dfrac{d}{dx}5 = \dfrac{d}{dx}(0)$ **7.** $y = 3x^2 - 5$

$\qquad 4yy' - 9x^2 - 0 = 0$ $\qquad \dfrac{dy}{dt} = \dfrac{d(3x^2)}{dt} - \dfrac{d(5)}{dt}$

$\qquad\qquad y' = \dfrac{9x^2}{4y}$ $\qquad \dfrac{dy}{dt} = 6x\dfrac{dx}{dt}$

$\qquad \dfrac{dy}{dx}\bigg|_{(1,\ 2)} = \dfrac{9 \cdot 1^2}{4 \cdot 2} = \dfrac{9}{8}$ $\qquad x = 12;\ \dfrac{dx}{dt} = 3$

$\qquad\qquad\qquad\qquad$ (5-4) $\qquad \dfrac{dy}{dt} = 6 \cdot 12 \cdot 3 = 216$ (5-4)

8. $y = 100e^{-0.1x}$

<u>Step 1. Analyze $f(x)$</u>:

(A) Domain: All real numbers, $(-\infty, \infty)$.

(B) Intercepts: y intercept: $f(0) = 100e^{-0.1(0)} = 100$

$\qquad\qquad\qquad\quad$ x intercept: Since $100e^{-0.1x} \neq 0$ for all x, there are no x intercepts.

(C) Asymptotes:

$\qquad \lim\limits_{x \to \infty} 100e^{-0.1x} = \lim\limits_{x \to \infty} \dfrac{100}{e^{0.1x}} = 0$

$\qquad \lim\limits_{x \to -\infty} 100e^{-0.1x}$ does not exist.

Thus, $y = 0$ is a horizontal asymptote. There are no vertical asymptotes.

<u>Step 2. Analyze $f'(x)$</u>:

$y' = 100e^{-0.1x}(-0.1)$

$\quad = -10e^{-0.1x} < 0$ on $(-\infty, \infty)$

Thus, y is decreasing on $(-\infty, \infty)$; there are no local extrema.

<u>Step 3. Analyze $f''(x)$</u>:

$y'' = -10e^{-0.1x}(-0.1)$

$\quad = e^{-0.1x} > 0$ on $(-\infty, \infty)$

Thus, the graph of f is concave upward on $(-\infty, \infty)$; there are no inflection points.

x	y
0	100
-1	~ 110
10	~ 37

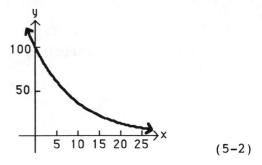

(5-2)

9. $\dfrac{d}{dz}[(\ln z)^7 + \ln z^7] = \dfrac{d}{dz}[\ln z]^7 + \dfrac{d}{dz}7\ln z$

$$= 7[\ln z]^6 \dfrac{d}{dz}\ln z + 7\dfrac{d}{dz}\ln z$$

$$= 7[\ln z]^6 \dfrac{1}{z} + \dfrac{7}{z}$$

$$= \dfrac{7(\ln z)^6 + 7}{z} = \dfrac{7[(\ln z)^6 + 1]}{z}$$ (5-3)

10. $\dfrac{d}{dx}x^6 \ln x = x^6 \dfrac{d}{dx}\ln x + (\ln x)\dfrac{d}{dx}x^6$

$$= x^6\left(\dfrac{1}{x}\right) + (\ln x)6x^5 = x^5(1 + 6\ln x)$$ (5-2)

11. $\dfrac{d}{dx}\left(\dfrac{e^x}{x^6}\right) = \dfrac{x^6 \dfrac{d}{dx}e^x - e^x \dfrac{d}{dx}x^6}{(x^6)^2} = \dfrac{x^6 e^x - 6x^5 e^x}{x^{12}} = \dfrac{xe^x - 6e^x}{x^7} = \dfrac{e^x(x - 6)}{x^7}$ (5-2)

12. $y = \ln(2x^3 - 3x)$

$y' = \dfrac{1}{2x^3 - 3x}(6x^2 - 3) = \dfrac{6x^2 - 3}{2x^3 - 3x}$

(5-3)

13. $f(x) = e^{x^3-x^2}$

$f'(x) = e^{x^3-x^2}(3x^2 - 2x)$

$= (3x^2 - 2x)e^{x^3-x^2}$ (5-3)

14. $y = e^{-2x}\ln 5x$

$\dfrac{dy}{dx} = e^{-2x}\left(\dfrac{1}{5x}\right)(5) + (\ln 5x)(e^{-2x})(-2)$

$$= e^{-2x}\left(\dfrac{1}{x} - 2\ln 5x\right) = \dfrac{1 - 2x\ln 5x}{xe^{2x}}$$ (5-3)

15. $f(x) = 1 + e^{-x}$

$f'(x) = e^{-x}(-1) = -e^{-x}$

An equation for the tangent line to the graph of f at $x = 0$ is:

$y - y_1 = m(x - x_1)$,

where $x_1 = 0$, $y_1 = f(0) = 1 + e^0 = 2$, and $m = f'(0) = -e^0 = -1$.

Thus, $y - 2 = -1(x - 0)$ or $y = -x + 2$.

An equation for the tangent line to the graph of f at $x = -1$ is

$y - y_1 = m(x - x_1)$,

where $x_1 = -1$, $y_1 = f(-1) = 1 + e$, and $m = f'(-1) = -e$. Thus,

$y - (1 + e) = -e[x - (-1)]$ or $y - 1 - e = -ex - e$ and $y = -ex + 1$.

(5-2)

16. $x^2 - 3xy + 4y^2 = 23$

Differentiate implicitly:

$2x - 3(xy' + y \cdot 1) + 8yy' = 0$

$\quad 2x - 3xy' - 3y + 8yy' = 0$

$\qquad\qquad 8yy' - 3xy' = 3y - 2x$

$\qquad\qquad (8y - 3x)y' = 3y - 2x$

$\qquad\qquad\qquad y' = \dfrac{3y - 2x}{8y - 3x}$

$\qquad y'\Big|_{(-1,\ 2)} = \dfrac{3 \cdot 2 - 2(-1)}{8 \cdot 2 - 3(-1)} = \dfrac{8}{19}$ [Slope at $(-1, 2)$] $\qquad$ (5-4)

17.
$$x^3 - 2t^2x + 8 = 0$$

$3x^2x' - (2t^2x' + x \cdot 4t) + 0 = 0$

$\qquad 3x^2x' - 2t^2x' - 4xt = 0$

$\qquad\quad (3x^2 - 2t^2)x' = 4xt$

$\qquad\qquad\qquad x' = \dfrac{4xt}{3x^2 - 2t^2}$

$\qquad x'\Big|_{(-2,\ 2)} = \dfrac{4 \cdot 2 \cdot (-2)}{3(2^2) - 2(-2)^2} = \dfrac{-16}{12 - 8} = \dfrac{-16}{4} = -4$ $\qquad$ (5-4)

18. $x - y^2 = e^y$

Differentiate implicitly:

$1 - 2yy' = e^y y'$

$\qquad 1 = e^y y' + 2yy'$

$\qquad 1 = y'(e^y + 2y)$

$\qquad y' = \dfrac{1}{e^y + 2y}$

$y'\Big|_{(1,\ 0)} = \dfrac{1}{e^0 + 2 \cdot 0} = 1$ $\quad$ (5-4)

19. $\ln y = x^2 - y^2$

Differentiate implicitly:

$\dfrac{y'}{y} = 2x - 2yy'$

$y'\left(\dfrac{1}{y} + 2y\right) = 2x$

$y'\left(\dfrac{1 + 2y^2}{y}\right) = 2x$

$\qquad y' = \dfrac{2xy}{1 + 2y^2}$

$\qquad y'\Big|_{(1,\ 1)} = \dfrac{2 \cdot 1 \cdot 1}{1 + 2(1)^2} = \dfrac{2}{3}$ $\qquad$ (5-4)

20. $y^2 - 4x^2 = 12$

Differentiate with respect to t:

$2y\dfrac{dy}{dt} - 8x\dfrac{dx}{dt} = 0$

Given: $\dfrac{dx}{dt} = -2$ when $x = 1$ and $y = 4$. Therefore,

$2 \cdot 4\dfrac{dy}{dt} - 8 \cdot 1 \cdot (-2) = 0$

$\qquad 8\dfrac{dy}{dt} + 16 = 0$

$\qquad\qquad \dfrac{dy}{dt} = -2.$

The y coordinate is decreasing at 2 units per second. $\qquad$ (5-5)

21. From the figure, $x^2 + y^2 = 17^2$.

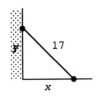

Differentiate with respect to t:

$$2x\frac{dx}{dt} + 2y\frac{dy}{dt} = 0 \quad \text{or} \quad x\frac{dx}{dt} + y\frac{dy}{dt} = 0$$

We are given $\frac{dx}{dt} = -0.5$ feet per second. Therefore,

$$x(-0.5) + y\frac{dy}{dt} = 0 \quad \text{or} \quad \frac{dy}{dt} = \frac{0.5x}{y} = \frac{x}{2y}$$

Now, when $x = 8$, we have: $8^2 + y^2 = 17^2$

$$y^2 = 289 - 64 = 225$$
$$y = 15$$

Therefore, $\left.\dfrac{dy}{dt}\right|_{(8,\ 15)} = \dfrac{8}{2(15)} = \dfrac{4}{15} \approx 0.27$ ft/sec.   (5-5)

22. $A = \pi R^2$. Given: $\frac{dA}{dt} = 24$ square inches per minute.

Differentiate with respect to t:

$$\frac{dA}{dt} = 2\pi R\frac{dR}{dt}$$

$$24 = 2\pi R\frac{dR}{dt}$$

Therefore, $\dfrac{dR}{dt} = \dfrac{24}{2\pi R} = \dfrac{12}{\pi R}$.

$$\left.\frac{dR}{dt}\right|_{(R\ =\ 12)} = \frac{12}{\pi \cdot 12} = \frac{1}{\pi} \approx 0.318 \text{ inches per minute}$$ (5-5)

23. $f(x) = 11x - 2x \ln x,\ x > 0$

$$f'(x) = 11 - 2x\left(\frac{1}{x}\right) - (\ln x)(2)$$

$$= 11 - 2 - 2\ln x = 9 - 2\ln x,\ x > 0$$

Critical value(s): $f'(x) = 9 - 2\ln x = 0$

$$2\ln x = 9$$
$$\ln x = \frac{9}{2}$$
$$x = e^{9/2}$$

$f''(x) = -\dfrac{2}{x}$ and $f''(e^{9/2}) = -\dfrac{2}{e^{9/2}} < 0$

Since $x = e^{9/2}$ is the only critical value, and $f''(e^{9/2}) < 0$, f has an absolute maximum at $x = e^{9/2}$. The absolute maximum is:

$$f(e^{9/2}) = 11e^{9/2} - 2e^{9/2}\ln(e^{9/2})$$
$$= 11e^{9/2} - 9e^{9/2}$$
$$= 2e^{9/2} \approx 180.03$$ (5-2)

24. $f(x) = 10xe^{-2x},\ x > 0$

$$f'(x) = 10xe^{-2x}(-2) + 10e^{-2x}(1) = 10e^{-2x}(1 - 2x),\ x > 0$$

Critical value(s): $f'(x) = 10e^{-2x}(1 - 2x) = 0$

$$1 - 2x = 0$$
$$x = \frac{1}{2}$$

$$f''(x) = 10e^{-2x}(-2) + 10(1 - 2x)e^{-2x}(-2)$$
$$= -20e^{-2x}(1 + 1 - 2x)$$
$$= -40e^{-2x}(1 - x)$$
$$f''\left(\frac{1}{2}\right) = -20e^{-1} < 0$$

Since $x = \frac{1}{2}$ is the only critical value, and $f''\left(\frac{1}{2}\right) = -20e^{-1} < 0$, f has an absolute maximum at $x = \frac{1}{2}$. The absolute maximum of f is:

$$f\left(\frac{1}{2}\right) = 10\left(\frac{1}{2}\right)e^{-2(1/2)}$$
$$= 5e^{-1} \approx 1.84 \qquad (5\text{-}3)$$

25. $f(x) = 3x - x^2 + e^{-x}, \ x > 0$

$f'(x) = 3 - 2x - e^{-x}, \ x > 0$

Critical value(s): $f'(x) = 3 - 2x - e^{-x} = 0$

$$x \approx 1.373$$

$f''(x) = -2 + e^{-x}$ and $f''(1.373) = -2 + e^{-1.373} < 0$

Since $x \approx 1.373$ is the only critical value, and $f''(1.373) < 0$, f has an absolute maximum at $x = 1.373$. The absolute maximum of f is:

$$f(1.373) = 3(1.373) - (1.373)^2 + e^{-1.373}$$
$$\approx 2.487 \qquad (5\text{-}2)$$

26. $f(x) = \dfrac{\ln x}{e^x}, \ x > 0$

$$f'(x) = \frac{e^x\left(\frac{1}{x}\right) - (\ln x)e^x}{(e^x)^2} = \frac{e^x\left(\frac{1}{x} - \ln x\right)}{e^{2x}}$$
$$= \frac{1 - x \ln x}{xe^x}, \ x > 0$$

Critical value(s): $f'(x) = \dfrac{1 - x \ln x}{xe^x} = 0$

$$1 - x \ln x = 0$$
$$x \ln x = 1$$
$$x \approx 1.763$$

$$f''(x) = \frac{xe^x[-1 - \ln x] - (1 - x \ln x)(xe^x + e^x)}{x^2 e^{2x}}$$
$$= \frac{-x(1 + \ln x) - (x + 1)(1 - x \ln x)}{x^2 e^x};$$
$$f''(1.763) \approx \frac{-1.763(1.567) - (2.763)(0.000349)}{(1.763)^2 e^{1.763}} < 0$$

Since $x = 1.763$ is the only critical value, and $f''(1.763) < 0$, f has an absolute maximum at $x = 1.763$. The absolute maximum of f is:

$$f(1.763) = \frac{\ln(1.763)}{e^{1.763}} \approx 0.097 \qquad (5\text{-}2)$$

27. $f(x) = 5 - 5e^{-x}$

Step 1. Analyze $f(x)$:

(A) Domain: All real numbers, $(-\infty, \infty)$.

(B) Intercepts: y intercept: $f(0) = 5 - 5e^{-0} = 0$

x intercepts: $5 - 5e^{-x} = 0$

$e^{-x} = 1$

$x = 0$

(C) Asymptotes:

$$\lim_{x \to \infty} (5 - 5e^{-x}) = \lim_{x \to \infty} \left(5 - \frac{5}{e^x}\right) = 5$$

$\lim_{x \to -\infty} (5 - 5e^{-x})$ does not exist.

Thus, $y = 5$ is a horizontal asymptote.

Since $f(x) = 5 - \dfrac{5}{e^x} = \dfrac{5e^x - 5}{e^x}$ and $e^x \neq 0$ for all x, there are no vertical asymptotes.

Step 2. Analyze $f'(x)$:

$f'(x) = -5e^{-x}(-1) = 5e^{-x} > 0$ on $(-\infty, \infty)$

Thus, f is increasing on $(-\infty, \infty)$; there are no local extrema.

Step 3. Analyze $f''(x)$:

$f''(x) = -5e^{-x} < 0$ on $(-\infty, \infty)$.

Thus, the graph of f is concave downward on $(-\infty, \infty)$; there are no inflection points.

Step 4. Sketch the graph of f:

x	$f(x)$
0	0
-1	-8.59
2	4.32

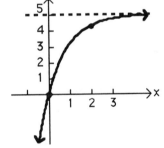

28. $f(x) = x^3 \ln x$ (5-2)

Step 1. Analyze $f(x)$:

(A) Domain: all positive real numbers, $(0, \infty)$.

(B) Intercepts: y intercept: Since $x = 0$ is not in the domain, there is no y intercept.

x intercepts: $x^3 \ln x = 0$

$\ln x = 0$

$x = 1$

(C) Asymptotes:

$\lim_{x \to \infty} (x^3 \ln x)$ does not exist.

It can be shown that $\lim_{x \to 0^+} (x^3 \ln x) = 0$. Thus, there are no horizontal or vertical asymptotes.

Step 2. Analyze $f'(x)$:

$$f'(x) = x^3\left(\frac{1}{x}\right) + (\ln x)3x^2$$

$$= x^2[1 + 3\ln x], \quad x > 0$$

Critical values: $x^2[1 + 3\ln x] = 0$

$$1 + 3\ln x = 0 \quad \text{(since } x > 0\text{)}$$

$$\ln x = -\frac{1}{3}$$

$$x = e^{-1/3} \approx 0.72$$

Partition numbers: $x = e^{-1/3}$

Sign chart for f':

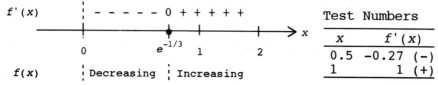

x	f'(x)
0.5	-0.27 (−)
1	1 (+)

Test Numbers

Thus, f is decreasing on $(0, e^{-1/3})$ and increasing on $(e^{-1/3}, \infty)$; f has a local minimum at $x = e^{-1/3}$.

Step 3. Analyze $f''(x)$:

$$f''(x) = x^2\left(\frac{3}{x}\right) + (1 + 3\ln x)2x$$

$$= x(5 + 6\ln x), \quad x > 0$$

Partition numbers: $x(5 + 6\ln x) = 0$

$$5 + 6\ln x = 0$$

$$\ln x = -\frac{5}{6}$$

$$x = e^{-5/6} \approx 0.43$$

Sign chart for f'':

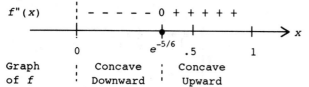

x	f''(x)
.2	-0.93 (−)
1	5 (+)

Test Numbers

Thus, the graph of f is concave downward on $(0, e^{-5/6})$ and concave upward on $(e^{-5/6}, \infty)$; the graph has an inflection point at $x = e^{-5/6}$.

Step 4. Sketch the graph of f:

x	f(x)
$e^{-5/6}$	-0.07
$e^{-1/3}$	-0.12
1	0

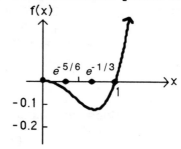

(5-2)

29. $y = w^3$, $w = \ln u$, $u = 4 - e^x$

   (A) $y = [\ln(4 - e^x)]^3$

   (B) $\dfrac{dy}{dx} = \dfrac{dy}{dw} \cdot \dfrac{dw}{du} \cdot \dfrac{du}{dx}$

$$= 3w^2 \cdot \dfrac{1}{u} \cdot (-e^x) = 3[\ln(4 - e^x)]^2 \left(\dfrac{1}{4 - e^x}\right)(-e^x)$$

$$= \dfrac{-3e^x[\ln(4 - e^x)]^2}{4 - e^x} \qquad\qquad (5\text{-}3)$$

30. $y = 5^{x^2-1}$

$$y' = 5^{x^2-1}(\ln 5)(2x) = 2x5^{x^2-1}(\ln 5) \qquad\qquad (5\text{-}3)$$

31. $\dfrac{d}{dx} \log_5(x^2 - x) = \dfrac{1}{x^2 - x} \cdot \dfrac{1}{\ln 5} \cdot \dfrac{d}{dx}(x^2 - x) = \dfrac{1}{\ln 5} \cdot \dfrac{2x - 1}{x^2 - x}$ $(5\text{-}3)$

32. $\dfrac{d}{dx}\sqrt{\ln(x^2 + x)} = \dfrac{d}{dx}[\ln(x^2 + x)]^{1/2} = \dfrac{1}{2}[\ln(x^2 + x)]^{-1/2}\dfrac{d}{dx}\ln(x^2 + x)$

$$= \dfrac{1}{2}[\ln(x^2 + x)]^{-1/2} \dfrac{1}{x^2 + x} \dfrac{d}{dx}(x^2 + x)$$

$$= \dfrac{1}{2}[\ln(x^2 + x)]^{-1/2} \cdot \dfrac{2x + 1}{x^2 + x} = \dfrac{2x + 1}{2(x^2 + x)[\ln(x^2 + x)]^{1/2}}$$

$$(5\text{-}3)$$

33. $e^{xy} = x^2 + y + 1$

Differentiate implicitly:

$$\dfrac{d}{dx}e^{xy} = \dfrac{d}{dx}x^2 + \dfrac{d}{dx}y + \dfrac{d}{dx}1$$

$$e^{xy}(xy' + y) = 2x + y'$$

$$xe^{xy}y' - y' = 2x - ye^{xy}$$

$$y' = \dfrac{2x - ye^{xy}}{xe^{xy} - 1}$$

$$y'\bigg|_{(0,\,0)} = \dfrac{2 \cdot 0 - 0 \cdot e^0}{0 \cdot e^0 - 1} = 0 \qquad\qquad (5\text{-}4)$$

34. $A = \pi r^2$, $r \geq 0$

Differentiate with respect to t:

$$\dfrac{dA}{dt} = 2\pi r\dfrac{dr}{dt} = 6\pi r \quad \text{since} \quad \dfrac{dr}{dt} = 3$$

The area increases at the rate $6\pi r$. This is smallest when $r = 0$; there is no largest value. $(5\text{-}5)$

35. $y = x^3$

Differentiate with respect to t:

$$\dfrac{dy}{dt} = 3x^2\dfrac{dx}{dt}$$

Solving for $\dfrac{dx}{dt}$, we get

$$\dfrac{dx}{dt} = \dfrac{1}{3x^2} \cdot \dfrac{dy}{dt} = \dfrac{5}{3x^2} \quad \text{since} \quad \dfrac{dy}{dt} = 5$$

To find where $\dfrac{dx}{dt} > \dfrac{dy}{dt}$, solve the inequality

$$\frac{5}{3x^2} > 5$$

$$\frac{1}{3x^2} > 1$$

$$3x^2 < 1$$

$$-\frac{1}{\sqrt{3}} < x < \frac{1}{\sqrt{3}} \quad \text{or} \quad \frac{-\sqrt{3}}{3} < x < \frac{\sqrt{3}}{3} \tag{5-5}$$

36. (A) The compound interest formula is: $A = P(1 + r)^t$. Thus, the time for P to double when $r = 0.05$ and interest is compounded annually can be found by solving

$$2P = P(1 + 0.05)^t \quad \text{or} \quad 2 = (1.05)^t \quad \text{for } t.$$
$$\ln(1.05)^t = \ln 2$$
$$t \ln(1.05) = \ln 2$$
$$t = \frac{\ln 2}{\ln(1.05)} \approx 14.2 \text{ years}$$

(B) The continuous compound interest formula is: $A = Pe^{rt}$. Proceeding as above, we have

$$2P = Pe^{0.05t} \quad \text{or} \quad e^{0.05t} = 2.$$
Therefore, $0.05t = \ln 2$ and
$$t = \frac{\ln 2}{.05} \approx 13.9 \text{ years} \tag{5-1}$$

37. $A(t) = 100e^{0.1t}$
$A'(t) = 100(0.1)e^{0.1t} = 10e^{0.1t}$
$A'(1) = 11.05 \quad \text{or} \quad \$11.05 \text{ per year}$
$A'(10) = 27.18 \quad \text{or} \quad \$27.18 \text{ per year}$ $\tag{5-1}$

38. $R(x) = xp(x) = 1000xe^{-0.02x}$
$R'(x) = 1000[xD_x e^{-0.02x} + e^{-0.02x}D_x x]$
$\qquad = 1000[x(-0.02)e^{-0.02x} + e^{-0.02x}]$
$\qquad = (1000 - 20x)e^{-0.02x}$ $\tag{5-3}$

39. From Problem 38,
$R'(x) = (1000 - 20x)e^{-0.02x}$
Critical value(s): $R'(x) = (1000 - 20x)e^{-0.02x} = 0$
$$1000 - 20x = 0$$
$$x = 50$$
$R''(x) = (1000 - 20x)e^{-0.02x}(-0.02) + e^{-0.02x}(-20)$
$\qquad = e^{-0.02x}[0.4x - 20 - 20]$
$\qquad = e^{-0.02x}(0.4x - 40)$
$R''(50) = e^{-0.02(50)}[0.4(50) - 40] = -20e^{-1} < 0$

Since $x = 50$ is the only critical value and $R''(50) < 0$, R has an absolute maximum at a production level of 50 units. The maximum revenue is
$R(50) = 1000(50)e^{-0.02(50)} = 50,000e^{-1} \approx 18,394$ or $18,394.
The price per unit at the production level of 50 units is
$p(50) = 1000e^{-0.02(50)} = 1000e^{-1} \approx 367.88$ or $367.88.   (5-3)

40. $R(x) = 1000xe^{-0.02x}$, $0 \le x \le 100$

Step 1. Analyze $R(x)$:
(A) Domain: $0 \le x \le 100$ or $[0, 100]$

(B) Intercepts: y intercept: $R(0) = 0$
x intercepts: $100xe^{-0.02x} = 0$
$x = 0$

(C) Asymptotes: There are no horizontal or vertical asymptotes.

Step 2. Analyze $R'(x)$:
From Problems 47 and 48, $R'(x) = (1000 - 20x)e^{-0.02x}$ and $x = 50$ is a critical value.
Sign chart for R':

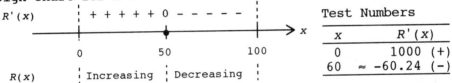

Test Numbers	
x	$R'(x)$
0	1000 (+)
60	$\approx$ -60.24 (-)

Thus, R is increasing on $(0, 50)$ and decreasing on $(50, 100)$; R has a maximum at $x = 50$.

Step 3. Analyze $R''(x)$:
$R''(x) = (0.4x - 40)e^{-0.02x} < 0$ on $(0, 100)$
Thus, the graph of R is concave downward on $(0, 100)$.

Step 4. Sketch the graph of R:

x	$R(x)$
0	0
50	18,394
100	13,533

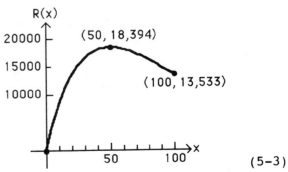

(5-3)

41. Cost: $C(x) = 200 + 50x - 50 \ln x$, $x \geq 1$

Average cost: $\overline{C} = \dfrac{C(x)}{x} = \dfrac{200}{x} + 50 - \dfrac{50}{x} \ln x$, $x \geq 1$

$\overline{C}'(x) = \dfrac{-200}{x^2} - \dfrac{50}{x}\left(\dfrac{1}{x}\right) + (\ln x)\dfrac{50}{x^2} = \dfrac{50(\ln x - 5)}{x^2}$, $x \geq 1$

Critical value(s): $\overline{C}'(x) = \dfrac{50(\ln x - 5)}{x^2} = 0$

$$\ln x = 5$$
$$x = e^5$$

Sign chart for $\overline{C}'$:

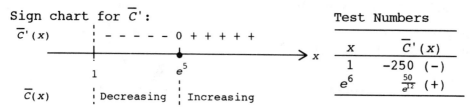

Test Numbers

x	$\overline{C}'(x)$
1	-250 $(-)$
e^6	$\dfrac{50}{e^{12}}$ $(+)$

By the first derivative test, $\overline{C}$ has a local minimum at $x = e^5$. Since this is the only critical value of $\overline{C}$, $\overline{C}$ has as absolute minimum at $x = e^5$. Thus, the minimal average cost is:

$\overline{C}(e^5) = \dfrac{200}{e^5} + 50 - \dfrac{50}{e^5} \ln(e^5)$

$\qquad = 50 - \dfrac{50}{e^5} \approx 49.66$ or \$49.66 (5-2)

42. $x = \sqrt{5000 - 2p^3} = (5000 - 2p^3)^{1/2}$

Differentiate implicitly with respect to x:

$1 = \dfrac{1}{2}(5000 - 2p^3)^{-1/2}(-6p^2)\dfrac{dp}{dx}$

$1 = \dfrac{-3p^2}{(5000 - 2p^3)^{1/2}}\dfrac{dp}{dx}$

$\dfrac{dp}{dx} = \dfrac{-(5000 - 2p^3)^{1/2}}{3p^2}$ (5-3)

43. Given: $R(x) = 36x - \dfrac{x^2}{20}$ and $\dfrac{dx}{dt} = 10$ when $x = 250$.

Differentiate with respect to t:

$\dfrac{dR}{dt} = 36\dfrac{dx}{dt} - \dfrac{1}{20}(2x)\dfrac{dx}{dt} = 36\dfrac{dx}{dt} - \dfrac{x}{10}\dfrac{dx}{dt}$

Thus, $\dfrac{dR}{dt}\bigg|_{x = 250 \text{ and } \frac{dx}{dt} = 10} = 36(10) - \dfrac{250}{10}(10)$

$= \$110$ per day (5-5)

44. $C(t) = 5e^{-0.3t}$

$C'(t) = 5e^{-0.3t}(-0.3) = -1.5e^{-0.3t}$

After one hour, the rate of change of concentration is

$C'(1) = -1.5e^{-0.3(1)} = -1.5e^{-0.3} \approx -1.111$ mg/ml per hour.

After five hours, the rate of change of concentration is

$C'(5) = -1.5e^{-0.3(5)} = -1.5e^{-1.5} \approx -0.335$ mg/ml per hour. (5-3)

45. Given: $A = \pi R^2$ and $\dfrac{dA}{dt} = -45$ mm^2 per day (negative because the area is decreasing).

Differentiate with respect to t:

$$\frac{dA}{dt} = \pi 2R \frac{dR}{dt}$$

$$-45 = 2\pi R \frac{dR}{dt}$$

$$\frac{dR}{dt} = -\frac{45}{2\pi R}$$

$$\left.\frac{dR}{dt}\right|_{R\,=\,15} = \frac{-45}{2\pi \cdot 15} = \frac{-3}{2\pi} \approx -0.477 \text{ mm per day} \qquad (5\text{-}5)$$

46. $N(t) = 10(1 - e^{-0.4t})$

(A) $N'(t) = -10e^{-0.4t}(-0.4) = 4e^{-0.4t}$

$N'(1) = 4e^{-0.4(1)} = 4e^{-0.4} \approx 2.68$.

Thus, learning is increasing at the rate of 2.68 units per day after 1 day.

$N'(5) = 4e^{-0.4(5)} = 4e^{-2} \approx 0.54$

Thus, learning is increasing at the rate of 0.54 units per day after 5 days.

(B) From (A), $N'(t) = 4e^{-0.4t} > 0$ on $(0, 10)$. Thus, N is increasing on $(0, 10)$.

$N''(t) = 4e^{-0.4t}(-0.4) = -1.6e^{-0.4t} < 0$ on $(0, 10)$.

Thus, the graph of N is concave downward on $(0, 10)$. The graph of N is:

t	$N(t)$
0	0
5	8.65
10	9.82

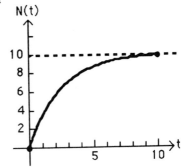

$(5\text{-}3)$

47. Given: $T = 2\left(1 + \dfrac{1}{x^{3/2}}\right) = 2 + 2x^{-3/2}$, and $\dfrac{dx}{dt} = 3$ when $x = 9$.

Differentiate with respect to t:

$$\frac{dT}{dt} = 0 + 2\left(-\frac{3}{2}x^{-5/2}\right)\frac{dx}{dt} = -3x^{-5/2}\frac{dx}{dt}$$

$$\left.\frac{dT}{dt}\right|_{x\,=\,9 \text{ and } \frac{dx}{dt}\,=\,3} = -3(9)^{-5/2}(3) = -3 \cdot 3^{-5} \cdot 3 = -3^{-3} = \frac{-1}{27}$$

$$\approx -0.037 \text{ minutes per operation per hour}$$

$(5\text{-}5)$

6 INTEGRATION

Things to remember:

1. A function $F(x)$ is an ANTIDERIVATIVE of $f(x)$ if $F'(x) = f(x)$.

2. THEOREM ON ANTIDERIVATIVES

    If the derivatives of two functions are equal on an open interval (a, b), then the functions can differ by at most a constant. Symbolically: If F and G are differentiable functions on the interval (a, b) and $F'(x) = G'(x)$ for all x in (a, b), then $F(x) = G(x) + k$ for some constant k.

3. The INDEFINITE INTEGRAL of $f(x)$, denoted

    $$\int f(x)\,dx,$$

    represents all antiderivatives of $f(x)$ and is given by

    $$\int f(x)\,dx = F(x) + C$$

    where $F(x)$ is any antiderivative of $f(x)$ and C is an arbitrary constant. The symbol $\int$ is called an INTEGRAL SIGN, the function $f(x)$ is called the INTEGRAND, and C is called the CONSTANT OF INTEGRATION.

4. Indefinite integration and differentiation are reverse operations (except for the addition of the constant of integration). This is expressed symbolically by:

    (a) $\dfrac{d}{dx}\left(\int f(x)\,dx\right) = f(x)$

    (b) $\int F'(x)\,dx = F(x) + C$

5. INDEFINITE INTEGRAL FORMULAS:

    (a) $\int k\,dx = kx + C,\ k$ constant

    (b) $\int x^n dx = \dfrac{x^{n+1}}{n+1} + C,\ n \neq -1$

    (c) $\int e^x dx = e^x + C$

    (d) $\int \dfrac{dx}{x} = \ln|x| + C,\ x \neq 0$

$\underline{6}$. INDEFINITE INTEGRATION PROPERTIES:

(a) $\displaystyle\int kf(x)\,dx = k \int f(x)\,dx$, k constant

(b) $\displaystyle\int [f(x) \pm g(x)]\,dx = \int f(x)\,dx \pm \int g(x)\,dx$

---

1. $\displaystyle\int 7\,dx = 7x + C$ [using $\underline{5}$(a)] Check: $(7x + C)' = 7$

3. $\displaystyle\int x^6\,dx = \frac{x^{6+1}}{6+1} + C$ [using $\underline{5}$(b)]

$\qquad\qquad = \frac{x^7}{7} + C$ Check: $\left(\frac{x^7}{7} + C\right)' = x^6$

5. $\displaystyle\int 8t^3\,dt = 8\int t^3\,dt$ [using $\underline{6}$(a)]

$\qquad\qquad = 8\,\frac{t^{3+1}}{3+1} + C = 2t^4 + C$ Check: $(2t^4 + C)' = 8t^3$

7. $\displaystyle\int (2u + 1)\,du = \int 2u\,du + \int 1\,du$ [using $\underline{6}$(b)]

$\qquad\qquad = 2\,\frac{u^2}{2} + u + C = u^2 + u + C$ Check: $(u^2 + u + C)' = 2u + 1$

9. $\displaystyle\int (3x^2 + 2x - 5)\,dx = \int 3x^2\,dx + \int 2x\,dx - \int 5\,dx$

$\qquad\qquad\qquad = 3\int x^2\,dx + 2\int x\,dx - \int 5\,dx$

$\qquad\qquad\qquad = 3\,\frac{x^3}{3} + 2\,\frac{x^2}{2} - 5x + C = x^3 + x^2 - 5x + C$

Check: $(x^3 + x^2 - 5x + C)' = 3x^2 + 2x - 5$

11. $\displaystyle\int (s^4 - 8s^5)\,ds = \int s^4\,ds - \int 8s^5\,ds = \int s^4\,ds - 8\int s^5\,ds$

$\qquad\qquad\qquad = \frac{s^5}{5} - 8\,\frac{s^6}{6} + C = \frac{s^5}{5} - \frac{4s^6}{3} + C$

Check: $\left(\frac{s^5}{5} - \frac{4s^6}{3} + C\right)' = s^4 - 8s^5$

13. $\displaystyle\int 3e^t\,dt = 3\int e^t\,dt$ [using $\underline{5}$(c)]

$\qquad\qquad = 3e^t + C$ Check: $(3e^t + C)' = 3e^t$

15. $\int 2z^{-1}dz = 2\int \frac{1}{z}dz = 2\ln|z| + C$ [using $\underline{5}$(d)]

Check: $(2\ln|z| + C)' = \frac{2}{z}$

17.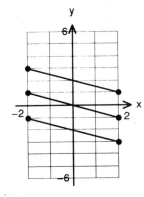

19.

21. $\frac{dy}{dx} = 200x^4$

$y = \int 200x^4 dx = 200 \int x^4 dx = 200\frac{x^5}{5} + C = 40x^5 + C$

23. $\frac{dP}{dx} = 24 - 6x$

$P = \int (24 - 6x)dx = \int 24\ dx - \int 6x\ dx = \int 24\ dx - 6\int x\ dx$

$= 24x - \frac{6x^2}{2} + C = 24x - 3x^2 + C$

25. $\frac{dy}{du} = 2u^5 - 3u^2 - 1$

$y = \int (2u^5 - 3u^2 - 1)du = \int 2u^5 du - \int 3u^2 du - \int 1\ du$

$= 2\int u^5 du - 3\int u^2 du - \int du$

$= \frac{2u^6}{6} - \frac{3u^3}{3} - u + C = \frac{u^6}{3} - u^3 - u + C$

27. $\frac{dy}{dx} = e^x + 3$

$y = \int (e^x + 3)dx = \int e^x dx + \int 3\ dx = e^x + 3x + C$

29. $\frac{dx}{dt} = 5t^{-1} + 1$

$x = \int (5t^{-1} + 1)dt = \int 5t^{-1}dt + \int 1\ dt = 5\int \frac{1}{t}dt + \int dt$

$= 5\ln|t| + t + C$

31. The graphs in this set ARE NOT graphs from a family of antiderivative functions since the graphs are not vertical translations of each other.

33. The graphs in this set could be graphs from a family of antiderivative functions since they appear to be vertical translations of each other.

35. $\int 6x^{1/2}dx = 6\int x^{1/2}dx = 6\,\dfrac{x^{(1/2)+1}}{\frac{1}{2}+1} + C = \dfrac{6x^{3/2}}{\frac{3}{2}} + C = 4x^{3/2} + C$

Check: $(4x^{3/2} + C)' = 4\left(\dfrac{3}{2}\right)x^{1/2} = 6x^{1/2}$

37. $\int 8x^{-3}dx = 8\int x^{-3}dx = 8\,\dfrac{x^{-3+1}}{-3+1} + C = \dfrac{8x^{-2}}{-2} + C = -4x^{-2} + C$

Check: $(-4x^{-2} + C)' = -4(-2)x^{-3} = 8x^{-3}$

39. $\int \dfrac{du}{\sqrt{u}} = \int \dfrac{du}{u^{1/2}} = \int u^{-1/2}du = \dfrac{u^{(-1/2)+1}}{-\frac{1}{2}+1} + C = \dfrac{u^{1/2}}{\frac{1}{2}} + C$

$$= 2u^{1/2} + C \text{ or } 2\sqrt{u} + C$$

Check: $(2u^{1/2} + C)' = 2\left(\dfrac{1}{2}\right)u^{-1/2} = \dfrac{1}{u^{1/2}} = \dfrac{1}{\sqrt{u}}$

41. $\int \dfrac{dx}{4x^3} = \dfrac{1}{4}\int x^{-3}dx = \dfrac{1}{4}\cdot\dfrac{x^{-2}}{-2} + C = \dfrac{-x^{-2}}{8} + C$

Check: $\left(\dfrac{-x^{-2}}{8} + C\right)' = \dfrac{1}{8}(-2)(-x^{-3}) = \dfrac{1}{4}x^{-3} = \dfrac{1}{4x^3}$

43. $\int \dfrac{du}{2u^5} = \dfrac{1}{2}\int u^{-5}du = \dfrac{1}{2}\cdot\dfrac{u^{-4}}{-4} + C = \dfrac{-u^{-4}}{8} + C$

Check: $\left(\dfrac{-u^{-4}}{8} + C\right)' = \dfrac{-1}{8}(-4)u^{-5} = \dfrac{u^{-5}}{2} = \dfrac{1}{2u^5}$

45. $\int \left(3x^2 - \dfrac{2}{x^2}\right)dx = \int 3x^2 dx - \int \dfrac{2}{x^2}dx$

$$= 3\int x^2 dx - 2\int x^{-2}dx = 3\cdot\dfrac{x^3}{3} - \dfrac{2x^{-1}}{-1} + C = x^3 + 2x^{-1} + C$$

Check: $(x^3 + 2x^{-1} + C)' = 3x^2 - 2x^{-2} = 3x^2 - \dfrac{2}{x^2}$

47. $\displaystyle\int \left(10x^4 - \frac{8}{x^5} - 2\right)dx = \int 10x^4\,dx - \int 8x^{-5}\,dx - \int 2\ dx$

$$= 10 \int x^4\,dx - 8\int x^{-5}\,dx - \int 2\ dx$$

$$= \frac{10x^5}{5} - \frac{8x^{-4}}{-4} - 2x + C = 2x^5 + 2x^{-4} - 2x + C$$

Check: $(2x^5 + 2x^{-4} - 2x + C)' = 10x^4 - 8x^{-5} - 2 = 10x^4 - \dfrac{8}{x^5} - 2$

49. $\displaystyle\int \left(3\sqrt{x} + \frac{2}{\sqrt{x}}\right)dx = 3\int x^{1/2}\,dx + 2\int x^{-1/2}\,dx$

$$= \frac{3x^{3/2}}{\frac{3}{2}} + \frac{2x^{1/2}}{\frac{1}{2}} + C = 2x^{3/2} + 4x^{1/2} + C$$

Check: $(2x^{3/2} + 4x^{1/2} + C)' = 2\left(\frac{3}{2}\right)x^{1/2} + 4\left(\frac{1}{2}\right)x^{-1/2}$

$$= 3x^{1/2} + 2x^{-1/2} = 3\sqrt{x} + \frac{2}{\sqrt{x}}$$

51. $\displaystyle\int \left(\sqrt[3]{x^2} - \frac{4}{x^3}\right)dx = \int x^{2/3}\,dx - 4\int x^{-3}\,dx = \frac{x^{5/3}}{\frac{5}{3}} - \frac{4x^{-2}}{-2} + C$

$$= \frac{3x^{5/3}}{5} + 2x^{-2} + C$$

Check: $\left(\frac{3}{5}x^{5/3} + 2x^{-2} + C\right)' = \frac{3}{5}\left(\frac{5}{3}\right)x^{2/3} + 2(-2)x^{-3}$

$$= x^{2/3} - 4x^{-3} = \sqrt[3]{x^2} - \frac{4}{x^3}$$

53. $\displaystyle\int \frac{e^x - 3x}{4}\,dx = \int \left(\frac{e^x}{4} - \frac{3x}{4}\right)dx = \frac{1}{4}\int e^x\,dx - \frac{3}{4}\int x\ dx$

$$= \frac{1}{4}e^x - \frac{3}{4}\cdot\frac{x^2}{2} + C = \frac{1}{4}e^x - \frac{3x^2}{8} + C$$

Check: $\left(\frac{1}{4}e^x - \frac{3x^2}{8} + C\right)' = \frac{1}{4}e^x - \frac{6x}{8} = \frac{1}{4}e^x - \frac{3}{4}x$

55. $\displaystyle\int (2z^{-3} + z^{-2} + z^{-1})\,dz = 2\int z^{-3}\,dz + \int z^{-2}\,dz + \int \frac{1}{z}\,dz$

$$= \frac{2z^{-2}}{-2} + \frac{z^{-1}}{-1} + \ln|z| + C = -z^{-2} - z^{-1} + \ln|z| + C$$

Check: $(-z^{-2} - z^{-1} + \ln|z| + C)' = 2z^{-3} + z^{-2} + \frac{1}{z}$

57.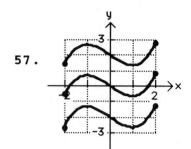

59.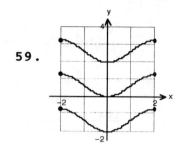

61. $\dfrac{dy}{dx} = 2x - 3$

$$y = \int (2x - 3)\,dx = 2\int x\,dx - \int 3\,dx = \dfrac{2x^2}{2} - 3x + C = x^2 - 3x + C$$

Given $y(0) = 5$: $5 = 0^2 - 3(0) + C$. Hence, $C = 5$ and $y = x^2 - 3x + 5$.

63. $C'(x) = 6x^2 - 4x$

$$C(x) = \int (6x^2 - 4x)\,dx = 6\int x^2\,dx - 4\int x\,dx = \dfrac{6x^3}{3} - \dfrac{4x^2}{2} + C = 2x^3 - 2x^2 + C$$

Given $C(0) = 3000$: $3000 = 2(0^3) - 2(0^2) + C$. Hence, $C = 3000$ and $C(x) = 2x^3 - 2x^2 + 3000$.

65. $\dfrac{dx}{dt} = \dfrac{20}{\sqrt{t}}$

$$x = \int \dfrac{20}{\sqrt{t}}\,dt = 20\int t^{-1/2}\,dt = 20\dfrac{t^{1/2}}{\frac{1}{2}} + C = 40\sqrt{t} + C$$

Given $x(1) = 40$: $40 = 40\sqrt{1} + C$ or $40 = 40 + C$. Hence, $C = 0$ and $x = 40\sqrt{t}$.

67. $\dfrac{dy}{dx} = 2x^{-2} + 3x^{-1} - 1$

$$y = \int (2x^{-2} + 3x^{-1} - 1)\,dx = 2\int x^{-2}\,dx + 3\int x^{-1}\,dx - \int dx$$

$$= \dfrac{2x^{-1}}{-1} + 3\ln|x| - x + C = \dfrac{-2}{x} + 3\ln|x| - x + C$$

Given $y(1) = 0$: $0 = -\dfrac{2}{1} + 3\ln|1| - 1 + C$. Hence, $C = 3$ and

$$y = -\dfrac{2}{x} + 3\ln|x| - x + 3.$$

69. $\dfrac{dx}{dt} = 4e^t - 2$

$$x = \int (4e^t - 2)\,dt = 4\int e^t\,dt - \int 2\,dt = 4e^t - 2t + C$$

Given $x(0) = 1$: $1 = 4e^0 - 2(0) + C = 4 + C$. Hence, $C = -3$ and $x = 4e^t - 2t - 3$.

71. $\dfrac{dy}{dx} = 4x - 3$

$y = \displaystyle\int (4x - 3)\,dx = 4\int x\,dx - \int 3\,dx = \dfrac{4x^2}{2} - 3x + C = 2x^2 - 3x + C$

Given $y(2) = 3$: $3 = 2 \cdot 2^2 - 3 \cdot 2 + C$. Hence, $C = 1$ and $y = 2x^2 - 3x + 1$.

73. $\displaystyle\int \dfrac{2x^4 - x}{x^3}\,dx = \int \left(\dfrac{2x^4}{x^3} - \dfrac{x}{x^3}\right)dx$

$\qquad = 2\displaystyle\int x\,dx - \int x^{-2}dx = \dfrac{2x^2}{2} - \dfrac{x^{-1}}{-1} + C = x^2 + x^{-1} + C$

75. $\displaystyle\int \dfrac{x^5 - 2x}{x^4}\,dx = \int \left(\dfrac{x^5}{x^4} - \dfrac{2x}{x^4}\right)dx$

$\qquad = \displaystyle\int x\,dx - 2\int x^{-3}dx = \dfrac{x^2}{2} - \dfrac{2x^{-2}}{-2} + C = \dfrac{x^2}{2} + x^{-2} + C$

77. $\displaystyle\int \dfrac{x^2 e^x - 2x}{x^2}\,dx = \int \left(\dfrac{x^2 e^x}{x^2} - \dfrac{2x}{x^2}\right)dx = \int e^x dx - 2\int x^{-1}dx = e^x - 2\ln|x| + C$

79. $\dfrac{dM}{dt} = \dfrac{t^2 - 1}{t^2}$

$M = \displaystyle\int \dfrac{t^2 - 1}{t^2}\,dt = \int \left(\dfrac{t^2}{t^2} - \dfrac{1}{t^2}\right)dt = \int dt - \int t^{-2}dt = t - \dfrac{t^{-1}}{-1} + C = t + \dfrac{1}{t} + C$

Given $M(4) = 5$: $5 = 4 + \dfrac{1}{4} + C$ or $C = 5 - \dfrac{17}{4} = \dfrac{3}{4}$. Hence, $M = t + \dfrac{1}{t} + \dfrac{3}{4}$.

81. $\dfrac{dy}{dx} = \dfrac{5x + 2}{\sqrt[3]{x}}$

$y = \displaystyle\int \dfrac{5x + 2}{\sqrt[3]{x}}\,dx = \int \left(\dfrac{5x}{x^{1/3}} + \dfrac{2}{x^{1/3}}\right)dx = 5\int x^{2/3}dx + 2\int x^{-1/3}dx$

$\qquad = \dfrac{5x^{5/3}}{\frac{5}{3}} + \dfrac{2x^{2/3}}{\frac{2}{3}} + C = 3x^{5/3} + 3x^{2/3} + C$

Given $y(1) = 0$: $0 = 3 \cdot 1^{5/3} + 3 \cdot 1^{2/3} + C$. Hence, $C = -6$ and
$y = 3x^{5/3} + 3x^{2/3} - 6$.

83. $p'(x) = -\dfrac{10}{x^2}$

$p(x) = \displaystyle\int -\dfrac{10}{x^2}\,dx = -10\int x^{-2}dx = \dfrac{-10x^{-1}}{-1} + C = \dfrac{10}{x} + C$

Given $p(1) = 20$: $20 = \dfrac{10}{1} + C = 10 + C$. Hence, $C = 10$ and

$p(x) = \dfrac{10}{x} + 10$.

85. $\overline{C}'(x) = -\dfrac{1{,}000}{x^2}$

$$\overline{C}(x) = \int \overline{C}'(x)\,dx = \int -\frac{1{,}000}{x^2}\,dx = -1{,}000\int x^{-2}\,dx$$

$$= -1{,}000\frac{x^{-1}}{-1} + C$$

$$= \frac{1{,}000}{x} + C$$

Given $\overline{C}(100) = 25$: $\quad \dfrac{1{,}000}{100} + C = 25$

$$C = 15$$

Thus, $\overline{C}(x) = \dfrac{1{,}000}{x} + 15.$

Cost function: $C(x) = x\overline{C}(x) = 15x + 1{,}000$

Fixed costs: $C(0) = \$1{,}000$

87. (A) The cost function increases from 0 to 8. The graph is concave downward from 0 to 4 and concave upward from 4 to 8. There is an inflection point at $x = 4$.

(B) $C(x) = \displaystyle\int C'(x)\,dx = \int (3x^2 - 24x + 53)\,dx$

$$= 3\int x^2\,dx - 24\int x\,dx + \int 53\,dx$$

$$= x^3 - 12x^2 + 53x + K$$

Since $C(0) = 30$, we have $K = 30$ and

$$C(x) = x^3 - 12x^2 + 53x + 30.$$

$C(4) = 4^3 - 12(4)^2 + 53(4) + 30 = \$114$ thousand

$C(8) = 8^3 - 12(8)^2 + 53(8) + 30 = \$198$ thousand

(C)
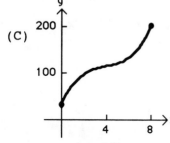

(D) Manufacturing plants are often inefficient at low and high levels of production.

89. $S'(t) = -25t^{2/3}$

$$S(t) = \int S'(t)\,dt = \int -25t^{2/3}\,dt = -25\int t^{2/3}\,dt = -25\,\frac{t^{5/3}}{5/3} + C = -15t^{5/3} + C$$

Given $S(0) = 2000$: $-15(0)^{5/3} + C = 2000$. Hence, $C = 2000$ and $S(t) = -15t^{5/3} + 2000$. Now, we want to find t such that $S(t) = 800$, that is: $\quad -15t^{5/3} + 2000 = 800$

$$-15t^{5/3} = -1200$$

$$t^{5/3} = 80$$

and $\qquad\qquad t = 80^{3/5} \approx 14$

Thus, the company should manufacture the computer for 14 months.

91. $S'(t) = -25t^{2/3} - 70$

$$S(t) = \int S'(t)\,dt = \int (-25t^{2/3} - 70)\,dt$$

$$= -25 \int t^{2/3}\,dt - \int 70\,dt$$

$$= -25 \frac{t^{5/3}}{5/3} - 70t + C$$

$$= -15t^{5/3} - 70t + C$$

Given $S(0) = 2{,}000$ implies $C = 2{,}000$ and
$$S(t) = 2{,}000 - 15t^{5/3} - 70t$$

Graphing $y_1 = 2{,}000 - 15t^{5/3} - 70t$, $y_2 = 800$ on $0 \le x \le 10$, $0 \le y \le 1000$, we see that the point of intersection is $x \approx 8.92066$, $y = 800$.
So we get $t \approx 8.92$ months.

93. $L'(x) = g(x) = 2400x^{-1/2}$

$$L(x) = \int g(x)\,dx = \int 2400x^{-1/2}\,dx = 2400 \int x^{-1/2}\,dx = 2400 \frac{x^{1/2}}{1/2} + C$$

$$= 4800\, x^{1/2} + C$$

Given $L(16) = 19{,}200$: $19{,}200 = 4800(16)^{1/2} + C = 19{,}200 + C$. Hence, $C = 0$ and $L(x) = 4800x^{1/2}$.
$L(25) = 4800(25)^{1/2} = 4800(5) = 24{,}000$ labor hours.

95. $\dfrac{dW}{dh} = 0.0015h^2$

$$W = \int 0.0015h^2\,dh = 0.0015 \int h^2\,dh = 0.0015 \frac{h^3}{3} + C = 0.0005h^3 + C$$

Given $W(60) = 108$: $108 = 0.0005(60)^3 + C$ or $108 = 108 + C$.
Hence, $C = 0$ and $W(h) = 0.0005h^3$. Now $5'10'' = 70''$ and
$W(70) = 0.0005(70)^3 = 171.5$ lb.

97. $\dfrac{dN}{dt} = 400 + 600\sqrt{t}$, $0 \le t \le 9$

$$N = \int (400 + 600\sqrt{t})\,dt = \int 400\,dt + 600 \int t^{1/2}\,dt$$

$$= 400t + 600 \frac{t^{3/2}}{3/2} + C = 400t + 400t^{3/2} + C$$

Given $N(0) = 5000$: $5000 = 400(0) + 400(0)^{3/2} + C$. Hence, $C = 5000$ and
$N(t) = 400t + 400t^{3/2} + 5000$.
$N(9) = 400(9) + 400(9)^{3/2} + 5000 = 3600 + 10{,}800 + 5000 = 19{,}400$

Things to remember:

<u>1</u>. REVERSING THE CHAIN RULE

The chain rule formula for differentiating a composite function:

$$\frac{d}{dx} f[g(x)] = f'[g(x)]g'(x),$$

yields the integral formula

$$\int f'[g(x)]g'(x)\,dx = f[g(x)] + C$$

<u>2</u>. GENERAL INDEFINITE INTEGRAL FORMULAS (Version 1)

(a) $\displaystyle\int [f(x)]^n f'(x)\,dx = \frac{[f(x)]^{n+1}}{n+1} + C, \ n \neq -1$

(b) $\displaystyle\int e^{f(x)} f'(x)\,dx = e^{f(x)} + C$

(c) $\displaystyle\int \frac{1}{f(x)} f'(x)\,dx = \ln|f(x)| + C$

<u>3</u>. DIFFERENTIALS

If $y = f(x)$ defines a differentiable function, then:

(a) The DIFFERENTIAL dx of the independent variable x is an arbitrary real number.

(b) The DIFFERENTIAL dy of the dependent variable y is defined as the product of $f'(x)$ and dx; that is: $dy = f'(x)\,dx$.

<u>4</u>. GENERAL INDEFINITE INTEGRAL FORMULAS (Version 2)

(a) $\displaystyle\int u^n\,du = \frac{u^{n+1}}{n+1} + C, \ n \neq -1$

(b) $\displaystyle\int e^u\,du = e^u + C$

(c) $\displaystyle\int \frac{1}{u}\,du = \ln|u| + C$

<u>5</u>. INTEGRATION BY SUBSTITUTION

(a) Select a substitution that appears to simplify the integrand. In particular, try to select u so that du is a factor in the integrand.

(b) Express the integrand entirely in terms of u and du, completely eliminating the original variable and its differential.

(c) Evaluate the new integral, if possible.

(d) Express the antiderivative found in Step (c) in terms of the original variable.

1. $\int (x^2 - 4)^5 2x \, dx$

Let $u = x^2 - 4$, then $du = 2x \, dx$ and

$$\int (x^2 - 4)^5 2x \, dx = \int u^5 du = \frac{u^6}{6} + C \quad \text{[using formula } \underline{4}(a)]$$

$$= \frac{(x^2 - 4)^6}{6} + C$$

Check: $\frac{d}{dx}\left[\frac{(x^2 - 4)^6}{6} + C\right] = \frac{1}{6}(6)(x^2 - 4)^5(2x) = (x^2 - 4)^5 2x$

3. $\int e^{4x} 4 \, dx$

Let $u = 4x$, then $du = 4 \, dx$ and

$$\int e^{4x} 4 \, dx = \int e^u du = e^u + C \quad \text{[using formula } \underline{4}(b)]$$

$$= e^{4x} + C$$

Check: $\frac{d}{dx}[e^{4x} + C] = e^{4x}(4)$

5. $\int \frac{1}{2t + 3} 2 \, dt$

Let $u = 2t + 3$, then $du = 2 \, dt$ and

$$\int \frac{1}{2t + 3} 2 \, dt = \int \frac{1}{u} du = \ln|u| + C \quad \text{[using formula } \underline{4}(c)]$$

$$= \ln|2t + 3| + C$$

Check: $\frac{d}{dt}[\ln|2t + 3| + C] = \frac{1}{2t + 3}(2)$

7. $\int (3x - 2)^7 dx$

Let $u = 3x - 2$, then $du = 3 \, dx$ and

$$\int (3x - 2)^7 dx = \int (3x - 2)^7 \frac{3}{3} dx = \frac{1}{3}\int (3x - 2)^7 3 \, dx = \frac{1}{3}\int u^7 du$$

$$= \frac{1}{3} \cdot \frac{u^8}{8} + C = \frac{(3x - 2)^8}{24} + C$$

Check: $\frac{d}{dx}\left[\frac{(3x - 2)^8}{24} + C\right] = \frac{1}{24}(8)(3x - 2)^7(3) = (3x - 2)^7$

9. Let $u = x^2 + 3$, then $du = 2x \, dx$.

$$\int (x^2 + 3)^7 x \, dx = \int (x^2 + 3)^7 \frac{2}{2} x \, dx = \frac{1}{2}\int (x^2 + 3)^7 2x \, dx$$

$$= \frac{1}{2}\int u^7 du = \frac{1}{2} \cdot \frac{u^8}{8} + C = \frac{u^8}{16} + C = \frac{(x^2 + 3)^8}{16} + C$$

Check: $\dfrac{d}{dx}\left[\dfrac{(x^2 + 3)^8}{16} + C\right] = \dfrac{1}{16}(8)(x^2 + 3)^7(2x) = (x^2 + 3)^7 x$

11. Let $u = -0.5t$, then $du = -0.5\ dt$.

$$\int 10e^{-0.5t}dt = 10\int e^{-0.5t}dt = 10\int e^{-0.5t}\dfrac{(-0.5)}{-0.5}\ dt$$

$$= \dfrac{10}{-0.5}\int e^u du = -20e^u + C = -20e^{-0.5t} + C$$

Check: $\dfrac{d}{dt}[-20e^{-0.5t} + C] = -20e^{-0.5t}(-0.5) = 10e^{-0.5t}$

13. Let $u = 10x + 7$, then $du = 10\ dx$.

$$\int \dfrac{1}{10x + 7}dx = \int \dfrac{1}{10x + 7}\cdot\dfrac{10}{10}dx = \dfrac{1}{10}\int\dfrac{1}{10x + 7}10\ dx$$

$$= \dfrac{1}{10}\int\dfrac{1}{u}du = \dfrac{1}{10}\ln|u| + C = \dfrac{1}{10}\ln|10x + 7| + C$$

Check: $\dfrac{d}{dx}\left[\dfrac{1}{10}\ln|10x + 7| + C\right] = \left(\dfrac{1}{10}\right)\dfrac{1}{10x + 7}(10) = \dfrac{1}{10x + 7}$

15. Let $u = 2x^2$, then $du = 4x\ dx$.

$$\int xe^{2x^2}dx = \int e^{2x^2}x\cdot\dfrac{4}{4}dx = \dfrac{1}{4}\int e^{2x^2}(4x)\,dx$$

$$= \dfrac{1}{4}\int e^u du = \dfrac{1}{4}e^u + C = \dfrac{1}{4}e^{2x^2} + C$$

Check: $\dfrac{d}{dx}\left[\dfrac{1}{4}e^{2x^2} + C\right] = \dfrac{1}{4}e^{2x^2}(4x) = xe^{2x^2}$

17. Let $u = x^3 + 4$, then $du = 3x^2 dx$.

$$\int\dfrac{x^2}{x^3 + 4}dx = \int\dfrac{1}{x^3 + 4}\cdot\dfrac{3}{3}x^2 dx = \dfrac{1}{3}\int\dfrac{1}{x^3 + 4}(3x^2)\,dx$$

$$= \dfrac{1}{3}\int\dfrac{1}{u}du = \dfrac{1}{3}\ln|u| + C = \dfrac{1}{3}\ln|x^3 + 4| + C$$

Check: $\dfrac{d}{dx}\left[\dfrac{1}{3}\ln|x^3 + 4| + C\right] = \left(\dfrac{1}{3}\right)\dfrac{1}{x^3 + 4}(3x^2) = \dfrac{x^2}{x^3 + 4}$

19. Let $u = 3t^2 + 1$, then $du = 6t\,dt$.

$$\int \frac{t}{(3t^2 + 1)^4}\,dt = \int (3t^2 + 1)^{-4}t\,dt = \int (3t^2 + 1)^{-4}\frac{6}{6}t\,dt$$

$$= \frac{1}{6}\int (3t^2 + 1)^{-4}6t\,dt = \frac{1}{6}\int u^{-4}\,du$$

$$= \frac{1}{6}\cdot\frac{u^{-3}}{-3} + C = \frac{-1}{18}(3t^2 + 1)^{-3} + C$$

Check: $\dfrac{d}{dt}\left[\dfrac{-1}{18}(3t^2 + 1)^{-3} + C\right] = \left(\dfrac{-1}{18}\right)(-3)(3t^2 + 1)^{-4}(6t) = \dfrac{t}{(3t^2 + 1)^4}$

21. Let $u = 4 - x^3$, then $du = -3x^2\,dx$.

$$\int \frac{x^2}{(4 - x^3)^2}\,dx = \int (4 - x^3)^{-2}x^2\,dx = \int (4 - x^3)^{-2}\left(\frac{-3}{-3}\right)x^2\,dx$$

$$= \frac{-1}{3}\int (4 - x^3)^{-2}(-3x^2)\,dx = \frac{-1}{3}\int u^{-2}\,du = \frac{-1}{3}\cdot\frac{u^{-1}}{-1} + C$$

$$= \frac{1}{3}(4 - x^3)^{-1} + C$$

Check: $\dfrac{d}{dx}\left[\dfrac{1}{3}(4 - x^3)^{-1} + C\right] = \dfrac{1}{3}(-1)(4 - x^3)^{-2}(-3x^2) = \dfrac{x^2}{(4 - x^3)^2}$

23. $\displaystyle\int x\sqrt{x + 4}\,dx$

Let $u = x + 4$, then $du = dx$ and $x = u - 4$.

$$\int x\sqrt{x + 4}\,dx = \int (u - 4)u^{1/2}\,du = \int (u^{3/2} - 4u^{1/2})\,du$$

$$= \frac{u^{5/2}}{\frac{5}{2}} - \frac{4u^{3/2}}{\frac{3}{2}} + C = \frac{2}{5}u^{5/2} - \frac{8}{3}u^{3/2} + C$$

$$= \frac{2}{5}(x + 4)^{5/2} - \frac{8}{3}(x + 4)^{3/2} + C \quad \text{(since } u = x + 4\text{)}$$

Check: $\dfrac{d}{dx}\left[\dfrac{2}{5}(x + 4)^{5/2} - \dfrac{8}{3}(x + 4)^{3/2} + C\right]$

$$= \frac{2}{5}\left(\frac{5}{2}\right)(x + 4)^{3/2}(1) - \frac{8}{3}\left(\frac{3}{2}\right)(x + 4)^{1/2}(1)$$

$$= (x + 4)^{3/2} - 4(x + 4)^{1/2} = (x + 4)^{1/2}[(x + 4) - 4] = x\sqrt{x + 4}$$

25. $\int \dfrac{x}{\sqrt{x - 3}}\, dx$

Let $u = x - 3$, then $du = dx$ and $x = u + 3$.

$\int \dfrac{x}{\sqrt{x - 3}}\, dx = \int \dfrac{u + 3}{u^{1/2}}\, du = \int (u^{1/2} + 3u^{-1/2})\, du = \dfrac{u^{3/2}}{3/2} + \dfrac{3u^{1/2}}{1/2} + C$

$\qquad = \dfrac{2}{3} u^{3/2} + 6u^{1/2} + C = \dfrac{2}{3}(x - 3)^{3/2} + 6(x - 3)^{1/2} + C$

$\qquad\qquad\qquad\qquad\qquad\qquad\qquad\qquad\text{(since } u = x - 3\text{)}$

Check: $\dfrac{d}{dx}\left[\dfrac{2}{3}(x - 3)^{3/2} + 6(x - 3)^{1/2} + C\right]$

$\qquad = \dfrac{2}{3}\left(\dfrac{3}{2}\right)(x - 3)^{1/2}(1) + 6\left(\dfrac{1}{2}\right)(x - 3)^{-1/2}(1)$

$\qquad = (x - 3)^{1/2} + \dfrac{3}{(x - 3)^{1/2}} = \dfrac{x - 3 + 3}{(x - 3)^{1/2}} = \dfrac{x}{\sqrt{x - 3}}$

27. $\int x(x - 4)^9\, dx$

Let $u = x - 4$, then $du = dx$ and $x = u + 4$.

$\int x(x - 4)^9\, dx = \int (u + 4)u^9\, du = \int (u^{10} + 4u^9)\, du$

$\qquad = \dfrac{u^{11}}{11} + \dfrac{4u^{10}}{10} + C = \dfrac{(x - 4)^{11}}{11} + \dfrac{2}{5}(x - 4)^{10} + C$

Check: $\dfrac{d}{dx}\left[\dfrac{(x - 4)^{11}}{11} + \dfrac{2}{5}(x - 4)^{10} + C\right]$

$\qquad = \dfrac{1}{11}(11)(x - 4)^{10}(1) + \dfrac{2}{5}(10)(x - 4)^9(1)$

$\qquad = (x - 4)^9[(x - 4) + 4] = x(x - 4)^9$

29. Let $u = 1 + e^{2x}$, then $du = 2e^{2x}\, dx$.

$\int e^{2x}(1 + e^{2x})^3\, dx = \int (1 + e^{2x})^3 \dfrac{2}{2} e^{2x}\, dx = \dfrac{1}{2}\int (1 + e^{2x})^3 2e^{2x}\, dx$

$\qquad = \dfrac{1}{2}\int u^3\, du = \dfrac{1}{2}\cdot\dfrac{u^4}{4} + C = \dfrac{1}{8}(1 + e^{2x})^4 + C$

Check: $\dfrac{d}{dx}\left[\dfrac{1}{8}(1 + e^{2x})^4 + C\right] = \left(\dfrac{1}{8}\right)(4)(1 + e^{2x})^3 e^{2x}(2) = e^{2x}(1 + e^{2x})^3$

31. Let $u = 4 + 2x + x^2$, then $du = (2 + 2x)\,dx = 2(1 + x)\,dx$.

$$\int \frac{1 + x}{4 + 2x + x^2}\,dx = \int \frac{1}{4 + 2x + x^2} \cdot \frac{2(1 + x)}{2}\,dx = \frac{1}{2}\int \frac{1}{4 + 2x + x^2}\,2(1 + x)\,dx$$

$$= \frac{1}{2}\int \frac{1}{u}\,du = \frac{1}{2}\ln|u| + C = \frac{1}{2}\ln|4 + 2x + x^2| + C$$

Check: $\dfrac{d}{dx}\left[\dfrac{1}{2}\ln|4 + 2x + x^2| + C\right] = \left(\dfrac{1}{2}\right)\dfrac{1}{4 + 2x + x^2}(2 + 2x) = \dfrac{1 + x}{4 + 2x + x^2}$

33. Let $u = x^2 + x + 1$, then $du = (2x + 1)\,dx$.

$$\int (2x + 1)e^{x^2+x+1}\,dx = \int e^u\,du = e^u + C = e^{x^2+x+1} + C$$

Check: $\dfrac{d}{dx}[e^{x^2+x+1} + C] = e^{x^2+x+1}(2x + 1)$

35. Let $u = e^x - 2x$, then $du = (e^x - 2)\,dx$.

$$\int (e^x - 2x)^3(e^x - 2)\,dx = \int u^3\,du = \frac{u^4}{4} + C = \frac{(e^x - 2x)^4}{4} + C$$

Check: $\dfrac{d}{dx}\left[\dfrac{(e^x - 2x)^4}{4} + C\right] = \dfrac{1}{4}(4)(e^x - 2x)^3(e^x - 2) = (e^x - 2x)^3(e^x - 2)$

37. Let $u = x^4 + 2x^2 + 1$, then $du = (4x^3 + 4x)\,dx = 4(x^3 + x)\,dx$.

$$\int \frac{x^3 + x}{(x^4 + 2x^2 + 1)^4}\,dx = \int (x^4 + 2x^2 + 1)^{-4}\frac{4}{4}(x^3 + x)\,dx$$

$$= \frac{1}{4}\int (x^4 + 2x^2 + 1)^{-4}4(x^3 + x)\,dx$$

$$= \frac{1}{4}\int u^{-4}\,du = \frac{1}{4}\cdot\frac{u^{-3}}{-3} + C = \frac{-u^{-3}}{12} + C$$

$$= \frac{-(x^4 + 2x^2 + 1)^{-3}}{12} + C$$

Check: $\dfrac{d}{dx}\left[-\dfrac{1}{12}(x^4 + 2x^2 + 1)^{-3} + C\right] = \left(-\dfrac{1}{12}\right)(-3)(x^4 + 2x^2 + 1)^{-4}(4x^3 + 4x)$

$$= (x^4 + 2x^2 + 1)^{-4}(x^3 + x)$$

39. (A) Differentiate $F(x) = \ln|2x - 3| + C$ to see if you get the integrand
$$f(x) = \frac{1}{2x - 3}$$

(B) Wrong; $\dfrac{d}{dx}[\ln|2x - 3| + C] = \dfrac{1}{2x - 3}\,(2) = \dfrac{2}{2x - 3} \neq \dfrac{1}{2x - 3}$

(C) Let $u = 2x - 3$, then $du = 2\,dx$

$$\int \frac{1}{2x - 3}\,dx = \int \frac{1}{2x - 3} \cdot \frac{2}{2}\,dx = \frac{1}{2}\int \frac{1}{2x - 3}\,2\,dx$$

$$= \frac{1}{2}\int \frac{1}{u}\,du$$

$$= \frac{1}{2}\ln|u| + C$$

$$= \frac{1}{2}\ln|2x - 3| + C$$

Check: $\dfrac{d}{dx}\left[\dfrac{1}{2}\ln|2x - 3| + C\right] = \dfrac{1}{2} \cdot \dfrac{1}{2x - 3} \cdot 2 = \dfrac{1}{2x - 3}$

41. (A) Differentiate $F(x) = e^{x^4} + C$ to see if you get the integrand $f(x) = x^3 e^{x^4}$.

(B) Wrong; $\dfrac{d}{dx}[e^{x^4} + c] = e^{x^4}(4x^3) = 4x^3 e^{x^4} \neq x^3 e^{x^4}$

(C) Let $u = x^4$, then $du = 4x^3\,dx$

$$\int x^3 e^{x^4}\,dx = \int \frac{4}{4}x^3 e^{x^4}\,dx = \frac{1}{4}\int 4\,x^3 e^{x^4}\,dx$$

$$= \frac{1}{4}\int e^u\,du$$

$$= \frac{1}{4}e^u + C$$

$$= \frac{1}{4}e^{x^4} + C$$

Check: $\dfrac{d}{dx}\left[\dfrac{1}{4}e^{x^4} + C\right] = \dfrac{1}{4}e^{x^4}(4x^3) = x^3 e^{x^4}$

43. (A) Differentiate $F(x) = \dfrac{(x^2 - 2)^2}{3x} + C$ to see if you get the integrand $f(x) = 2(x^2 - 2)^2$

(B) Wrong; $\dfrac{d}{dx}\left[\dfrac{(x^2 - 2)^2}{3x} + C\right] = \dfrac{3x \cdot 2(x^2 - 2)(2x) - (x^2 - 2)^2 \cdot 3}{9x^2}$

$$= \frac{(x^2 - 2)[9x^2 + 6]}{9x^2} = \frac{9x^4 - 12x^2 - 12}{9x^2}$$

$$= \frac{3x^4 - 4x^2 - 4}{3x^2}$$

$$\neq 2(x^2 - 2)^2$$

(C) $\displaystyle\int 2(x^2 - 2)^2\,dx = 2\int (x^4 - 4x^2 + 4)\,dx$

$$= 2 \cdot \left[\frac{1}{5}x^5 - \frac{4}{3}x^3 + 4x\right] + C$$

$$= \frac{2}{5}x^5 - \frac{8}{3}x^3 + 8x + C$$

Check: $\dfrac{d}{dx}\left[\dfrac{2}{5}x^5 - \dfrac{8}{3}x^3 + 8x + C\right] = 2x^4 - 8x^2 + 8 = 2[x^4 - 4x^2 + 4]$

$$= 2(x^2 - 2)^2$$

45. Let $u = 3x^2 + 7$, then $du = 6x\ dx$.

$$\int x\sqrt{3x^2 + 7}\ dx = \int (3x^2 + 7)^{1/2}x\ dx = \int (3x^2 + 7)^{1/2}\dfrac{6}{6}x\ dx$$

$$= \dfrac{1}{6}\int u^{1/2}\,du = \dfrac{1}{6}\cdot\dfrac{u^{3/2}}{\dfrac{3}{2}} + C = \dfrac{1}{9}(3x^2 + 7)^{3/2} + C$$

Check: $\dfrac{d}{dx}\left[\dfrac{1}{9}(3x^2 + 7)^{3/2} + C\right] = \dfrac{1}{9}\left(\dfrac{3}{2}\right)(3x^2 + 7)^{1/2}(6x) = x(3x^2 + 7)^{1/2}$

47. $\displaystyle\int x(x^3 + 2)^2\,dx = \int x(x^6 + 4x^3 + 4)\,dx = \int (x^7 + 4x^4 + 4x)\,dx$

$$= \dfrac{x^8}{8} + \dfrac{4}{5}x^5 + 2x^2 + C$$

Check: $\dfrac{d}{dx}\left[\dfrac{x^8}{8} + \dfrac{4}{5}x^5 + 2x^2 + C\right] = x^7 + 4x^4 + 4x$

$$= x(x^6 + 4x^3 + 4) = x(x^3 + 2)^2$$

49. $\displaystyle\int x^2(x^3 + 2)^2\,dx$

Let $u = x^3 + 2$, then $du = 3x^2\,dx$.

$$\int x^2(x^3 + 2)^2\,dx = \int (x^3 + 2)^2\dfrac{3x^2}{3}\,dx = \dfrac{1}{3}\int (x^3 + 2)^2 3x^2\,dx$$

$$= \dfrac{1}{3}\int u^2\,du = \dfrac{1}{3}\cdot\dfrac{u^3}{3} + C = \dfrac{1}{9}u^3 + C = \dfrac{1}{9}(x^3 + 2)^3 + C$$

Check: $\dfrac{d}{dx}\left[\dfrac{1}{9}(x^3 + 2)^3 + C\right] = \dfrac{1}{9}(3)(x^3 + 2)^2(3x^2) = x^2(x^3 + 2)^2$

51. Let $u = 2x^4 + 3$, then $du = 8x^3\,dx$.

$$\int \dfrac{x^3}{\sqrt{2x^4 + 3}}\,dx = \int (2x^4 + 3)^{-1/2}x^3\,dx = \int (2x^4 + 3)^{-1/2}\dfrac{8}{8}x^3\,dx$$

$$= \dfrac{1}{8}\int u^{-1/2}\,du = \dfrac{1}{8}\cdot\dfrac{u^{1/2}}{\dfrac{1}{2}} + C = \dfrac{1}{4}(2x^4 + 3)^{1/2} + C$$

Check: $\dfrac{d}{dx}\left[\dfrac{1}{4}(2x^4 + 3)^{1/2} + C\right] = \dfrac{1}{4}\left(\dfrac{1}{2}\right)(2x^4 + 3)^{-1/2}(8x^3) = \dfrac{x^3}{(2x^4 + 3)^{1/2}}$

53. Let $u = \ln x$, then $du = \dfrac{1}{x}dx$.

$$\int \frac{(\ln x)^3}{x}dx = \int u^3 du = \frac{u^4}{4} + C = \frac{(\ln x)^4}{4} + C$$

Check: $\dfrac{d}{dx}\left[\dfrac{(\ln x)^4}{4} + C\right] = \dfrac{1}{4}(4)(\ln x)^3 \cdot \dfrac{1}{x} = \dfrac{(\ln x)^3}{x}$

55. Let $u = \dfrac{-1}{x} = -x^{-1}$, then $du = \dfrac{1}{x^2}dx$.

$$\int \frac{1}{x^2}e^{-1/x}dx = \int e^u du = e^u + C = e^{-1/x} + C$$

Check: $\dfrac{d}{dx}[e^{-1/x} + C] = e^{-1/x}\left(\dfrac{1}{x^2}\right) = \dfrac{1}{x^2}e^{-1/x}$

57. $\dfrac{dx}{dt} = 7t^2(t^3 + 5)^6$

Let $u = t^3 + 5$, then $du = 3t^2 dt$.

$$x = \int 7t^2(t^3 + 5)^6 dt = 7\int t^2(t^3 + 5)^6 dt = 7\int (t^3 + 5)^6 \frac{3}{3}t^2 dt$$

$$= \frac{7}{3}\int u^6 du = \frac{7}{3}\cdot\frac{u^7}{7} + C = \frac{1}{3}(t^3 + 5)^7 + C$$

59. $\dfrac{dy}{dt} = \dfrac{3t}{\sqrt{t^2 - 4}}$

Let $u = t^2 - 4$, then $du = 2t\, dt$.

$$y = \int \frac{3t}{(t^2 - 4)^{1/2}}dt = 3\int (t^2 - 4)^{-1/2}t\, dt = 3\int (t^2 - 4)^{-1/2}\frac{2}{2}t\, dt$$

$$= \frac{3}{2}\int u^{-1/2}du = \frac{3}{2}\cdot\frac{u^{1/2}}{\frac{1}{2}} + C = 3(t^2 - 4)^{1/2} + C$$

61. $\dfrac{dp}{dx} = \dfrac{e^x + e^{-x}}{(e^x - e^{-x})^2}$

Let $u = e^x - e^{-x}$, then $du = (e^x + e^{-x})dx$.

$$p = \int \frac{e^x + e^{-x}}{(e^x - e^{-x})^2}dx = \int (e^x - e^{-x})^{-2}(e^x + e^{-x})dx = \int u^{-2}du$$

$$= \frac{u^{-1}}{-1} + C = -(e^x - e^{-x})^{-1} + C$$

63. Let $v = au$, then $dv = a\ du$.

$$\int e^{au}du = \int e^{au}\frac{a}{a}\ du = \frac{1}{a}\int e^{au}a\ du = \frac{1}{a}\int e^v dv = \frac{1}{a}e^v + C = \frac{1}{a}e^{au} + C$$

<u>Check</u>: $\dfrac{d}{du}\left[\dfrac{1}{a}e^{au} + C\right] + \dfrac{1}{a}e^{au}(a) = e^{au}$

65. $p'(x) = \dfrac{-6000}{(3x + 50)^2}$

Let $u = 3x + 50$, then $du = 3\ dx$.

$$p(x) = \int \frac{-6000}{(3x + 50)^2}\ dx = -6000\int (3x + 50)^{-2}dx = -6000\int (3x + 50)^{-2}\frac{3}{3}\ dx$$

$$= -2000\int u^{-2}du = -2000 \cdot \frac{u^{-1}}{-1} + C = \frac{2000}{3x + 50} + C$$

Given $p(150) = 4$:

$4 = \dfrac{2000}{(3 \cdot 150 + 50)} + C$

$4 = \dfrac{2000}{500} + C$

$C = 0$

Thus, $p(x) = \dfrac{2000}{3x + 50}$.

Now, $2.50 = \dfrac{2000}{3x + 50}$

$2.50(3x + 50) = 2000$

$7.5x + 125 = 2000$

$7.5x = 1875$

$x = 250$

Thus, the demand is 250 bottles when the price is \$2.50.

67. $C'(x) = 12 + \dfrac{500}{x + 1}$, $x > 0$

$$C(x) = \int \left(12 + \frac{500}{x + 1}\right)dx = \int 12\ dx + 500\int \frac{1}{x + 1}dx \qquad (u = x + 1,\ du = dx)$$

$$= 12x + 500\ \ln(x + 1) + C$$

Now, $C(0) = 2000$. Thus, $C(x) = 12x + 500\ \ln(x + 1) + 2000$. The average cost is:

$\overline{C}(x) = 12 + \dfrac{500}{x}\ln(x + 1) + \dfrac{2000}{x}$

and

$\overline{C}(1000) = 12 + \dfrac{500}{1000}\ln(1001) + \dfrac{2000}{1000} = 12 + \dfrac{1}{2}\ln(1001) + 2$

≈ 17.45 or \$17.45 per pair of shoes

69. $S'(t) = 10 - 10e^{-0.1t}$, $0 \leq t \leq 24$

(A) $S(t) = \int (10 - 10e^{-0.1t})\,dt = \int 10\,dt - 10\int e^{-0.1t}\,dt$

$$= 10t - \frac{10}{-0.1}e^{-0.1t} + C = 10t + 100e^{-0.1t} + C$$

Given $S(0) = 0$: $\quad 0 + 100e^0 + C = 0$
$$100 + C = 0$$
$$C = -100$$

Total sales at time t:
$$S(t) = 10t + 100e^{-0.1t} - 100$$

(B) $S(12) = 10(12) + 100e^{-0.1(12)} - 100$
$$= 20 + 100e^{-1.2} \approx 50$$

Total estimated sales for the first twelve months: $50 million.

(C) On a graphing utility, solve
$$10t + 100e^{-0.1t} - 100 = 100$$
or $\qquad 10t + 100e^{-0.1t} = 200$

The result is: $t \approx 18.41$ months.

71. $Q(t) = \int R(t)\,dt = \int \left(\frac{100}{t+1} + 5\right)dt = 100\int \frac{1}{t+1}\,dt + \int 5\,dt$

$$= 100\ln(t+1) + 5t + C$$

Given $Q(0) = 0$:
$0 = 100\ln(1) + 0 + C$
Thus, $C = 0$ and $Q(t) = 100\ln(t+1) + 5t$, $0 \leq t \leq 20$.
$Q(9) = 100\ln(9+1) + 5(9) = 100\ln 10 + 45 \approx 275$ thousand barrels.

73. $W(t) = \int W'(t)\,dt = \int 0.2e^{0.1t}\,dt = \frac{0.2}{0.1}\int e^{0.1t}(0.1)\,dt = 2e^{0.1t} + C$

Given $W(0) = 2$:
$2 = 2e^0 + C$.
Thus, $C = 0$ and $W(t) = 2e^{0.1t}$.
The weight of the culture after 8 hours is given by:
$W(8) = 2e^{0.1(8)} = 2e^{0.8} \approx 4.45$ grams.

75. $\dfrac{dN}{dt} = \dfrac{-2000t}{1+t^2}$, $0 \leq t \leq 10$

Let $u = 1 + t^2$, then $du = 2t\,dt$.

$$N = \int \frac{-2000t}{1+t^2}\,dt = \frac{-2000}{2}\int \frac{2t}{1+t^2}\,dt = -1000\int \frac{1}{u}\,du$$

$$= -1000\ln|u| + C = -1000\ln(1+t^2) + C$$

Given $N(0) = 5000$:

$5000 = -1000 \ln(1) + C$

Hence, $C = 5000$ and $N(t) = 5000 - 1000 \ln(1 + t^2)$, $0 \le t \le 10$.

Now, $N(10) = 5000 - 1000 \ln(1 + 10^2)$

$= 5000 - 1000 \ln(101)$

≈ 385 bacteria per milliliter.

77. $N'(t) = 6e^{-0.1t}$, $0 \le t \le 15$

$$N(t) = \int N'(t)\,dt = \int 6e^{-0.1t}\,dt = 6\int e^{-0.1t}\,dt$$

$$= \frac{6}{-0.1}\int e^{-0.1t}(-0.1)\,dt = -60e^{-0.1t} + C$$

Given $N(0) = 40$:

$40 = -60e^0 + C$

Hence, $C = 100$ and $N(t) = 100 - 60e^{-0.1t}$, $0 \le t \le 15$.

The number of words per minute after completing the course is:

$N(15) = 100 - 60e^{-0.1(15)} = 100 - 60e^{-1.5} \approx 87$ words per minute.

79. $\dfrac{dE}{dt} = 5000(t + 1)^{-3/2}$, $t \ge 0$

Let $u = t + 1$, then $du = dt$

$$E = \int 5000(t + 1)^{-3/2}\,dt = 5000\int (t + 1)^{-3/2}\,dt = 5000\int u^{-3/2}\,du$$

$$= 5000\frac{u^{-1/2}}{-\dfrac{1}{2}} + C = -10{,}000(t + 1)^{-1/2} + C$$

$$= \frac{-10{,}000}{\sqrt{t + 1}} + C$$

Given $E(0) = 2000$:

$2000 = \dfrac{-10{,}000}{\sqrt{1}} + C$

Hence, $C = 12{,}000$ and $E(t) = 12{,}000 - \dfrac{10{,}000}{\sqrt{t + 1}}$.

The projected enrollment 15 years from now is:

$E(15) = 12{,}000 - \dfrac{10{,}000}{\sqrt{15 + 1}} = 12{,}000 - \dfrac{10{,}000}{\sqrt{16}} = 12{,}000 - \dfrac{10{,}000}{4}$

$= 9500$ students

EXERCISE 6-3

Things to remember:

1. A DIFFERENTIAL EQUATION is an equation that involves an unknown function and one or more of its derivatives. The ORDER of a differential equation is the order of the highest derivative of the unknown function.

<u>2.</u>   A SLOPE FIELD for a first-order differential equation is obtained by drawing tangent line segments determined by the equation at each point in a grid.

<u>3.</u>   EXPONENTIAL GROWTH LAW

If $\dfrac{dQ}{dt} = rQ$ and $Q(0) = Q_0$, then $Q(t) = Q_0 e^{rt}$, where

    Q_0 = Amount at $t = 0$

    r = Continuous compound growth rate (expressed as a decimal)

    t = Time

    Q = Quantity at time t

<u>4.</u>   COMPARISON OF EXPONENTIAL GROWTH PHENOMENA

DESCRIPTION	MODEL	SOLUTION	GRAPH	USES
Unlimited growth: Rate of growth is proportional to the amount present	$\dfrac{dy}{dt} = ky$ $k, t > 0$ $y(0) = c$	$y = ce^{kt}$		• Short-term population growth (people, bacteria, etc.) • Growth of money at continuous compound interest • Price-supply curves • Depletion of natural resources
Exponential decay: Rate of growth is proportional to the amount present	$\dfrac{dy}{dt} = -ky$ $k, t > 0$ $y(0) = c$	$y = ce^{-kt}$		• Radioactive decay • Light absorption in water • Price-demand curves • Atmospheric pressure (t is altitude)
Limited growth: Rate of growth is proportional to the difference between the amount present and a fixed limit	$\dfrac{dy}{dt} = k(M - y)$ $k, t > 0$ $y(0) = 0$	$y = M(1 - e^{-kt})$		• Sales fads (e.g., skateboards) • Depreciation of equipment • Company growth • Learning
Logistic growth: Rate of growth is proportional to the amount present and to the difference between the amount present and a fixed limit	$\dfrac{dy}{dt} = ky(M - y)$ $k, t > 0$ $y(0) = \dfrac{M}{1 + c}$	$y = \dfrac{M}{1 - ce^{-kMt}}$		• Long-term population growth • Epidemics • Sales of new products • Rumor spread • Company growth

---

1. $\dfrac{dy}{dx} = e^{0.5x}$; $\displaystyle\int \dfrac{dy}{dx}\, dx = \int e^{0.5x}\, dx$

$$y = \dfrac{e^{0.5x}}{0.5} + C$$

General solution: $y = 2e^{0.5x} + C$

3. $\dfrac{dy}{dx} = x^2 - x; \quad y(0) = 0$

$$\int \frac{dy}{dx}\, dx = \int (x^2 - x)\, dx$$

$$y = \frac{1}{3}x^3 - \frac{1}{2}x^2 + C$$

Given $y(0) = 0$: $\frac{1}{3}(0)^3 - \frac{1}{2}(0)^2 + C = 0$

$$C = 0$$

Particular solution: $y = \frac{1}{3}x^3 - \frac{1}{2}x^2$

5. $\dfrac{dy}{dx} = -2xe^{-x^2}; \quad y(0) = 3$

$$\int \frac{dy}{dx}\, dx = \int -2xe^{-x^2}\, dx$$

$$y = \int -2xe^{-x^2}\, dx$$

Let $u = -x^2$, then $du = -2x\, dx$ and

$$\int -2xe^{-x^2}\, dx = \int e^u\, du = e^u + c = e^{-x^2} + c$$

Thus, $\quad y = e^{-x^2} + c$

Given $y(0) = 3$: $3 = e^0 + c$

$$3 = 1 + c$$

$$c = 2$$

Particular solution: $y = e^{-x^2} + 2$

7. Figure (b). When $x = 1$, $\dfrac{dy}{dx} = 1 - 1 = 0$ for any y. When $x = 0$, $\dfrac{dy}{dx} = 0 - 1 = -1$ for any y. When $x = 2$, $\dfrac{dy}{dx} = 2 - 1 = 1$ for any y; and so on. These facts are consistent with the slope-field in Figure (b); they are not consistent with the slope-field in Figure (a).

9. $\dfrac{dy}{dx} = x - 1$

$$\int \frac{dy}{dx}\, dx = \int (x - 1)\, dx$$

General solution: $y = \frac{1}{2}x^2 - x + c$

Given $y(0) = -2$: $\frac{1}{2}(0)^2 - 0 + c = -2$

$$c = -2$$

Particular solution: $y = \frac{1}{2}x^2 - x - 2$

11.

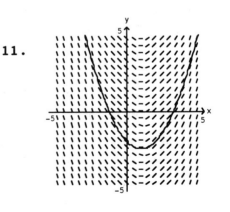

13. $\dfrac{dy}{dx} = -0.8y$

$$\int \frac{1}{y} \frac{dy}{dx}\, dx = \int -0.8\, dx$$

$$\int \frac{1}{y}\, dy = \int -0.8\, dx$$

$$\ln |y| = -0.8x + K \quad (K \text{ an arbitrary constant})$$
$$|y| = e^{-0.8x+K} = e^K e^{-0.8x}$$
$$|y| = Ce^{-0.8x} \quad \text{where} \quad C = e^K$$

If we assume $y > 0$, we get:
    General solution: $y = Ce^{-0.8x}$

Note: The differential equation $\dfrac{dy}{dx} = -0.8x$ is the model for exponential decay with decay rate $r = 0.8$ (see 3). Thus,
    $y = Ce^{-0.8x}$

15. $\dfrac{dy}{dx} = 0.07y, \ y(0) = 1{,}000$

$$\int \frac{1}{y} \frac{dy}{dx}\, dx = \int 0.07\, dx$$

$$\int \frac{1}{y}\, dy = \int 0.07\, dx$$

$$\ln |y| = 0.07x + K \quad (K \text{ an arbitrary constant})$$
$$|y| = e^{0.07x+K} = e^K e^{0.07x}$$
$$|y| = Ce^{0.07x} \quad (C = e^K)$$

If we assume $y > 0$, we get
    General solution: $y = Ce^{0.07x}$

Given $y(0) = 1{,}000$: $1000 = Ce^0$
                           $C = 1000$
    Particular solution: $y = 1{,}000e^{0.07x}$

17. $\dfrac{dx}{dt} = -x$

$$\int \frac{1}{x} \frac{dx}{dt}\, dt = -\int dt$$

$$\int \frac{1}{x}\, dx = -\int dt$$

$$\ln |x| = -t + K \quad (K \text{ an arbitrary constant})$$
$$|x| = e^{-t+K} = e^K e^{-t}$$
$$|x| = Ce^{-t}, \quad (C = e^K)$$

If we assume $x > 0$, we get
    General solution: $x = Ce^{-t}$

19. Figure (c). When $y = 1$, $\frac{dy}{dx} = 1 - 1 = 0$ for any x.

When $y = 2$, $\frac{dy}{dx} = 1 - 2 = -1$ for any x; and so on. This is consistent with the slope-field in Figure (c); it is not consistent with the slope-field in Figure (d).

21. $y = 1 - Ce^{-x}$

$\frac{dy}{dx} = \frac{d}{dx}[1 - Ce^{-x}] = Ce^{-x}$

From the original equation,

$\quad Ce^{-x} = 1 - y$

Thus, we have

$\quad \frac{dy}{dx} = 1 - y$

and $y = 1 - Ce^{-x}$ is a solution of the differential equation for any number c.

Given $y(0) = 0$: $0 = 1 - Ce^0 = 1 - c$

$\qquad\qquad\qquad c = 1$

Particular solution: $y = 1 - e^{-x}$

23.

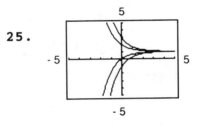

25.

27. $y = 1,000e^{0.08t}$

$\quad 0 \le t \le 15,\ 0 \le y \le 3,500$

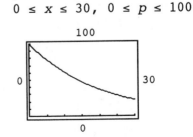

29. $p = 100e^{-0.05x}$

$\quad 0 \le x \le 30,\ 0 \le p \le 100$

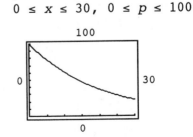

31. $N = 100(1 - e^{-0.05t})$
$0 \leq t \leq 100,\ 0 \leq N \leq 100$

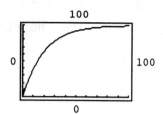

33. $N = \dfrac{1,000}{1 + 999e^{-0.4t}}$
$0 \leq t \leq 40,\ 0 \leq N \leq 1,000$

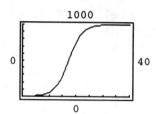

35. $\dfrac{dA}{dt} = 0.08A,\ A(0) = 1,000.$

This is an unlimited growth model. From $\underline{4}$, $A(t) = 1,000e^{0.08t}$.

37. $\dfrac{dA}{dt} = rA,\ A(0) = 8,000$

is an unlimited growth model. From $\underline{4}$, $A(t) = 8,000e^{rt}$.
Since $A(2) = 9,020$, we solve $8,000e^{2r} = 9,020$ for r.
$$8000e^{2r} = 9,020$$
$$e^{2r} = \frac{902}{800}$$
$$2r = \ln(902/800)$$
$$r = \frac{\ln(902/800)}{2} \approx 0.06$$

Thus, $A(t) = 8,000e^{0.06t}$.

39. (A) $\dfrac{dp}{dx} = rp,\ p(0) = 100$

This is an UNLIMITED GROWTH MODEL. From $\underline{4}$, $p(x) = 100e^{rx}$.
Since $p(5) = 77.88$, we have
$$77.88 = 100e^{5r}$$
$$e^{5r} = 0.7788$$
$$5r = \ln(0.7788)$$
$$r = \frac{\ln(0.7788)}{5} \approx -0.05$$

Thus, $p(x) = 100e^{-0.05x}$.

(B) $p(10) = 100e^{-0.05(10)} = 100e^{-0.5}$
$\approx \$60.65$ per unit

(C)

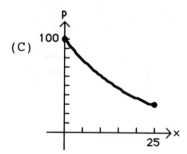

41. (A) $\dfrac{dN}{dt} = k(L - N)$; $N(0) = 0$

This is a LIMITED GROWTH MODEL. From $\underline{4}$, $N(t) = L(1 - e^{-kt})$.
Since $N(10) = 0.4L$, we have

$$0.4L = L(1 - e^{-10k})$$
$$1 - e^{-10k} = 0.4$$
$$e^{-10k} = 0.6$$
$$-10k = \ln(0.6)$$
$$k = \frac{\ln(0.6)}{-10} \approx 0.051$$

Thus, $N(t) = L(1 - e^{-0.051t})$.

(B) $N(5) = L[1 - e^{-0.051(5)}] = L[1 - e^{-0.255}] \approx 0.225L$

Approximately 22.5% of the possible viewers will have been exposed after 5 days.

(C) Solve $L(1 - e^{-0.051t}) = 0.8L$ for t:

$$1 - e^{-0.051t} = 0.8$$
$$e^{-0.051t} = 0.2$$
$$-0.051t = \ln(0.2)$$
$$t = \frac{\ln(0.2)}{-0.051} \approx 31.56$$

It will take 32 days for 80% of the possible viewers to be exposed.

(D)

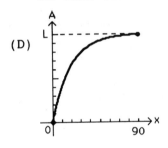

43. $\dfrac{dI}{dx} = -kI$, $I(0) = I_0$

This is an exponential decay model. From $\underline{4}$, $I(x) = I_0 e^{-kx}$ with

$k = 0.00942$, we have
$$I(x) = I_0 e^{-0.00942x}$$

To find the depth at which the light is reduced to half of that at the surface, solve,

$$I_0 e^{-0.00942x} = \frac{1}{2} I_0$$

for x:

$$e^{-0.00942x} = 0.5$$
$$-0.00942x = \ln(0.5)$$
$$x = \frac{\ln(0.5)}{-0.00942} \approx 74 \text{ feet}$$

45. $\frac{dQ}{dt} = -0.04Q$, $Q(0) = Q_0$.

(A) This is a model for exponential decay. From 4,
$$Q(t) = Q_0 e^{-0.04t}$$
With $Q_0 = 3$, we have
$$Q(t) = 3e^{-0.04t}$$

(B) $Q(10) = 3e^{-0.04(10)} = 3e^{-0.4} \approx 2.01$.
There are approximately 2.01 milliliters in the body after 10 hours.

(C) $3e^{-0.04t} = 1$

$e^{-0.04t} = \frac{1}{3}$

$-0.04t = \ln(1/3)$

$t = \frac{\ln(1/3)}{-0.04} \approx 27.47$

It will take approximately 27.47 hours
for Q to decrease to 1 milliliter.

(D)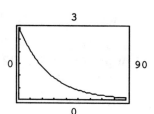

47. Using the exponential decay model, we have $\frac{dy}{dt} = -ky$, $y(0) = 100$, $k > 0$
where $y = y(t)$ is the amount of cesium-137 present at time t. From 4,
$$y(t) = 100e^{-kt}$$
Since $y(3) = 93.3$, we solve $93.3 = 100e^{-3k}$ for k to find the continuous
compound decay rate:
$$93.3 = 100e^{-3k}$$
$$e^{-3k} = 0.933$$
$$-3k = \ln(0.933)$$
$$k = \frac{\ln(0.933)}{-3} \approx 0.023117$$

49. From Example 3: $Q = Q_0 e^{-0.0001238t}$

Now, the amount of radioactive carbon-14 present is 5% of the original
amount. Thus, $0.05Q_0 = Q_0 e^{-0.0001238t}$ or $e^{-0.0001238t} = 0.05$.

Therefore, $-0.0001238t = \ln(0.05) \approx -2.9957$ and $t \approx 24{,}200$ years.

51. $N(k) = 180e^{-0.11(k-1)}$, $1 \le k \le 10$
Thus, $N(6) = 180e^{-0.11(6-1)} = 180e^{-0.55} \approx 104$ times
and $N(10) = 180e^{-0.11(10-1)} = 180e^{-0.99} \approx 67$ times.

53. (A) $x(t) = \dfrac{400}{1 + 399e^{-0.4t}}$

$x(5) = \dfrac{400}{1 + 399e^{(-0.4)5}} = \dfrac{400}{1 + 399e^{-2}} \approx \dfrac{400}{55} \approx 7$ people

$x(20) = \dfrac{400}{1 + 399e^{(-0.4)20}} = \dfrac{400}{1 + 399e^{-8}} \approx 353$ people

(B) $\lim\limits_{t \to \infty} x(t) = 400$.

(C)

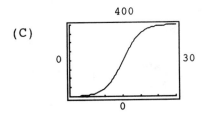

400

0   30

0

EXERCISE 6-4

Things to remember:

1. APPROXIMATING AREA UNDER A GRAPH; LEFT SUMS AND RIGHT SUMS

   Let $f(x)$ be defined and positive on the interval $[a, b]$. Divide the interval into n equal subintervals of length $\Delta x = \dfrac{b - a}{n}$, with x_1, x_2, x_3, ..., x_{n-1}, the points of subdivision.

   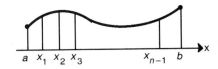

   Then
   $$L_n = f(a)\Delta x + f(x_1)\Delta x + f(x_2)\Delta x + \ldots + f(x_{n-1})\Delta x$$

   is called a LEFT SUM;
   $$R_n = f(x_1)\Delta x + f(x_2)\Delta x + f(x_3)\Delta x + \ldots + f(x_{n-1})\Delta x + f(x_n)\Delta x$$

   is called a RIGHT SUM;
   $$A_n = \frac{L_n + R_n}{2} \text{ is the average of } L_n \text{ and } R_n.$$

   Left and right sums and their averages are used to approximate the area under the graph of f. The exact area under the graph of $y = f(x)$ from $x = a$ to $x = b$ is denoted by the DEFINITE INTEGRAL SYMBOL
   $$\int_a^b f(x)\,dx = \begin{pmatrix}\text{area under the graph} \\ \text{from } x = a \text{ to } x = b\end{pmatrix}$$

2. MONOTONE FUNCTIONS AND AREA UNDER THE GRAPH

   A function f is MONOTONE over an interval $[a, b]$ if it is either increasing over $[a, b]$ or decreasing over $[a, b]$.

   If f is a monotone function, then the area under the graph of f always lies between the left sum L_n and the right sum R_n for any integer n.

3. ERROR BOUNDS FOR LEFT AND RIGHT SUMS AND THEIR AVERAGE
(MONOTONE FUNCTIONS)

If $f(x)$ is monotonic on the interval $[a, b]$ and $I = \int_a^b f(x)\,dx$,

L_n = left sum, R_n = right sum, and $A_n = \dfrac{L_n + R_n}{2}$

then the following error bounds hold:

$$|I - L_n| \le |f(b) - f(a)| \, \frac{b - a}{n}$$

$$|I - R_n| \le |f(b) - f(a)| \, \frac{b - a}{n}$$

$$|I - A_n| \le |f(b) - f(a)| \, \frac{b - a}{2n}$$

4. DEFINITE INTEGRAL SYMBOL FOR FUNCTIONS WITH NEGATIVE VALUES

If $f(x)$ is positive for some values of x on $[a, b]$ and negative for
others, then the DEFINITE INTEGRAL SYMBOL

$$\int_a^b f(x)\,dx$$

represents the cumulative sum of the signed areas between the curve
$y = f(x)$ and the x axis where the areas above the x axis are counted
positively and the areas below the x axis are counted negatively
(see the figure where A and B are actual areas of the indicated
regions).

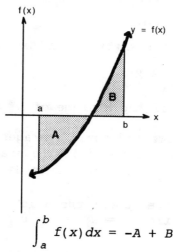

$$\int_a^b f(x)\,dx = -A + B$$

5. RATE, AREA, AND DISTANCE

If $r = r(t)$ is a positive rate function for an object moving on a
line, then

$$\int_a^b r(t)\,dt = \begin{pmatrix} \text{Net distance traveled} \\ \text{from } t = a \text{ to } t = b \end{pmatrix}$$

$$= \begin{pmatrix} \text{Total Change in Position} \\ \text{from } t = a \text{ to } t = b \end{pmatrix}$$

6

RATE, AREA, AND TOTAL CHANGE

If $y = F'(x)$ is a rate function (derivative), then the cumulated sum of the signed areas between the curve $y = F'(x)$ and the x axis from $x = a$ to $x = b$ represents the total net change in $F(x)$ from $x = a$ to $x = b$. Symbolically,

$$\int_a^b F'(x)\,dx = \begin{pmatrix} \text{Total net change in } F(x) \\ \text{from } x = a \text{ to } x = b \end{pmatrix}$$

1.
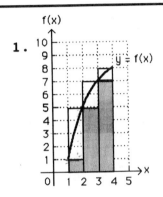
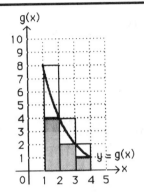

3. For Figure (a):

$$L_3 = f(1) \cdot 1 + f(2) \cdot 1 + f(3) \cdot 1$$
$$= 1 + 5 + 7 = 13$$

$$R_3 = f(2) \cdot 1 + f(3) \cdot 1 + f(4) \cdot 1$$
$$= 5 + 7 + 8 = 20$$

$$A_3 = \frac{L_3 + R_3}{2} = \frac{13 + 20}{2} = \frac{33}{2} = 16.5$$

For Figure (b):

$$L_3 = g(1) \cdot 1 + g(2) \cdot 1 + g(3) \cdot 1$$
$$= 8 + 4 + 2 = 14$$

$$R_3 = g(2) \cdot 1 + g(3) \cdot 1 + g(4) \cdot 1$$
$$= 4 + 2 + 1 = 7$$

$$A_3 = \frac{L_3 + R_3}{2} = \frac{14 + 7}{2} = 10.5$$

5. $L_3 \le \int_1^4 f(x)\,dx \le R_3$, $R_3 \le \int_1^4 g(x)\,dx \le L_3$; since f is increasing on $[1, 4]$, L_3 underestimates the area and R_3 overestimates the area; since g is decreasing on $[1, 4]$, L_3 overestimates the area and R_3 underestimates the area.

7. For Figure (a).

Error bound for L_3 and R_3:

$$\text{Error} \le |f(4) - f(1)| \left(\frac{4-1}{3}\right) = |8 - 1| = 7$$

Error bound for A_3:

$$\text{Error} \le \frac{7}{2} = 3.5$$

For Figure (b).

Error bound for L_3 and R_3:

$$\text{Error} \le |f(4) - f(1)| \left(\frac{4-1}{3}\right) = |1 - 8| = |-7| = 7$$

Error bound for A_3:

$$\text{Error} \le \frac{7}{2} = 3.5$$

9. The exact area under the graph of $y = f(x)$ is within 3.5 units of the average of the left sum and right sum estimates, $A_3 = 16.5$.

11. $r(t)$ is a decreasing function.

(A) $L_4 = r(0) \cdot 1 + r(1) \cdot 1 + r(2) \cdot 1 + r(3) \cdot 1$

$\qquad = 128 + 96 + 64 + 32 = 320$

$R_4 = r(1) \cdot 1 + r(2) \cdot 1 + r(3) \cdot 1 + r(4) \cdot 1$

$\qquad = 96 + 64 + 32 + 0 = 192$

$$A_4 = \frac{L_4 + R_4}{2} = \frac{320 + 192}{2} = 256$$

Error bound for L_4 and R_4:

$$\text{Error} \le |f(4) - f(0)| \left(\frac{4-0}{4}\right) = 128$$

Error bound for A_4:

$$\text{Error} \le \frac{128}{2} = 64$$

(B) The height of each rectangle represents an instantaneous rate and the base of each rectangle is a time interval; rate *times* time *equals* distance.

(C) We want to find n such that $|I - A_n| < 1$, that is:

$$|f(4) - f(0)| \left(\frac{4-0}{2n}\right) < 1$$

$$|0 - 128| \left(\frac{2}{n}\right) < 1$$

$$\frac{256}{n} < 1 \quad \text{or} \quad n > 256$$

13. $h(x)$ is an increasing function; $\Delta x = 100$

$L_{10} = h(0)100 + h(100)100 + h(200)100 + \ldots + h(900)(100)$

$\quad = [0 + 183 + 235 + 245 + 260 + 286 + 322 + 388 + 453 + 489]100$

$\quad = (2,861)100 = 286,100$ sq ft

$R_{10} = h(100)100 + h(200)100 + h(300)100 + \ldots + h(1,000)100$

$\quad = [183 + 235 + 245 + 260 + 286 + 322 + 388 + 453 + 489 + 500]100$

$\quad = (3,361)100 = 336,100$ sq ft

$A_{10} = \dfrac{L_{10} + R_{10}}{2} = \dfrac{286,100 + 336,100}{2} = 311,100$ sq ft

Error bound for A_{10}:

$$\text{Error} \le |h(1,000) - h(0)| \left(\frac{1000 - 0}{2 \cdot 10}\right) = 500(50) = 25,000 \text{ sq ft}$$

We want to find n such that $|I - A_n| \le 2,500$:

$$|h(1000) - h(0)| \left(\frac{1000 - 0}{2n}\right) \le 2,500$$

$$500\left(\frac{500}{n}\right) \le 2,500$$

$$250,000 \le 2,500n$$

$$n \ge 100$$

15. $r(t)$ is a decreasing function; $\Delta t = 1$

(A) $L_7 = r(0)1 + r(1)1 + r(2)1 + r(3)1 + r(4)1 + r(5)1 + r(6)1$

$\quad = 110 + 85 + 63 + 45 + 29 + 16 + 5 = 353$

$R_7 = r(1)1 + r(2)1 + r(3)1 + r(4)1 + r(5)1 + r(6)1 + r(7)1$

$\quad = 85 + 63 + 45 + 29 + 16 + 5 + 0 = 243$

$A_7 = \dfrac{L_7 + R_7}{2} = \dfrac{353 + 243}{2} = 298$ ft

Error bound for A_7:

$$\text{Error} \le |r(7) - r(0)| \left(\frac{7 - 0}{2 \cdot 7}\right) = |0 - 110|\, \frac{1}{2} = 55 \text{ ft}$$

(B) We want to find n such that $|I - A_n| \le 5$, that is:

$$|r(7) - r(0)| \left(\frac{7 - 0}{2n}\right) < 5$$

$$\frac{(110)7}{2n} < 5$$

$$10n > 770$$

$$n > 77$$

17. (A) $P = 2$

(B)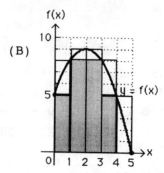

(C) To the left of $P = 2$, the left rectangles underestimate the true area and the right rectangles overestimate the true area. To the right of $P = 2$, the left rectangles overestimate the true area and the right rectangles underestimate the true area.

(D) N_1 = sum of areas of rectangles below the graph of f:
$$= 5 \cdot 1 + 8 \cdot 1 + 8 \cdot 1 + 5 \cdot 1 + 0 \cdot 1 = 26$$
N_2 = sum of areas of rectangles above the graph of f:
$$= 8 \cdot 1 + 9 \cdot 1 + 9 \cdot 1 + 8 \cdot 1 + 5 \cdot 1 = 39$$
Thus, $26 \le \displaystyle\int_0^5 f(x)\,dx \le 39$.

19. $f(x) = 0.25x^2 - 4$ on $[2, 5]$

$L_6 = f(2)\Delta x + f(2.5)\Delta x + f(3)\Delta x + f(3.5)\Delta x + f(4)\Delta x + f(4.5)\Delta x$
where $\Delta x = 0.5$
Thus,
$L_6 = [-3 - 2.44 - 1.75 - 0.94 + 0 + 1.06](0.5) = -3.53$
$R_6 = f(2.5)\Delta x + f(3)\Delta x + f(3.5)\Delta x + f(4)\Delta x + f(4.5)\Delta x + f(5)\Delta x$
where $\Delta x = 0.5$
Thus,
$R_6 = [-2.44 - 1.75 - 0.94 + 0 + 1.06 + 2.25](0.5) = -0.91$
$$A_6 = \frac{L_6 + R_6}{2} = \frac{-3.53 - 0.91}{2} = -2.22$$
Error bound for L_6 and R_6:
$$\text{Error} \le |f(5) - f(2)| \left(\frac{5 - 2}{6}\right) = |2.25 - (-3)|\,(0.5) = 2.63$$
Error bound for A_6:
$$\text{Error} \le \frac{2.63}{2} = 1.32$$

Geometrically, the definite integral over the interval $[2, 5]$ is the area of the region which lies above the x-axis minus the area of the region which lies below the x-axis. From the figure, if R_1 represents the region bounded by the graph of f and the x-axis for $2 \le x \le 4$ and R_2 represents the region bounded by the graph of f and the x-axis for $4 \le x \le 5$, then

$$\int_2^5 f(x)\,dx = \text{area}(R_2) - \text{area}(R_1)$$

21. $\displaystyle\int_1^2 x^x\,dx; \quad f(x) = x^x$

$$|I - R_n| \le |f(2) - f(1)|\left(\frac{2-1}{n}\right) \le 0.05$$

$$|4 - 1|\left(\frac{1}{n}\right) \le 0.05$$

$$\frac{3}{n} \le 0.05$$

$$n \ge \frac{3}{0.05} = 60$$

23. $\displaystyle\int_0^2 e^{-x^2}\,dx; \quad f(x) = e^{-x^2}$

$$|I - L_n| = |f(2) - f(0)|\left(\frac{2-0}{n}\right) \le 0.005$$

$$|e^{-4} - e^0|\left(\frac{2}{n}\right) \le 0.005$$

$$|0.018316 - 1|\left(\frac{2}{n}\right) \le 0.005$$

$$(0.981684)\frac{2}{n} \le 0.005$$

$$n \ge \frac{(0.981684)2}{0.005} \approx 393$$

25. Suppose f is monotonic on $[a, b]$. Let $I = \displaystyle\int_a^b f(x)\,dx$. Then

$$|I - L_n| \le |f(b) - f(a)|\left(\frac{b-a}{n}\right)$$

Now, $\displaystyle\lim_{n\to\infty} |f(b) - f(a)|\left(\frac{b-a}{n}\right) = 0$

Thus, $\displaystyle\lim_{n\to\infty} |I - L_n| = 0$ which implies $I = \displaystyle\lim_{n\to\infty} L_n$

27. Let $a = 300$, $b = 900$, $n = 2$. Then $\Delta x = \dfrac{900 - 300}{2} = 300$

$$L_2 = f(300)(300) + f(600)(300)$$
$$= [400 + 300]300 = 210,000$$

$$R_2 = f(600)300 + f(900)(300)$$
$$= [300 + 200]300 = 150,000$$

$$A_2 = \frac{L_2 + R_2}{2} = \frac{210,000 + 150,000}{2} = \$180,000$$

Error bound for A_2:

$$\text{Error} \le |f(900) - f(300)|\left(\frac{900 - 300}{2 \cdot 2}\right)$$

$$= |200 - 400|\,(150)$$
$$= 200(150) = \$30,000$$

29. Let $A'(t) = 800e^{0.08t}$, $a = 2$, $b = 6$, $n = 4$. Then $\Delta t = \dfrac{6 - 2}{4} = 1$.

$L_4 = A(2)1 + A(3)1 + A(4)1 + A(5)1$

$\quad = 939 + 1017 + 1102 + 1193 = \$4251$

$R_4 = A(3)1 + A(4)1 + A(5)1 + A(6)1$

$\quad = 1017 + 1102 + 1193 + 1293 = \$4605$

Since $A'(t)$ is increasing on $[2, 6]$,

$$\$4251 \le \int_0^2 800e^{0.08t}\, dt \le \$4605$$

31. First 60 days:

$L_3 = N(0)20 + N(20)20 + N(40)20$

$\quad = (10 + 51 + 68)20 = 2580$

$R_3 = N(20)20 + N(40)20 + N(60)20$

$\quad = (51 + 68 + 76)20 = 3900$

$A_3 = \dfrac{2580 + 3900}{2} = 3240$

Error bound for A_3:

$$\text{Error} \le |N(60) - N(0)| \left(\frac{60 - 0}{2 \cdot 3}\right) = (76 - 10)(10) = 660 \text{ units}$$

Second 60 days:

$L_3 = N(60)20 + N(80)20 + N(100)20$

$\quad = (76 + 81 + 84)20 = 4820$

$R_3 = N(80)20 + N(100)20 + N(120)20$

$\quad = (81 + 84 + 86)20 = 5020$

$A_3 = \dfrac{4820 + 5020}{2} = 4920$

Error bound for A_3:

$$\text{Error} \le |N(120) - N(60)| \left(\frac{120 - 60}{2 \cdot 3}\right) = (86 - 76)(10) = 100 \text{ units}$$

33. (A) Geometrically, $\displaystyle\int_{100}^{200} R'(x)\, dx$ is the area under the graph of $R'(x)$ on the interval $[100, 200]$. $\displaystyle\int_{100}^{200} R'(x)\, dx$ also represents the total change in revenue going from sales of 100 six packs per day to 200 six packs per day.

(B) $R'(x) = 8 - \dfrac{x}{25}$, $a = 100$, $b = 200$, $n = 4$; $\quad \Delta x = \dfrac{200 - 100}{4} = 25$

$\quad L_4 = R'(100)25 + R'(125)25 + R'(150)25 + R'(175)25$

$\qquad = (4 + 3 + 2 + 1)25 = 250$

$\quad R_4 = R'(125)25 + R'(150)25 + R'(175)25 + R'(200)25$

$\qquad = (3 + 2 + 1 + 0)25 = 150$

$\quad A_4 = \dfrac{250 + 150}{2} = \$200$

Error bound for A_4:

$\qquad \text{Error} \leq |R'(200 - R'(100)| \left(\dfrac{200 - 100}{2 \cdot 4}\right)$

$\qquad\qquad = |0 - 4|(12.5) = 4(12.5) = \$50$

(C) $R(200) - R(100) = 8(200) - \dfrac{(200)^2}{50} - \left[8(100) - \dfrac{(100)^2}{50}\right]$

$\qquad\qquad = 1600 - 800 - (800 - 200)$

$\qquad\qquad = \$200$

$R(200) - R(100)$ and $\displaystyle\int_{100}^{200} R'(x)\,dx$ each represents the total change in revenue going from sales of 100 six packs per day to 200 six packs per day. This suggests that

$$\int_{100}^{200} R'(x)\,dx = R(200) - R(100)$$

35. (A) $L_5 = A'(0)1 + A'(1)1 + A'(2)1 + A'(3)1 + A'(4)1$

$\qquad = 0.90 + 0.81 + 0.74 + 0.67 + 0.60$

$\qquad = 3.72$ sq cm

$\quad R_5 = A'(1)1 + A'(2)1 + A'(3)1 + A'(4)1 + A'(5)1$

$\qquad = (0.81 + 0.74 + 0.67 + 0.60 + 0.55)$

$\qquad = 3.37$ sq cm

(B) Since $A'(t)$ is a decreasing function

$\qquad 3.37 \leq \displaystyle\int_0^5 A'(t)\,dt \leq 3.72$

37. $L_3 = N'(6)2 + N'(8)2 + N'(10)2$

$\quad = (21 + 19 + 17)2 = 114$

$R_3 = N'(8)2 + N'(10)2 + N'(12)2$

$\quad = (19 + 17 + 15)2 = 102$

$A_3 = \dfrac{L_3 + R_3}{2} = \dfrac{114 + 102}{2} = 108$ code symbols

Error bound for A_3:

$\qquad \text{Error} \leq |N'(12) - N'(6)| \left(\dfrac{12 - 6}{2 \cdot 3}\right)$

$\qquad |15 - 21|(1) = 6$ code symbols

Things to remember:

1. LEFT SUM, RIGHT SUM, MIDPOINT SUM

   Let f be defined on the interval $[a, b]$. Partition $[a, b]$ into n equal subintervals of length $\Delta x = \dfrac{b - a}{n}$ with endpoints $a = x_0, x_1, x_2, \ldots, x_{n-1}, x_n = b$.

   Then,

   (a) LEFT SUM $= L_n = \displaystyle\sum_{k=1}^{n} f(x_{k-1})\Delta x$

   $\qquad\qquad\qquad = f(x_0)\Delta x + f(x_1)\Delta x + \ldots + f(x_{n-1})\Delta x$

   (b) RIGHT SUM $= R_n = \displaystyle\sum_{k=1}^{n} f(x_k)\Delta x$

   $\qquad\qquad\qquad\ = f(x_1)\Delta x + f(x_2)\Delta x + \ldots + f(x_n)\Delta x$

   (c) MIDPOINT SUM $= M_n = \displaystyle\sum_{k=1}^{n} f\left(\dfrac{x_{k-1} + x_k}{2}\right)\Delta x$

   $\qquad\qquad\qquad\ = f\left(\dfrac{x_0 + x_1}{2}\right)\Delta x + f\left(\dfrac{x_1 + x_2}{2}\right)\Delta x + \ldots$

   $\qquad\qquad\qquad\qquad\qquad\qquad\qquad + f\left(\dfrac{x_{n-1} + x_n}{2}\right)\Delta x$

2. DEFINITION OF A DEFINITE INTEGRAL

   Let f be a continuous function defined on the closed interval $[a, b]$, and let

   1. $a = x_0 < x_1 < \ldots < x_{n-1} < x_n = b$
   2. $\Delta x_k = x_k - x_{k-1}$ for $k = 1, 2, \ldots, n$
   3. $\Delta x_k \to 0$ as $n \to \infty$
   4. $x_{k-1} \le c_k \le x_k$ for $k = 1, 2, \ldots, n$

   Then,

   $$\int_a^b f(x)\,dx = \lim_{n \to \infty} \sum_{k=1}^{n} f(c_k)\Delta x_k$$
   $$= \lim_{n \to \infty} [f(c_1)\Delta x_1 + f(c_2)\Delta x_2 + \ldots + f(c_n)\Delta x_n]$$

   is called a DEFINITE INTEGRAL of f from a to b. The INTEGRAND is $f(x)$, the LOWER LIMIT is a and the UPPER LIMIT is b.

<u>3</u>. ERROR BOUNDS FOR L_n, R_n, M_n

Let

$$I = \int_a^b f(x)\,dx \qquad L_n = \text{Left Sum} \qquad R_n = \text{Right Sum}$$

$$M_n = \text{Midpoint Sum}$$

LEFT AND RIGHT SUM ERROR BOUND:

If $|f'(x)| \le B_1$ for all x on $[a, b]$, then

$$|I - L_n| \le \frac{B_1(b - a)^2}{2n}, \qquad |I - R_n| \le \frac{B_1(b - a)^2}{2n}$$

MIDPOINT ERROR BOUND:

If $|f''(x)| \le B_2$ for all x on $[a, b]$, then

$$|I - M_n| \le \frac{B_2(b - a)^3}{24n^2}$$

<u>4</u>. DEFINITE INTEGRAL PROPERTIES

(a) $\displaystyle\int_a^a f(x)\,dx = 0$

(b) $\displaystyle\int_a^b f(x)\,dx = -\int_b^a f(x)\,dx$

(c) $\displaystyle\int_a^b Kf(x)\,dx = K\int_a^b f(x)\,dx \qquad K$ is a constant

(d) $\displaystyle\int_a^b [f(x) \pm g(x)]\,dx = \int_a^b f(x)\,dx \pm \int_a^b g(x)\,dx$

(e) $\displaystyle\int_a^b f(x)\,dx = \int_a^c f(x)\,dx + \int_c^b f(x)\,dx$

<u>5</u>. FUNDAMENTAL THEOREM OF CALCULUS

If f is a continuous function on the closed interval $[a, b]$ and F is any antiderivative of f, then

$$\int_a^b f(x)\,dx = F(x)\,\Big|_a^b = F(b) - F(a);$$
$$F'(x) = f(x)$$

<u>6</u>. AVERAGE VALUE OF A CONTINUOUS FUNCTION OVER $[a, b]$

Let f be continuous on $[a, b]$. Then the AVERAGE VALUE of f over $[a, b]$ is:

$$\frac{1}{b - a} \int_a^b f(x)\,dx$$

1. $\int_a^0 f(x)\,dx = -2.33$

3. $\int_a^b f(x)\,dx = -2.33 + 10.67 = 8.34$

5. $\int_0^b \dfrac{f(x)}{10}\,dx = \dfrac{1}{10}\int_0^b f(x)\,dx = \dfrac{1}{10}(10.67) = 1.067$

7. $\int_2^3 2x\,dx = 2 \cdot \dfrac{x^2}{2}\Big|_2^3 = 3^2 - 2^2 = 5$

9. $\int_3^4 5\,dx = 5x\Big|_3^4 = 5\cdot 4 - 5\cdot 3 = 5$

11. $\int_1^3 (2x - 3)\,dx = (x^2 - 3x)\Big|_1^3 = (3^2 - 3\cdot 3) - (1^2 - 3\cdot 1) = 2$

13. $\int_0^4 (3x^2 - 4)\,dx = (x^3 - 4x)\Big|_0^4 = (4^3 - 4\cdot 4) - (0^3 - 4\cdot 0) = 48$

15. $\int_{-3}^4 (4 - x^2)\,dx = \left(4x - \dfrac{x^3}{3}\right)\Big|_{-3}^4 = \left(4\cdot 4 - \dfrac{4^3}{3}\right) - \left(4(-3) - \dfrac{(-3)^3}{3}\right)$

$$= 16 - \dfrac{64}{3} + 3 = -\dfrac{7}{3}$$

17. $\int_0^1 24x^{11}\,dx = 24\dfrac{x^{12}}{12}\Big|_0^1 = 2x^{12}\Big|_0^1 = 2\cdot 1^{12} - 2\cdot 0^{12} = 2$

19. $\int_0^1 e^{2x}\,dx = \dfrac{1}{2}e^{2x}\Big|_0^1 = \dfrac{1}{2}e^{2\cdot 1} - \dfrac{1}{2}e^{2\cdot 0} = \dfrac{1}{2}(e^2 - 1)$

21. $\int_1^{3.5} 2x^{-1}\,dx = 2\ln x\Big|_1^{3.5} = 2\ln 3.5 - 2\ln 1$

$$= 2\ln 3.5 \quad (\text{Recall: } \ln 1 = 0)$$

23. $\int_b^0 f(x)\,dx = -\int_0^b f(x)\,dx = -10.67$

25. $\int_c^0 f(x)\,dx = -\int_0^c f(x)\,dx = -\left[\int_0^b f(x)\,dx + \int_b^c f(x)\,dx\right]$

$$= -[10.67 - 5.63]$$
$$= -5.04$$

27. $\int_1^4 32t\,dt = 16t^2\Big|_1^4 = 16(4)^2 - 16(1)^2 = 256 - 16 = 240 \text{ ft}$

29. $\int_1^2 (2x^{-2} - 3)\,dx = (-2x^{-1} - 3x)\Big|_1^2 = \left(-\dfrac{2}{x} - 3x\right)\Big|_1^2$

$$= -\dfrac{2}{2} - 3\cdot 2 - \left(-\dfrac{2}{1} - 3\cdot 1\right) = -7 - (-5) = -2$$

31. $\int_1^4 3\sqrt{x}\,dx = 3\int_1^4 x^{1/2}\,dx = 3\cdot\dfrac{2}{3}x^{3/2}\Big|_1^4 = 2x^{3/2}\Big|_1^4$

$$= 2\cdot 4^{3/2} - 2\cdot 1^{3/2} = 16 - 2 = 14$$

33. $\int_2^3 12(x^2 - 4)^5 x \, dx$. Consider the indefinite integral $\int 12(x^2 - 4)^5 x \, dx$.

Let $u = x^2 - 4$, then $du = 2x \, dx$.

$$\int 12(x^2 - 4)^5 x \, dx = 6\int (x^2 - 4)^5 2x \, dx = 6\int u^5 du$$

$$= 6\frac{u^6}{6} + C = u^6 + C = (x^2 - 4)^6 + C$$

Thus,

$$\int_2^3 12(x^2 - 4)^5 x \, dx = (x^2 - 4)^6 \Big|_2^3 = (3^2 - 4)^6 - (2^2 - 4)^6 = 5^6 = 15{,}625.$$

35. $\int_3^9 \frac{1}{x-1} \, dx$

Let $u = x - 1$. Then $du = dx$ and $u = 8$ when $x = 9$, $u = 2$ when $x = 3$.

Thus,

$$\int_3^9 \frac{1}{x-1} \, dx = \int_2^8 \frac{1}{u} \, du = \ln u \Big|_2^8 = \ln 8 - \ln 2 = \ln 4 \approx 1.386.$$

37. $\int_{-5}^{10} e^{-0.05x} dx$

Let $u = -0.05x$. Then $du = -0.05 \, dx$ and $u = -0.5$ when $x = 10$, $u = 0.25$ when $x = -5$. Thus,

$$\int_{-5}^{10} e^{-0.05x} dx = -\frac{1}{0.05} \int_{-5}^{10} e^{-0.05x}(-0.05) \, dx = -\frac{1}{0.05} \int_{0.25}^{-0.5} e^u du$$

$$= -\frac{1}{0.05} e^u \Big|_{0.25}^{-0.5} = -\frac{1}{0.05} [e^{-0.5} - e^{0.25}]$$

$$= 20(e^{0.25} - e^{-0.5}) \approx 13.550$$

39. $\int_{-6}^0 \sqrt{4 - 2x} \, dx$

Consider the indefinite integral $\int \sqrt{4 - 2x} \, dx = \int (4 - 2x)^{1/2} dx$.

Let $u = 4 - 2x$, then $du = -2 \, dx$.

$$\int (4 - 2x)^{1/2} dx = -\frac{1}{2} \int (4 - 2x)^{1/2}(-2) \, dx = -\frac{1}{2} \int u^{1/2} du$$

$$= -\frac{1}{2} \cdot \frac{u^{3/2}}{\frac{3}{2}} + C = -\frac{1}{3} u^{3/2} + C = -\frac{1}{3}(4 - 2x)^{3/2} + C$$

Thus,

$$\int_{-6}^0 (4 - 2x)^{1/2} dx = -\frac{1}{3}(4 - 2x)^{3/2} \Big|_{-6}^0 = -\frac{1}{3}[4^{3/2} - 16^{3/2}]$$

$$= -\frac{1}{3}(8 - 64) = \frac{56}{3} \approx 18.667$$

41. $\int_{-1}^{7} \dfrac{x}{\sqrt{x + 2}}\, dx$

Consider the indefinite integral $\int \dfrac{x}{\sqrt{x + 2}}\, dx = \int x(x + 2)^{-1/2} dx$.

Let $u = x + 2$, then $du = dx$ and $x = u - 2$.

$$\int x(x + 2)^{-1/2} dx = \int (u - 2) u^{-1/2} du = \int (u^{1/2} - 2u^{-1/2})\, du$$

$$= \dfrac{u^{3/2}}{\dfrac{3}{2}} - \dfrac{2u^{1/2}}{\dfrac{1}{2}} + C = \dfrac{2}{3}(x + 2)^{3/2} - 4(x + 2)^{1/2} + C$$

Thus,

$$\int_{-1}^{7} \dfrac{x}{\sqrt{x + 2}}\, dx = \left[\dfrac{2}{3}(x + 2)^{3/2} - 4(x + 2)^{1/2}\right]\Bigg|_{-1}^{7}$$

$$= \dfrac{2}{3}(9)^{3/2} - 4(9)^{1/2} - \left(\dfrac{2}{3}(1)^{3/2} - 4(1)^{1/2}\right)$$

$$= \dfrac{2}{3}(27) - 12 - \left(\dfrac{2}{3} - 4\right) = 6 + \dfrac{10}{3} = \dfrac{28}{3} \approx 9.333.$$

43. $\int_{0}^{1} (e^{2x} - 2x)^2 (e^{2x} - 1)\, dx = \dfrac{1}{2}\int_{0}^{1} (e^{2x} - 2x)^2 (2e^{2x} - 2)\, dx$

$$= \dfrac{1}{2} \cdot \dfrac{(e^{2x} - 2x)^3}{3}\Bigg|_{0}^{1}$$

$$= \dfrac{1}{6}(e^{2x} - 2x)^3 \Bigg|_{0}^{1}$$

$$= \dfrac{1}{6}[(e^2 - 2)^3 - 1]$$

$$\approx 25.918$$

$\left[\underline{\text{Note}} : \text{The integrand}\right.$
has the form $u^2 du$; an
antiderivative is
$\left.\dfrac{u^3}{3} = \dfrac{(e^{2x} - 2x)^3}{3}.\right]$

45. $\int_{-2}^{-1} (x^{-1} + 2x)\, dx = (\ln|x| + x^2)\Big|_{-2}^{-1}$

$$= \ln|-1| + (-1)^2 - [\ln|-2| + (-2)^2]$$
$$= 1 - \ln 2 - 4$$
$$= -3 - \ln 2 \approx -3.693$$

47. $f(x) = 500 - 50x$ on $[0, 10]$

(A) Ave $f(x) = \dfrac{1}{10 - 0}\int_{0}^{10} (500 - 50x)\, dx$

$$= \dfrac{1}{10}(500x - 25x^2)\Big|_{0}^{10}$$

$$= \dfrac{1}{10}[5{,}000 - 2{,}500] = 250$$

(B)

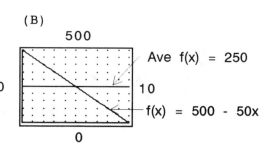

49. $f(t) = 3t^2 - 2t$ on $[-1, 2]$

(A) Ave $f(t) = \dfrac{1}{2 - (-1)} \displaystyle\int_{-1}^{2} (3t^2 - 2t)\,dt$ **(B)**

$$= \frac{1}{3}(t^3 - t^2)\Big|_{-1}^{2}$$

$$= \frac{1}{3}[4 - (-2)] = 2$$

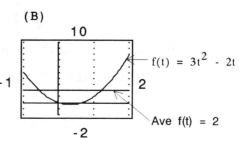

$f(t) = 3t^2 - 2t$

Ave f(t) = 2

51. $f(x) = \sqrt[3]{x} = x^{1/3}$ on $[1, 8]$

(A) Ave $f(x) = \dfrac{1}{8 - 1} \displaystyle\int_{1}^{8} x^{1/3}\,dx$ **(B)**

$$= \frac{1}{7}\left(\frac{3}{4}x^{4/3}\right)\Big|_{1}^{8}$$

$$= \frac{3}{28}(16 - 1) = \frac{45}{28} \approx 1.61$$

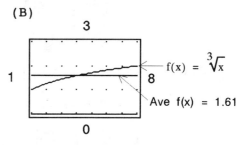

$f(x) = \sqrt[3]{x}$

Ave f(x) = 1.61

53. $f(x) = 4e^{-0.2x}$ on $[0, 10]$

(A) Ave $f(x) = \dfrac{1}{10 - 0} \displaystyle\int_{0}^{10} 4e^{-0.2x}\,dx$ **(B)**

$$= \frac{1}{10}(-20e^{-0.2x})\Big|_{0}^{10}$$

$$= \frac{1}{10}(20 - 20e^{-2}) \approx 1.73$$

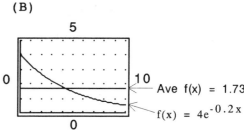

Ave f(x) = 1.73

$f(x) = 4e^{-0.2x}$

55. $f(x) = 0.25x^2 - 4$ on $[0, 8]$, $a = 0$, $b = 8$, $n = 4$;
$$\Delta x = \frac{8 - 0}{4} = 2$$

$$M_4 = f(1)\Delta x + f(3)\Delta x + f(5)\Delta x + f(7)\Delta x$$
$$= (-3.75 - 1.75 + 2.25 + 8.25)2 = 10$$

Thus, $I = \displaystyle\int_{0}^{8} (0.25x^2 - 4)\,dx \approx 10$.

Error bound:
$\quad f'(x) = 0.5x,\ \ f''(x) = 0.5$

From 3,

$$|I - M_4| \leq \frac{0.5(8 - 0)^3}{24(4)^2} = \frac{256}{384} = 0.67$$

Thus, $I = 10 \pm 0.67$

57. $I = \int_0^8 (0.25x^2 - 4)\,dx = \int_0^8 \left(\frac{1}{4}x^2 - 4\right)dx$

$$= \left(\frac{1}{12}x^3 - 4x\right)\Big|_0^8 = \frac{512}{12} - 32 = 10.67$$

$|I - M_4| = |10.67 - 10| = 0.67$

This error does lie within the error bound calculated in Problem 55.

59. $f''(x) = 0.5$ on $[0, 8]$

$$|I - M_n| \le \frac{0.5(8 - 0)^3}{24n^2} = \frac{256}{24n^2} = \frac{32}{3n^2};$$

$$\frac{32}{3n^2} \le 0.005$$

$$n^2 \ge \frac{32}{0.015} \approx 2133.33$$

$$n \ge 47$$

61. $\lim\limits_{n \to \infty}[(1 - c_1^2)\Delta x + (1 - c_2^2)\Delta x + \ldots + (1 - c_n^2)\Delta x]$ where $\Delta x = \dfrac{5 - 2}{n}$ and

$c_k = 2 + k\cdot\dfrac{3}{n}$, $k = 1, 2, \ldots, n$ is a Riemann sum for $\int_2^5 (1 - x^2)\,dx$.

$$\int_2^5 (1 - x^2)\,dx = \left(x - \frac{1}{3}x^3\right)\Big|_2^5 = \left[5 - \frac{125}{3} - \left(2 - \frac{8}{3}\right)\right] = -36$$

63. $\lim\limits_{n \to \infty}[(3c_1^2 - 2c_1 + 3)\Delta x + (3c_2^2 - 2c_2 + 3)\Delta x + \ldots + (3c_n^2 - 2c_n + 3)\Delta x]$,

where $\Delta x = \dfrac{12 - 2}{n}$ and $c_k = 2 + k\cdot\dfrac{10}{n}$, $k = 1, 2, \ldots, n$ is a Riemann sum

for $\int_2^{12} (3x^2 - 2x + 3)\,dx$.

$$\int_2^{12} (3x^2 - 2x + 3)\,dx = (x^3 - x^2 + 3x)\Big|_2^{12} = (12)^3 - (12)^2 + 36 - (10)$$

$$= 1{,}610$$

65. $\int_2^3 x\sqrt{2x^2 - 3}\,dx = \int_2^3 x(2x^2 - 3)^{1/2}\,dx$

$$= \frac{1}{4}\int_2^3 (2x^2 - 3)^{1/2}4x\,dx$$

$$= \frac{1}{4}\left(\frac{2}{3}\right)(2x^2 - 3)^{3/2}\Big|_2^3$$

$\left[\begin{array}{l}\underline{\text{Note}}:\text{ The integrand has the}\\ \text{form } u^{1/2}du;\text{ the antiderivative}\\ \text{is }\frac{2}{3}u^{3/2} = \frac{2}{3}(2x^2 - 3)^{3/2}.\end{array}\right.$

$$= \frac{1}{6}[2(3)^2 - 3]^{3/2} - \frac{1}{6}[2(2)^2 - 3]^{3/2}$$

$$= \frac{1}{6}(15)^{3/2} - \frac{1}{6}(5)^{3/2} = \frac{1}{6}[15^{3/2} - 5^{3/2}] \approx 7.819$$

67. $\int_0^1 \dfrac{x - 1}{x^2 - 2x + 3}\, dx$

Consider the indefinite integral and let $u = x^2 - 2x + 3$.
Then $du = (2x - 2)\, dx = 2(x - 1)\, dx$.

$\int \dfrac{x - 1}{x^2 - 2x + 3}\, dx = \dfrac{1}{2} \int \dfrac{2(x - 1)}{x^2 - 2x + 3}\, dx = \dfrac{1}{2} \int \dfrac{1}{u}\, du = \dfrac{1}{2} \ln|u| + C$

Thus,

$\int_0^1 \dfrac{x - 1}{x^2 - 2x + 3}\, dx = \dfrac{1}{2} \ln|x^2 - 2x + 3| \,\Big|_0^1$

$\qquad\qquad = \dfrac{1}{2} \ln 2 - \dfrac{1}{2} \ln 3 = \dfrac{1}{2}(\ln 2 - \ln 3) \approx -0.203$

69. $\int_{-1}^1 \dfrac{e^{-x} - e^x}{(e^{-x} + e^x)^2}\, dx$

Consider the indefinite integral and let $u = e^{-x} + e^x$.
Then $du = (-e^{-x} + e^x)\, dx = -(e^{-x} - e^x)\, dx$.

$\int \dfrac{e^{-x} - e^x}{(e^{-x} + e^x)^2}\, dx = -\int \dfrac{-(e^{-x} - e^x)}{(e^{-x} + e^x)^2}\, dx = -\int u^{-2}\, du = \dfrac{-u^{-1}}{-1} + C = \dfrac{1}{u} + C$

Thus,

$\int_{-1}^1 \dfrac{e^{-x} - e^x}{(e^{-x} + e^x)^2}\, dx = \dfrac{1}{e^{-x} + e^x} \,\Big|_{-1}^1 = \dfrac{1}{e^{-1} + e^1} - \dfrac{1}{e^{-(-1)} + e^{-1}}$

$\qquad\qquad = \dfrac{1}{e^{-1} + e} - \dfrac{1}{e^{-1} + e} = 0$

71. $f(t) = \dfrac{1}{t}$ on $[1, 2]$

(A) $a = 1$, $b = 2$, $n = 5$, $\Delta x = \dfrac{2 - 1}{5} = 0.2$

partition $\{1, 1.2, 1.4, 1.6, 1.8, 2\}$
midpoints $\{1.1, 1.3, 1.5, 1.7, 1.9\}$

$M_5 = f(1.1)\Delta x + f(1.3)\Delta x + f(1.5)\Delta x + f(1.7)\Delta x + f(1.9)\Delta x$

$\qquad \approx [0.9091 + 0.7692 + 0.6667 + 0.5882 + 0.5263]0.2$

$\qquad = (3.4595)0.2 = 0.6919$

Error bound:

$\qquad f'(t) = -\dfrac{1}{t^2}, \quad f''(t) = \dfrac{2}{t^3}$

Max $|f''(t)|$ on $[1, 2]$ = max $\dfrac{2}{t^3}$ on $[1, 2]$ = 2

$|I - M_5| \le \dfrac{2(2 - 1)^3}{24(5)^2} = \dfrac{1}{300} = 0.0033$

Thus, $\ln 2 = 0.6919 \pm 0.0033$

(B) $\ln 2 \approx 0.6931$

(C) Error $= |\ln 2 - M_5| = |0.6931 - 0.6919| = 0.0012$

This error is within the bound determined in part (A).

73. $f(t) = \frac{1}{t}$ on $[1, 2]$; $f'(t) = -\frac{1}{t^2}$, $f''(t) = \frac{2}{t^3}$

Max $f''(t)$ on $[1, 2]$ = max $\frac{2}{t^3}$ on $[1, 2]$ = 2

$$|I - M_n| \le \frac{2(2 - 1)^3}{24n^2} = \frac{1}{12n^2};$$

$$\frac{1}{12n^2} \le 0.0005$$

$$n^2 \ge \frac{1}{12(0.0005)} \approx 166.67$$

$$n \ge 13$$

75. $C'(x) = 500 - \frac{x}{3}$ on $[300, 900]$

The increase in cost from a production level of 300 bikes per month to a production level of 900 bikes per month is given by:

$$\int_{300}^{900} \left(500 - \frac{x}{3}\right) dx = \left(500x - \frac{1}{6}x^2\right)\Big|_{300}^{900}$$

$$= 315,000 - (135,000)$$
$$= \$180,000$$

77. Total loss in value in the first 5 years:

$$V(5) - V(0) = \int_0^5 V'(t)\,dt = \int_0^5 500(t - 12)\,dt = 500\left(\frac{t^2}{2} - 12t\right)\Big|_0^5$$

$$= 500\left(\frac{25}{2} - 60\right) = -\$23,750$$

Total loss in value in the second 5 years:

$$V(10) - V(5) = \int_5^{10} V'(t)\,dt = \int_5^{10} 500(t - 12)\,dt = 500\left(\frac{t^2}{2} - 12t\right)\Big|_5^{10}$$

$$= 500\left[(50 - 120) - \left(\frac{25}{2} - 60\right)\right] = -\$11,250$$

79. $C(x) = 1 + 12x - x^2$, $0 \le x \le 12$

(A) For the average cash reserve for the first quarter, take $a = 0$, $b = 3$

$$\text{Ave } C(x) = \frac{1}{3 - 0}\int_0^3 (1 + 12x - x^2)\,dx$$

$$= \frac{1}{3}\left(x + 6x^2 - \frac{1}{3}x^3\right)\Big|_0^3$$

$$= \frac{1}{3}(48) = 16$$

Thus, Ave $C(x) = \$16,000$.

(B)

$C(x) = 1 + 12x - x^2$

Ave $C(x) = 16$
(Cash reserves in
thousands of dollars.)

81. (A) To find the useful life, set $C'(t) = R'(t)$ and solve for t.

$$\frac{1}{11}t = 5te^{-t^2}$$

$$e^{t^2} = 55$$

$$t^2 = \ln 55$$

$$t = \sqrt{\ln 55} \approx 2 \text{ years}$$

(B) The total profit accumulated during the useful life is:

$$P(2) - P(0) = \int_0^2 [R'(t) - C'(t)]dt = \int_0^2 \left(5te^{-t^2} - \frac{1}{11}t\right)dt$$

$$= \int_0^2 5te^{-t^2}dt - \int_0^2 \frac{1}{11}t\,dt$$

$$= -\frac{5}{2}\int_0^2 e^{-t^2}(-2t)dt - \frac{1}{11}\int_0^2 t\,dt$$

[**Note:** In the first
integral, the integrand
has the form $e^u du$, where
$u = -t^2$; an antiderivative
is $e^u = e^{-t^2}$.]

$$= -\frac{5}{2}e^{-t^2}\Big|_0^2 - \frac{1}{22}t^2\Big|_0^2$$

$$= -\frac{5}{2}e^{-4} + \frac{5}{2} - \frac{4}{22} = \frac{51}{22} - \frac{5}{2}e^{-4} \approx 2.272$$

Thus, the total profit is approximately $2,272.

83. $C(x) = 60,000 + 300x$

(A) Average cost per unit:

$$\overline{C}(x) = \frac{C(x)}{x} = \frac{60,000}{x} + 300$$

$$\overline{C}(500) = \frac{60,000}{500} + 300 = \$420$$

(B) Ave $C(x) = \frac{1}{500}\int_0^{500}(60,000 + 300x)\,dx$

$$= \frac{1}{500}(60,000x + 150x^2)\Big|_0^{500}$$

$$= \frac{1}{500}(30,000,000 + 37,500,000) = \$135,000$$

(C) $\overline{C}(500)$ is the average cost per unit at a production level of 500
units; Ave $C(x)$ is the average value of the total cost as production
increases from 0 units to 500 units.

85. $A'(t) = 800e^{0.08t}$, $t \geq 0$

The change in the account from the end of the second year to the end of the sixth year is given by:

$$\int_2^6 A'(t)\,dt) = \int_2^6 800e^{0.08t}\,dt$$

$$= \left(\frac{800}{0.08} e^{0.08t}\right)\Big|_2^6$$

$$= 10{,}000(e^{0.48} - e^{0.16})$$

$$\approx \$4{,}425.64$$

87. Average price:

$$\text{Ave } S(x) = \frac{1}{30 - 20}\int_{20}^{30} 10(e^{0.02x} - 1)\,dx = \int_{20}^{30}(e^{0.02x} - 1)\,dx$$

$$= \int_{20}^{30} e^{0.02x}\,dx - \int_{20}^{30} dx$$

$$= \frac{1}{0.02}\int_{20}^{30} e^{0.02x}(0.02)\,dx - x\Big|_{20}^{30}$$

$$= 50e^{0.02x}\Big|_{20}^{30} - (30 - 20)$$

$$= 50e^{0.6} - 50e^{0.4} - 10$$

$$\approx 6.51 \text{ or } \$6.51$$

89. $g(x) = 2400x^{-1/2}$ and $L'(x) = g(x)$.

The number of labor hours to assemble the 17th through the 25th control units is:

$$L(25) - L(16) = \int_{16}^{25} g(x)\,dx = \int_{16}^{25} 2400x^{-1/2}\,dx = 2400(2)x^{1/2}\Big|_{16}^{25}$$

$$= 4800x^{1/2}\Big|_{16}^{25} = 4800[25^{1/2} - 16^{1/2}] = 4800 \text{ labor hours}$$

91. (A) The inventory function is obtained by finding the equation of the line joining (0, 600) and (3, 0).

Slope: $m = \dfrac{0 - 600}{3 - 0} = -200$, y intercept: $b = 600$

Thus, the equation of the line is: $I = -200t + 600$

(B) The average of I over [0, 3] is given by:

$$\text{Ave } I(t) = \frac{1}{3 - 0}\int_0^3 I(t)\,dt = \frac{1}{3}\int_0^3 (-200t + 600)\,dt$$

$$= \frac{1}{3}(-100t^2 + 600t)\Big|_0^3$$

$$= \frac{1}{3}[-100(3^2) + 600(3) - 0]$$

$$= \frac{900}{3} = 300 \text{ units}$$

93. Rate of production: $R(t) = \dfrac{100}{t + 1} + 5$, $0 \le t \le 20$

Total production from year N to year M is given by:

$$P = \int_{N}^{M} R(t)\,dt = \int_{N}^{M} \left(\frac{100}{t + 1} + 5\right) dt = 100 \int_{N}^{M} \frac{1}{t + 1}\,dt + \int_{N}^{M} 5\,dt$$

$$= 100\,\ln|t + 1|\Big|_{N}^{M} + 5t\Big|_{N}^{M}$$

$$= 100\,\ln(M + 1) - 100\,\ln(N + 1) + 5(M - N)$$

Thus, for total production during the first 10 years, let $M = 10$ and $N = 0$.

$P = 100\,\ln 11 - 100\,\ln 1 + 5(10 - 0)$

$ = 100\,\ln 11 + 50 \approx 290$ thousand barrels

For the total production from the end of the 10th year to the end of the 20th year, let $M = 20$ and $N = 10$.

$P = 100\,\ln 21 - 100\,\ln 11 + 5(20 - 10)$

$ = 100\,\ln 21 - 100\,\ln 11 + 50 \approx 115$ thousand barrels

95. Let $P(t) = R(t) - C(t)$. Then the total accumulated profits over the five-year period are given by:

$$P(5) - P(0) = \int_{0}^{5} P'(t)\,dt = \int_{0}^{5} [R'(t) - C'(t)]\,dt$$

$$= \int_{0}^{5} R'(t)\,dt - \int_{0}^{5} C'(t)\,dt$$

Now, $C'(t) = 1{,}500$ (constant). Therefore,

$$\int_{0}^{5} C'(t)\,dt = \int_{0}^{5} 1{,}500\,dt = 1{,}500t\Big|_{0}^{5} = 7{,}500$$

Using a midpoint sum with $n = 5$ to approximate

$$\int_{0}^{5} R'(t)\,dt,$$

we have $\Delta x = 1$ and

$$M_5 = R'\left(\frac{1}{2}\right)1 + R'\left(\frac{3}{2}\right)1 + R'\left(\frac{5}{2}\right)1 + R'\left(\frac{7}{2}\right)1 + R'\left(\frac{9}{2}\right)1$$

$$= 5{,}000 + 4{,}500 + 3{,}500 + 2{,}500 + 2{,}000$$

$$= 17{,}500$$

Therefore, the total accumulated profits are (approximately):

$$P(5) - P(0) = \int_{0}^{5} R'(t)\,dt - \int_{0}^{5} C'(t)\,dt \approx 17{,}500 - 7{,}500 = \$10{,}000$$

97.

x	300	900	1,500	2,100
$f(x)$	900	1,700	1,700	900

Using a midpoint sum with $n = 4$ and $\Delta x = 600$, we have

$$\int_{0}^{2{,}400} f(x)\,dx \approx M_4 = f(300)\Delta x + f(900)\Delta x + f(1{,}500)\Delta x + f(2{,}100)\Delta x$$

$$= [900 + 1{,}700 + 1{,}700 + 900]600$$

$$= (5{,}200)600 = 3{,}120{,}000 \text{ sq ft}$$

99. $W'(t) = 0.2e^{0.1t}$

The weight increase during the first eight hours is given by:

$$W(8) - W(0) = \int_0^8 W'(t)\,dt = \int_0^8 0.2e^{0.1t}\,dt = 0.2\int_0^8 e^{0.1t}\,dt$$

$$= \frac{0.2}{0.1}\int_0^8 e^{0.1t}(0.1)\,dt \qquad \text{(Let } u = 0.1t, \text{ then } du = 0.1dt.)$$

$$= 2e^{0.1t}\Big|_0^8 = 2e^{0.8} - 2 \approx 2.45 \text{ grams}$$

The weight increase during the second eight hours, i.e., from the 8th hour through the 16th hour, is given by:

$$W(16) - W(8) = \int_8^{16} W'(t)\,dt = \int_8^{16} 0.2e^{0.1t}\,dt = 2e^{0.1t}\Big|_8^{16}$$

$$= 2e^{1.6} - 2e^{0.8} \approx 5.45 \text{ grams}$$

101. Average temperature over time period [0, 2] is given by:

$$\frac{1}{2-0}\int_0^2 C(t)\,dt = \frac{1}{2}\int_0^2 (t^3 - 2t + 10)\,dt = \frac{1}{2}\left(\frac{t^4}{4} - \frac{2t^2}{2} + 10t\right)\Big|_0^2$$

$$= \frac{1}{2}(4 - 4 + 20) = 10° \text{ Celsius}$$

103. Using a midpoint sum with $n = 3$ and $\Delta t = 1$, and estimating the values of $R(t)$ from the graph, we have

$$\int_0^3 R(t)\,dt \approx M_3 = R\left(\frac{1}{2}\right)1 + R\left(\frac{3}{2}\right)1 + R\left(\frac{5}{2}\right)1$$

$$= 0.3 + 0.5 + 0.3 = 1.1$$

Thus, the total volume of air inhaled is approximately 1.1 liters.

105. $P(t) = \dfrac{8.4t}{t^2 + 49} + 0.1, \; 0 \le t \le 24$

(A) Average fraction of people during the first seven months:

$$\frac{1}{7-0}\int_0^7 \left[\frac{8.4t}{t^2 + 49} + 0.1\right]dt = \frac{4.2}{7}\int_0^7 \frac{2t}{t^2 + 49}\,dt + \frac{1}{7}\int_0^7 0.1\,dt$$

$$= 0.6 \ln(t^2 + 49)\Big|_0^7 + \frac{0.1}{7}t\Big|_0^7$$

$$= 0.6[\ln 98 - \ln 49] + 0.1$$

$$= 0.6 \ln 2 + 0.1 \approx 0.516$$

(B) Average fraction of people during the first two years:

$$\frac{1}{24-0}\int_0^{24} \left[\frac{8.4t}{t^2 + 49} + 0.1\right]dt = \frac{4.2}{24}\int_0^{24} \frac{2t}{t^2 + 49}\,dt + \frac{1}{24}\int_0^{24} 0.1\,dt$$

$$= 0.175 \ln(t^2 + 49)\Big|_0^{24} + \frac{0.1}{24}t\Big|_0^{24}$$

$$= 0.175[\ln 625 - \ln 49] + 0.1 \approx 0.546$$

1. $\int (3t^2 - 2t)\,dt = 3\int t^2 dt - 2\int t\ dt = 3 \cdot \dfrac{t^3}{3} - 2 \cdot \dfrac{t^2}{2} + C = t^3 - t^2 + C$ (6-1)

2. $\int_2^5 (2x - 3)\,dx = 2\int_2^5 x\ dx - 3\int_2^5 dx = x^2 \Big|_2^5 - 3x \Big|_2^5$

$\qquad\qquad = (25 - 4) - (15 - 6) = 12$ (6-5)

3. $\int (3t^{-2} - 3)\,dt = 3\int t^{-2} dt - 3\int dt = 3 \cdot \dfrac{t^{-1}}{-1} - 3t + C = -3t^{-1} - 3t + C$ (6-1)

4. $\int_1^4 x\ dx = \dfrac{x^2}{2}\Big|_1^4 = \dfrac{16}{2} - \dfrac{1}{2} = \dfrac{15}{2}$ (6-5)

5. $\int e^{-0.5x} dx = \dfrac{e^{-0.5x}}{-0.5} + C = -2e^{-0.5x} + C$ (6-2)

6. $\int_1^5 \dfrac{2}{u}\,du = 2\int_1^5 \dfrac{du}{u} = 2\ \ln u\Big|_1^5 = 2\ \ln 5 - 2\ \ln 1 = 2\ \ln 5$ (6-5)

7. $\dfrac{dy}{dx} = 3x^2 - 2$

$\qquad y = f(x) = \int (3x^2 - 2)\,dx$

$\qquad\quad f(x) = x^3 - 2x + C$

$\qquad\quad f(0) = C = 4$

$\qquad\quad f(x) = x^3 - 2x + 4$ (6-3)

8. The graph of an antiderivative function f is increasing on $(0, 2)$, decreasing on $(2, 4)$, concave down on $(0, 4)$; f has a local maximum at $x = 2$. The graphs of the antiderivative functions differ by a vertical translation. (6-1)

9.

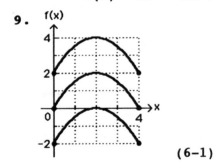

(6-1)

10. (A) $\int (8x^3 - 4x - 1)\,dx = 8\int x^3 dx - 4\int x\ dx - \int dx$

$\qquad\qquad\qquad = 8 \cdot \dfrac{1}{4}x^4 - 4\ \dfrac{1}{2}x^2 - x + C$

$\qquad\qquad\qquad = 2x^4 - 2x^2 - x + C$ (6-1)

$\quad$ (B) $\int (e^t - 4t^{-1})\,dt = \int e^t - 4\int \dfrac{1}{t}\,dt$

$\qquad\qquad\qquad = e^t - 4\ \ln |t| + C$ (6-1)

11. $f(x) = x^2 + 1$, $a = 1$, $b = 5$, $n = 2$, $\Delta x = \dfrac{5 - 1}{2} = 2$;

$$M_2 = f(2)\Delta x + f(4)\Delta x$$
$$= 5 \cdot 2 + 17 \cdot 2 = 44$$

Error bound for M_2:

$$f'(x) = 2x, \quad f''(x) = 2$$
$$|I - M_2| \le \frac{2(5 - 1)^3}{24(2)^2} = \frac{128}{96} \approx 1.333$$

Thus, $\displaystyle\int_1^5 (x^2 + 1)\,dx = 44 \pm 1.333$ (6-5)

12. $\displaystyle\int_1^5 (x^2 + 1)\,dx = \left(\frac{1}{3}x^3 + x\right)\Big|_1^5 = \frac{125}{3} + 5 - \left(\frac{1}{3} + 1\right) \approx 45.333$

$|I - M_2| = 1.333$ (6-5)

13. Using the values of f in the table with $a = 1$, $b = 17$, $n = 4$,

$\Delta x = \dfrac{17 - 1}{4} = 4$;

$$M_4 = f(3)\Delta x + f(7)\Delta x + f(11)\Delta x + f(15)\Delta x$$
$$= [1.2 + 3.4 + 2.6 + 0.5]4 = 30.8$$ (6-5)

14. $f(x) = 6x^2 + 2x$ on $[-1, 2]$;

$$\text{Ave } f(x) = \frac{1}{2 - (-1)}\int_{-1}^2 (6x^2 + 2x)\,dx$$
$$= \frac{1}{3}(2x^3 + x^2)\Big|_{-1}^2 = \frac{1}{3}[20 - (-1)] = 7$$ (6-5)

15. width $= 2 - (-1) = 3$, height $=$ Ave $f(x) = 7$ (6-5)

16. $\displaystyle\int_a^b 5f(x)\,dx = 5\int_a^b f(x)\,dx = 5(-2) = -10$ (6-4, 6-5)

17. $\displaystyle\int_b^c \frac{f(x)}{5}\,dx = \frac{1}{5}\int_b^c f(x)\,dx = \frac{1}{5}(2) = \frac{2}{5} = 0.4$ (6-4, 6-5)

18. $\displaystyle\int_b^d f(x)\,dx = \int_b^c f(x)\,dx + \int_c^d f(x)\,dx = 2 - 0.6 = 1.4$ (6-4, 6-5)

19. $\displaystyle\int_a^c f(x)\,dx = \int_a^b f(x)\,dx + \int_b^c f(x)\,dx = -2 + 2 = 0$ (6-4, 6-5)

20. $\displaystyle\int_0^d f(x)\,dx = \int_0^a f(x)\,dx + \int_a^b f(x)\,dx + \int_b^c f(x)\,dx + \int_c^d f(x)\,dx$
$$= 1 - 2 + 2 - 0.6 = 0.4$$ (6-4, 6-5)

21. $\displaystyle\int_b^a f(x)\,dx = -\int_a^b f(x)\,dx = -(-2) = 2$ (6-4, 6-5)

22. $\int_c^b f(x)\,dx = -\int_b^c f(x)\,dx = -2$ (6-4, 6-5)

23. $\int_d^0 f(x)\,dx = -\int_0^d f(x)\,dx = -0.4$ (from Problem 20) (6-4, 6-5)

24. Let f be an antiderivative function. Then:
f is increasing on $[0, 1]$ and $[3, 4]$ ($f'(x) > 0$);
f is decreasing on $[1, 3]$ ($f'(x) < 0$);
the graph of f is concave down on $[0, 2]$ (f' is decreasing); the graph of f is concave up on $[2, 4]$ (f' is increasing); f has a local maximum at $x = 1$; f has a local minimum at $x = 3$; there is an inflection point at $x = 2$. The graphs of antiderivative functions differ by a vertical translation. (6-1)

25.

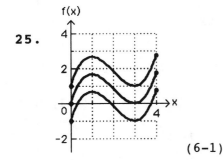

(6-1)

26. (A) $\dfrac{dy}{dx} = \dfrac{2y}{x}$; $\dfrac{dy}{dx}\bigg|_{(2,\,1)} = \dfrac{2(1)}{2} = 1$, $\dfrac{dy}{dx}\bigg|_{(-2,\,-1)} = \dfrac{2(-1)}{-2} = 1$

(B) $\dfrac{dy}{dx} = \dfrac{2x}{y}$; $\dfrac{dy}{dx}\bigg|_{(2,\,1)} = \dfrac{2(2)}{1} = 4$, $\dfrac{dy}{dx}\bigg|_{(-2,\,-1)} = \dfrac{2(-2)}{-1} = 4$ (6-3)

27. $\dfrac{dy}{dx} = \dfrac{2y}{x}$; from the figure, the slopes at $(2, 1)$ and $(-2, -1)$ are approximately equal to 1 as computed in Problem 26(A), not 4 as computed in Problem 26(B). (6-3)

28. Let $y = Cx^2$. Then $\dfrac{dy}{dx} = 2Cx$. From the original equation, $C = \dfrac{y}{x^2}$ so

$$\frac{dy}{dx} = 2x\left(\frac{y}{x^2}\right) = \frac{2y}{x}$$

(6-3)

29. Letting $x = 2$ and $y = 1$ in $y = Cx^2$, we get
$$1 = 4C \quad \text{so} \quad C = \frac{1}{4} \quad \text{and} \quad y = \frac{1}{4}x^2$$
Letting $x = -2$ and $y = -1$ in $y = Cx^2$, we get
$$-1 = 4C \quad \text{so} \quad C = -\frac{1}{4} \quad \text{and} \quad y = -\frac{1}{4}x^2$$

(6-3)

30.

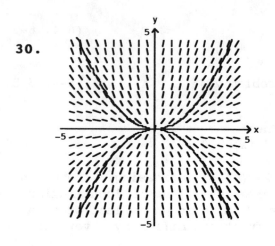

(6-3)

31.

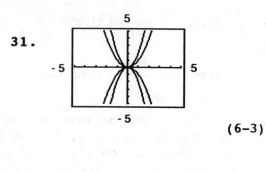

(6-3)

32. $\int \sqrt[3]{6x - 5}\, dx = \int (6x - 5)^{1/3} dx = \frac{1}{6} \int (6x - 5)^{1/3} 6\ dx$

$$= \frac{1}{6} \frac{(6x - 5)^{4/3}}{\frac{4}{3}} + C$$

$$= \frac{1}{8} (6x - 5)^{4/3} + C \qquad\qquad (6\text{-}1,\ 6\text{-}2)$$

33. $\int_0^1 10(2x - 1)^4 dx = 5 \int_0^1 (2x - 1)^4 2\ dx = \left. \frac{5(2x - 1)^5}{5} \right|_0^1$

$$= (2x - 1)^5 \Big|_0^1 = 1 - (-1)^5 = 2 \qquad\qquad (6\text{-}5)$$

34. $\int \left(\frac{2}{x^2} - 2xe^{x^2} \right) dx = 2 \int x^{-2} dx - \int 2xe^{x^2} dx = \frac{2x^{-1}}{-1} - e^{x^2} + C$

$$= -2x^{-1} - e^{x^2} + C \qquad\qquad (6\text{-}2)$$

35. $\int_0^4 \sqrt{x^2 + 4}\, x\ dx = \int_0^4 (x^2 + 4)^{1/2} x\ dx = \frac{1}{2} \int_0^4 (x^2 + 4)^{1/2} 2x\ dx$

$$= \frac{1}{2} \cdot \left. \frac{(x^2 + 4)^{3/2}}{\frac{3}{2}} \right|_0^4 = \left. \frac{(x^2 + 4)^{3/2}}{3} \right|_0^4 = \frac{(20)^{3/2} - 8}{3} \approx 27.148$$

$$(6\text{-}5)$$

36. $\int (e^{-2x} + x^{-1})\, dx = \int e^{-2x} dx + \int \frac{1}{x} dx = -\frac{1}{2} \int e^{-2x}(-2)\, dx + \ln|x| + C$

$$= -\frac{1}{2} e^{-2x} + \ln|x| + C \qquad\qquad (6\text{-}2)$$

37. $\displaystyle\int_0^{10} 10e^{-0.02x}dx = 10\int_0^{10} e^{-0.02x}dx = \frac{10}{-0.02}\int_0^{10} e^{-0.02x}(-0.02)\,dx$

$$= -500e^{-0.02x}\Big|_0^{10} = -500e^{-0.2} + 500 \approx 90.635 \qquad\qquad (6-5)$$

38. Let $u = 1 + x^2$, then $du = 2x\ dx$.

$\displaystyle\int_0^3 \frac{x}{1 + x^2}dx = \int_0^3 \frac{1}{1 + x^2}\frac{2}{2}x\ dx$

$$= \frac{1}{2}\int_0^3 \frac{1}{1 + x^2}2x\ dx = \frac{1}{2}\ \ln(1 + x^2)\ \Big|_0^3$$

$$= \frac{1}{2}\ \ln\ 10 - \frac{1}{2}\ \ln\ 1 = \frac{1}{2}\ \ln\ 10 \approx 1.151 \qquad\qquad (6-5)$$

39. Let $u = 1 + x^2$, then $du = 2x\ dx$.

$\displaystyle\int_0^3 \frac{x}{(1 + x^2)^2}dx = \int_0^3 (1 + x^2)^{-2}\frac{2}{2}x\ dx = \frac{1}{2}\int_0^3 (1 + x^2)^{-2}2x\ dx$

$$= \frac{1}{2}\cdot\frac{(1 + x^2)^{-1}}{-1}\Big|_0^3 = \frac{-1}{2(1 + x^2)}\Big|_0^3 = -\frac{1}{20} + \frac{1}{2} = \frac{9}{20} = 0.45$$

$$\qquad\qquad (6-5)$$

40. Let $u = 2x^4 + 5$, then $du = 8x^3dx$.

$\displaystyle\int x^3(2x^4 + 5)^5dx = \int(2x^4 + 5)^5\frac{8}{8}x^3dx = \frac{1}{8}\int u^5\,du$

$$= \frac{1}{8}\cdot\frac{u^6}{6} + C = \frac{(2x^4 + 5)^6}{48} + C \qquad\qquad (6-2)$$

41. Let $u = e^{-x} + 3$, then $du = -e^{-x}dx$.

$\displaystyle\int\frac{e^{-x}}{e^{-x} + 3}dx = \int\frac{1}{e^{-x} + 3}\cdot\frac{(-1)}{(-1)}e^{-x}dx = -\int\frac{1}{u}\,du$

$$= -\ln|u| + C = -\ln|e^{-x} + 3| + C = -\ln(e^{-x} + 3) + C$$

[Note: Absolute value not needed since $e^{-x} + 3 > 0$.] $\qquad\qquad (6-2)$

42. Let $u = e^x + 2$, then $du = e^xdx$.

$\displaystyle\int\frac{e^x}{(e^x + 2)^2}dx = \int(e^x + 2)^{-2}e^xdx = \int u^{-2}du$

$$= \frac{u^{-1}}{-1} + C = -(e^x + 2)^{-1} + C = \frac{-1}{(e^x + 2)} + C \qquad\qquad (6-2)$$

43. $\dfrac{dy}{dx} = 3x^{-1} - x^{-2}$

$\displaystyle y = \int(3x^{-1} - x^{-2})dx = 3\int\frac{1}{x}dx - \int x^{-2}dx$

$$= 3\ \ln|x| - \frac{x^{-1}}{-1} + C = 3\ \ln|x| + x^{-1} + C$$

Given $y(1) = 5$:

$5 = 3\ \ln\ 1 + 1 + C$ and $C = 4$

Thus, $y = 3\ \ln|x| + x^{-1} + 4$. $\qquad\qquad (6-2,\ 6-3)$

44. $\dfrac{dy}{dx} = 6x + 1$

$f(x) = y = \displaystyle\int (6x + 1)\,dx = \dfrac{6x^2}{2} + x + C = 3x^2 + x + C$

We have $y = 10$ when $x = 2$: $3(2)^2 + 2 + C = 10$

$\qquad\qquad\qquad\qquad\qquad\qquad\qquad C = 10 - 12 - 2 = -4$

Thus, the equation of the curve is $y = 3x^2 + x - 4$. $\qquad\qquad$ (6-3)

45. $r(t)$ given in the figure, $a = 0$, $b = 1$, $n = 5$, $\Delta t = \dfrac{5 - 0}{5} = 1$.

$L_5 = r(0)\Delta t + r(1)\Delta t + r(2)\Delta t + r(3)\Delta t + r(4)\Delta t$

$\quad\ = 160 + 128 + 96 + 64 + 32 = 480$ ft

$R_5 = r(1)\Delta t + r(2)\Delta t + r(3)\Delta t + r(4)\Delta t + r(5)\Delta t$

$\quad\ = 128 + 96 + 64 + 32 + 0 = 320$ ft

$A_5 = \dfrac{L_5 + R_5}{2} = \dfrac{480 + 320}{2} = 400$ ft

Error bound for L_5 and R_5.

Error $\le |r(5) - r(0)|\left(\dfrac{5 - 0}{5}\right) = |0 - 160| = 160$ ft

Error bound for A_5:

Error $\le |r(5) - r(0)|\left(\dfrac{5 - 0}{2 \cdot 5}\right) = |0 - 160|\dfrac{1}{2} = 80$ ft $\qquad$ (6-4)

46. The height of each rectangle represents an instantaneous rate and the base of each rectangle is a time interval; and rate *times* time *equals* distance. $\qquad\qquad$ (6-4)

47. We want to find n such that

$$|I - A_n| < 1:$$

$$|r(5) - r(0)|\left(\dfrac{5 - 0}{2n}\right) < 1$$

$$|0 - 160|\dfrac{5}{2n} < 1$$

$$\dfrac{400}{n} < 1$$

$$n > 400 \qquad\qquad (6\text{-}4)$$

48. The graph of r is a straight line.

Slope: $\dfrac{0 - 160}{5 - 0} = -32$; y-intercept: 160; equation: $r = -32t + 160$

Therefore, Height $= \displaystyle\int_0^5 (-32t + 160)\,dt$

$$= (-16t^2 + 160t)\,\Big|_0^5$$

$$= 400 \text{ ft} \qquad\qquad (6\text{-}4)$$

49. (A) $f(x) = 3\sqrt{x} = 3x^{1/2}$ on $[1, 9]$

$$\text{Ave } f(x) = \frac{1}{9-1}\int_1^9 3x^{1/2}\,dx$$

$$= \frac{3}{8}\cdot\frac{x^{3/2}}{\frac{3}{2}}\Big|_1^9 = \frac{1}{4}x^{3/2}\Big|_1^9 = \frac{27}{4} - \frac{1}{4} = \frac{26}{4} = 6.5$$

(B)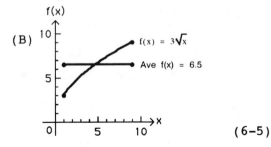

(6-5)

50. Let $u = \ln x$, then $du = \frac{1}{x}dx$.

$$\int\frac{(\ln x)^2}{x}\,dx = \int(\ln x)^2\frac{1}{x}\,dx = \int u^2\,du = \frac{u^3}{3} + C = \frac{(\ln x)^3}{3} + C \qquad (6-2)$$

51. $\int x(x^3 - 1)^2\,dx = \int x(x^6 - 2x^3 + 1)\,dx$ (square $x^3 - 1$)

$$= \int(x^7 - 2x^4 + x)\,dx = \frac{x^8}{8} - \frac{2x^5}{5} + \frac{x^2}{2} + C \qquad (6-2)$$

52. Let $u = 6 - x$, then $x = 6 - u$ and $dx = -du$.

$$\int\frac{x}{\sqrt{6-x}}\,dx = -\int\frac{(6-u)\,du}{u^{1/2}} = \int(u^{1/2} - 6u^{-1/2})\,du$$

$$= \frac{u^{3/2}}{\frac{3}{2}} - \frac{6u^{1/2}}{\frac{1}{2}} + C = \frac{2}{3}u^{3/2} - 12u^{1/2} + C$$

$$= \frac{2}{3}(6-x)^{3/2} - 12(6-x)^{1/2} + C \qquad (6-2)$$

53. $\int_0^7 x\sqrt{16 - x}\,dx.$ First consider the indefinite integral:

Let $u = 16 - x$, then $x = 16 - u$ and $dx = -du$.

$$\int x\sqrt{16-x}\,dx = -\int(16-u)u^{1/2}\,du = \int(u^{3/2} - 16u^{1/2})\,du = \frac{u^{5/2}}{\frac{5}{2}} - \frac{16u^{3/2}}{\frac{3}{2}} + C$$

$$= \frac{2}{5}u^{5/2} - \frac{32}{3}u^{3/2} + C = \frac{2(16-x)^{5/2}}{5} - \frac{32(16-x)^{3/2}}{3} + C$$

$$\int_0^7 x\sqrt{16-x}\,dx = \left[\frac{2(16-x)^{5/2}}{5} - \frac{32(16-x)^{3/2}}{3}\right]\Bigg|_0^7$$

$$= \frac{2\cdot 9^{5/2}}{5} - \frac{32\cdot 9^{3/2}}{3} - \left(\frac{2\cdot 16^{5/2}}{5} - \frac{32\cdot 16^{3/2}}{3}\right)$$

$$= \frac{2\cdot 3^5}{5} - \frac{32\cdot 3^3}{3} - \left(\frac{2\cdot 4^5}{5} - \frac{32\cdot 4^3}{3}\right)$$

$$= \frac{486}{5} - 288 - \left(\frac{2048}{5} - \frac{2048}{3}\right) = \frac{1234}{15} \approx 82.267 \qquad (6\text{-}5)$$

54. Let $u = x + 1$, then $x = u - 1$, $dx = du$; and $u = 0$ when $x = -1$, $u = 2$ when $x = 1$.

$$\int_{-1}^1 x(x+1)^4 dx = \int_0^2 (u-1)u^4 du = \int_0^2 (u^5 - u^4)\,du$$

$$= \left[\frac{u^6}{6} - \frac{u^5}{5}\right]\Bigg|_0^2 = \frac{2^6}{6} - \frac{2^5}{5} = \frac{32}{3} - \frac{32}{5} = \frac{160 - 96}{15} = \frac{64}{15} \approx 4.267$$

$$(6\text{-}5)$$

55. $\dfrac{dy}{dx} = 9x^2 e^{x^3}$, $f(0) = 2$

Let $u = x^3$, then $du = 3x^2 dx$.

$$y = \int 9x^2 e^{x^3} dx = 3\int e^{x^3}\cdot 3x^2 dx = 3\int e^u du = 3e^u + C = 3e^{x^3} + C$$

Given $f(0) = 2$:
$$2 = 3e^0 + C = 3 + C$$
Hence, $C = -1$ and $y = f(x) = 3e^{x^3} - 1$. $\qquad (6\text{-}3)$

56. $\dfrac{dN}{dt} = 0.06N$, $N(0) = 800$, $N > 0$

From the differential equation, $N(t) = Ce^{0.06t}$, where C is an arbitrary constant. Since $N(0) = 800$, we have
$$800 = Ce^0 = C.$$
Hence, $C = 800$ and $N(t) = 800e^{0.06t}$. $\qquad (6\text{-}3)$

57.

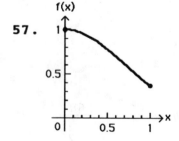

$(6\text{-}5)$

58. $f(x) = e^{-x^2}$ on $[0, 1]$; $a = 0$, $b = 1$, $n = 5$, $\Delta x = \dfrac{1-0}{5} = \dfrac{1}{5} = 0.2$

$M_5 = f(0.1)\Delta x + f(0.3)\Delta x + f(0.5)\Delta x + f(0.7)\Delta x + f(0.9)\Delta x$

$\quad = [0.9900 + 0.9139 + 0.7788 + 0.6126 + 0.4449]0.2$

$\quad = 0.74805 \qquad (6\text{-}5)$

59. $f(x) = e^{-x^2}$
$f'(x) = -2xe^{-x^2}$
$f''(x) = -2e^{-x^2} + 4x^2 e^{-x^2} = (4x^2 - 2)e^{-x^2}$
The graph of f'' is shown at the right.

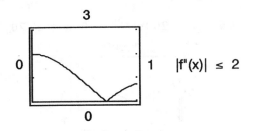

$|f''(x)| \leq 2$

$|f''(x)| \leq 2$ (6-5)

60. From Problems 58 and 59, $M_5 \approx 0.74805$ and $|f''(x)| \leq 2$.

Error $= |I - M_5| \leq \dfrac{B_2(b - a)^3}{24n^2} = \dfrac{2(1 - 0)^3}{24(5)^2} = \dfrac{1}{300} < 0.00334$

Therefore,
$I = M_5 \pm 0.00334 = 0.74805 \pm 0.00334$ (6-5)

61. We want to find n such that
$|I - M_n| \leq 0.0005$:

$$|I - M_n| \leq \frac{2(1 - 0)^3}{24n^2} \leq 0.0005$$

$$\frac{1}{12n^2} \leq 0.0005$$

$$n^2 \geq \frac{1}{0.006} \approx 166.67$$

$$n \geq 12.9$$

Take $n \geq 13$.   (6-5)

62. $N = 50(1 - e^{-0.07t})$, $0 \leq t \leq 80$, $0 \leq N \leq 60$

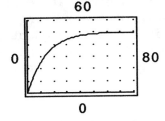

Limited growth

(6-3)

63. $p = 500e^{-0.03x}$, $0 \leq x \leq 100$, $0 \leq p \leq 500$

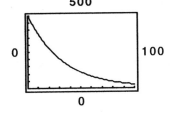

Exponential decay

(6-3)

64. $A = 200e^{0.08t}$, $0 \le t \le 20$, $0 \le A \le 1{,}000$

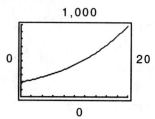

Unlimited growth

(6-3)

65. $N = \dfrac{100}{1 + 9e^{-0.3t}}$, $0 \le t \le 25$, $0 \le N \le 100$

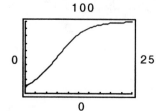

Logistic growth

(6-3)

66. $a = 200$, $b = 600$, $n = 2$, $\Delta x = \dfrac{600 - 200}{2} = 200$

$L_2 = C'(200)\Delta x + C'(400)\Delta x$

$\quad = [500 + 400]200 = \$180{,}000$

$R_2 = C'(400)\Delta x + C'(600)\Delta x$

$\quad = [400 + 300]200 = \$140{,}000$

$\quad 140{,}000 \le \displaystyle\int_{200}^{600} C'(x)\,dx \le 180{,}000$

(6-4)

67. The height of the rectangle, $C'(x)$, represents the marginal cost at a production level of x units; that is, $C'(x)$ is the approximate cost per unit at the production level of x units. The width of the rectangle represents the number of units involved in the increase in production. In the case of Problem 66, the width is 200. Thus, the cost per unit *times* the number of units *equals* the increase in production costs.

Note: The approximation improves as n increases.

(6-4)

68. The graph of $C'(x)$ is a straight line with y-intercept = 600 and

$\quad$ slope $= \dfrac{300 - 600}{600 - 0} = -\dfrac{1}{2}$

Thus, $C'(x) = -\dfrac{1}{2}x + 600$

Increase in costs:

$\displaystyle\int_{200}^{600} \left(600 - \dfrac{1}{2}x\right)dx = \left(600x - \dfrac{1}{4}x^2\right)\Big|_{200}^{600}$

$\qquad\qquad\qquad\qquad = 270{,}000 - 110{,}000$

$\qquad\qquad\qquad\qquad = \$160{,}000$

(6-5)

69. The total change in profit for a production change from 10 units per week to 40 units per week is given by:

$$\int_{10}^{40}\left(150 - \frac{x}{10}\right)dx = \left(150x - \frac{x^2}{20}\right)\Big|_{10}^{40}$$

$$= \left(150(40) - \frac{40^2}{20}\right) - \left(150(10) - \frac{10^2}{20}\right)$$

$$= 5920 - 1495 = \$4425 \qquad\qquad (6\text{-}5)$$

70. $P'(x) = 100 - 0.02x$

$$P(x) = \int(100 - 0.02x)\,dx = 100x - 0.02\frac{x^2}{2} + C = 100x - 0.01x^2 + C$$

$$P(0) = 0 - 0 + C = 0$$
$$C = 0$$

Thus, $P(x) = 100x - 0.01x^2$.
The profit on 10 units of production is given by:
$P(10) = 100(10) - 0.01(10)^2 = \$999 \qquad\qquad (6\text{-}3)$

71. The required definite integral is:

$$\int_{0}^{15}(60 - 4t)\,dt = (60t - 2t^2)\Big|_{0}^{15}$$

$$= 60(15) - 2(15)^2 = 450 \text{ or } 450{,}000 \text{ barrels}$$

The total production in 15 years is 450,000 barrels. $\qquad\qquad (6\text{-}5)$

72. Average inventory from $t = 3$ to $t = 6$:

$$\text{Ave } I(t) = \frac{1}{6-3}\int_{3}^{6}(10 + 36t - 3t^2)\,dt$$

$$= \frac{1}{3}[10t + 18t^2 - t^3]\Big|_{3}^{6}$$

$$= \frac{1}{3}[60 + 648 - 216 - (30 + 162 - 27)]$$

$$= 109 \text{ items} \qquad\qquad (6\text{-}5)$$

73. $S(x) = 8(e^{0.05x} - 1)$
Average price over the interval [40, 50]:

$$\text{Ave } S(x) = \frac{1}{50-40}\int_{40}^{50} 8(e^{0.05x} - 1)\,dx = \frac{8}{10}\int_{40}^{50}(e^{0.05x} - 1)\,dx$$

$$= \frac{4}{5}\left[\frac{e^{0.05x}}{0.05} - x\right]\Big|_{40}^{50}$$

$$= \frac{4}{5}[20e^{2.5} - 50 - (20e^2 - 40)]$$

$$= 16e^{2.5} - 16e^2 - 8 \approx \$68.70 \qquad (6\text{-}5)$$

74. From the table, $a = 0$, $b = 50$, $n = 5$, $\Delta t = 10$.

$L_5 = N(0)\Delta t + N(10)\Delta t + N(20)\Delta t + N(30)\Delta t + N(40)\Delta t$

$\qquad = [5 + 10 + 14 + 17 + 19]10 = 650$

$R_5 = N(10)\Delta t + N(20)\Delta t + N(30)\Delta t + N(40)\Delta t + N(50)\Delta t$

$\qquad = [10 + 14 + 17 + 19 + 20]10 = 800$

$A_5 = \dfrac{L_5 + R_5}{2} = \dfrac{650 + 800}{2} = 725$ components

Error bound for A_5:

$\text{Error} \le |N(50) - N(0)| \dfrac{(50 - 0)}{2 \cdot 5} = |20 - 5|5 = 75$ $\hspace{3cm}$ (6-4)

75. To find the useful life, set $R'(t) = C'(t)$:

$20e^{-0.1t} = 3$

$e^{-0.1t} = \dfrac{3}{20}$

$-0.1t = \ln\left(\dfrac{3}{20}\right) \approx -1.897$

$\qquad t = 18.97$ or 19 years

$\text{Total profit} = \displaystyle\int_0^{19} [R'(t) - C'(t)]dt = \int_0^{19} (20e^{-0.1t} - 3)dt$

$\qquad = 20\displaystyle\int_0^{19} e^{-0.1t}dt - \int_0^{19} 3\,dt = \dfrac{20}{-0.1}\int_0^{19} e^{-0.1t}(-0.1)dt - \int_0^{19} 3\,dt$

$\qquad = -200e^{-0.1t}\Big|_0^{19} - 3t\Big|_0^{19}$

$\qquad = -200e^{-1.9} + 200 - 57 \approx 113.086$ or $\$113{,}086$ $\hspace{2cm}$ (6-5)

76. $S'(t) = 4e^{-0.08t}$, $0 \le t \le 24$. Therefore,

$S(t) = \displaystyle\int 4e^{-0.08t}dt = \dfrac{4e^{-0.08t}}{-0.08} + C = -50e^{-0.08t} + C.$

Now, $S(0) = 0$, so

$0 = -50e^{-0.08(0)} + C = -50 + C.$

Thus, $C = 50$, and $S(t) = 50(1 - e^{-0.08t})$ gives the total sales after t months.

Estimated sales after 12 months:

$S(12) = 50(1 - e^{-0.08(12)}) = 50(1 - e^{-0.96}) \approx 31$ or $\$31$ million.

To find the time to reach $\$40$ million in sales, solve

$40 = 50(1 - e^{-0.08t})$

for t.

$0.8 = 1 - e^{-0.08t}$

$e^{-0.08t} = 0.2$

$-0.08t = \ln(0.2)$

$\qquad t = \dfrac{\ln(0.2)}{-0.08} \approx 20$ months $\hspace{4cm}$ (6-3)

77. Using a midpoint sum with $n = 6$, $\Delta t = \dfrac{12 - 0}{6} = 2$ and the values of C from the graph, we have:

$$\text{Ave } C(t) = \frac{1}{12 - 0} \int_0^{12} C(t)\,dt = \frac{1}{12} \int_0^{12} C(t)\,dt$$

$$\approx \frac{1}{12} M_6 = \frac{1}{12} [C(1)\Delta t + C(3)\Delta t + C(5)\Delta t + C(7)\Delta t + C(9)\Delta t + C(11)\Delta t]$$

$$= \frac{2}{12} [4 + 6 + 8 + 9 + 8 + 4] = \frac{39}{6} = 6.5 \text{ parts per million} \qquad (6\text{-}5)$$

78. $\dfrac{dA}{dt} = -5t^{-2}$, $1 \le t \le 5$

$$A = \int -5t^{-2}\,dt = -5 \int t^{-2}\,dt = -5 \cdot \frac{t^{-1}}{-1} + C = \frac{5}{t} + C$$

Now $A(1) = \dfrac{5}{1} + C = 5$. Therefore, $C = 0$ and

$$A(t) = \frac{5}{t}$$

$$A(5) = \frac{5}{5} = 1$$

The area of the wound after 5 days is 1 cm^2. $\qquad (6\text{-}3)$

79. The total amount of seepage during the first four years is given by:

$$T = \int_0^4 R(t)\,dt = \int_0^4 \frac{1000}{(1 + t)^2}\,dt = 1000 \int_0^4 (1 + t)^{-2}\,dt = 1000 \frac{(1 + t)^{-1}}{-1} \Big|_0^4$$

$$[\text{Let } u = 1 + t, \text{ then } du = dt.] = \frac{-1000}{1 + t} \Big|_0^4 = \frac{-1000}{5} + 1000 = 800 \text{ gallons}$$

$$(6\text{-}5)$$

80. (A) $A(t) = 770e^{0.01t}$, $t \ge 0$ (1995 is $t = 0$)

Population in year 2030: $t = 35$
$$A(35) = 770e^{0.01(35)} = 770e^{0.35} \approx 1093 \text{ million}$$

(B) Time to double:
$$770e^{0.01t} = 1540$$
$$e^{0.01t} = 2$$
$$0.01t = \ln 2$$
$$t = \frac{\ln 2}{0.01} \approx 69.3$$

It will take approximately 70 years for the population to double.

$$(6\text{-}3)$$

81. Let $Q = Q(t)$ be the amount of carbon-14 present in the bone at time t. Then,

$$\frac{dQ}{dt} = -0.0001238Q \quad \text{and} \quad Q(t) = Q_0 e^{-0.0001238t},$$

where Q_0 is the amount present originally (i.e., at the time the animal died). We want to find t such that $Q(t) = 0.04Q_0$.

$$0.04Q_0 = Q_0 e^{-0.0001238t}$$
$$e^{-0.0001238t} = 0.04$$
$$-0.0001238t = \ln 0.04$$
$$t = \frac{\ln 0.04}{-0.0001238} \approx 26{,}000 \text{ years} \tag{6-3}$$

82. $N'(t) = 7e^{-0.1t}$ and $N(0) = 25$.

$$N(t) = \int 7e^{-0.1t} dt = 7\int e^{-0.1t} dt = \frac{7}{-0.1}\int e^{-0.1t}(-0.1)\, dt$$

$$= -70e^{-0.1t} + C, \quad 0 \le t \le 15$$

Given $N(0) = 25$: $25 = -70e^0 + C = -70 + C$

Hence, $C = 95$ and $N(t) = 95 - 70e^{-0.1t}$. The student would be expected to type $N(15) = 95 - 70e^{-0.1(15)} = 95 - 70e^{-1.5} \approx 79$ words per minute after completing the course. $\tag{6-3}$

7 ADDITIONAL INTEGRATION TOPICS

Things to remember:

1. **AREA UNDER A CURVE**

   If f is continuous and $f(x) \geq 0$ over the interval $[a, b]$, then the area between $y = f(x)$ and the x-axis from $x = a$ to $x = b$ is given by the definite integral:

   $$A = \int_a^b f(x)\,dx$$

   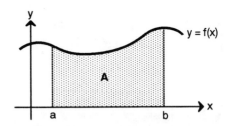

   If $f(x) \leq 0$ over the interval $[a, b]$, then the area between $y = f(x)$ and the x-axis from $x = a$ to $x = b$ is given by

   $$\int_a^b [-f(x)]\,dx.$$

   Finally, if $f(x)$ is positive for some values of x and negative for others, the area between the graph of f and the x axis can be obtained by dividing $[a, b]$ into subintervals on which f is always positive or always negative, finding the area over each subinterval, and then summing these areas.

2. **AREA BETWEEN TWO CURVES**

   If f and g are continuous and $f(x) \geq g(x)$ over the interval $[a, b]$, then the area bounded by $y = f(x)$ and $y = g(x)$, for $a \leq x \leq b$, is given exactly by:

   $$A = \int_a^b [f(x) - g(x)]\,dx.$$

   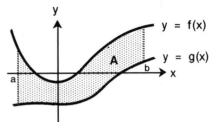

<u>3.</u> INDEX OF INCOME CONCENTRATION

If $y = f(x)$ is the equation of a Lorenz curve, then the

Index of Income Concentration $= 2\int_0^1 [x - f(x)]dx$.

---

1. $A = \int_a^b g(x)\,dx$

3. $A = \int_a^b [-h(x)]\,dx$

5. Since the shaded region in Figure (c) is below the x-axis, $h(x) \le 0$.
Thus, $\int_a^b h(x)\,dx$ represents the negative of the area of the region.

7. $A = \int_1^2 3x^2\,dx = \left.\frac{3x^3}{3}\right|_1^2 = \left.x^3\right|_1^2$
$= 2^3 - 1 = 7$

9. $A = \int_0^4 -[-2x - 1]\,dx = \int_0^4 [2x + 1]\,dx$
$= \left.(x^2 + x)\right|_0^4 = 20$

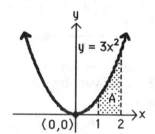

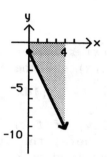

11. $A = \int_{-1}^0 (x^2 + 2)\,dx = \left.\left(\frac{x^3}{3} + 2x\right)\right|_{-1}^0$
$= 0 - \left(-\frac{1}{3} - 2\right) = \frac{7}{3} \approx 2.333$

13. $A = \int_{-1}^2 -[x^2 - 4]\,dx = \int_{-1}^2 [4 - x^2]\,dx$
$= \left.\left(4x - \frac{x^3}{3}\right)\right|_{-1}^2 = \left(8 - \frac{8}{3}\right) - \left(-4 + \frac{1}{3}\right)$
$= 9$

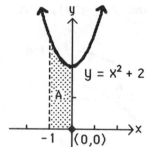

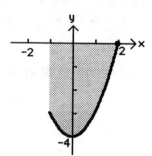

15. $A = \int_{-1}^{2} e^x dx = e^x \Big|_{-1}^{2}$

$= e^2 - e^{-1} \approx 7.021$

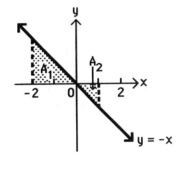

17. $A = \int_{0.5}^{1} -\left[-\frac{1}{t}\right] dt = \int_{0.5}^{1} \frac{1}{t} dt = \ln t \Big|_{0.5}^{1} = \ln 1 - \ln(0.5) \approx 0.693$

19. $A = \int_{a}^{b} [-f(x)] dx$ 　　　 **21.** $a = \int_{b}^{c} f(x) dx + \int_{c}^{d} [-f(x)] dx$

23. $A = \int_{c}^{d} [f(x) - g(x)] dx$ 　 **25.** $A = \int_{a}^{b} [f(x) - g(x)] dx + \int_{b}^{c} [g(x) - f(x)] dx$

27. Find the x-coordinates of the points of intersection of the two curves on [a, d] by solving the equation $f(x) = g(x)$, $a \le x \le d$, to find $x = b$ and $x = c$. Then note that $f(x) \ge g(x)$ on [a, b], $g(x) \ge f(x)$ on [b, c] and $f(x) \ge g(x)$ on [c, d].

Thus,

$$\text{Area} = \int_{a}^{b} [f(x) - g(x)] dx + \int_{b}^{c} [g(x) - f(x)] dx + \int_{c}^{d} [f(x) - g(x)] dx$$

29. $A = A_1 + A_2 = \int_{-2}^{0} -x \, dx + \int_{0}^{1} -(-x) dx$

$= -\int_{-2}^{0} x \, dx + \int_{0}^{1} x \, dx$

$= -\frac{x^2}{2}\Big|_{-2}^{0} + \frac{x^2}{2}\Big|_{0}^{1}$

$= -\left(0 - \frac{(-2)^2}{2}\right) + \left(\frac{1^2}{2} - 0\right)$

$= 2 + \frac{1}{2} = \frac{5}{2} = 2.5$

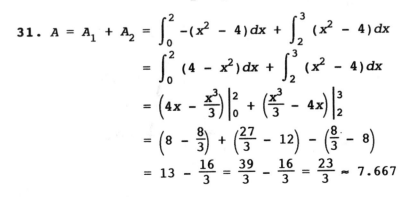

31. $A = A_1 + A_2 = \int_{0}^{2} -(x^2 - 4) dx + \int_{2}^{3} (x^2 - 4) dx$

$= \int_{0}^{2} (4 - x^2) dx + \int_{2}^{3} (x^2 - 4) dx$

$= \left(4x - \frac{x^3}{3}\right)\Big|_{0}^{2} + \left(\frac{x^3}{3} - 4x\right)\Big|_{2}^{3}$

$= \left(8 - \frac{8}{3}\right) + \left(\frac{27}{3} - 12\right) - \left(\frac{8}{3} - 8\right)$

$= 13 - \frac{16}{3} = \frac{39}{3} - \frac{16}{3} = \frac{23}{3} \approx 7.667$

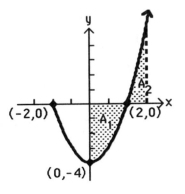

33. $A = A_1 + A_2 + A_3 = \int_{-3}^{-2} -[4 - x^2]\,dx + \int_{-2}^{2} (4 - x^2)\,dx + \int_{2}^{4} -[4 - x^2]\,dx$

$$= \int_{-3}^{-2} (x^2 - 4)\,dx + \int_{-2}^{2} (4 - x^2)\,dx + \int_{2}^{4} (x^2 - 4)\,dx$$

$$= \left(\frac{x^3}{3} - 4x\right)\Big|_{-3}^{-2} + \left(4x - \frac{x^3}{3}\right)\Big|_{-2}^{2} + \left(\frac{x^3}{3} - 4x\right)\Big|_{2}^{4}$$

$$= \left(-\frac{8}{3} + 8\right) - (-9 + 12) + \left(8 - \frac{8}{3}\right) - \left(-8 + \frac{8}{3}\right)$$

$$+ \left(\frac{64}{3} - 16\right) - \left(\frac{8}{3} - 8\right)$$

$$= 13 + \frac{32}{3} \approx 23.667$$

35. $A = \int_{-1}^{2} [12 - (-2x + 8)]\,dx = \int_{-1}^{2} (2x + 4)\,dx$

$$= \left(\frac{2x^2}{2} + 4x\right)\Big|_{-1}^{2} = (x^2 + 4x)\Big|_{-1}^{2}$$

$$= (4 + 8) - (1 - 4)$$

$$= 12 + 3 = 15$$

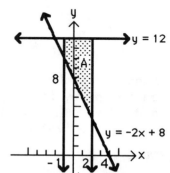

37. $A = \int_{-2}^{2} (12 - 3x^2)\,dx = \left(12x - \frac{3x^3}{3}\right)\Big|_{-2}^{2}$

$$= (12x - x^3)\Big|_{-2}^{2}$$

$$= (12 \cdot 2 - 2^3) - [12 \cdot (-2) - (-2)^3]$$

$$= 16 - (-16) = 32$$

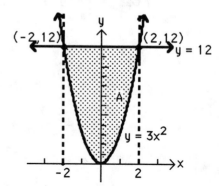

39. $(3, -5)$ and $(-3, -5)$ are the points of intersection.

$$A = \int_{-3}^{3} [4 - x^2 - (-5)]\,dx$$

$$= \int_{-3}^{3} (9 - x^2)\,dx = \left(9x - \frac{x^3}{3}\right)\Big|_{-3}^{3}$$

$$= \left(9 \cdot 3 - \frac{3^3}{3}\right) - \left(9(-3) - \frac{(-3)^3}{3}\right)$$

$$= 18 + 18 = 36$$

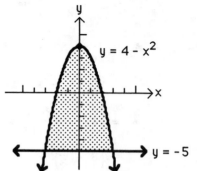

41. $A = \int_{-1}^{2} [(x^2 + 1) - (2x - 2)]dx$

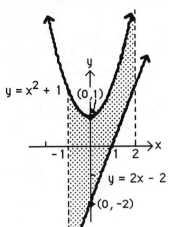

$= \int_{-1}^{2} (x^2 - 2x + 3)dx = \left(\dfrac{x^3}{3} - x^2 + 3x\right)\Big|_{-1}^{2}$

$= \left(\dfrac{8}{3} - 4 + 6\right) - \left(-\dfrac{1}{3} - 1 - 3\right)$

$= 3 - 4 + 6 + 1 + 3 = 9$

43. $A = \int_{1}^{2} \left[e^{0.5x} - \left(-\dfrac{1}{x}\right)\right]dx$

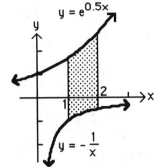

$= \int_{1}^{2} \left(e^{0.5x} + \dfrac{1}{x}\right)dx$

$= \left(\dfrac{e^{0.5x}}{0.5} + \ln|x|\right)\Big|_{1}^{2}$

$= 2e + \ln 2 - 2e^{0.5}$

≈ 2.832

45. The graphs of $y = 3 - 5x - 2x^2$ and $y = 2x^2 + 3x - 2$ are shown at the right. The x-coordinates of the points of intersection are: $x_1 = -2.5$, $x_2 = 0.5$.

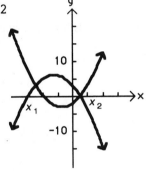

$A = \int_{-2.5}^{0.5} [(3 - 5x - 2x^2) - (2x^2 + 3x - 2)]dx$

$= \int_{-2.5}^{0.5} (5 - 8x - 4x^2)dx = \left(5x - 4x^2 - \dfrac{4}{3}x^3\right)\Big|_{-2.5}^{0.5}$

$= 1.333 + 16.667 = 18$

47. The graphs of $y = -0.5x + 2.25$ and $y = \dfrac{1}{x}$ are shown at the right. The x-coordinates of the points of intersection are: $x_1 = 0.5$, $x_2 = 4$.

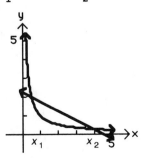

$A = \int_{0.5}^{4} \left[(-0.5x + 2.25) - \left(\dfrac{1}{x}\right)\right]dx$

$= \left(-\dfrac{1}{4}x^2 + \dfrac{9}{4}x - \ln x\right)\Big|_{0.5}^{4}$

$= [-4 + 9 - \ln 4] - [-0.0625 + 1.125 - \ln(0.5)]$

≈ 1.858

49. The graphs of $y = 10 - 2x$ and $y = 4 + 2x$, $0 \le x \le 4$, are shown at the right.

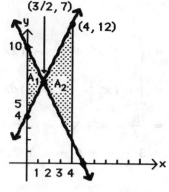

To find the point of intersection of the two lines, solve:

$$10 - 2x = 4 + 2x$$
$$-4x = -6$$
$$x = \frac{3}{2}$$

Substituting $x = \frac{3}{2}$ into either equation, we find $y = 7$. Now we have:

$$A = A_1 + A_2$$
$$= \int_0^{3/2} [(10 - 2x) - (4 + 2x)]dx + \int_{3/2}^4 [(4 + 2x) - (10 - 2x)]dx$$
$$= \int_0^{3/2} (6 - 4x)dx + \int_{3/2}^4 (4x - 6)dx$$
$$= (6x - 2x^2)\Big|_0^{3/2} + (2x^2 - 6x)\Big|_{3/2}^4$$
$$= 9 - \frac{9}{2} + (32 - 24) - \left(\frac{9}{2} - 9\right) = 17$$

51. The graphs are given at the right. To find the points of intersection, solve:

$$x^3 = 4x$$
$$x^3 - 4x = 0$$
$$x(x^2 - 4) = 0$$
$$x(x + 2)(x - 2) = 0$$

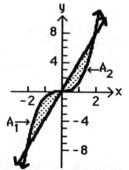

Thus, the points of intersection are $(-2, -8)$, $(0, 0)$, and $(2, 8)$.

$$A = A_1 + A_2 = \int_{-2}^0 (x^3 - 4x)dx + \int_0^2 (4x - x^3)dx$$

$$= \left(\frac{x^4}{4} - 2x^2\right)\Big|_{-2}^0 + \left(2x^2 - \frac{x^4}{4}\right)\Big|_0^2$$
$$= 0 - \left[\frac{(-2)^4}{4} - 2(-2)^2\right] + \left[2(2^2) - \frac{2^4}{4}\right] - 0$$
$$= -4 + 8 + 8 - 4 = 8$$

53. The graphs are given at the right. To find the points of intersection, solve:

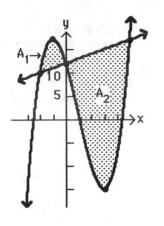

$$x^3 - 3x^2 - 9x + 12 = x + 12$$
$$x^3 - 3x^2 - 10x = 0$$
$$x(x^2 - 3x - 10) = 0$$
$$x(x - 5)(x + 2) = 0$$
$$x = -2, \ x = 0, \ x = 5$$

Thus, $(-2, 10)$, $(0, 12)$, and $(5, 17)$ are the points of intersection.

$$A = A_1 + A_2$$

$$= \int_{-2}^{0} [x^3 - 3x^2 - 9x + 12 - (x + 12)]\,dx$$

$$\qquad + \int_{0}^{5} [x + 12 - (x^3 - 3x^2 - 9x + 12)]\,dx$$

$$= \int_{-2}^{0} (x^3 - 3x^2 - 10x)\,dx + \int_{0}^{5} (-x^3 + 3x^2 + 10x)\,dx$$

$$= \left(\frac{x^4}{4} - x^3 - 5x^2\right)\Big|_{-2}^{0} + \left(-\frac{x^4}{4} + x^3 + 5x^2\right)\Big|_{0}^{5}$$

$$= -\left[\frac{(-2)^4}{4} - (-2)^3 - 5(-2)^2\right] + \left(\frac{-5^4}{4} + 5^3 + 5 \cdot 5^2\right)$$

$$= 8 + \frac{375}{4} = \frac{407}{4} = 101.75$$

55. The graphs are given at the right. To find the points of intersection, solve:

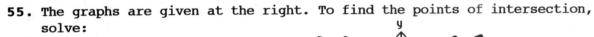

$$x^4 - 4x^2 + 1 = x^2 - 3$$
$$x^4 - 5x^2 + 4 = 0$$
$$(x^2 - 4)(x^2 - 1) = 0$$
$$x = -2, \ -1, \ 1, \ 2$$

$$A = A_1 + A_2 + A_3$$

$$= \int_{-2}^{-1} [(x^2 - 3) - (x^4 - 4x^2 + 1)]\,dx + \int_{-1}^{1} [(x^4 - 4x^2 + 1) - (x^2 - 3)]\,dx$$

$$\qquad + \int_{1}^{2} [(x^2 - 3) - (x^4 - 4x^2 + 1)]\,dx$$

$$= \int_{-2}^{-1} (-x^4 + 5x^2 - 4)\,dx + \int_{-1}^{1} (x^4 - 5x^2 + 4)\,dx + \int_{1}^{2} (-x^4 + 5x^2 - 4)\,dx$$

$$= \left(-\frac{x^5}{5} + \frac{5}{3}x^3 - 4x\right)\Big|_{-2}^{-1} + \left(\frac{x^5}{5} - \frac{5}{3}x^3 + 4x\right)\Big|_{-1}^{1} + \left(-\frac{x^5}{5} + \frac{5}{3}x^3 - 4x\right)\Big|_{1}^{2}$$

$$= \left(\frac{1}{5} - \frac{5}{3} + 4\right) - \left(\frac{32}{5} - \frac{40}{3} + 8\right) + \left(\frac{1}{5} - \frac{5}{3} + 4\right) - \left(-\frac{1}{5} + \frac{5}{3} - 4\right)$$

$$\qquad + \left(-\frac{32}{5} + \frac{40}{3} - 8\right) - \left(-\frac{1}{5} + \frac{5}{3} - 4\right) = 8$$

57. The graphs are given below. The x-coordinates of the points of intersection are: $x_1 = -2$, $x_2 = 0.5$, $x_3 = 2$

$A = A_1 + A_2$

$$= \int_{-2}^{0.5} [(x^3 - x^2 + 2) - (-x^3 + 8x - 2)]dx$$

$$+ \int_{0.5}^{2} [(-x^3 + 8x - 2) - (x^3 - x^2 + 2)]dx$$

$$= \int_{-2}^{0.5} (2x^3 - x^2 - 8x + 4)dx + \int_{0.5}^{2} (-2x^3 + x^2 + 8x - 4)dx$$

$$= \left(\frac{1}{2}x^4 - \frac{1}{3}x^3 - 4x^2 + 4x\right)\Big|_{-2}^{0.5} + \left(-\frac{1}{2}x^4 + \frac{1}{3}x^3 + 4x^2 - 4x\right)\Big|_{0.5}^{2}$$

$$= \left(\frac{1}{32} - \frac{1}{24} - 1 + 2\right) - \left(8 + \frac{8}{3} - 16 - 8\right)$$

$$+ \left(-8 + \frac{8}{3} + 16 - 8\right) - \left(-\frac{1}{32} + \frac{1}{24} + 1 - 2\right)$$

$$= 18 + \frac{1}{16} - \frac{1}{12} \approx 17.979$$

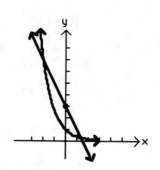

59. The graphs are given at the right. The x-coordinates of the points of intersection are: $x_1 \approx -1.924$, $x_2 \approx 1.373$

$$A = \int_{-1.924}^{1.373} [(3 - 2x) - e^{-x}]dx$$

$$= (3x - x^2 + e^{-x})\Big|_{-1.924}^{1.373}$$

$$\approx 2.487 - (-2.626) = 5.113$$

61. The graphs are given at the right. The x-coordinates of the points of intersection are: $x_1 \approx -2.247$, $x_2 \approx 0.264$, $x_3 \approx 1.439$

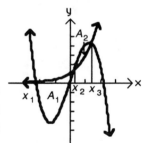

$$A = A_1 + A_2 = \int_{-2.247}^{0.264} [e^x - (5x - x^3)]\,dx$$

$$+ \int_{0.264}^{1.439} [5x - x^3 - e^x]\,dx$$

$$= \left(e^x - \frac{5}{2}x^2 + \frac{1}{4}x^4\right)\Big|_{-2.247}^{0.264} + \left(\frac{5}{2}x^2 - \frac{1}{4}x^4 - e^x\right)\Big|_{0.264}^{1.439}$$

$$\approx (1.129) - (-6.144) + (-0.112) - (-1.129) = 8.290$$

63. $\displaystyle\int_5^{10} R(t)\,dt = \int_5^{10} \left(\frac{100}{t + 10} + 10\right)dt = 100\int_5^{10} \frac{1}{t + 10}\,dt + \int_5^{10} 10\ dt$

$$= 100\ \ln(t + 10)\Big|_5^{10} + 10t\Big|_5^{10}$$

$$= 100\ \ln 20 - 100\ \ln 15 + 10(10 - 5)$$

$$= 100\ \ln 20 - 100\ \ln 15 + 50 \approx 79$$

The total production from the end of the fifth year to the end of the tenth year is approximately 79 thousand barrels.

65. To find the useful life, set $R'(t) = C'(t)$ and solve for t:

$$9e^{-0.3t} = 2$$

$$e^{-0.3t} = \frac{2}{9}$$

$$-0.3t = \ln\frac{2}{9}$$

$$-0.3t \approx -1.5$$

$$t \approx 5 \text{ years}$$

$$\int_0^5 [R'(t) - C'(t)]\,dt] = \int_0^5 [9e^{-0.3t} - 2]\,dt$$

$$= 9\int_0^5 e^{-0.3t}\,dt - \int_0^5 2\ dt = \frac{9}{-0.3}e^{-0.3t}\Big|_0^5 - 2t\Big|_0^5$$

$$= -30e^{-1.5} + 30 - 10$$

$$= 20 - 30e^{-1.5} \approx 13.306$$

The total profit over the useful life of the game is approximately $13,306.

67. For 1935: $f(x) = x^{2.4}$

Index of Income Concentration $= 2\displaystyle\int_0^1 [x - f(x)] = 2\int_0^1 (x - x^{2.4})\,dx$

$$= 2\left(\frac{x^2}{2} - \frac{x^{3.4}}{3.4}\right)\Big|_0^1$$

$$= 2\left(\frac{1}{2} - \frac{1}{3.4}\right) \approx 0.412$$

For 1947: $g(x) = x^{1.6}$

Index of Income Concentration $= 2\int_0^1 [x - g(x)]\,dx = 2\int_0^1 (x - x^{1.6})\,dx$

$$= 2\left(\frac{x^2}{2} - \frac{x^{2.6}}{2.6}\right)\Big|_0^1$$

$$= 2\left(\frac{1}{2} - \frac{1}{2.6}\right) \approx 0.231$$

Interpretation: Income was more equally distributed in 1947.

69. For 1963: $f(x) = x^{10}$

Index of Income Concentration $= 2\int_0^1 [x - f(x)]\,dx = 2\int_0^1 (x - x^{10})\,dx$

$$= 2\left(\frac{x^2}{2} - \frac{x^{11}}{11}\right)\Big|_0^1$$

$$= 2\left(\frac{1}{2} - \frac{1}{11}\right) \approx 0.818$$

For 1983: $g(x) = x^{12}$

Index of Income Concentration $= 2\int_0^1 [x - g(x)]\,dx = 2\int_0^1 (x - x^{12})\,dx$

$$= 2\left(\frac{x^2}{2} - \frac{x^{13}}{13}\right)\Big|_0^1$$

$$= 2\left(\frac{1}{2} - \frac{1}{13}\right) \approx 0.846$$

Interpretation: Total assets were less equally distributed in 1983.

71. $W(t) = \int_0^{10} W'(t)\,dt = \int_0^{10} 0.3e^{0.1t}\,dt = 0.3\int_0^{10} e^{0.1t}\,dt$

$$= \frac{0.3}{0.1}e^{0.1t}\Big|_0^{10} = 3e^{0.1t}\Big|_0^{10} = 3e - 3 \approx 5.15$$

Total weight gain during the first 10 hours is approximately 5.15 grams.

73. $V = \int_2^4 \frac{15}{t}\,dt = 15\int_2^4 \frac{1}{t}\,dt = 15\ln t\Big|_2^4$

$$= 15\ln 4 - 15\ln 2 = 15\ln\left(\frac{4}{2}\right) = 15\ln 2 \approx 10$$

Average number of words learned during the second 2 hours is 10.

Things to remember:

1. PROBABILITY DENSITY FUNCTION

    A function f which satisfies the following three conditions:

    a. $f(x) \geq 0$ for all real x.

    b. The area under the graph of f over the interval $(-\infty, \infty)$ is exactly 1.

    c. If $[c, d]$ is a subinterval of $(-\infty, \infty)$, then the probability that the outcome x of an experiment will be in the interval $[c, d]$, denoted Probability $(c \leq x \leq d)$, is given by

$$\text{Probability } (c \leq x \leq d) = \int_c^d f(x)\,dx$$

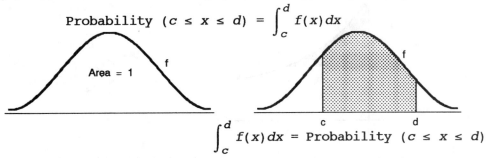

$$\int_c^d f(x)\,dx = \text{Probability } (c \leq x \leq d)$$

2. TOTAL INCOME FOR A CONTINUOUS INCOME STREAM

    If $f(t)$ is the rate of flow of a continuous income stream, then the TOTAL INCOME produced during the time period from $t = a$ to $t = b$ is:

$$\text{Total income} = \int_a^b f(t)\,dt$$

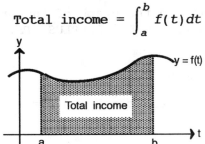

3. FUTURE VALUE OF A CONTINUOUS INCOME STREAM

    If $f(t)$ is the rate of flow of a continuous income stream, $0 \leq t \leq T$, and if the income is continuously invested at a rate r, compounded continuously, then the FUTURE VALUE, FV, at the end of T years is given by:

$$FV = \int_0^T f(t)e^{r(T-t)}\,dt = e^{rT}\int_0^T f(t)e^{-rt}\,dt$$

    The future value of a continuous income stream is the total value of all money produced by the continuous income stream (income and interest) at the end of T years.

<u>4</u>. CONSUMERS' SURPLUS

If $(\overline{x}, \overline{p})$ is a point on the graph of the price-demand equation $p = D(x)$ for a particular product, then the CONSUMERS' SURPLUS, CS, at a price level of $\overline{p}$ is

$$CS = \int_0^{\overline{x}} [D(x) - \overline{p}]dx$$

which is the area between $p = \overline{p}$ and $p = D(x)$ from $x = 0$ to $x = \overline{x}$.

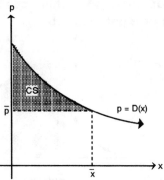

Consumer's surplus represents the total savings to consumers who are willing to pay more than $\overline{p}$ for the product but are still able to buy the product for $\overline{p}$.

<u>5</u>. PRODUCERS' SURPLUS

If $(\overline{x}, \overline{p})$ is a point on the graph of the price-supply equation $p = S(x)$, then the PRODUCERS' SURPLUS, PS, at a price level of $\overline{p}$ is

$$PS = \int_0^{\overline{x}} [\overline{p} - S(x)]dx$$

which is the area between $p = \overline{p}$ and $p = S(x)$ from $x = 0$ to $x = \overline{x}$.

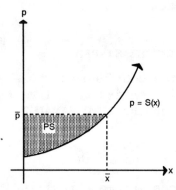

Producers' surplus represents the total gain to producers who are willing to supply units at a lower price than $\bar{p}$ but are still able to supply units at $\bar{p}$.

6. EQUILIBRIUM PRICE AND EQUILIBRIUM QUANTITY

If $p = D(x)$ and $p = S(x)$ are the price-demand and the price-supply equations, respectively, for a product and if $(\bar{x}, \bar{p})$ is the point of intersection of these equations, then $\bar{p}$ is called the EQUILIBRIUM PRICE and $\bar{x}$ is called the EQUILIBRIUM QUANTITY.

---

1. $f(x) = \begin{cases} \dfrac{2}{(x + 2)^2}, & x \geq 0 \\ 0 & x < 0 \end{cases}$

(A) Probability $(0 \leq x \leq 6) = \displaystyle\int_0^6 f(x)\,dx = \int_0^6 \frac{2}{(x + 2)^2}\,dx$

$$= 2\frac{(x + 2)^{-1}}{-1}\bigg|_0^6 = \frac{-2}{(x + 2)}\bigg|_0^6$$

$$= -\frac{1}{4} + 1 = \frac{3}{4} = 0.75$$

Thus, Probability $(0 \leq x \leq 6) = 0.75$

(B) Probability $(6 \leq x \leq 12) = \displaystyle\int_6^{12} f(x)\,dx = \int_6^{12} \frac{2}{(x + 2)^2}\,dx$

$$= \frac{-2}{x + 2}\bigg|_6^{12} = -\frac{1}{7} + \frac{1}{4} = \frac{3}{28} \approx 0.11$$

(C)
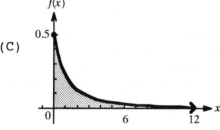

3. We want to find d such that

Probability $(0 \leq x \leq d) = \displaystyle\int_0^d f(x)\,dx = 0.8$:

$$\int_0^d f(x)\,dx = \int_0^d \frac{2}{(x + 2)^2}\,dx = -\frac{2}{x + 2}\bigg|_0^d = \frac{-2}{d + 2} + 1 = \frac{d}{d + 2}$$

Now, $\dfrac{d}{d + 2} = 0.8$

$$d = 0.8d + 1.6$$
$$0.2d = 1.6$$
$$d = 8 \text{ years}$$

5. $f(t) = \begin{cases} 0.01e^{-0.01t} & \text{if } t \geq 0 \\ 0 & \text{otherwise} \end{cases}$

    (A) Since t is in months, the probability of failure during the warranty period of the first year is

$$\text{Probability } (0 \leq t \leq 12) = \int_0^{12} f(t)\,dt = \int_0^{12} 0.01e^{-0.01t}\,dt$$

$$= \frac{0.01}{-0.01}e^{-0.01t}\Big|_0^{12} = -1(e^{-0.12} - 1) \approx 0.11$$

    (B) $\text{Probability } (12 \leq t \leq 24 = \int_{12}^{24} 0.01e^{-0.01t}\,dt = -1e^{-0.01t}\Big|_{12}^{24}$

$$= -1(e^{-0.24} - e^{-0.12}) \approx 0.10$$

7. $\text{Probability } (0 \leq t \leq \infty) = 1 = \int_0^{\infty} f(t)\,dt$

But, $\int_0^{\infty} f(t)\,dt = \int_0^{12} f(t)\,dt + \int_{12}^{\infty} f(t)\,dt$

Thus, $\text{Probability } (t \geq 12) = 1 - \text{Probability } (0 \leq t \leq 12)$
$$\approx 1 - 0.11 = 0.89$$

9. $f(t) = 2500$

    $\text{Total income} = \int_0^5 2500\,dt = 2500t\Big|_0^5 = \$12{,}500$

11.

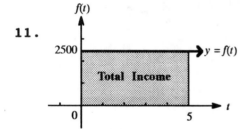

If $f(t)$ is the rate of flow of a continuous income stream, then the total income produced from 0 to 5 years is the area under the curve $y = f(t)$ from $t = 0$ to $t = 5$.

13. $f(t) = 400e^{0.05t}$

    $\text{Total income} = \int_0^3 400e^{0.05t}\,dt = \frac{400}{0.05}e^{0.05t}\Big|_0^3 = 8000(e^{0.15} - 1) \approx \$1295$

15.

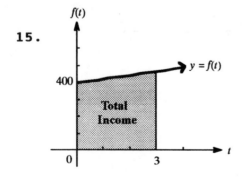

If $f(t)$ is the rate of flow of a continuous income stream, then the total income produced from 0 to 3 years is the area under the curve $y = f(t)$ from $t = 0$ to $t = 3$.

17. $f(t) = 2,000e^{0.05t}$

The amount in the account after 40 years is given by:

$$\int_0^{40} 2,000e^{0.05t}\, dt = 40,000e^{0.05t}\Big|_0^{40} = 295,562.24 - 40,000 \approx \$255,562$$

Since $\$2,000 \times 40 = \$80,000$ was deposited into the account, the interest earned is:

$$\$255,562 - \$80,000 = \$175,562$$

19. $f(t) = 1500e^{-0.02t}$, $r = 0.1$, $T = 4$

$$FV = e^{0.1(4)}\int_0^4 1500e^{-0.02t}e^{-0.1t}dt = 1500e^{0.4}\int_0^4 e^{-0.12t}dt$$

$$= \frac{1500e^{0.4}}{-0.12}e^{-0.12t}\Big|_0^4 = -12,500e^{0.4}(e^{-0.48} - 1)$$

$$= 12,500(e^{0.4} - e^{-0.08}) \approx \$7,109$$

21. Total Income $= \displaystyle\int_0^4 1,500e^{-0.02t}\, dt = -75,000e^{-0.02t}\Big|_0^4$

$$= 75,000 - 69,233.73 \approx \$5,766$$

From Problem 19,

Interest Earned $= \$7,109 - \$5,766 = \$1,343$

23. Clothing store: $f(t) = 12,000$, $r = 0.1$, $T = 5$.

$$FV = e^{0.1(5)}\int_0^5 12,000e^{-0.1t}dt = 12,000e^{0.5}\int_0^5 e^{-0.1t}dt$$

$$= \frac{12,000e^{0.5}}{-0.1}e^{-0.1t}\Big|_0^5 = -120,000e^{0.5}(e^{-0.5} - 1)$$

$$= 120,000(e^{0.5} - 1) \approx \$77,847$$

Computer store: $g(t) = 10,000e^{0.05t}$, $r = 0.1$, $T = 5$.

$$FV = e^{0.1(5)}\int_0^5 10,000e^{0.05t}e^{-0.1t}dt = 10,000e^{0.5}\int_0^5 e^{-0.05t}dt$$

$$= \frac{10,000e^{0.5}}{-0.05}e^{-0.05t}\Big|_0^5 = -200,000e^{0.5}(e^{-0.25} - 1)$$

$$= 200,000(e^{0.5} - e^{0.25}) \approx \$72,939$$

The clothing store is the better investment.

25. Bond: $P = \$10,000$, $r = 0.08$, $t = 5$.

$FV = 10,000e^{0.08(5)} = 10,000e^{0.4} \approx \$14,918$

Business: $f(t) = 2000$, $r = 0.08$, $T = 5$.

$$FV = e^{0.08(5)}\int_0^5 2000e^{-0.08t}dt = 2000e^{0.4}\int_0^5 e^{-0.08t}dt$$

$$= \frac{2000e^{0.4}}{-0.08}e^{-0.08t}\Big|_0^5 = -25,000e^{0.4}(e^{-0.4} - 1)$$

$$= 25,000(e^{0.4} - 1) \approx \$12,296$$

The bond is the better investment.

27. $f(t) = 9000$, $r = 0.12$, $T = 8$

$$FV = e^{0.12(8)} \int_0^8 9000e^{-0.12t}dt = 9000e^{0.96} \int_0^8 e^{-0.12t}dt$$

$$= \frac{9000e^{0.96}}{-0.12} e^{-0.12t} \Big|_0^8 = -75,000e^{0.96}(e^{-0.96} - 1)$$

$$= 75,000(e^{0.96} - 1) \approx \$120,877$$

The relationship between present value (*PV*) and future value (*FV*) at a continuously compounded interest rate r (expressed as a decimal) for t years is:

$FV = PVe^{rt}$ or $PV = FVe^{-rt}$

Thus, we have:

$PV = 120,877e^{-0.12(8)} = 120,877e^{-0.96} \approx 46,283$

Thus, the single deposit should be $46,283.

29. $f(t) = k$, rate r (expressed as a decimal), years T:

$$FV = e^{rT} \int_0^T ke^{-rt}dt = ke^{rT} \int_0^T e^{-rt}dt = \frac{ke^{rT}}{-r} e^{-rt} \Big|_0^T$$

$$= -\frac{k}{r} e^{rT} (e^{-rT} - 1) = \frac{k}{r} (e^{rT} - 1)$$

31. $D(x) = 400 - \frac{1}{20}x$, $\overline{p} = 150$

First, find $\overline{x}$: $150 = 400 - \frac{1}{20}\overline{x}$

$$\overline{x} = 5000$$

$$CS = \int_0^{5000} \left[400 - \frac{1}{20}x - 150 \right] dx = \int_0^{5000} \left(250 - \frac{1}{20}x \right)dx$$

$$= \left(250x - \frac{1}{40}x^2 \right) \Big|_0^{5000} = \$625,000$$

33.

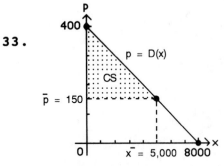

The shaded area is the consumers' surplus and represents the total savings to consumers who are willing to pay more than $150 for a product but are still able to buy the product for $150.

35. $p = S(x) = 10 + 0.1x + 0.0003x^2$, $\overline{p} = 67$.

First find $\overline{x}$: $67 = 10 + 0.1\overline{x} + 0.0003\overline{x}^2$

$$0.0003\overline{x}^2 + 0.1\overline{x} - 57 = 0$$

$$\overline{x} = \frac{-0.1 + \sqrt{0.01 + 0.0684}}{0.0006}$$

$$= \frac{-0.1 + 0.28}{0.0006} = 300$$

$$PS = \int_0^{300} [67 - (10 + 0.1x + 0.0003x^2)]\,dx$$

$$= \int_0^{300} (57 - 0.1x - 0.0003x^2)\,dx$$

$$= (57x - 0.05x^2 - 0.0001x^3) \Big|_0^{300} = \$9,900$$

37.

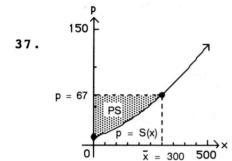

The area of the region PS is the producers' surplus and represents the total gain to producers who are willing to supply units at a lower price than \$67 but are still able to supply the product at \$67.

39. $p = D(x) = 50 - 0.1x;\ p = S(x) = 11 + 0.05x$

Equilibrium price: $D(x) = S(x)$

$$50 - 0.1x = 11 + 0.05x$$
$$39 = 0.15x$$
$$x = 260$$

Thus, $\bar{x} = 260$ and $\bar{p} = 50 - 0.1(260) = 24.$

$$CS = \int_0^{260} [(50 - 0.1x) - 24]\,dx = \int_0^{260} (26 - 0.1x)\,dx$$

$$= (26x - 0.05x^2) \Big|_0^{260}$$

$$= \$3,380$$

$$PS = \int_0^{260} [24 - (11 + 0.05x)]\,dx = \int_0^{260} [13 - 0.05x]\,dx$$

$$= (13x - 0.025x^2) \Big|_0^{260}$$

$$= \$1,690$$

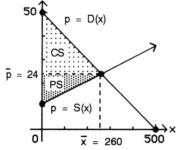

41. $D(x) = 80e^{-0.001x}$ and $S(x) = 30e^{0.001x}$
Equilibrium price: $D(x) = S(x)$

$$80e^{-0.001x} = 30e^{0.001x}$$

$$e^{0.002x} = \frac{8}{3}$$

$$0.002x = \ln\left(\frac{8}{3}\right)$$

$$\overline{x} = \frac{\ln\left(\frac{8}{3}\right)}{0.002} \approx 490$$

Thus, $\overline{p} = 30e^{0.001(490)} \approx 49$.

$$CS = \int_0^{490} [80e^{-0.001x} - 49]dx = \left(\frac{80e^{-0.001x}}{-0.001} - 49x\right)\Bigg|_0^{490}$$

$$= -80,000e^{-0.49} + 80,000 - 24,010 \approx \$6,980$$

$$PS = \int_0^{490} [49 - 30e^{0.001x}]dx = \left(49x - \frac{30e^{0.001x}}{0.001}\right)\Bigg|_0^{490}$$

$$= 24,010 - 30,000(e^{0.49} - 1) \approx \$5,041$$

43. $D(x) = 80 - 0.04x;\ S(x) = 30e^{0.001x}$
Equilibrium price: $D(x) = S(x)$

$$80 - 0.04x = 30e^{0.001x}$$

Using a graphing utility, we find that

$$\overline{x} \approx 614$$

Thus, $\overline{p} = 80 - (0.04)614 \approx 55$

$$CS = \int_0^{614} [80 - 0.04x - 55]dx = \int_0^{614} (25 - 0.04x)dx$$

$$= (25x - 0.02x^2)\Bigg|_0^{614}$$

$$\approx \$7,810$$

$$PS = \int_0^{614} [55 - (30e^{0.001x})]dx = \int_0^{614} (55 - 30e^{0.001x})dx$$

$$= (55x - 30,000e^{0.001x})\Bigg|_0^{614}$$

$$\approx \$8,336$$

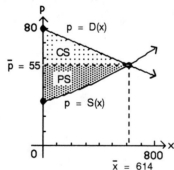

45. $D(x) = 80e^{-0.001x}$; $S(x) = 15 + 0.0001x^2$
Equilibrium price: $D(x) = S(x)$
Using a graphing utility, we find that
$$\overline{x} \approx 556$$
Thus, $\overline{p} = 15 + 0.0001(556)^2 \approx 46$

$$CS = \int_0^{556} [80e^{-0.001x} - 46]dx = (-80,000e^{-0.001x} - 46x)\Big|_0^{556}$$

$$\approx \$8,544$$

$$PS = \int_0^{556} [46 - (15 + 0.0001x^2)]dx = \int_0^{556} (31 - 0.0001x^2)dx$$

$$= \left(31x - \frac{0.0001}{3}x^3\right)\Big|_0^{556}$$

$$\approx \$11,507$$

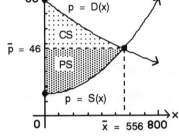

EXERCISE 7-3

Things to remember:

1. INTEGRATION-BY-PARTS FORMULA

$$\int u \; dv = uv - \int v \; du$$

2. INTEGRATION-BY-PARTS: SELECTION OF u AND dv

   (a) The product udv must equal the original integrand.

   (b) It must be possible to integrate dv (preferably by using standard formulas or simple substitutions.)

   (c) The new integral, $\int v \, du$, should not be any more involved than the original integral $\int u \, dv$.

   (d) For integrals involving $x^p e^{ax}$, try
   $u = x^p$; $dv = e^{ax}dx$.

   (e) For integrals involving $x^p (\ln x)^q$, try
   $u = (\ln x)^q$; $dv = x^p dx$.

1. $\int xe^{3x}dx$

Let $u = x$ and $dv = e^{3x}dx$. Then $du = dx$ and $v = \dfrac{e^{3x}}{3}$.

$\int xe^{3x}dx = \dfrac{xe^{3x}}{3} - \int \dfrac{e^{3x}}{3}dx = \dfrac{1}{3}xe^{3x} - \dfrac{1}{3}\int e^{3x}dx = \dfrac{1}{3}xe^{3x} - \dfrac{1}{9}e^{3x} + C$

3. $\int x^2 \ln x\ dx$

Let $u = \ln x$ and $dv = x^2 dx$. Then $du = \dfrac{dx}{x}$ and $v = \dfrac{x^3}{3}$.

$\int x^2 \ln x\ dx = (\ln x)\left(\dfrac{x^3}{3}\right) - \int \dfrac{x^3}{3}\cdot\dfrac{dx}{x} = \dfrac{1}{3}x^3 \ln x - \dfrac{1}{3}\int x^2 dx$

$\qquad = \dfrac{x^3 \ln x}{3} - \dfrac{1}{3}\cdot\dfrac{x^3}{3} + C = \dfrac{x^3 \ln x}{3} - \dfrac{x^3}{9} + C$

5. $\int xe^{-x}dx$

Let $u = x$ and $dv = e^{-x}dx$. Then $du = dx$ and $v = -e^{-x}$.

$\int xe^{-x}dx = x(-e^{-x}) - \int(-e^{-x})dx = -xe^{-x} + \int e^{-x}dx = -xe^{-x} - e^{-x} + C$

7. $\int xe^{x^2}dx = \int e^{x^2}\dfrac{2}{2}x\ dx = \dfrac{1}{2}\int e^{x^2}2x\ dx = \dfrac{1}{2}\int e^u du$

Let $u = x^2$,
then $du = 2x\ dx$. $\qquad = \dfrac{1}{2}e^u + C = \dfrac{1}{2}e^{x^2} + C$

9. $\int_0^1 (x - 3)e^x dx$

Let $u = (x - 3)$ and $dv = e^x dx$. Then $du = dx$ and $v = e^x$.

$\int (x - 3)e^x dx = (x - 3)e^x - \int e^x dx = (x - 3)e^x - e^x + C$

$\qquad = xe^x - 4e^x + C.$

Thus, $\int_0^1 (x - 3)e^x dx = (xe^x - 4e^x)\Big|_0^1 = (e - 4e) - (-4)$

$\qquad\qquad = -3e + 4 \approx -4.1548.$

11. $\int_1^3 \ln 2x\ dx$

Let $u = \ln 2x$ and $dv = dx$. Then $du = \dfrac{dx}{x}$ and $v = x$.

$\int \ln 2x\ dx = (\ln 2x)(x) - \int x\cdot\dfrac{dx}{x} = x \ln 2x - x + C$

Thus, $\int_1^3 \ln 2x\ dx = (x \ln 2x - x)\Big|_1^3 = (3 \ln 6 - 3) - (\ln 2 - 1) \approx 2.6821.$

13. $\int \dfrac{2x}{x^2 + 1}\,dx = \int \dfrac{1}{u}\,du = \ln|u| + C = \ln(x^2 + 1) + C$

Substitution: $u = x^2 + 1$
$\qquad\qquad\quad du = 2x\,dx$

[<u>Note</u>: Absolute value not needed, since $x^2 + 1 \geq 0$.]

15. $\int \dfrac{\ln\,x}{x}\,dx = \int u\,du = \dfrac{u^2}{2} + C = \dfrac{(\ln\,x)^2}{2} + C$

Substitution: $u = \ln\,x$
$\qquad\qquad\quad du = \dfrac{1}{x}\,dx$

17. $\int \sqrt{x}\,\ln\,x\,dx = \int x^{1/2}\,\ln\,x\,dx$

Let $u = \ln\,x$ and $dv = x^{1/2}dx$. Then $du = \dfrac{dx}{x}$ and $v = \dfrac{2}{3}x^{3/2}$.

$\int x^{1/2}\,\ln\,x\,dx = \dfrac{2}{3}x^{3/2}\,\ln\,x - \int \dfrac{2}{3}x^{3/2}\dfrac{dx}{x} = \dfrac{2}{3}x^{3/2}\,\ln\,x - \dfrac{2}{3}\int x^{1/2}dx$

$\qquad\qquad\qquad\qquad\qquad = \dfrac{2}{3}x^{3/2}\,\ln\,x - \dfrac{4}{9}x^{3/2} + C$

19.

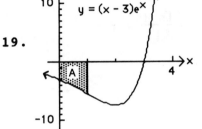

Since $f(x) = (x - 3)e^x < 0$ on $[0, 1]$, the integral represents the negative of the area between the graph of f and the x-axis from $x = 0$ to $x = 1$.

21.

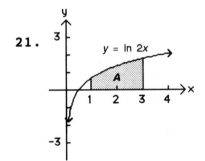

The integral represents the area between the curve $y = \ln\,2x$ and the x-axis from $x = 1$ to $x = 3$.

23. $\int x^2 e^x dx$

Let $u = x^2$ and $dv = e^x dx$. Then $du = 2x\,dx$ and $v = e^x$.

$\int x^2 e^x dx = x^2 e^x - \int e^x (2x)\,dx = x^2 e^x - 2\int xe^x dx$

$\int xe^x dx$ can be computed by using integration-by-parts again.

Let $u = x$ and $dv = e^x dx$. Then $du = dx$ and $v = e^x$.

$$\int xe^x dx = xe^x - \int e^x dx = xe^x - e^x + C$$

and

$$\int x^2 e^x dx = x^2 e^x - 2(xe^x - e^x) + C = x^2 e^x - 2xe^x + 2e^x + C$$

$$= (x^2 - 2x + 2)e^x + C$$

25. $\int xe^{ax} dx$

Let $u = x$ and $dv = e^{ax} dx$. Then $du = dx$ and $v = \dfrac{e^{ax}}{a}$.

$$\int xe^{ax} dx = x \cdot \frac{e^{ax}}{a} - \int \frac{e^{ax}}{a} dx = \frac{xe^{ax}}{a} - \frac{e^{ax}}{a^2} + C$$

27. $\int_1^e \dfrac{\ln x}{x^2} dx$

Let $u = \ln x$ and $dv = \dfrac{dx}{x^2}$. Then $du = \dfrac{dx}{x}$ and $v = \dfrac{-1}{x}$.

$$\int \frac{\ln x}{x^2} dx = (\ln x)\left(-\frac{1}{x}\right) - \int -\frac{1}{x} \cdot \frac{dx}{x} = -\frac{\ln x}{x} + \int \frac{dx}{x^2} = -\frac{\ln x}{x} - \frac{1}{x} + C$$

Thus, $\displaystyle\int_1^e \frac{\ln x}{x^2} dx = \left(-\frac{\ln x}{x} - \frac{1}{x}\right)\Big|_1^e = -\frac{\ln e}{e} - \frac{1}{e} - \left(-\frac{\ln 1}{1} - \frac{1}{1}\right)$

$$= -\frac{2}{e} + 1 \approx 0.2642.$$

[<u>Note</u>: $\ln e = 1$.]

29. $\int_0^2 \ln(x + 4) dx$

Let $t = x + 4$. Then $dt = dx$ and

$$\int \ln(x + 4) dx = \int \ln t \, dt.$$

Now, let $u = \ln t$ and $dv = dt$. Then $du = \dfrac{dt}{t}$ and $v = t$.

$$\int \ln t \, dt = t \ln t - \int t\left(\frac{1}{t}\right) dt = t \ln t - \int dt = t \ln t - t + C$$

Thus, $\displaystyle\int \ln(x + 4) dx = (x + 4) \ln(x + 4) - (x + 4) + C$

and

$$\int_0^2 \ln(x + 4) dx = [(x + 4) \ln(x + 4) - (x + 4)]\Big|_0^2$$

$$= 6 \ln 6 - 6 - (4 \ln 4 - 4) = 6 \ln 6 - 4 \ln 4 - 2 \approx 3.205.$$

31. $\int xe^{x-2} dx$

Let $u = x$ and $dv = e^{x-2} dx$. Then $du = dx$ and $v = e^{x-2}$.

$$\int xe^{x-2} dx = xe^{x-2} - \int e^{x-2} dx = xe^{x-2} - e^{x-2} + C$$

33. $\int x \ln(1 + x^2)\,dx$

Let $t = 1 + x^2$. Then $dt = 2x\,dx$ and

$$\int x \ln(1 + x^2)\,dx = \int \ln(1 + x^2)x\,dx = \int \ln t \frac{dt}{2} = \frac{1}{2}\int \ln t\,dt.$$

Now, for $\int \ln t\,dt$, let $u = \ln t$, $dv = dt$. Then $du = \frac{dt}{t}$ and $v = t$.

$$\int \ln t\,dt = t \ln t - \int t\left(\frac{1}{t}\right)dt = t \ln t - \int dt = t \ln t - t + C$$

Therefore,

$$\int x \ln(1 + x^2)\,dx = \frac{1}{2}(1 + x^2)\ln(1 + x^2) - \frac{1}{2}(1 + x^2) + C.$$

35. $\int e^x \ln(1 + e^x)\,dx$

Let $t = 1 + e^x$. Then $dt = e^x dx$ and

$$\int e^x \ln(1 + e^x)\,dx = \int \ln t\,dt.$$

Now, as shown in Problems 29 and 33,

$$\int \ln t\,dt = t \ln t - t + C.$$

Thus, $\int e^x \ln(1 + e^x)\,dx = (1 + e^x)\ln(1 + e^x) - (1 + e^x) + C.$

37. $\int (\ln x)^2\,dx$

Let $u = (\ln x)^2$ and $dv = dx$. Then $du = \frac{2 \ln x}{x}dx$ and $v = x$.

$$\int (\ln x)^2\,dx = x(\ln x)^2 - \int x \cdot \frac{2 \ln x}{x}dx = x(\ln x)^2 - 2\int \ln x\,dx$$

$\int \ln x\,dx$ can be computed by using integration-by-parts again.

As shown in Problems 29 and 33,

$$\int \ln x\,dx = x \ln x - x + C.$$

Thus, $\int (\ln x)^2\,dx = x(\ln x)^2 - 2(x \ln x - x) + C$

$$= x(\ln x)^2 - 2x \ln x + 2x + C.$$

39. $\int (\ln x)^3\,dx$

Let $u = (\ln x)^3$ and $dv = dx$. Then $du = 3(\ln x)^2 \cdot \frac{1}{x}dx$ and $v = x$.

$$\int (\ln x)^3\,dx = x(\ln x)^3 - \int x \cdot 3(\ln x)^2 \cdot \frac{1}{x}dx = x(\ln x)^3 - 3\int (\ln x)^2\,dx$$

Now, using Problem 37,

$$\int (\ln x)^2 dx = x(\ln x)^2 - 2x \ln x + 2x + C.$$

Therefore, $\int (\ln x)^3 dx = x(\ln x)^3 - 3[x(\ln x)^2 - 2x \ln x + 2x] + C$

$$= x(\ln x)^3 - 3x(\ln x)^2 + 6x \ln x - 6x + C.$$

41. $y = x - 2 - \ln x$, $1 \le x \le 4$

$y = 0$ at $x \approx 3.146$

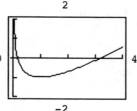

$$A = \int_1^{3.146} [-(x - 2 - \ln x)]dx + \int_{3.146}^4 (x - 2 - \ln x)dx$$

$$= \int_1^{3.146} (\ln x + 2 - x)dx + \int_{3.146}^4 (x - 2 - \ln x)dx$$

Now, $\int \ln x \, dx$ is found using integration-by-parts. Let $u = \ln x$ and

$dv = dx$. Then $du = \frac{1}{x} dx$ and $v = x$.

$$\int \ln x \, dx = x \ln x - \int x\left(\frac{1}{x}\right)dx = x \ln x - \int dx = x \ln x - x + C$$

Thus,

$$A = \left(x \ln x - x + 2x - \frac{1}{2}x^2\right)\Big|_1^{3.146} + \left(\frac{1}{2}x^2 - 2x - x \ln x + x\right)\Big|_{3.146}^4$$

$$= \left(x \ln x + x - \frac{1}{2}x^2\right)\Big|_1^{3.146} + \left(\frac{1}{2}x^2 - x - x \ln x\right)\Big|_{3.146}^4$$

$$\approx (1.803 - 0.5) + (-1.545 + 1.803) = 1.561$$

43. $y = 5 - xe^x$, $0 \le x \le 3$

$y = 0$ at $x \approx 1.327$

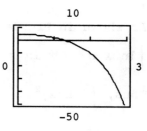

$$A = \int_0^{1.327} (5 - xe^x)dx + \int_{1.327}^3 [-(5 - xe^x)]dx$$

$$= \int_0^{1.327} (5 - xe^x)dx + \int_{1.327}^3 (xe^x - 5)dx$$

Now, $\int xe^x dx$ is found using integration-by-parts. Let $u = x$ and

$dv = e^x dx$. Then, $du = dx$ and $v = e^x$.

$$\int xe^x dx = xe^x - \int e^x dx = xe^x - e^x + C$$

Thus,

$$A = (5x - [xe^x - e^x]) \Big|_0^{1.327} + (xe^x - e^x - 5x) \Big|_{1.327}^3$$

$$\approx (5.402 - 1) + (25.171 - [-5.402]) \approx 34.98$$

45. Marginal profit: $P'(t) = 2t - te^{-t}$.
The total profit over the first 5 years is given by the definite integral:

$$\int_0^5 (2t - te^{-t})\,dt = \int_0^5 2t\,dt - \int_0^5 te^{-t}\,dt$$

We calculate the second integral using integration-by-parts. Let $u = t$ and $dv = e^{-t}\,dt$. Then $du = dt$ and $v = -e^{-t}$

$$\int te^{-t}\,dt = -te^{-t} - \int -e^{-t}\,dt = -te^{-t} - e^{-t} + C = -e^{-t}[t + 1] + C$$

Thus,

$$\text{Total profit} = t^2 \Big|_0^5 + (e^{-t}[t + 1]) \Big|_0^5$$

$$\approx 25 + (0.040 - 1) = 24.040$$

To the nearest million, the total profit is \$24 million.

47.

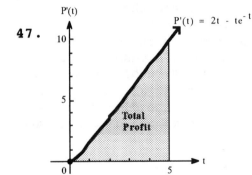

The total profit for the first five years (in millions of dollars) is the same as the area under the marginal profit function, $P'(t) = 2t - te^{-t}$, from $t = 0$ to $t = 5$.

49. From Exercise 7-2, Future Value $= e^{rT} \int_0^T f(t)e^{-rt}dt$. Now $r = 0.08$, $T = 5$, $f(t) = 1000 - 200t$. Thus,

$$FV = e^{(0.08)5} \int_0^5 (1000 - 200t)e^{-0.08t}dt$$

$$= 1000e^{0.4} \int_0^5 e^{-0.08t}dt - 200e^{0.4} \int_0^5 te^{-0.08t}dt.$$

We calculate the second integral using integration-by-parts.

Let $u = t$, $dv = e^{-0.08t}dt$. Then $du = dt$ and $v = \dfrac{e^{-0.08t}}{-0.08}$.

$$\int te^{-0.08t}dt = \frac{te^{-0.08t}}{-0.08} - \int \frac{e^{-0.08t}}{-0.08}\,dt = -12.5te^{-0.08t} - \frac{e^{-0.08t}}{0.0064} + C$$

$$= -12.5te^{-0.08t} - 156.25e^{-0.08t} + C$$

Thus, we have:

$$FV = 1000e^{0.4}\frac{e^{-0.08t}}{-0.08}\Big|_0^5 - 200e^{0.4}[-12.5te^{-0.08t} - 156.25e^{-0.08t}]\Big|_0^5$$

$$= -12{,}500 + 12{,}500e^{0.4} - 200e^{0.4}[-62.5e^{-0.4} - 156.25e^{-0.4} + 156.25]$$

$$= -12{,}500 + 12{,}500e^{0.4} + 43{,}750 - 31{,}250e^{0.4}$$

$$= 31{,}250 - 18{,}750e^{0.4} \approx 3{,}278 \text{ or } \$3{,}278$$

51. Index of Income Concentration $= 2\int_0^1 (x - xe^{x-1})\,dx$

$$= 2\int_0^1 x\,dx - 2\int_0^1 xe^{x-1}dx$$

We calculate the second integral using integration-by-parts.
Let $u = x$, $dv = e^{x-1}dx$. Then $du = dx$, $v = e^{x-1}$.

$$\int xe^{x-1}dx = xe^{x-1} - \int e^{x-1}dx = xe^{x-1} - e^{x-1} + C$$

Therefore, $2\int_0^1 x\,dx - 2\int_0^1 xe^{x-1}dx = x^2\Big|_0^1 - 2[xe^{x-1} - e^{x-1}]\Big|_0^1$

$$= 1 - 2[1 - 1 + (e^{-1})]$$

$$= 1 - 2e^{-1} \approx 0.264.$$

53.

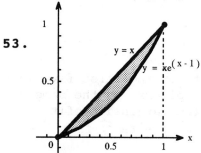

The area bounded by $y = x$ and the Lorenz curve $y = xe^{(x-1)}$ divided by the area under the curve $y = x$ from $x = 0$ to $x = 1$ is the index of income concentration, in this case 0.264. It is a measure of the concentration of income—the closer to zero, the closer to all the income being equally distributed; the closer to one, the closer to all the income being concentrated in a few hands.

55. $p = D(x) = 9 - \ln(x + 4)$; $\overline{p} = \$2.089$. To find $\overline{x}$, solve

$$9 - \ln(\overline{x} + 4) = 2.089$$

$$\ln(\overline{x} + 4) = 6.911$$

$$\overline{x} + 4 = e^{6.911} \quad (\text{take the exponential of both sides})$$

$$\overline{x} \approx 1{,}000$$

Now,

$$CS = \int_0^{1{,}000} (D(x) - \overline{p})\,dx = \int_0^{1{,}000} [9 - \ln(x + 4) - 2.089]\,dx$$

$$= \int_0^{1{,}000} 6.911\,dx - \int_0^{1{,}000} \ln(x + 4)\,dx$$

To calculate the second integral, we first let $z = x + 4$ and $dz = dx$ to get

$$\int \ln(x + 4)\,dx = \int \ln z\,dz$$

Then we use integration-by-parts. Let $u = \ln z$ and $dv = dz$. Then $du = \frac{1}{z}dz$ and $v = z$.

$$\int \ln z \, dz = z \ln z - \int z \cdot \frac{1}{z} dz = z \ln z - z + C$$

Therefore,

$$\int \ln(x + 4)\,dx = (x + 4)\ln(x + 4) - (x + 4) + C$$

and

$$CS = 6.911x \Big|_0^{1,000} - [(x + 4)\ln(x + 4) - (x + 4)]\Big|_0^{1,000}$$

$$\approx 6911 - (5935.39 - 1.55) \approx \$977$$

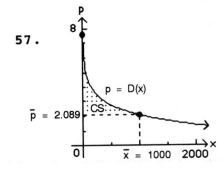

57.

The area bounded by the price-demand equation, $p = 9 - \ln(x + 4)$, and the price equation, $y = \bar{p} = 2.089$, from $x = 0$ to $x = \bar{x} = 1,000$, represents the consumers' surplus. This is the amount saved by consumer's who are willing to pay more than $2.089.

59. Average concentration: $= \dfrac{1}{5 - 0}\displaystyle\int_0^5 \dfrac{20\,\ln(t + 1)}{(t + 1)^2}\,dt = 4\displaystyle\int_0^5 \dfrac{\ln(t + 1)}{(t + 1)^2}\,dt$

$\displaystyle\int \dfrac{\ln(t + 1)}{(t + 1)^2}\,dt$ is found using integration-by-parts.

Let $u = \ln(t + 1)$ and $dv = (t + 1)^{-2}dt$.

Then $du = \dfrac{1}{t + 1}dt = (t + 1)^{-1}dt$ and $v = -(t + 1)^{-1}$.

$$\int \frac{\ln(t + 1)}{(t + 1)^2}\,dt = -\frac{\ln(t + 1)}{t + 1} - \int -(t + 1)^{-1}(t + 1)^{-1}dt$$

$$= -\frac{\ln(t + 1)}{t + 1} + \int(t + 1)^{-2}dt = -\frac{\ln(t + 1)}{t + 1} - \frac{1}{t + 1} + C$$

Therefore, the average concentration is:

$$\frac{1}{5}\int_0^5 \frac{20\,\ln(t + 1)}{(t + 1)^2}\,dt = 4\left[-\frac{\ln(t + 1)}{t + 1} - \frac{1}{t + 1}\right]\Big|_0^5 = 4\left(-\frac{\ln 6}{6} - \frac{1}{6}\right) - 4(-\ln 1 - 1)$$

$$= 4 - \frac{2}{3}\ln 6 - \frac{2}{3} = \frac{1}{3}(10 - 2\ln 6) \approx 2.1388 \text{ ppm}$$

61. Average number of voters $= \dfrac{1}{5}\displaystyle\int_0^5 (20 + 4t - 5te^{-0.1t})\,dt$

$$= \frac{1}{5}\int_0^5 (20 + 4t)\,dt - \int_0^5 te^{-0.1t}dt$$

$\int te^{-0.1t}dt$ is found using integration-by-parts.

Let $u = t$ and $dv = e^{-0.1t}dt$. Then $du = dt$ and $v = \dfrac{e^{-0.1t}}{-0.1} = -10e^{-0.1t}$.

$$\int te^{-0.1t}dt = -10te^{-0.1t} - \int -10e^{-0.1t}dt = -10te^{-0.1t} + 10\int e^{-0.1t}dt$$

$$= -10te^{-0.1t} + \dfrac{10e^{-0.1t}}{-0.1} + C = -10te^{-0.1t} - 100e^{-0.1t} + C$$

Therefore, the average number of voters is:

$$\dfrac{1}{5}\int_0^5 (20 + 4t)\,dt - \int_0^5 te^{-0.1t}dt$$

$$= \dfrac{1}{5}(20t + 2t^2)\Big|_0^5 - (-10te^{-0.1t} - 100e^{-0.1t})\Big|_0^5$$

$$= \dfrac{1}{5}(100 + 50) + (10te^{-0.1t} + 100e^{-0.1t})\Big|_0^5$$

$$= 30 + (50e^{-0.5} + 100e^{-0.5}) - 100$$

$$= 150e^{-0.5} - 70$$

$$\approx 20.98 \text{ (thousands) or } 20{,}980$$

EXERCISE 7-4

1. Use Formula 9 with $a = b = 1$.

$$\int \dfrac{1}{x(1 + x)}\,dx = \dfrac{1}{1}\ln\left|\dfrac{x}{1 + x}\right| + C = \ln\left|\dfrac{x}{x + 1}\right| + C$$

3. Use Formula 18 with $a = 3$, $b = 1$, $c = 5$, $d = 2$:

$$\int \dfrac{1}{(3 + x)^2(5 + 2x)}\,dx = \dfrac{1}{3\cdot 2 - 5\cdot 1}\cdot\dfrac{1}{3 + x} + \dfrac{2}{(3\cdot 2 - 5\cdot 1)^2}\ln\left|\dfrac{5 + 2x}{3 + x}\right| + C$$

$$= \dfrac{1}{3 + x} + 2\ln\left|\dfrac{5 + 2x}{3 + x}\right| + C$$

5. Use Formula 25 with $a = 16$ and $b = 1$:

$$\int \dfrac{x}{\sqrt{16 + x}}\,dx = \dfrac{2(x - 2\cdot 16)}{3\cdot 1^2}\sqrt{16 + x} + C = \dfrac{2(x - 32)}{3}\sqrt{16 + x} + C$$

7. Use Formula 37 with $a = 2$ ($a^2 = 4$):

$$\int \dfrac{1}{x\sqrt{x^2 + 4}}\,dx = \dfrac{1}{2}\ln\left|\dfrac{x}{2 + \sqrt{x^2 + 4}}\right| + C$$

9. Use Formula 51 with $n = 2$:

$$\int x^2 \ln x\,dx = \dfrac{x^{2+1}}{2 + 1}\ln x - \dfrac{x^{2+1}}{(2 + 1)^2} + C = \dfrac{x^3}{3}\ln x - \dfrac{x^3}{9} + C$$

11. First use Formula 5 with $a = 3$ and $b = 1$ to find the indefinite integral.

$$\int \frac{x^2}{3 + x}\,dx = \frac{(3 + x)^2}{2 \cdot 1^3} - \frac{2 \cdot 3(3 + x)}{1^3} + \frac{3^2}{1^3}\ln|3 + x| + C$$

$$= \frac{(3 + x)^2}{2} - 6(3 + x) + 9\ln|3 + x| + C$$

Thus, $\displaystyle\int_1^3 \frac{x^2}{3 + x}\,dx = \left[\frac{(3 + x)^2}{2} - 6(3 + x) + 9\ln|3 + x|\right]\Bigg|_1^3$

$$= \frac{(3 + 3)^2}{2} - 6(3 + 3) + 9\ln|3 + 3|$$

$$- \left[\frac{(3 + 1)^2}{2} - 6(3 + 1) + 9\ln|3 + 1|\right]$$

$$= 9\ln\frac{3}{2} - 2 \approx 1.6492.$$

13. First use Formula 15 with $a = 3$, $b = c = d = 1$ to find the indefinite integral.

$$\int \frac{1}{(3 + x)(1 + x)}\,dx = \frac{1}{3 \cdot 1 - 1 \cdot 1}\ln\left|\frac{1 + x}{3 + x}\right| + C = \frac{1}{2}\ln\left|\frac{1 + x}{3 + x}\right| + C$$

Thus, $\displaystyle\int_0^7 \frac{1}{(3 + x)(1 + x)}\,dx = \frac{1}{2}\ln\left|\frac{1 + x}{3 + x}\right|\Bigg|_0^7 = \frac{1}{2}\ln\left|\frac{1 + 7}{3 + 7}\right| - \frac{1}{2}\ln\left|\frac{1}{3}\right|$

$$= \frac{1}{2}\ln\left|\frac{4}{5}\right| - \frac{1}{2}\ln\left|\frac{1}{3}\right| = \frac{1}{2}\ln\frac{12}{5} \approx 0.4377.$$

15. First use Formula 36 with $a = 3$ $(a^2 = 9)$ to find the indefinite integral:

$$\int \frac{1}{\sqrt{x^2 + 9}}\,dx = \ln\left|x + \sqrt{x^2 + 9}\right| + C$$

Thus, $\displaystyle\int_0^4 \frac{1}{\sqrt{x^2 + 9}}\,dx = \ln\left|x + \sqrt{x^2 + 9}\right|\Bigg|_0^4 = \ln\left|4 + \sqrt{16 + 9}\right| - \ln\left|\sqrt{9}\right|$

$$= \ln 9 - \ln 3 = \ln 3 \approx 1.0986.$$

17. Consider Formula 35. Let $u = 2x$. Then $u^2 = 4x^2$, $x = \frac{u}{2}$, and $dx = \frac{du}{2}$.

$$\int \frac{\sqrt{4x^2 + 1}}{x^2}\,dx = \int \frac{\sqrt{u^2 + 1}}{\frac{u^2}{4}}\,\frac{du}{2} = 2\int \frac{\sqrt{u^2 + 1}}{u^2}\,du$$

$$= 2\left[-\frac{\sqrt{u^2 + 1}}{u} + \ln\left|u + \sqrt{u^2 + 1}\right|\right] + C$$

$$= 2\left[-\frac{\sqrt{4x^2 + 1}}{2x} + \ln\left|2x + \sqrt{4x^2 + 1}\right|\right] + C$$

$$= -\frac{\sqrt{4x^2 + 1}}{x} + 2\ln\left|2x + \sqrt{4x^2 + 1}\right| + C$$

19. Let $u = x^2$. Then $du = 2x \, dx$.

$$\int \frac{x}{\sqrt{x^4 - 16}} \, dx = \frac{1}{2} \int \frac{1}{\sqrt{u^2 - 16}} \, du$$

Now use Formula 43 with $a = 4$ ($a^2 = 16$):

$$\frac{1}{2} \int \frac{1}{\sqrt{u^2 - 16}} \, du = \frac{1}{2} \ln \left| u + \sqrt{u^2 - 16} \right| + C = \frac{1}{2} \left| \ln x^2 + \sqrt{x^4 - 16} \right| + C$$

21. Let $u = x^3$. Then $du = 3x^2 dx$.

$$\int x^2 \sqrt{x^6 + 4} \, dx = \frac{1}{3} \int \sqrt{u^2 + 4} \, du$$

Now use Formula 32 with $a = 2$ ($a^2 = 4$):

$$\frac{1}{3} \int \sqrt{u^2 + 4} \, du = \frac{1}{3} \cdot \frac{1}{2} \left[u \sqrt{u^2 + 4} + 4 \ln \left| u + \sqrt{u^2 + 4} \right| \right] + C$$

$$= \frac{1}{6} \left[x^3 \sqrt{x^6 + 4} + 4 \ln \left| x^3 + \sqrt{x^6 + 4} \right| \right] + C$$

23. $\displaystyle \int \frac{1}{x^3 \sqrt{4 - x^4}} \, dx = \int \frac{x}{x^4 \sqrt{4 - x^4}} \, dx$

Let $u = x^2$. Then $du = 2x \, dx$.

$$\int \frac{x}{x^4 \sqrt{4 - x^4}} \, dx = \frac{1}{2} \int \frac{1}{u^2 \sqrt{4 - u^2}} \, du$$

Now use Formula 30 with $a = 2$ ($a^2 = 4$):

$$\frac{1}{2} \int \frac{1}{u^2 \sqrt{4 - u^2}} \, du = -\frac{1}{2} \cdot \frac{\sqrt{4 - u^2}}{4u} + C = \frac{-\sqrt{4 - x^4}}{8x^2} + C$$

25. $\displaystyle \int \frac{e^x}{(2 + e^x)(3 + 4e^x)} \, dx = \int \frac{1}{(2 + u)(3 + 4u)} \, du$

Substitution: $u = e^x$, $du = e^x dx$.

Now use Formula 15 with $a = 2$, $b = 1$, $c = 3$, $d = 4$:

$$\int \frac{1}{(2 + u)(3 + 4u)} \, du = \frac{1}{2 \cdot 4 - 3 \cdot 1} \ln \left| \frac{3 + 4u}{2 + u} \right| + C = \frac{1}{5} \ln \left| \frac{3 + 4e^x}{2 + e^x} \right| + C$$

27. $\displaystyle \int \frac{\ln x}{x\sqrt{4 + \ln x}} \, dx = \int \frac{u}{\sqrt{4 + u}} \, du$

Substitution: $u = \ln x$, $du = \frac{1}{x} dx$.

Use Formula 25 with $a = 4$, $b = 1$:

$$\int \frac{u}{\sqrt{4 + u}} \, du = \frac{2(u - 2 \cdot 4)}{3 \cdot 1^2} \sqrt{4 + u} + C = \frac{2(u - 8)}{3} \sqrt{4 + u} + C$$

$$= \frac{2(\ln x - 8)}{3} \sqrt{4 + \ln x} + C$$

29. Use Formula 47 with $n = 2$ and $a = 5$:

$$\int x^2 e^{5x} dx = \frac{x^2 e^{5x}}{5} - \frac{2}{5} \int x e^{5x} dx$$

To find $\int xe^{5x}dx$, use Formula 47 with $n = 1$, $a = 5$:

$$\int xe^{5x}dx = \frac{xe^{5x}}{5} - \frac{1}{5}\int e^{5x}dx = \frac{xe^{5x}}{5} - \frac{1}{5} \cdot \frac{e^{5x}}{5}$$

Thus, $\int x^2 e^{5x}dx = \frac{x^2 e^{5x}}{5} - \frac{2}{5}\left[\frac{xe^{5x}}{5} - \frac{1}{25}e^{5x}\right] + C = \frac{x^2 e^{5x}}{5} - \frac{2xe^{5x}}{25} + \frac{2e^{5x}}{125} + C.$

31. Use Formula 47 with $n = 3$ and $a = -1$.

$$\int x^3 e^{-x}dx = \frac{x^3 e^{-x}}{-1} - \frac{3}{-1}\int x^2 e^{-x}dx = -x^3 e^{-x} + 3\int x^2 e^{-x}dx$$

Now $\int x^2 e^{-x}dx = \frac{x^2 e^{-x}}{-1} - \frac{2}{-1}\int xe^{-x}dx = -x^2 e^{-x} + 2\int xe^{-x}dx$

and $\int xe^{-x}dx = \frac{xe^{-x}}{-1} - \frac{1}{-1}\int e^{-x}dx = -xe^{-x} - e^{-x}$, using Formula 47.

Thus, $\int x^3 e^{-x}dx = -x^3 e^{-x} + 3[-x^2 e^{-x} + 2(-xe^{-x} - e^{-x})] + C$

$$= -x^3 e^{-x} - 3x^2 e^{-x} - 6xe^{-x} - 6e^{-x} + C.$$

33. Use Formula 52 with $n = 3$:

$$\int (\ln x)^3 dx = x(\ln x)^3 - 3\int (\ln x)^2 dx$$

Now $\int (\ln x)^2 dx = x(\ln x)^2 - 2\int \ln x\, dx$ using Formula 52 again, and

$\int \ln x\, dx = x \ln x - x$ by Formula 49.

Thus, $\int (\ln x)^3 dx = x(\ln x)^3 - 3[x(\ln x)^2 - 2(x \ln x - x)] + C$

$$= x(\ln x)^3 - 3x(\ln x)^2 + 6x \ln x - 6x + C.$$

35. $\int_3^5 x\sqrt{x^2 - 9}\, dx$. First consider the indefinite integral.

Let $u = x^2 - 9$. Then $du = 2x\, dx$ or $x\, dx = \frac{1}{2}du$. Thus,

$$\int x\sqrt{x^2 - 9}\, dx = \frac{1}{2}\int u^{1/2}du = \frac{1}{2} \cdot \frac{u^{3/2}}{\frac{3}{2}} + C = \frac{1}{3}(x^2 - 9)^{3/2} + C.$$

Now, $\int_3^5 x\sqrt{x^2 - 9}\, dx = \frac{1}{3}(x^2 - 9)^{3/2}\Big|_3^5 = \frac{1}{3} \cdot 16^{3/2} = \frac{64}{3}.$

37. $\int_2^4 \frac{1}{x^2 - 1}dx$. Consider the indefinite integral:

$$\int \frac{1}{x^2 - 1}dx = \frac{1}{2 \cdot 1}\ln\left|\frac{x - 1}{x + 1}\right| + C, \text{ using Formula 13 with } a = 1.$$

Thus, $\int_2^4 \frac{1}{x^2 - 1}dx = \frac{1}{2}\ln\left|\frac{x - 1}{x + 1}\right|\Big|_2^4 = \frac{1}{2}\ln\left|\frac{3}{5}\right| - \frac{1}{2}\ln\left|\frac{1}{3}\right| = \frac{1}{2}\ln\frac{9}{5} \approx 0.2939.$

39. $\displaystyle\int \frac{x + 1}{x^2 + 2x}\,dx = \frac{1}{2}\int \frac{du}{u}$

$\qquad$ Substitution: $u = x^2 + 2x$
$$du = (2x + 2)\,dx$$
$$= 2(x + 1)\,dx$$
$$\frac{1}{2}\,du = (x + 1)\,dx$$

$\qquad\quad = \dfrac{1}{2}\,\ln|u| + C$

$\qquad\quad = \dfrac{1}{2}\,\ln|x^2 + 2x| + C$

41. $\displaystyle\int \frac{x + 1}{x^2 + 3x}\,dx = \int \frac{x}{x(x + 3)}\,dx + \int \frac{1}{x(x + 3)}\,dx = \int \frac{1}{x + 3}\,dx + \int \frac{1}{x(x + 3)}\,dx$

Now $\displaystyle\int \frac{1}{x + 3}\,dx = \ln|x + 3|.$ $\qquad$ Substitution: $u = x + 3,\ du = dx.$

Use Formula 15 with $a = 0,\ b = 1,\ c = 3,\ d = 1$ on the second integral:

$$\int \frac{1}{x(x + 3)}\,dx = \frac{1}{0\cdot 1 - 1\cdot 3}\,\ln\left|\frac{x + 3}{x}\right| = -\frac{1}{3}\,\ln\left|\frac{x + 3}{x}\right|$$

Thus, $\displaystyle\int \frac{x + 1}{x^2 + 3x}\,dx = \ln|x + 3| - \frac{1}{3}\,\ln\left|\frac{x + 3}{x}\right| + C$

$$= \ln|x + 3| - \frac{1}{3}\,\ln|x + 3| + \frac{1}{3}\,\ln|x| + C$$

$$= \frac{2}{3}\,\ln|x + 3| + \frac{1}{3}\,\ln|x| + C$$

43. $f(x) = \dfrac{10}{\sqrt{x^2 + 1}},\quad g(x) = x^2 + 3x$

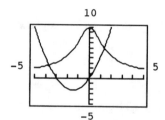

The graphs of f and g are shown at the right.
The x-coordinates of the points of intersection
are: $x_1 \approx -3.70,\ x_2 \approx 1.36$

$$A = \int_{-3.70}^{1.36}\left[\frac{10}{\sqrt{x^2 + 1}} - (x^2 + 3x)\right]dx$$

$$= 10\int_{-3.70}^{1.36}\frac{1}{\sqrt{x^2 + 1}}\,dx - \int_{-3.70}^{1.36}(x^2 + 3x)\,dx$$

For the first integral, use Formula 36 with $a = 1$:

$$A = (10\,\ln|x + \sqrt{x^2 + 1}|)\,\Big|_{-3.70}^{1.36} - \left(\frac{1}{3}x^3 + \frac{3}{2}x^2\right)\Big|_{-3.70}^{1.36}$$

$$\approx [11.15 - (-20.19)] - [3.61 - (3.65)] = 31.38$$

45. $f(x) = x\sqrt{x + 4},\quad g(x) = 1 + x$

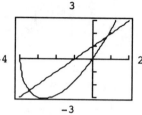

The graphs of f and g are shown at the right.
The x-coordinates of the points of intersection
are: $x_1 \approx -3.49,\ x_2 \approx 0.83$

$$A = \int_{-3.49}^{0.83}[1 + x - x\sqrt{x + 4}]\,dx = \int_{-3.49}^{0.83}(1 + x)\,dx - \int_{-3.49}^{0.83}x\sqrt{x + 4}\,dx$$

For the second integral, use Formula 22 with $a = 4$ and $b = 1$:

$$A = \left(x + \frac{1}{2}x^2\right)\Big|_{-3.49}^{0.83} - \left(\frac{2[3x - 8]}{15}\sqrt{(x + 4)^3}\right)\Big|_{-3.49}^{0.83}$$

$$\approx (1.17445 - 2.60005) - (-7.79850 + 0.89693) \approx 5.48$$

47. Find $\overline{x}$, the demand when the price $\overline{p} = 15$:

$$15 = \frac{7500 - 30\overline{x}}{300 - \overline{x}}$$

$$4500 - 15\overline{x} = 7500 - 30\overline{x}$$

$$15\overline{x} = 3000$$

$$\overline{x} = 200$$

Consumers' surplus:

$$CS = \int_0^{\overline{x}} [D(x) - \overline{p}]dx = \int_0^{200}\left[\frac{7500 - 30x}{300 - x} - 15\right]dx = \int_0^{200}\left[\frac{3000 - 15x}{300 - x}\right]dx$$

Use Formula 20 with $a = 3000$, $b = -15$, $c = 300$, $d = -1$:

$$CS = \left[\frac{-15x}{-1} + \frac{3000(-1) - (-15)(300)}{(-1)^2}\ln|300 - x|\right]\Big|_0^{200}$$

$$= [15x + 1500 \ln|300 - x|]\Big|_0^{200}$$

$$= 3000 + 1500 \ln(100) - 1500 \ln(300)$$

$$= 3000 + 1500 \ln\left(\frac{1}{3}\right) \approx 1352$$

Thus, the consumers' surplus is $1352.

49.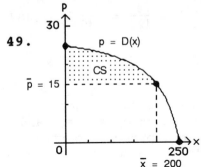

The shaded region represents the consumers' surplus.

51. $FV = e^{rT}\int_0^T f(t)e^{-rt}dt$

Now, $r = 0.1$, $T = 10$, $f(t) = 50t^2$.

$$FV = e^{(0.1)10}\int_0^{10} 50t^2 e^{-0.1t}dt = 50e\int_0^{10} t^2 e^{-0.1t}dt$$

To evaluate the integral, use Formula 47 with $n = 2$ and $a = -0.1$:

$$\int t^2 e^{-0.1t}dt = \frac{t^2 e^{-0.1t}}{-0.1} - \frac{2}{-0.1}\int te^{-0.1t}dt = -10t^2 e^{-0.1t} + 20\int te^{-0.1t}dt$$

Now, using Formula 47 again:

$$\int te^{-0.1t}dt = \frac{te^{-0.1t}}{-0.1} - \frac{1}{-0.1}\int e^{-0.1t}dt = -10te^{-0.1t} + 10\frac{e^{-0.1t}}{-0.1}$$

$$= -10te^{-0.1t} - 100e^{-0.1t}$$

Thus, $\int t^2 e^{-0.1t}dt = -10t^2 e^{-0.1t} - 200te^{-0.1t} - 2000e^{-0.1t} + C.$

$$FV = 50e[-10t^2 e^{-0.1t} - 200te^{-0.1t} - 2000e^{-0.1t}]\Big|_0^{10}$$

$$= 50e[-1000e^{-1} - 2000e^{-1} - 2000e^{-1} + 2000] = 100,000e - 250,000$$
$$\approx 21,828 \text{ or } \$21,828$$

53. Index of Income Concentration:

$$2\int_0^1 [x - f(x)]dx = 2\int_0^1 \left[x - \frac{1}{2}x\sqrt{1 + 3x}\right]dx = \int_0^1 [2x - x\sqrt{1 + 3x}]dx$$

$$= \int_0^1 2x\,dx - \int_0^1 x\sqrt{1 + 3x}\,dx$$

For the second integral, use Formula 22 with $a = 1$ and $b = 3$:

$$= x^2\Big|_0^1 - \frac{2(3\cdot 3x - 2\cdot 1)}{15(3)^2}\sqrt{(1 + 3x)^3}\Big|_0^1$$

$$= 1 - \frac{2(9x - 2)}{135}\sqrt{(1 + 3x)^3}\Big|_0^1$$

$$= 1 - \frac{14}{135}\sqrt{4^3} - \frac{4}{135}\sqrt{1^3} = 1 - \frac{112}{135} - \frac{4}{135} = \frac{19}{135} \approx 0.1407$$

55.

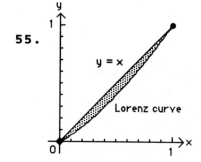

As the area bounded by the two curves gets smaller, the Lorenz curve approaches $y = x$ and the distribution of income approaches perfect equality—all individuals share equally in the income.

57. $S'(t) = \frac{t^2}{(1 + t)^2}$; $S(t) = \int \frac{t^2}{(1 + t)^2}dt$

Use Formula 7 with $a = 1$ and $b = 1$:

$$S(t) = \frac{1 + t}{1^3} - \frac{1^2}{1^3(1 + t)} - \frac{2(1)}{1^3}\ln|1 + t| + C$$

$$= 1 + t - \frac{1}{1 + t} - 2\ln|1 + t| + C$$

Since $S(0) = 0$, we have $0 = 1 - 1 - 2\ln 1 + C$ and $C = 0$. Thus,

$$S(t) = 1 + t - \frac{1}{1 + t} - 2\ln|1 + t|.$$

Now, the total sales during the first two years (= 24 months) is given by:

$$S(24) = 1 + 24 - \frac{1}{1 + 24} - 2 \ln|1 + 24| = 24.96 - 2 \ln 25 \approx 18.5$$

Thus, total sales during the first two years is approximately $18.5 million.

59.

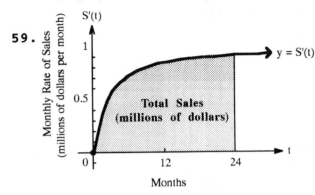

The total sales, in millions of dollars, over the first two years (24 months) is the area under the curve $y = S'(t)$ from $t = 0$ to $t = 24$.

61. $\frac{dR}{dt} = \frac{100}{\sqrt{t^2 + 9}}$. Therefore,

$$R = \int \frac{100}{\sqrt{t^2 + 9}} dt = 100 \int \frac{1}{\sqrt{t^2 + 9}} dt$$

Using Formula 36 with $a = 3$ ($a^2 = 9$), we have:

$$R = 100 \ln\left| t + \sqrt{t^2 + 9} \right| + C$$

Now $R(0) = 0$, so $0 = 100 \ln|3| + C$ or $C = -100 \ln 3$. Thus,

$$R(t) = 100 \ln\left| t + \sqrt{t^2 + 9} \right| - 100 \ln 3$$

and

$$\begin{aligned} R(4) &= 100 \ln(4 + \sqrt{4^2 + 9}) - 100 \ln 3 \\ &= 100 \ln 9 - 100 \ln 3 \\ &= 100 \ln 3 \approx 110 \text{ feet} \end{aligned}$$

63. $N'(t) = \frac{60}{\sqrt{t^2 + 25}}$

The number of items learned in the first twelve hours of study is given by:

$$\begin{aligned} N &= \int_0^{12} \frac{60}{\sqrt{t^2 + 25}} dt = 60 \int_0^{12} \frac{1}{\sqrt{t^2 + 25}} dt \\ &= 60 \left(\ln\left| t + \sqrt{t^2 + 25} \right| \right) \Big|_0^{12}, \text{ using Formula 36} \\ &= 60 \left[\ln\left| 12 + \sqrt{12^2 + 25} \right| - \ln\sqrt{25} \right] \\ &= 60(\ln 25 - \ln 5) \\ &= 60 \ln 5 \approx 96.57 \text{ or } 97 \text{ items} \end{aligned}$$

65.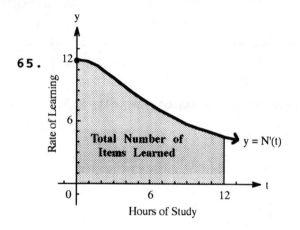

The area under the rate of learning curve, $y = N'(t)$, from $t = 0$ to $t = 12$ represents the total number of items learned in that time interval.

CHAPTER 7 REVIEW

1. $A = \displaystyle\int_a^b f(x)\,dx$ (7-1)   **2.** $A = \displaystyle\int_b^c [-f(x)]\,dx$ (7-1)

3. $A = \displaystyle\int_a^b f(x)\,dx + \int_b^c [-f(x)]\,dx$ (7-1)

4. $A = \displaystyle\int_{0.5}^1 [-\ln x]\,dx + \int_1^e \ln x\,dx$

We evaluate the integral using integration-by-parts. Let $u = \ln x$, $dv = dx$.

Then $du = \dfrac{1}{x}\,dx$, $v = x$, and $\displaystyle\int \ln x\,dx =$

$x \ln x - \displaystyle\int x\left(\dfrac{1}{x}\right) dx = x \ln x - x + C$

Thus,

$A = -\displaystyle\int_{0.5}^1 \ln x\,dx + \int_1^e \ln x\,dx$

$= (-x \ln x + x)\Big|_{0.5}^{1} + (x \ln x - x)\Big|_{1}^{e}$

$\approx (1 - 0.847) + (1) = 1.153$

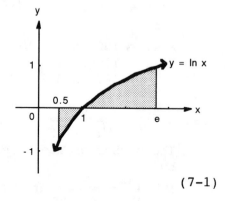

(7-1)

5. $\displaystyle\int xe^{4x}\,dx$.   Use integration-by-parts:

Let $u = x$ and $dv = e^{4x}\,dx$.   Then $du = dx$ and $v = \dfrac{e^{4x}}{4}$.

$\displaystyle\int xe^{4x}\,dx = \dfrac{xe^{4x}}{4} - \int \dfrac{e^{4x}}{4}\,dx = \dfrac{xe^{4x}}{4} - \dfrac{e^{4x}}{16} + C$ (7-3, 7-4)

6. $\int x \ln x \, dx.$ Use integration-by-parts:

Let $u = \ln x$ and $dv = x \, dx.$ Then $du = \frac{1}{x} dx$ and $v = \frac{x^2}{2}.$

$\int x \ln x \, dx = \frac{x^2 \ln x}{2} - \int \frac{1}{x} \cdot \frac{x^2}{2} dx = \frac{x^2 \ln x}{2} - \frac{1}{2} \int x \, dx = \frac{x^2 \ln x}{2} - \frac{x^2}{4} + C$

$(7\text{-}3, \; 7\text{-}4)$

7. Use Formula 11 with $a = 1$ and $b = 1.$

$\int \frac{1}{x(1 + x)^2} dx = \frac{1}{1(1 + x)} + \frac{1}{1^2} \ln \left| \frac{x}{1 + x} \right| + C = \frac{1}{1 + x} + \ln \left| \frac{x}{1 + x} \right| + C$ $(7\text{-}4)$

8. Use Formula 28 with $a = 1$ and $b = 1.$

$\int \frac{1}{x^2 \sqrt{1 + x}} dx = -\frac{\sqrt{1 + x}}{1 \cdot x} - \frac{1}{2 \cdot 1 \sqrt{1}} \ln \left| \frac{\sqrt{1 + x} - \sqrt{1}}{\sqrt{1 + x} + \sqrt{1}} \right| + C$

$= -\frac{\sqrt{1 + x}}{x} - \frac{1}{2} \ln \left| \frac{\sqrt{1 + x} - 1}{\sqrt{1 + x} + 1} \right| + C$ $(7\text{-}4)$

9. $A = \int_a^b [f(x) - g(x)] dx$ $(7\text{-}1)$ **10.** $A = \int_b^c [g(x) - f(x)] dx$ $(7\text{-}1)$

11. $A = \int_b^c [g(x) - f(x)] dx + \int_c^d [f(x) - g(x)] dx$ $(7\text{-}1)$

12. $A = \int_a^b [f(x) - g(x)] dx + \int_b^c [g(x) - f(x)] dx + \int_c^d [f(x) - g(x)] dx$ $(7\text{-}1)$

13. $A = \int_0^5 [(9 - x) - (x^2 - 6x + 9)] dx$

$= \int_0^5 (5x - x^2) \, dx$

$= \left(\frac{5}{2} x^2 - \frac{1}{3} x^3 \right) \Big|_0^5$

$= \frac{125}{2} - \frac{125}{3} = \frac{125}{6} \approx 20.833$

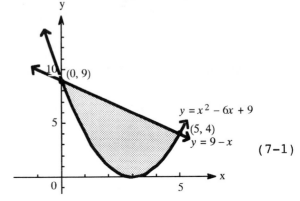

$(7\text{-}1)$

14. $\int_0^1 x e^x dx.$ Use integration-by-parts.

Let $u = x$ and $dv = e^x dx.$ Then $du = dx$ and $v = e^x.$

$\int x e^x dx = x e^x - \int e^x dx = x e^x - e^x + C$

Therefore, $\int_0^1 x e^x dx = (x e^x - e^x) \Big|_0^1 = 1 \cdot e - e - (0 \cdot 1 - 1)$

$= 1$ $(7\text{-}3, \; 7\text{-}4)$

15. Use Formula 38 with $a = 4$

$$\int_0^3 \frac{x^2}{\sqrt{x^2 + 16}}\,dx = \frac{1}{2}\left[\,x\sqrt{x^2 + 16} - 16\ln\left|\,x + \sqrt{x^2 + 16}\,\right|\,\right]\Big|_0^3$$

$$= \frac{1}{2}\left[\,3\sqrt{25} - 16\ln(3 + \sqrt{25})\,\right] - \frac{1}{2}(-16\ln\sqrt{16})$$

$$= \frac{1}{2}[15 - 16\ln 8] + 8\ln 4$$

$$= \frac{15}{2} - 8\ln 8 + 8\ln 4 \approx 1.955 \qquad\qquad (7\text{-}4)$$

16. Let $u = 3x$, then $du = 3\,dx$. Now, use Formula 40 with $a = 7$.

$$\int \sqrt{9x^2 - 49}\,dx = \frac{1}{3}\int \sqrt{u^2 - 49}\,du$$

$$= \frac{1}{3}\cdot\frac{1}{2}\left(u\sqrt{u^2 - 49} - 49\ln\left|\,u + \sqrt{u^2 - 49}\,\right|\right) + C$$

$$= \frac{1}{6}\left(3x\sqrt{9x^2 - 49} - 49\ln\left|\,3x + \sqrt{9x^2 - 49}\,\right|\right) + C \qquad (7\text{-}4)$$

17. $\int te^{-0.5t}\,dt$. Use integration-by-parts.

Let $u = t$ and $dv = e^{-0.5t}\,dt$. Then $du = dt$ and $v = \dfrac{e^{-0.5t}}{-0.5}$.

$$\int te^{-0.5t}\,dt = \frac{-te^{-0.5t}}{0.5} + \int \frac{e^{-0.5t}}{0.5}\,dt = \frac{-te^{-0.5t}}{0.5} + \frac{e^{-0.5t}}{-0.25} + C$$

$$= -2te^{-0.5t} - 4e^{-0.5t} + C \qquad\qquad (7\text{-}3,\ 7\text{-}4)$$

18. $\int x^2 \ln x\,dx$. Use integration-by-parts.

Let $u = \ln x$ and $dv = x^2\,dx$. Then $du = \dfrac{1}{x}\,dx$ and $v = \dfrac{x^3}{3}$.

$$\int x^2 \ln x\,dx = \frac{x^3 \ln x}{3} - \int \frac{1}{x}\cdot\frac{x^3}{3}\,dx = \frac{x^3 \ln x}{3} - \frac{1}{3}\int x^2\,dx$$

$$= \frac{x^3 \ln x}{3} - \frac{x^3}{9} + C \qquad\qquad (7\text{-}3,\ 7\text{-}4)$$

19. Use Formula 48 with $a = 1$, $c = 1$, and $d = 2$.

$$\int \frac{1}{1 + 2e^x}\,dx = \frac{x}{1} - \frac{1}{1\cdot 1}\ln\left|\,1 + 2e^x\,\right| + C = x - \ln\left|\,1 + 2e^x\,\right| + C \qquad (7\text{-}4)$$

20. (A)

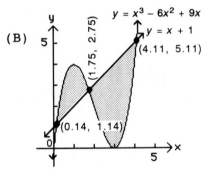

$$A = \int_0^2 [(x^3 - 6x^2 + 9x) - x]dx + \int_2^4 [x - (x^3 - 6x^2 + 9x)]dx$$

$$= \int_0^2 (x^3 - 6x^2 + 8x)dx + \int_2^4 (-x^3 + 6x^2 - 8x)dx$$

$$= \left(\frac{1}{4}x^4 - 2x^3 + 4x^2\right)\Big|_0^2 + \left(-\frac{1}{4}x^4 + 2x^3 - 4x^2\right)\Big|_2^4$$

$$= 4 + 4 = 8$$

(B)

The x-coordinates of the points of intersection are: $x_1 \approx 0.14$, $x_2 \approx 1.75$, $x_3 \approx 4.11$.

$$A = \int_{0.14}^{1.75} [(x^3 - 6x^2 + 9x) - (x + 1)]dx$$

$$+ \int_{1.75}^{4.11} [(x + 1) - (x^3 - 6x^2 + 9x)]dx$$

$$= \int_{0.14}^{1.75} (x^3 - 6x^2 + 8x - 1)dx + \int_{1.75}^{4.11} (1 - x^3 + 6x^2 - 8x)dx$$

$$= \left(\frac{1}{4}x^4 - 2x^3 + 4x^2 - x\right)\Big|_{0.14}^{1.75} + \left(x - \frac{1}{4}x^4 + 2x^3 - 4x^2\right)\Big|_{1.75}^{4.11}$$

$$= [2.126 - (-0.066)] + [4.059 - (-2.126)] \approx 8.38 \tag{7-1}$$

21. $\int \frac{(\ln x)^2}{x}\, dx = \int u^2 du = \frac{u^3}{3} + C = \frac{(\ln x)^3}{3} + C$ Substitution: $u = \ln x$

$$\qquad\qquad\qquad\qquad\qquad\qquad\qquad\qquad du = \frac{1}{x}\, dx$$

$$(6\text{-}2)$$

22. $\int x(\ln x)^2 dx.$ Use integration-by-parts.

Let $u = (\ln x)^2$ and $dv = x\, dx.$ Then $du = 2(\ln x)\frac{1}{x}\, dx$ and $v = \frac{x^2}{2}.$

$\int x(\ln x)^2 dx = \frac{x^2(\ln x)^2}{2} - \int 2(\ln x)\frac{1}{x}\cdot\frac{x^2}{2}\, dx = \frac{x^2(\ln x)^2}{2} - \int x \ln x\, dx$

Let $u = \ln x$ and $dv = x\, dx.$ Then $du = \frac{1}{x}\, dx$ and $v = \frac{x^2}{2}.$

$\int x \ln x\, dx = \frac{x^2 \ln x}{2} - \int \frac{x^2}{2}\cdot\frac{1}{x}\, dx = \frac{x^2 \ln x}{2} - \frac{1}{2}\int x\, dx = \frac{x^2 \ln x}{2} - \frac{x^2}{4} + C$

Thus, $\int x(\ln x)^2 dx = \frac{x^2(\ln x)^2}{2} - \left[\frac{x^2 \ln x}{2} - \frac{x^2}{4}\right] + C$

$$= \frac{x^2(\ln x)^2}{2} - \frac{x^2 \ln x}{2} + \frac{x^2}{4} + C. \qquad\qquad (7\text{-}3,\ 7\text{-}4)$$

23. Let $u = x^2 - 36.$ Then $du = 2x\, dx.$

$\int \frac{x}{\sqrt{x^2 - 36}}\, dx = \int \frac{x}{(x^2 - 36)^{1/2}}\, dx = \frac{1}{2}\int \frac{1}{u^{1/2}}\, du = \frac{1}{2}\int u^{-1/2} du$

$$= \frac{1}{2}\cdot\frac{u^{1/2}}{\frac{1}{2}} + C = u^{1/2} + C = \sqrt{x^2 - 36} + C \qquad\qquad (6\text{-}2)$$

24. Let $u = x^2,\ du = 2x\, dx.$
Then use Formula 43 with $a = 6.$

$\int \frac{x}{\sqrt{x^4 - 36}}\, dx = \frac{1}{2}\int \frac{du}{\sqrt{u^2 - 36}} = \frac{1}{2} \ln \left| u + \sqrt{u^2 - 36} \right| + C$

$$= \frac{1}{2} \ln \left| x^2 + \sqrt{x^4 - 36} \right| + C \qquad\qquad (7\text{-}4)$$

25. $\int_0^4 x \ln(10 - x)\, dx$ Substitution: $t = 10 - x$

Consider $\int x \ln(10 - x)\, dx = \int (10 - t)\ln t(-dt)$ $dt = -dx$

$$\qquad\qquad\qquad\qquad\qquad\qquad x = 10 - t$$

$$= \int t \ln t\, dt - 10\int \ln t\, dt.$$

Now use integration-by-parts on the two integrals.

Let $u = \ln t,\ dv = t\, dt.$ Then $du = \frac{1}{t}\, dt,\ v = \frac{t^2}{2}.$

$\int t \ln t\, dt = \frac{t^2}{2} \ln t - \int \frac{t^2}{2}\cdot\frac{1}{t}\, dt = \frac{t^2 \ln t}{2} - \frac{t^2}{4} + C$

Let $u = \ln t,\ dv = dt.$ Then $du = \frac{1}{t}\, dt,\ v = t.$

$$\int \ln \, t \, dt = t \, \ln \, t - \int t \cdot \frac{1}{t} dt = t \, \ln \, t - t + C$$

Thus, $\displaystyle\int_0^4 x \, \ln(10 - x) \, dx = \left[\frac{(10 - x)^2 \, \ln(10 - x)}{2} - \frac{(10 - x)^2}{4}\right.$

$$\left.- 10(10 - x) \, \ln(10 - x) + 10(10 - x)\right]\Big|_0^4$$

$$= \frac{36 \, \ln \, 6}{2} - \frac{36}{4} - 10(6) \, \ln \, 6 + 10(6)$$

$$- \left[\frac{100 \, \ln \, 10}{2} - \frac{100}{4} - 10(10) \, \ln \, 10 + 10(10)\right]$$

$$= 18 \, \ln \, 6 - 9 - 60 \, \ln \, 6 + 60 - 50 \, \ln \, 10 + 25$$
$$+ \, 100 \, \ln \, 10 - 100$$

$$= 50 \, \ln \, 10 - 42 \, \ln \, 6 - 24 \approx 15.875. \qquad (7\text{-}3, \; 7\text{-}4)$$

26. Use Formula 52 with $n = 2$.

$$\int (\ln \, x)^2 dx = x(\ln \, x)^2 - 2 \int \ln \, x \, dx$$

Now use integration-by-parts to calculate $\displaystyle\int \ln \, x \, dx$.

Let $u = \ln \, x$, $dv = dx$. Then $du = \frac{1}{x} dx$, $v = x$.

$$\int \ln \, x \, dx = x \, \ln \, x - \int x \cdot \frac{1}{x} dx = x \, \ln \, x - x + C$$

Therefore, $\displaystyle\int (\ln \, x)^2 dx = x(\ln \, x)^2 - 2[x \, \ln \, x - x] + C$

$$= x(\ln \, x)^2 - 2x \, \ln \, x + 2x + C. \qquad (7\text{-}3, \; 7\text{-}4)$$

27. $\displaystyle\int xe^{-2x^2} \, dx$

Let $u = -2x^2$. Then $du = -4x \, dx$.

$$\int xe^{-2x^2} \, dx = -\frac{1}{4} \int e^u \, du = -\frac{1}{4} e^u + C$$

$$= -\frac{1}{4} e^{-2x^2} + C \qquad\qquad (6\text{-}2)$$

28. $\displaystyle\int x^2 e^{-2x} \, dx$. Use integration-by-parts. Let $u = x^2$ and $dv = e^{-2x} \, dx$.

Then $du = 2x \, dx$ and $v = -\frac{1}{2} e^{-2x}$.

$$\int x^2 e^{-2x} \, dx = -\frac{1}{2} x^2 e^{-2x} + \int xe^{-2x} \, dx$$

Now use integration-by-parts again. Let $u = x$ and $dv = e^{-2x} \, dx$.
Then $du = dx$ and $v = -\frac{1}{2} e^{-2x}$.

$$\int xe^{-2x} \, dx = -\frac{1}{2} xe^{-2x} + \frac{1}{2} \int e^{-2x} \, dx$$

$$= -\frac{1}{2} xe^{-2x} - \frac{1}{4} e^{-2x} + C$$

Thus,

$$\int x^2 e^{-2x}\,dx = -\frac{1}{2}x^2 e^{-2x} + \left[-\frac{1}{2}xe^{-2x} - \frac{1}{4}e^{-2x}\right] + C$$

$$= -\frac{1}{2}x^2 e^{-2x} - \frac{1}{2}xe^{-2x} - \frac{1}{4}e^{-2x} + C \qquad (7\text{-}3,\ 7\text{-}4)$$

29. (A) Probability $(0 \le t \le 1) = \displaystyle\int_0^1 0.21e^{-0.21t}\,dt$

$$= -e^{-0.21t}\Big|_0^1$$

$$= -e^{-0.21} + 1 \approx 0.189$$

(B) Probability $(1 \le t \le 2) = \displaystyle\int_1^2 0.21e^{-0.21t}\,dt$

$$= -e^{-0.21t}\Big|_1^2$$

$$= e^{-0.21} - e^{-0.42} \approx 0.154 \qquad (7\text{-}2)$$

30.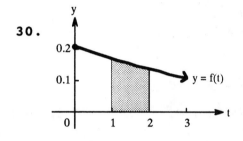

The probability that the product will fail during the second year of warranty is the area under the probability density function $y = f(t)$ from $t = 1$ to $t = 2$.

$(7\text{-}2)$

31. (A)

(B) Total income $= \displaystyle\int_1^4 2{,}500e^{0.05t}\,dt$

$$= 50{,}000e^{0.05t}\Big|_1^4$$

$$= 50{,}000[e^{0.2} - e^{0.05}] \approx \$8{,}507 \qquad (7\text{-}2)$$

32. $f(t) = 2,500e^{0.05t}$, $r = 0.15$, $T = 5$

(A) $FV = e^{(0.15)5} \displaystyle\int_0^5 2,500e^{0.05t} e^{-0.15t} \, dt = 2,500e^{0.75} \displaystyle\int_0^5 e^{-0.1t} \, dt$

$$= -25,000e^{0.75} e^{-0.1t} \Big|_0^5$$

$$= 25,000[e^{0.75} - e^{0.25}] \approx \$20,824$$

(B) Total income $= \displaystyle\int_0^5 2,500e^{0.05t} \, dt = 50,000e^{0.05t} \Big|_0^5$

$$= 50,000[e^{0.25} - 1]$$

$$\approx \$14,201$$

Interest $= FV$ − Total income $= \$20,824 - \$14,201 = \$6,623$ (7-2)

33. (A)

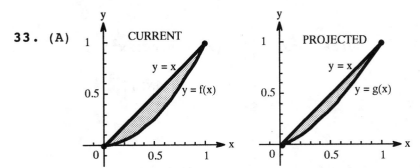

(B) The income will be more equally distributed 10 years from now since the area between $y = x$ and the projected Lorenz curve is less than the area between $y = x$ and the current Lorenz curve.

(C) Current:

Index of income concentration $= 2 \displaystyle\int_0^1 [x - (0.1x + 0.9x^2)] \, dx$

$= 2 \displaystyle\int_0^1 (0.9x - 0.9x^2) \, dx = 2(0.45x^2 - 0.3x^3) \Big|_0^1 = 0.30$

Projected:

Index of Income Concentration $= 2 \displaystyle\int_0^1 (x - x^{1.5}) \, dx$

$= 2 \displaystyle\int_0^1 (x - x^{3/2}) \, dx = 2\left(\frac{1}{2}x^2 - \frac{2}{5}x^{5/2}\right) \Big|_0^1 = 2\left(\frac{1}{10}\right) = 0.2$

Thus, income will be more equally distributed 10 years from now, as indicated in part (B).   (7-1)

34. (A) $p = D(x) = 70 - 0.2x$, $p = S(x) = 13 + 0.0012x^2$

Equilibrium price: $D(x) = S(x)$

$$70 - 0.2x = 13 + 0.0012x^2$$

$$0.0012x^2 + 0.2x - 57 = 0$$

$$x = \frac{-0.2 \pm \sqrt{0.04 + 0.2736}}{0.0024}$$

$$= \frac{-0.2 \pm 0.56}{0.0024}$$

Therefore, $\overline{x} = \dfrac{-0.2 + 0.56}{0.0024} = 150$, and $\overline{p} = 70 - 0.2(150) = 40$.

$$CS = \int_0^{150} (70 - 0.2x - 40)\,dx = \int_0^{150} (30 - 0.2x)\,dx$$

$$= (30x - 0.1x^2) \Big|_0^{150}$$

$$= \$2{,}250$$

$$PS = \int_0^{150} [40 - (13 + 0.0012x^2)]\,dx = \int_0^{150} (27 - 0.0012x^2)\,dx$$

$$= (27x - 0.0004x^3) \Big|_0^{150}$$

$$= \$2{,}700$$

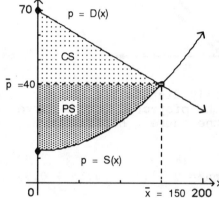

(B) $p = D(x) = 70 - 0.2x$, $p = S(x) = 13e^{0.006x}$

Equilibrium price: $D(x) = S(x)$

$$70 - 0.2x = 13e^{0.006x}$$

Using a graphing utility to solve for x, we get $\overline{x} \approx 170$ and $\overline{p} = 70 - 0.2(170) \approx 36$.

$$CS = \int_0^{170} (70 - 0.2x - 36)\,dx = \int_0^{170} (34 - 0.2x)\,dx$$

$$= (34x - 0.1x^2) \Big|_0^{170}$$

$$= \$2{,}890$$

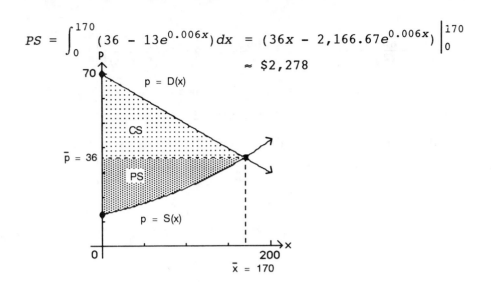

$$PS = \int_0^{170} (36 - 13e^{0.006x})\,dx = (36x - 2{,}166.67e^{0.006x})\Big|_0^{170}$$

$$\approx \$2{,}278$$

(7-2)

35. $R(t) = \dfrac{60t}{(t+1)^2(t+2)}$

The amount of the drug eliminated during the first hour is given by

$$A = \int_0^1 \frac{60t}{(t+1)^2(t+2)}\,dt$$

We will use the Table of Integration Formulas to calculate this integral. First, let $u = t + 2$. Then $t = u - 2$, $t + 1 = u - 1$, $du = dt$ and

$$\int \frac{60t}{(t+1)^2(t+2)}\,dt = 60 \int \frac{u-2}{(u-1)^2 \cdot u}\,du$$

$$= 60 \int \frac{1}{(u-1)^2}\,du - 120 \int \frac{1}{u(u-1)^2}\,du$$

In the first integral, let $v = u - 1$, $dv = du$. Then

$$60 \int \frac{1}{(u-1)^2}\,du = 60 \int v^{-2}\,dv = -60v^{-1} = \frac{-60}{u-1}$$

For the second integral, use Formula 11 with $a = -1$, $b = 1$:

$$-120 \int \frac{1}{u(u-1)^2}\,du = -120 \left[\frac{-1}{u-1} + \ln\left|\frac{u}{u-1}\right|\right]$$

Combining these results and replacing u by $t + 2$, we have:

$$\int \frac{60t}{(t+1)^2(t+2)}\,dt = \frac{-60}{t+1} + \frac{120}{t+1} - 120 \ln\left|\frac{t+2}{t+1}\right| + C$$

$$= \frac{60}{t+1} - 120 \ln\left|\frac{t+2}{t+1}\right| + C$$

Now,

$$A = \int_0^1 \frac{60t}{(t+1)^2(t+2)}\,dt = \left[\frac{60}{t+1} - 120\,\ln\left(\frac{t+2}{t+1}\right)\right]\Big|_0^1$$

$$= 30 - 120\,\ln\left(\frac{3}{2}\right) - 60 + 120\,\ln 2$$

$$\approx 4.522 \text{ milliliters}$$

The amount of drug eliminated during the 4th hour is given by:

$$A = \int_3^4 \frac{60t}{(t+1)^2(t+2)}\,dt = \left[\frac{60}{t+1} - 120\,\ln\left(\frac{t+2}{t+1}\right)\right]\Big|_3^4$$

$$= 12 - 120\,\ln\left(\frac{6}{5}\right) - 15 + 120\,\ln\left(\frac{5}{4}\right)$$

$$\approx 1.899 \text{ milliliters} \qquad\qquad (6\text{-}5,\ 7\text{-}4)$$

36.

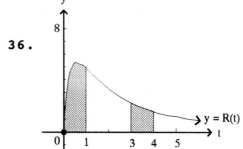

37. $f(t) = \begin{cases} \dfrac{4/3}{(t+1)^2} & 0 \le t \le 3 \\ 0 & \text{otherwise} \end{cases}$

(A) Probability $(0 \le t \le 1) = \int_0^1 \dfrac{4/3}{(t+1)^2}\,dt$

To calculate the integral, let $u = t + 1$, $du = dt$. Then,

$$\int \frac{4/3}{(t+1)^2}\,dt = \frac{4}{3}\int u^{-2}\,du = \frac{4}{3}\frac{u^{-1}}{-1} = -\frac{4}{3u} + C = \frac{-4}{3(t+1)} + C$$

Thus,

$$\int_0^1 \frac{4/3}{(t+1)^2}\,dt = \frac{-4}{3(t+1)}\Big|_0^1 = -\frac{2}{3} + \frac{4}{3} = \frac{2}{3} \approx 0.667$$

(B) Probability $(t \ge 1) = \int_1^3 \dfrac{4/3}{(t+1)^2}\,dt$

$$= \frac{-4}{3(t+1)}\Big|_1^3$$

$$= -\frac{1}{3} + \frac{2}{3} = \frac{1}{3} \approx 0.333 \qquad\qquad (7\text{-}2)$$

38.

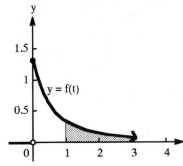

The probability that the doctor will spend more than an hour with a randomly selected patient is the area under the probability density function $y = f(t)$ from $t = 1$ to $t = 3$.

(7-2)

39. $N'(t) = \dfrac{100t}{(1 + t^2)^2}$. To find $N(t)$, we calculate

$$\int \frac{100t}{(1 + t^2)^2}\, dt$$

Let $u = 1 + t^2$. Then $du = 2t\, dt$, and

$$N(t) = \int \frac{100t}{(1 + t^2)^2}\, dt = 50 \int \frac{1}{u^2}\, du = 50 \int u^{-2}\, du$$

$$= -50\,\frac{1}{u} + C$$

$$= \frac{-50}{1 + t^2} + C$$

At $t = 0$, we have
$$N(0) = -50 + C$$

Therefore, $C = N(0) + 50$ and
$$N(t) = \frac{-50}{1 + t^2} + 50 + N(0)$$

Now,
$$N(3) = \frac{-5}{1 + 3^2} + 50 + N(0) = 45 + N(0)$$

Thus, the population will increase by 45 thousand during the next 3 years. (6-5, 7-1)

40. We want to find Probability $(t \geq 2) = \displaystyle\int_2^\infty f(t)\, dt$

Since
$$\int_{-\infty}^\infty f(t)\, dt = \int_{-\infty}^2 f(t)\, dt + \int_2^\infty f(t)\, dt = 1,$$

$$\int_2^\infty f(t)\, dt = 1 - \int_{-\infty}^2 f(t)\, dt = 1 - \int_0^2 f(t)\, dt \quad (\text{since } f(t) = 0 \text{ for } t \leq 0)$$

$$= 1 - \text{Probability } (0 \leq t \leq 2)$$

Now, Probability $(0 \le t \le 2) = \int_0^2 0.5e^{-0.5t}\, dt$

$$= -e^{-0.5t} \Big|_0^2$$

$$= -e^{-1} + 1 \approx 0.632$$

Therefore, Probability $(t \ge 2) = 1 - 0.632 = 0.368$ \hfill (7-2)

8 MULTIVARIABLE CALCULUS

Things to remember:

1. An equation of the form $z = f(x, y)$ describes a FUNCTION OF TWO INDEPENDENT VARIABLES if for each ordered pair (x, y) in the domain of f there is one and only one value of z determined by $f(x, y)$.

2. An equation of the form $w = f(x, y, z)$ describes a FUNCTION OF THREE INDEPENDENT VARIABLES if for each ordered triple (x, y, z) in the domain of f there is one and only one value of w determined by $f(x, y, z)$.

3. Functions of more than three independent variables are defined similarly.

---

1. $f(x, y) = 10 + 2x - 3y$
   $f(0, 0) = 10 + 2 \cdot 0 - 3 \cdot 0 = 10$
   $(x = 0$ and $y = 0)$

3. $f(x, y) = 10 + 2x - 3y$
   $f(-3, 1) = 10 + 2(-3) - 3(1) = 1$
   $(x = -3$ and $y = 1)$

5. $g(x, y) = x^2 - 3y^2$
   $g(0, 0) = 0^2 - 3 \cdot 0^2 = 0$
   $(x = 0$ and $y = 0)$

7. $g(x, y) = x^2 - 3y^2$
   $g(2, -1) = 2^2 - 3(-1)^2 = 1$
   $(x = 2$ and $y = -1)$

9. $A(x, y) = xy$

   $A(2, 3) = 2 \cdot 3 = 6$

   $(x = 2$ and $y = 3)$

11. $Q(M, C) = \dfrac{M}{C}(100)$

    $Q(12, 8) = \dfrac{12}{8}(100) = 150$

    $(M = 12$ and $C = 8)$

13. $V(r, h) = \pi r^2 h$
    $V(2, 4) = \pi \cdot 2^2 \cdot 4 = 16\pi$
    $(r = 2$ and $h = 4)$

15. $R(x, y) = -5x^2 + 6xy - 4y^2 + 200x + 300y$
    $R(1, 2) = -5(1)^2 + 6 \cdot 1 \cdot 2 - 4 \cdot 2^2 + 200 \cdot 1 + 300 \cdot 2$
    $\qquad = -5 + 12 - 16 + 200 + 600$
    $\qquad = 791$
    $(x = 1$ and $y = 2)$

17. $R(L, r) = .002\dfrac{L}{r^4}$

    $R(6, 0.5) = 0.002\dfrac{6}{(0.5)^4} = \dfrac{0.012}{0.0625} = 0.192$

    $(L = 6$ and $r = 0.5)$

19. $A(P, r, t) = P + Prt$
$A(100, 0.06, 3) = 100 + 100(0.06)3 = 118$
($P = 100$, $r = 0.06$, and $t = 3$)

21. $A(P, r, t) = Pe^{rt}$
$A(100, 0.08, 10) = 100e^{0.08 \cdot 10} = 100e^{0.8} \approx 222.55$
($P = 100$, $r = 0.08$, and $t = 10$)

23. $F(x, y) = x^2 + e^x y - y^2$; $F(x, 2) = x^2 + 2e^x - 4$.

We use a graphing utility to solve $F(x, 2) = 0$.
The graph of $u = F(x, 2)$ is shown at the right.

The solutions of $F(x, 2) = 0$ are: $x_1 \approx -1.926$,
$x_2 \approx 0.599$

25. $f(x, y) = x^2 + 2y^2$
$$\frac{f(x + h, y) - f(x, y)}{h} = \frac{(x + h)^2 + 2y^2 - (x^2 + 2y^2)}{h}$$
$$= \frac{x^2 + 2xh + h^2 + 2y^2 - x^2 - 2y^2}{h}$$
$$= \frac{2xh + h^2}{h} = \frac{h(2x + h)}{h} = 2x + h, \ h \neq 0$$

27. $f(x, y) = 2xy^2$
$$\frac{f(x + h, y) - f(x, y)}{h} = \frac{2(x + h)y^2 - 2xy^2}{h}$$
$$= \frac{2xy^2 + 2hy^2 - 2xy^2}{h} = \frac{2hy^2}{h} = 2y^2, \ h \neq 0$$

29. Coordinates of point $E = E(0, 0, 3)$.
Coordinates of point $F = F(2, 0, 3)$.

31. $f(x, y) = x^2$

(A) In the plane $y = c$, c any constant, the graph of $z = x^2$ is a parabola.

(B) Cross-section corresponding to $x = 0$: the y-axis

Cross-section corresponding to $x = 1$: the line passing through $(1, 0, 1)$ parallel to the y-axis.

Cross-section corresponding to $x = 2$: the line passing through $(2, 0, 4)$ parallel to the y-axis.

(C) The surface $z = x^2$ is a parabolic trough lying on the y-axis.

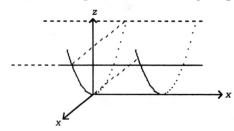

33. Monthly cost function = $C(x, y) = 2000 + 70x + 100y$

$$C(20, 10) = 2000 + 70 \cdot 20 + 100 \cdot 10 = \$4400$$
$$C(50, 5) = 2000 + 70 \cdot 50 + 100 \cdot 5 = \$6000$$
$$C(30, 30) = 2000 + 70 \cdot 30 + 100 \cdot 30 = \$7100$$

35. $R(p, q) = p \cdot x + q \cdot y = 200p - 5p^2 + 4pq + 300q - 4q^2 + 2pq$ or
$R(p, q) = -5p^2 + 6pq - 4q^2 + 200p + 300q$
$R(2, 3) = -5 \cdot 2^2 + 6 \cdot 2 \cdot 3 - 4 \cdot 3^2 + 200 \cdot 2 + 300 \cdot 3 = 1280$ or \$1280
$R(3, 2) = -5 \cdot 3^2 + 6 \cdot 3 \cdot 2 - 4 \cdot 2^2 + 200 \cdot 3 + 300 \cdot 2 = 1175$ or \$1175

37. $f(x, y) = 20x^{0.4}y^{0.6}$
$f(1250, 1700) = 20(1250)^{0.4}(1700)^{0.6}$
$$\approx 20(17.3286)(86.7500) \approx 30,065 \text{ units}$$

39. $FV = F(P, i, n) = P\dfrac{(1 + i)^n - 1}{i}$

$F(2000, .09, 30) = 2000\dfrac{(1 + .09)^{30} - 1}{.09} \approx \$272,615.08$

41. $T(V, x) = \dfrac{33V}{x + 33}$

$T(70, 47) = \dfrac{33 \cdot 70}{47 + 33} = \dfrac{33 \cdot 70}{80} = 28.8 \approx 29$ minutes

$T(60, 27) = \dfrac{33 \cdot 60}{27 + 33} = 33$ minutes

43. $C(W, L) = 100\dfrac{W}{L}$

$C(6, 8) = 100\dfrac{6}{8} = 75$

$C(8.1, 9) = 100\dfrac{8.1}{9} = 90$

45. $Q(M, C) = \dfrac{M}{C}100$

$Q(12, 10) = \dfrac{12}{10}100 = 120$

$Q(10, 12) = \dfrac{10}{12}100 = 83.33 \approx 83$

EXERCISE 8-2

Things to remember:

1. Let $z = f(x, y)$ be a function of two independent variables. The PARTIAL DERIVATIVE OF z WITH RESPECT TO x, denoted by $\dfrac{\partial z}{\partial x}$, f_x, or $f_x(x, y)$, is given by

$$\frac{\partial z}{\partial x} = \lim_{h \to 0} \frac{f(x + h, y) - f(x, y)}{h}$$

provided this limit exists. Similarly, the PARTIAL DERIVATIVE OF z WITH RESPECT TO y, denoted by $\dfrac{\partial z}{\partial y}$, f_y, or $f_y(x, y)$, is given by

$$\frac{\partial z}{\partial y} = \lim_{k \to 0} \frac{f(x, y + k) - f(x, y)}{k}$$

provided this limit exists.

SECOND-ORDER PARTIAL DERIVATIVES

If $z = f(x, y)$, then:

$$\frac{\partial^2 z}{\partial x^2} = \frac{\partial\left(\frac{\partial z}{\partial x}\right)}{\partial x} = f_{xx}(x, y) = f_{xx}$$

$$\frac{\partial^2 z}{\partial x \partial y} = \frac{\partial\left(\frac{\partial z}{\partial y}\right)}{\partial x} = f_{yx}(x, y) = f_{yx}$$

$$\frac{\partial^2 z}{\partial y \partial x} = \frac{\partial\left(\frac{\partial z}{\partial x}\right)}{\partial y} = f_{xy}(x, y) = f_{xy}$$

$$\frac{\partial^2 z}{\partial y^2} = \frac{\partial\left(\frac{\partial z}{\partial y}\right)}{\partial y} = f_{yy}(x, y) = f_{yy}$$

Note: For the functions being considered in this text, the mixed partial derivatives f_{xy} and f_{yx} are equal, i.e.,
$$\frac{\partial^2 z}{\partial x \partial y} = \frac{\partial^2 z}{\partial y \partial x}.$$

---

1. $z = f(x, y) = 10 + 3x + 2y$
$\frac{\partial z}{\partial x} = 0 + 3 + 0 = 3$

3. $z = f(x, y) = 10 + 3x + 2y$
$f_y(x, y) = 0 + 0 + 2 = 2$
$f_y(1, 2) = 2$

5. $z = f(x, y) = 3x^2 - 2xy^2 + 1$
$\frac{\partial z}{\partial y} = 0 - 2x(2y) + 0 = -4xy$

7. $z = f(x, y) = 3x^2 - 2xy^2 + 1$
$f_x(x, y) = 6x - 2y^2 + 0 = 6x - 2y^2$
$f_x(2, 3) = 6\cdot 2 - 2\cdot 3^2 = -6$

9. $S(x, y) = 5x^2 y^3$
$S_x(x, y) = 5y^3 2x = 10xy^3$

11. $S(x, y) = 5x^2 y^3$
$S_y(x, y) = 5x^2 \cdot 3y^2 = 15x^2 y^2$
$S_y(2, 1) = 15 \cdot 2^2 \cdot 1^2 = 60$

13. $C(x, y) = x^2 - 2xy + 2y^2 + 6x - 9y + 5$
$C_x(x, y) = 2x - 2y + 0 + 6 - 0 + 0 = 2x - 2y + 6$

15. $C_x(x, y) = 2x - 2y + 6$ (from Problem 13)
$C_x(2, 2) = 2\cdot 2 - 2\cdot 2 + 6 = 6$

17. $C_x(x, y) = 2x - 2y + 6$ (from Problem 13)
$C_{xy}(x, y) = 0 - 2 + 0 = -2$

19. $C_x(x, y) = 2x - 2y + 6$ (from Problem 13)
$C_{xx}(x, y) = 2 - 0 + 0 = 2$

21. $z = f(x, y) = e^{(2x+3y)}$

$\dfrac{\partial z}{\partial x} = e^{(2x+3y)} \dfrac{\partial}{\partial x} (2x + 3y) = e^{(2x+3y)}2 = 2e^{(2x+3y)}$

23. $z = f(x, y) = e^{(2x+3y)}$

$\dfrac{\partial z}{\partial y} = e^{(2x+3y)} \dfrac{\partial}{\partial y} (2x + 3y) = e^{(2x+3y)}3 = 3e^{(2x+3y)}$

$\dfrac{\partial^2 z}{\partial x \partial y} = \dfrac{\partial \left(\frac{\partial z}{\partial y}\right)}{\partial x} = \dfrac{\partial (3e^{(2x+3y)})}{\partial x} = 3e^{(2x+3y)} \dfrac{\partial}{\partial x} (2x + 3y) = 3e^{(2x+3y)}2 = 6e^{(2x+3y)}$

25. $z = f(x, y) = e^{(2x+3y)}$

$f_x = e^{(2x+3y)} \dfrac{\partial}{\partial x} (2x + 3y) = 2e^{(2x+3y)}$

$f_{xy} = \dfrac{\partial}{\partial y} \left(\dfrac{\partial z}{\partial x}\right) = 2e^{(2x+3y)} \dfrac{\partial}{\partial y} (2x + 3y) = 6e^{(2x+3y)}$

$f_{xy}(1, 0) = 6e^{(2 \cdot 1 + 0)} = 6e^2$

27. $f_x = 2e^{(2x+3y)}$ (from Problem 21)

$f_{xx} = \dfrac{\partial}{\partial x} (2e^{(2x+3y)}) = 2e^{(2x+3y)} \dfrac{\partial}{\partial x} (2x + 3y) = 4e^{(2x+3y)}$

$f_{xx}(0, 1) = 4e^{(0+3 \cdot 1)} = 4e^3$

29. $f(x, y) = (x^2 - y^3)^3$

$f_x(x, y) = 3(x^2 - y^3)^2 \dfrac{\partial (x^2 - y^3)}{\partial x} = 3(x^2 - y^3)^2 2x = 6x(x^2 - y^3)^2$

$f_y(x, y) = 3(x^2 - y^3)^2 \dfrac{\partial (x^2 - y^3)}{\partial y} = 3(x^2 - y^3)^2 (-3y^2) = -9y^2(x^2 - y^3)^2$

31. $f(x, y) = (3x^2 y - 1)^4$

$f_x(x, y) = 4(3x^2 y - 1)^3 \dfrac{\partial (3x^2 y - 1)}{\partial x} = 4(3x^2 y - 1)^3 6xy = 24xy(3x^2 y - 1)^3$

$f_y(x, y) = 4(3x^2 y - 1)^3 \dfrac{\partial (3x^2 y - 1)}{\partial y} = 4(3x^2 y - 1)^3 3x^2 = 12x^2(3x^2 y - 1)^3$

33. $f(x, y) = \ln(x^2 + y^2)$

$f_x(x, y) = \dfrac{1}{x^2 + y^2} \cdot \dfrac{\partial (x^2 + y^2)}{\partial x} = \dfrac{2x}{x^2 + y^2}$

$f_y(x, y) = \dfrac{1}{x^2 + y^2} \cdot \dfrac{\partial (x^2 + y^2)}{\partial y} = \dfrac{2y}{x^2 + y^2}$

35. $f(x, y) = y^2 e^{xy^2}$

$f_x(x, y) = y^2 e^{xy^2} \dfrac{\partial (xy^2)}{\partial x} = y^2 e^{xy^2} y^2 = y^4 e^{xy^2}$

$f_y(x, y) = y^2 \dfrac{\partial (e^{xy^2})}{\partial y} + e^{xy^2} \dfrac{\partial (y^2)}{\partial y}$ (Product rule)

$\qquad = y^2 e^{xy^2} 2yx + y^2 e^{xy^2} 2y = 2xy^3 e^{xy^2} + 2y e^{xy^2}$

37. $f(x, y) = \dfrac{x^2 - y^2}{x^2 + y^2}$

Applying the quotient rule:

$$f_x(x, y) = \frac{(x^2 + y^2)\dfrac{\partial(x^2 - y^2)}{\partial x} - (x^2 - y^2)\dfrac{\partial(x^2 + y^2)}{\partial x}}{(x^2 + y^2)^2}$$

$$= \frac{(x^2 + y^2)(2x) - (x^2 - y^2)(2x)}{(x^2 + y^2)^2}$$

$$= \frac{2x^3 + 2y^2x - 2x^3 + 2y^2x}{(x^2 + y^2)^2} = \frac{4xy^2}{(x^2 + y^2)^2}$$

Again, applying the quotient rule:

$$f_y(x, y) = \frac{(x^2 + y^2)(-2y) - (x^2 - y^2)(2y)}{(x^2 + y^2)^2} = \frac{-4x^2y}{(x^2 + y^2)^2}$$

39. (A) $f(x, y) = y^3 + 4y^2 - 5y + 3$

Since f is independent of x, $\dfrac{\partial f}{\partial x} = 0$

(B) If $g(x, y)$ depends on y only, that is, if $g(x, y) = G(y)$ is independent of x, then

$$\frac{\partial g}{\partial x} = 0$$

Clearly there are an infinite number of such functions.

41. $f(x, y) = x^2y^2 + x^3 + y$

$f_x(x, y) = 2xy^2 + 3x^2$ $\qquad\qquad$ $f_y(x, y) = 2x^2y + 1$

$f_{xx}(x, y) = 2y^2 + 6x$ $\qquad\qquad$ $f_{yx}(x, y) = 4xy$

$f_{xy}(x, y) = 4xy$ $\qquad\qquad$ $f_{yy}(x, y) = 2x^2$

43. $f(x, y) = \dfrac{x}{y} - \dfrac{y}{x}$

$f_x(x, y) = \dfrac{1}{y} + \dfrac{y}{x^2}$ $\qquad\qquad$ $f_y(x, y) = -\dfrac{x}{y^2} - \dfrac{1}{x}$

$f_{xx}(x, y) = -\dfrac{2y}{x^3}$ $\qquad\qquad$ $f_{yx}(x, y) = -\dfrac{1}{y^2} + \dfrac{1}{x^2}$

$f_{xy}(x, y) = -\dfrac{1}{y^2} + \dfrac{1}{x^2}$ $\qquad\qquad$ $f_{yy}(x, y) = \dfrac{2x}{y^3}$

45. $f(x, y) = xe^{xy}$

$f_x(x, y) = xye^{xy} + e^{xy}$ $\qquad\qquad$ $f_y(x, y) = x^2e^{xy}$

$f_{xx}(x, y) = xy^2e^{xy} + 2ye^{xy}$ $\qquad\qquad$ $f_{yx}(x, y) = x^2ye^{xy} + 2xe^{xy}$

$f_{xy}(x, y) = x^2ye^{xy} + 2xe^{xy}$ $\qquad\qquad$ $f_{yy}(x, y) = x^3e^{xy}$

47. $P(x, y) = -x^2 + 2xy - 2y^2 - 4x + 12y - 5$

$P_x(x, y) = -2x + 2y - 0 - 4 + 0 - 0 = -2x + 2y - 4$

$P_y(x, y) = 0 + 2x - 4y - 0 + 12 - 0 = 2x - 4y + 12$

$P_x(x, y) = 0$ and $P_y(x, y) = 0$ when

$$-2x + 2y - 4 = 0 \qquad (1)$$
$$2x - 4y + 12 = 0 \qquad (2)$$

Add equations (1) and (2): $-2y + 8 = 0$
$$y = 4$$

Substitute $y = 4$ into (1): $-2x + 2\cdot4 - 4 = 0$
$$-2x + 4 = 0$$
$$x = 2$$

Thus, $P_x(x, y) = 0$ and $P_y(x, y) = 0$ when $x = 2$ and $y = 4$.

49. $F(x, y) = x^3 - 2x^2y^2 - 2x - 4y + 10$;
$F_x(x, y) = 3x^2 - 4xy^2 - 2$; $F_y(x, y) = -4x^2y - 4$.

Set $F_x(x, y) = 0$ and $F_y(x, y) = 0$ and solve simultaneously:

$$3x^2 - 4xy^2 - 2 = 0 \qquad (1)$$
$$-4x^2y - 4 = 0 \qquad (2)$$

From (2), $y = -\dfrac{1}{x^2}$. Substituting this into (1),

$$3x^2 - 4x\left(-\frac{1}{x^2}\right)^2 - 2 = 0$$

$$3x^2 - 4x\left(\frac{1}{x^4}\right) - 2 = 0$$

$$3x^5 - 2x^3 - 4 = 0$$

Using a graphing utility, we find that
$$x \approx 1.200$$
Then, $y \approx -0.694$.

51. $f(x, y) = \ln(x^2 + y^2)$ (see Problem 33)

$$f_x(x, y) = \frac{2x}{x^2 + y^2} \qquad\qquad f_y(x, y) = \frac{2y}{x^2 + y^2}$$

$$f_{xx}(x, y) = \frac{(x^2 + y^2)2 - 2x(2x)}{(x^2 + y^2)^2} \qquad f_{yy}(x, y) = \frac{(x^2 + y^2)2 - 2y(2y)}{(x^2 + y^2)^2}$$

$$= \frac{2(y^2 - x^2)}{(x^2 + y^2)^2} \qquad\qquad = \frac{2(x^2 - y^2)}{(x^2 + y^2)^2} = \frac{-2(y^2 - x^2)}{(x^2 + y^2)^2}$$

$$f_{xx}(x, y) + f_{yy}(x, y) = \frac{2(y^2 - x^2)}{(x^2 + y^2)^2} + \frac{-2(y^2 - x^2)}{(x^2 + y^2)^2} = 0$$

53. $f(x, y) = x^2 + 2y^2$

(A) $\displaystyle\lim_{h\to 0} \frac{f(x + h, y) - f(x, y)}{h} = \lim_{h\to 0} \frac{(x + h)^2 + 2y^2 - (x^2 + 2y^2)}{h}$

$$= \lim_{h\to 0} \frac{x^2 + 2xh + h^2 + 2y^2 - x^2 - 2y^2}{h}$$

$$= \lim_{h\to 0} \frac{h(2x + h)}{h} = \lim_{h\to 0} (2x + h)$$

$$= 2x$$

(B) $\displaystyle\lim_{k \to 0} \frac{f(x, y + k) - f(x, y)}{k} = \lim_{k \to 0} \frac{x^2 + 2(y + k)^2 - (x^2 + 2y^2)}{k}$

$$= \lim_{k \to 0} \frac{x^2 + 2(y^2 + 2yk + k^2) - x^2 - 2y^2}{k}$$

$$= \lim_{k \to 0} \frac{4yk + 2k^2}{k} = \lim_{k \to 0} (4y + 2k)$$

$$= 4y$$

55. $R(x, y) = 80x + 90y + 0.04xy - 0.05x^2 - 0.05y^2$
$C(x, y) = 8x + 6y + 20,000$
The profit $P(x, y)$ is given by:
$P(x, y) = R(x, y) - C(x, y)$
$\qquad = 80x + 90y + 0.04xy - 0.05x^2 - 0.05y^2 - (8x + 6y + 20,000)$
$\qquad = 72x + 84y + 0.04xy - 0.05x^2 - 0.05y^2 - 20,000$

Now
$P_x(x, y) = 72 + 0.04y - 0.1x$
and
$P_x(1200, 1800) = 72 + 0.04(1800) - 0.1(1200)$
$\qquad\qquad\qquad = 72 + 72 - 120 = 24;$
$P_y(x, y) = 84 + 0.04x - 0.1y$
and
$P_y(1200, 1800) = 84 + 0.04(1200) - 0.1(1800)$
$\qquad\qquad\qquad = 84 + 48 - 180 = -48.$
Thus, at the (1200, 1800) output level, profit will increase approximately \$24 per unit increase in production of type A calculators; and profit will decrease \$48 per unit increase in production of type B calculators.

57. $x = 200 - 5p + 4q$
$y = 300 - 4q + 2p$
$\dfrac{\partial x}{\partial p} = -5, \dfrac{\partial y}{\partial p} = 2$
A \$1 increase in the price of brand A will decrease the demand for brand A by 5 pounds at any price level (p, q).

A \$1 increase in the price of brand A will increase the demand for brand B by 2 pounds at any price level (p, q).

59. $f(x, y) = 10x^{0.75}y^{0.25}$
  (A) $f_x(x, y) = 10(0.75)x^{-0.25}y^{0.25} = 7.5x^{-0.25}y^{0.25}$
  $\quad f_y(x, y) = 10(0.25)x^{0.75}y^{-0.75} = 2.5x^{0.75}y^{-0.75}$

  (B) Marginal productivity of labor $= f_x(600, 100)$
  $$= 7.5(600)^{-0.25}(100)^{0.25} \approx 4.79$$

  Marginal productivity of capital $= f_y(600, 100)$
  $$= 2.5(600)^{0.75}(100)^{-0.75} \approx 9.58$$

  (C) The government should encourage the increased use of capital.

61. $x = f(p, q) = 8000 - 0.09p^2 + 0.08q^2$ (Butter)

$y = g(p, q) = 15,000 + 0.04p^2 - 0.3q^2$ (Margarine)

$f_q(p, q) = 0.08(2)q = 0.16q > 0$

$g_p(p, q) = 0.04(2)p = 0.08p > 0$

Thus, the products are competitive.

63. $x = f(p, q) = 800 - 0.004p^2 - 0.003q^2$ (Skis)

$y = g(p, q) = 600 - 0.003p^2 - 0.002q^2$ (Ski boots)

$f_q(p, q) = -0.003(2)q = -0.006q < 0$

$g_p(p, q) = -0.003(2)p = -0.006p < 0$

Thus, the products are complementary.

65. $A = f(w, h) = 15.64w^{0.425}h^{0.725}$

(A) $f_w(w, h) = 15.64(0.425)w^{-0.575}h^{0.725} \approx 6.65w^{-0.575}h^{0.725}$

     $f_h(w, h) = 15.64(0.725)w^{0.425}h^{-0.275} \approx 11.34w^{0.425}h^{-0.275}$

(B) $f_w(65, 57) = 6.65(65)^{-0.575}(57)^{0.725} \approx 11.31$

For a 65 pound child 57 inches tall, the rate of change of surface area is approximately 11.31 square inches for a one-pound gain in weight, height held fixed.

$f_h(65, 57) = 11.34(65)^{0.425}(57)^{-0.275} \approx 21.99$

For a 65 pound child 57 inches tall, the rate of change of surface area is approximately 21.99 square inches for a one-inch gain in height, weight held fixed.

67. $C(W, L) = 100\dfrac{W}{L}$ $C_L(W, L) = -\dfrac{100W}{L^2}$

$C_W(W, L) = \dfrac{100}{L}$ $C_L(6, 8) = -\dfrac{100 \times 6}{8^2}$

$C_W(6, 8) = \dfrac{100}{8} = 12.5$ $= -\dfrac{600}{64} = -9.38$

The index increases 12.5 units per 1-inch increase in the width of the head (length held fixed) when $W = 6$ and $L = 8$.

The index decreases 9.38 units per 1-inch increase in length (width held fixed) when $W = 6$ and $L = 8$.

EXERCISE 8-3

Things to remember:

1. $f(a, b)$ is a LOCAL MAXIMUM if there exists a circular region in the domain of $f(x, y)$ with (a, b) as the center, such that $f(a, b) \geq f(x, y)$ for all (x, y) in the region. Similarly, $f(a, b)$ is a LOCAL MINIMUM if $f(a, b) \leq f(x, y)$ for all (x, y) in the region.

2. If $f(a, b)$ is either a local maximum of a local minimum for the function $f(x, y)$, and if $f_x(a, b)$ and $f_y(a, b)$ exist, then

$$f_x(a, b) = 0 \quad \text{and} \quad f_y(a, b) = 0.$$

3. SECOND-DERIVATIVE TEST FOR LOCAL EXTREMA FOR $z = f(x, y)$

Given:

(a) $f_x(a, b) = 0$ and $f_y(a, b) = 0$ [(a, b) is a critical point].

(b) All second-order partial derivatives of f exist in some circular region containing (a, b) as center.

(c) $A = f_{xx}(a, b)$, $B = f_{xy}(a, b)$, $C = f_{yy}(a, b)$.

Then:

   i) If $AC - B^2 > 0$ and $A < 0$, then $f(a, b)$ is a local maximum.

   ii) If $AC - B^2 > 0$ and $A > 0$, then $f(a, b)$ is a local minimum.

   iii) If $AC - B^2 < 0$, then f has a saddle point at (a, b).

   iv) If $AC - B^2 = 0$, then the test fails.

---

1. $f(x, y) = 6 - x^2 - 4x - y^2$
$f_x(x, y) = -2x - 4 = 0$
$$x = -2$$
$f_y(x, y) = -2y = 0$
$$y = 0$$
Thus, $(-2, 0)$ is a critical point.

$f_{xx} = -2$, $f_{xy} = 0$, $f_{yy} = -2$,
$f_{xx}(-2, 0) \cdot f_{yy}(-2, 0) - [f_{xy}(-2, 0)]^2 = (-2)(-2) - 0^2 = 4 > 0$

and $\qquad\qquad\qquad\qquad f_{xx}(-2, 0) = -2 < 0.$

Thus, $f(-2, 0) = 6 - (-2)^2 - 4(-2) - 0^2 = 10$ is a local maximum (using $\underline{3}$).

3. $f(x, y) = x^2 + y^2 + 2x - 6y + 14$
$f_x(x, y) = 2x + 2 = 0$
$$x = -1$$
$f_y(x, y) = 2y - 6 = 0$
$$y = 3$$
Thus, $(-1, 3)$ is a critical point.

$f_{xx} = 2 \qquad\qquad\qquad f_{xy} = 0 \qquad\qquad\qquad f_{yy} = 2$
$f_{xx}(-1, 3) = 2 > 0 \qquad\quad f_{xy}(-1, 3) = 0 \qquad\quad f_{yy}(-1, 3) = 2$
$f_{xx}(-1, 3) \cdot f_{yy}(-1, 3) - [f_{xy}(-1, 3)]^2 = 2 \cdot 2 - 0^2 = 4 > 0$
Thus, using $\underline{3}$, $f(-1, 3) = 4$ is a local minimum.

5. $f(x, y) = xy + 2x - 3y - 2$

$f_x = y + 2 = 0$

$\qquad y = -2$

$f_y = x - 3 = 0$

$\qquad x = 3$

Thus, $(3, -2)$ is a critical point.

$f_{xx} = 0 \qquad\qquad f_{xy} = 1 \qquad\qquad f_{yy} = 0$

$f_{xx}(3, -2) = 0 \qquad f_{xy}(3, -2) = 1 \qquad f_{yy}(3, -2) = 0$

$f_{xx}(3, -2) \cdot f_{yy}(3, -2) - [f_{xy}(3, -2)]^2 = 0 \cdot 0 - [1]^2 = -1 < 0$

Thus, using <u>3</u>, f has a saddle point at $(3, -2)$.

7. $f(x, y) = -3x^2 + 2xy - 2y^2 + 14x + 2y + 10$

$f_x = -6x + 2y + 14 = 0 \qquad (1)$

$f_y = 2x - 4y + 2 = 0 \qquad (2)$

Solving (1) and (2) for x and y, we obtain $x = 3$ and $y = 2$. Thus, $(3, 2)$ is a critical point.

$f_{xx} = -6 \qquad\qquad f_{xy} = 2 \qquad\qquad f_{yy} = -4$

$f_{xx}(3, 2) = -6 < 0 \qquad f_{xy}(3, 2) = 2 \qquad f_{yy}(3, 2) = -4$

$f_{xx}(3, 2) \cdot f_{yy}(3, 2) - [f_{xy}(3, 2)]^2 = (-6)(-4) - 2^2 = 20 > 0$

Thus, using <u>3</u>, $f(3, 2)$ is a local maximum and

$f(3, 2) = -3 \cdot 3^2 + 2 \cdot 3 \cdot 2 - 2 \cdot 2^2 + 14 \cdot 3 + 2 \cdot 2 + 10 = 33.$

9. $f(x, y) = 2x^2 - 2xy + 3y^2 - 4x - 8y + 20$

$f_x = 4x - 2y - 4 = 0 \qquad (1)$

$f_y = -2x + 6y - 8 = 0 \qquad (2)$

Solving (1) and (2) for x and y, we obtain $x = 2$ and $y = 2$. Thus, $(2, 2)$ is a critical point.

$f_{xx} = 4 \qquad\qquad f_{xy} = -2 \qquad\qquad f_{yy} = 6$

$f_{xx}(2, 2) = 4 > 0 \qquad f_{xy}(2, 2) = -2 \qquad f_{yy}(2, 2) = 6$

$f_{xx}(2, 2) \cdot f_{yy}(2, 2) - [f_{xy}(2, 2)]^2 = 4 \cdot 6 - [-2]^2 = 20 > 0$

Thus, using <u>3</u>, $f(2, 2)$ is a local minimum and

$f(2, 2) = 2 \cdot 2^2 - 2 \cdot 2 \cdot 2 + 3 \cdot 2^2 - 4 \cdot 2 - 8 \cdot 2 + 20 = 8.$

11. $f(x, y) = e^{xy}$

$f_x = e^{xy} \dfrac{\partial (xy)}{\partial x} \qquad\qquad f_y = e^{xy} \dfrac{\partial (xy)}{\partial y}$

$\quad = e^{xy} y = 0 \qquad\qquad\quad = e^{xy} x = 0$

$\quad y = 0 \ (e^{xy} \neq 0) \qquad\qquad x = 0 \ (e^{xy} \neq 0)$

Thus, $(0, 0)$ is a critical point.

$f_{xx} = y e^{xy} \dfrac{\partial (xy)}{\partial x} \qquad f_{xy} = e^{xy} \cdot 1 + y e^{xy} x \qquad f_{yy} = x e^{xy} \dfrac{\partial (xy)}{\partial y}$

$\quad = y e^{xy} y \qquad\qquad\quad = e^{xy} + xy e^{xy} \qquad\qquad = x^2 e^{xy}$

$\quad = y^2 e^{xy}$

$f_{xx}(0, 0) = 0 \qquad f_{xy}(0, 0) = 1 + 0 = 1 \qquad f_{yy}(0, 0) = 0$

$f_{xx}(0, 0) \cdot f_{yy}(0, 0) - [f_{xy}(0, 0)]^2 = 0 - [1]^2 = -1 < 0$

Thus, using <u>3</u>, $f(x, y)$ has a saddle point at $(0, 0)$.

13. $f(x, y) = x^3 + y^3 - 3xy$

$f_x = 3x^2 - 3y = 3(x^2 - y) = 0$

Thus, $y = x^2$.   (1)

$f_y = 3y^2 - 3x = 3(y^2 - x) = 0$

Thus, $y^2 = x$.   (2)

Combining (1) and (2), we obtain $x = x^4$ or $x(x^3 - 1) = 0$. Therefore, $x = 0$ or $x = 1$, and the critical points are $(0, 0)$ and $(1, 1)$.

$f_{xx} = 6x$ $f_{xy} = -3$ $f_{yy} = 6y$

For the critical point $(0, 0)$:

$f_{xx}(0, 0) = 0$ $f_{xy}(0, 0) = -3$ $f_{yy}(0, 0) = 0$

$f_{xx}(0, 0) \cdot f_{yy}(0, 0) - [f_{xy}(0, 0)]^2 = 0 - (-3)^2 = -9 < 0$

Thus, using $\underline{3}$, $f(x, y)$ has a saddle point at $(0, 0)$.

For the critical point $(1, 1)$:

$f_{xx}(1, 1) = 6$ $f_{xy}(1, 1) = -3$ $f_{yy}(1, 1) = 6$

$f_{xx}(1, 1) \cdot f_{yy}(1, 1) - [f_{xy}(1, 1)]^2 = 6 \cdot 6 - (-3)^2 = 27 > 0$

$f_{xx}(1, 1) > 0$

Thus, using $\underline{3}$, $f(1, 1)$ is a local minimum and

$f(1, 1) = 1^3 + 1^3 - 3 \cdot 1 \cdot 1 = 2 - 3 = -1$.

15. $f(x, y) = 2x^4 + y^2 - 12xy$

$f_x = 8x^3 - 12y = 0$

Thus, $y = \frac{2}{3}x^3$.

$f_y = 2y - 12x = 0$

Thus, $y = 6x$

Therefore, $6x = \frac{2}{3}x^3$

$x^3 - 9x = 0$

$x(x^2 - 9) = 0$

         $x = 0, \ x = 3, \ x = -3$

Thus, the critical points are $(0, 0)$, $(3, 18)$, $(-3, -18)$. Now,

$f_{xx} = 24x^2$ $f_{xy} = -12$ $f_{yy} = 2$

For the critical point $(0, 0)$:

$f_{xx}(0, 0) = 0$ $f_{xy}(0, 0) = -12$ $f_{yy}(0, 0) = 2$

and

$f_{xx}(0, 0) \cdot f_{yy}(0, 0) - [f_{xy}(0, 0)]^2 = 0 \cdot 2 - (-12)^2 = -144$.

Thus, $f(x, y)$ has a saddle point at $(0, 0)$.

For the critical point $(3, 18)$:

$f_{xx}(3, 18) = 24 \cdot 3^2 = 216 > 0$ $f_{xy}(3, 18) = -12$ $f_{yy}(3, 18) = 2$

and

$f_{xx}(3, 18) \cdot f_{yy}(3, 18) \ - [f_{xy}(3, 18)]^2 = 216 \cdot 2 - (-12)^2 = 288 > 0$

Thus, $f(3, 18) = -162$ is a local minimum.

For the critical point $(-3, -18)$:

$f_{xx}(-3, -18) = 216 > 0$ $\quad$ $f_{xy}(-3, -18) = -12$ $\quad$ $f_{yy}(-3, -18) = 2$

and

$f_{xx}(-3, -18) \cdot f_{yy}(-3, -18) - [f_{xy}(-3, -18)]^2 = 288 > 0$

Thus, $f(-3, -18) = -162$ is a local minimum.

17. $f(x, y) = x^3 - 3xy^2 + 6y^2$

$f_x = 3x^2 - 3y^2 = 0$

Thus, $y^2 = x^2$ or $y = \pm x$.

$f_y = -6xy + 12y = 0$ or $-6y(x - 2) = 0$

Thus, $y = 0$ or $x = 2$.

Therefore, the critical points are $(0, 0)$, $(2, 2)$, and $(2, -2)$. Now,

$f_{xx} = 6x$ $\quad$ $f_{xy} = -6y$ $\quad$ $f_{yy} = -6x + 12$

For the critical point $(0, 0)$:

$f_{xx}(0, 0) \cdot f_{yy}(0, 0) - [f_{xy}(0, 0]^2 = 0 \cdot 12 - 0^2 = 0$

Thus, the second-derivative test fails.

For the critical point $(2, 2)$:

$f_{xx}(2, 2) \cdot f_{yy}(2, 2) - [f_{xy}(2, 2)]^2 = 12 \cdot 0 - (-12)^2 = -144 < 0$

Thus, $f(x, y)$ has a saddle point at $(2, 2)$.

For the critical point $(2, -2)$:

$f_{xx}(2, -2) \cdot f_{yy}(2, -2) - [f_{xy}(2, -2)]^2 = 12 \cdot 0 - (12)^2 = -144 < 0$

Thus, $f(x, y)$ has a saddle point at $(2, -2)$.

19. $f(x, y) = y^3 + 2x^2y^2 - 3x - 2y + 8$;

$f_x = 4xy^2 - 3$; $f_y = 3y^2 + 4x^2y - 2$

Set $f_x = 0$ and $f_y = 0$ to find the critical points:

$$4xy^2 - 3 = 0 \quad\quad (1)$$
$$3y^2 + 4x^2y - 2 = 0 \quad\quad (2)$$

From (1) $x = \dfrac{3}{4y^2}$. Substituting this into (2), we have

$$3y^2 + 4\left(\frac{3}{4y^2}\right)^2 y - 2 = 0$$

$$3y^2 + 4\left(\frac{9}{16y^4}\right)y - 2 = 0$$

$$12y^5 - 8y^3 + 9 = 0$$

Using a graphing utility, we find that $y \approx -1.105$ and $x \approx 0.614$.

Now, $f_{xx} = 4y^2$ and $f_{xx}(0.614, -1.105) \approx 4.884$

$\quad\quad f_{xy} = 8xy$ and $f_{xy}(0.614, -1.105) \approx -5.428$

$\quad\quad f_{yy} = 6y + 4x^2$ and $f_{yy}(0.614, -1.105) \approx -5.122$

$\quad\quad f_{xx}(0.614, -1.105) \cdot f_{yy}(0.614, -1.105) - [f_{xy}(0.614, -1.105)]^2$

$\quad\quad\quad \approx -54.479 < 0$

Thus, $f(x, y)$ has a saddle point at $(0.614, -1.105)$.

21. $f(x, y) = x^2 \geq 0$ for all (x, y) and $f(x, y) = 0$ when $x = 0$. Thus, f has a local minimum at each point $(0, y, 0)$ on the y-axis.

23. $P(x, y) = R(x, y) - C(x, y)$

$$= 2x + 3y - (x^2 - 2xy + 2y^2 + 6x - 9y + 5)$$

$$= -x^2 + 2xy - 2y^2 - 4x + 12y - 5$$

$P_x = -2x + 2y - 4 = 0 \qquad (1)$

$P_y = 2x - 4y + 12 = 0 \qquad (2)$

Solving (1) and (2) for x and y, we obtain $x = 2$ and $y = 4$. Thus, $(2, 4)$ is a critical point.

$P_{xx} = -2$ and $P_{xx}(2, 4) = -2 < 0$

$P_{xy} = 2$ and $P_{xy}(2, 4) = 2$

$P_{yy} = -4$ and $P_{yy}(2, 4) = -4$

$P_{xx}(2, 4) \cdot P_{yy}(2, 4) - [P_{xy}(2, 4)]^2 = (-2)(-4) - [2]^2 = 4 > 0$

The maximum occurs when 2000 type A and 4000 type B calculators are produced. The maximum profit is given by $P(2, 4)$. Hence,

max $P = P(2, 4) = -(2)^2 + 2 \cdot 2 \cdot 4 - 2 \cdot 4^2 - 4 \cdot 2 + 12 \cdot 4 - 5$

$$= -4 + 16 - 32 - 8 + 48 - 5 = \$15 \text{ million.}$$

25. $x = 116 - 30p + 20q$ (Brand A)

$y = 144 + 16p - 24q$ (Brand B)

(A)

p	q	x	y
10	12	56	16
11	11	6	56

(B) In terms of p and q, the cost function C is given by:

$C = 6x + 8y = 6(116 - 30p + 20q) + 8(144 + 16p - 24q)$

$$= 1848 - 52p - 72q$$

The revenue function R is given by:

$R = px + qy = p(116 - 30p + 20q) + q(144 + 16p - 24q)$

$$= 116p - 30p^2 + 20pq + 144q + 16pq - 24q^2$$

$$= -30p^2 + 36pq - 24q^2 + 116p + 144q$$

Thus, the profit $P = R - C$ is given by:

$P = -30p^2 + 36pq - 24q^2 + 116p + 144q - (1848 - 52p - 72q)$

$$= -30p^2 + 36pq - 24q^2 + 168p + 216q - 1848$$

Now, calculating P_p and P_q and setting these equal to 0, we have:

$P_p = -60p + 36q + 168 = 0 \qquad (1)$

$P_q = 36p - 48q + 216 = 0 \qquad (2)$

Solving (1) and (2) for p and q, we get $p = 10$ and $q = 12$. Thus, $(10, 12)$ is a critical point of the profit function P.

$P_{pp} = -60$ and $P_{pp}(10, 12) = -60$

$P_{pq} = 36$ and $P_{pq}(10, 12) = 36$

$P_{qq} = -48$ and $P_{qq}(10, 12) = -48$

$P_{pp} \cdot P_{qq} - [P_{pq}]^2 = (-60)(-48) - (36)^2 = 1584 > 0$

Since $P_{pp}(10, 12) = -60 < 0$, we conclude that the maximum profit occurs when $p = \$10$ and $q = \$12$. The maximum profit is:
$$P(10, 12) = -30(10)^2 + 36(10)(12) - 24(12)^2 + 168(10) + 216(12) - 1848$$
$$= \$288$$

27. The square of the distance from P to A is: $x^2 + y^2$
The square of the distance from P to B is:
$$(x - 2)^2 + (y - 6)^2 = x^2 - 4x + y^2 - 12y + 40$$
The square of the distance from P to C is:
$$(x - 10)^2 + y^2 = x^2 - 20x + y^2 + 100$$
Thus, we have:
$$P(x, y) = 3x^2 - 24x + 3y^2 - 12y + 140$$

$P_x = 6x - 24 = 0$ $\qquad\qquad$ $P_y = 6y - 12 = 0$
$\qquad\qquad x = 4$ $\qquad\qquad\qquad\qquad y = 2$
Therefore, $(4, 2)$ is a critical point.

$P_{xx} = 6$ and $P_{xx}(4, 2) = 6 > 0$
$P_{xy} = 0$ and $P_{xy}(4, 2) = 0$
$P_{yy} = 6$ and $P_{yy}(4, 2) = 6$
$P_{xx} \cdot P_{yy} - [P_{xy}]^2 = 6 \cdot 6 - 0 = 36 > 0$
Therefore, P has a minimum at the point $(4, 2)$.

29. Let $x =$ length, $y =$ width, and $z =$ height. Then $V = xyz = 64$ or $z = \frac{64}{xy}$.
The surface area of the box is:

$$S = xy + 2xz + 4yz \quad \text{or} \quad S(x, y) = xy + \frac{128}{y} + \frac{256}{x}, \ x > 0, \ y > 0$$

$S_x = y - \dfrac{256}{x^2} = 0 \quad$ or $\quad y = \dfrac{256}{x^2} \qquad$ (1)

$S_y = x - \dfrac{128}{y^2} = 0 \quad$ or $\quad x = \dfrac{128}{y^2}$

Thus, $y = \dfrac{256}{\dfrac{(128)^2}{y^4}} \quad$ or $\quad y^4 - 64y = 0$

$\qquad\qquad\qquad\qquad y(y^3 - 64) = 0 \qquad$ (Since $y > 0$, $y = 0$ does not
$\qquad\qquad$ and $\qquad\qquad y = 4 \qquad$ yield a critical point.)

Setting $y = 4$ in (1), we find $x = 8$. Therefore, the critical point is $(8, 4)$.
Now we have:
$S_{xx} = \dfrac{512}{x^3}$ and $S_{xx}(8, 4) = 1 > 0$
$S_{xy} = 1$
$S_{yy} = \dfrac{256}{y^3}$ and $S_{yy}(8, 4) = 4$
$S_{xx}(8, 4) \cdot S_{yy}(8, 4) - [S_{xy}(8, 4)]^2 = 1 \cdot 4 - 1^2 = 3 > 0$

Thus, the dimensions that will require the least amount of material are:
Length $x = 8$ inches; Width $y = 4$ inches; Height $z = \dfrac{64}{8(4)} = 2$ inches.

31. Let x = length of the package, y = width, and z = height. Then
$x + 2y + 2z = 120$ (1)
Volume = $V = xyz$.

From (1), $z = \dfrac{120 - x - 2y}{2}$. Thus, we have:

$$V(x, y) = xy\left(\frac{120 - x - 2y}{2}\right) = 60xy - \frac{x^2 y}{2} - xy^2, \quad x > 0, \ y > 0$$

$V_x = 60y - xy - y^2 = 0$

    $y(60 - x - y) = 0$

        $60 - x - y = 0$ (2)   (Since $y > 0$, $y = 0$ does not yield a critical point.)

$V_y = 60x - \dfrac{x^2}{2} - 2xy = 0$

    $x\left(60 - \dfrac{x}{2} - 2y\right) = 0$

        $120 - x - 4y = 0$ (3)   (Since $x > 0$, $x = 0$ does not yield a critical point.)

Solving (2) and (3) for x and y, we obtain $x = 40$ and $y = 20$. Thus, (40, 20) is the critical point.

$V_{xx} = -y$ and $V_{xx}(40, 20) = -20 < 0$
$V_{xy} = 60 - x - 2y$ and $V_{xy}(40, 20) = 60 - 40 - 40 = -20$
$V_{yy} = -2x$ and $V_{yy}(40, 20) = -80$

$$V_{xx}(40, 20) \cdot V_{yy}(40, 20) - [V_{xy}(40, 20)]^2 = (-20)(-80) - [-20]^2$$
$$= 1600 - 400$$
$$= 1200 > 0$$

Thus, the maximum volume of the package is obtained when $x = 40$, $y = 20$, and $z = \dfrac{120 - 40 - 2 \cdot 20}{2} = 20$ inches. The package has dimensions:

Length $x = 40$ inches; Width $y = 20$ inches; Height $z = 20$ inches.

EXERCISE 8-4

Things to remember:

1. Any local maxima or minima of the function $z = f(x, y)$ subject to the constraint $g(x, y) = 0$ will be among those points (x_0, y_0) for which (x_0, y_0, λ_0) is a solution to the system:

        $F_x(x, y, \lambda) = 0$
        $F_y(x, y, \lambda) = 0$
        $F_\lambda(x, y, \lambda) = 0$

where $F(x, y, \lambda) = f(x, y) + \lambda g(x, y)$, provided all the partial derivatives exist.

2. METHOD OF LAGRANGE MULTIPLIERS

[Note: Although stated for functions of two variables, the method also applies to functions of three or more variables.]

(a) Formulate the problem in the form:
   Maximize (or Minimize) $z = f(x, y)$
   Subject to: $g(x, y) = 0$

(b) Form the function F:
   $$F(x, y, \lambda) = f(x, y) + \lambda g(x, y)$$

(c) Find the critical points (x_0, y_0, λ_0) for F, that is, solve the system:
   $$F_x(x, y, \lambda) = 0$$
   $$F_y(x, y, \lambda) = 0$$
   $$F_\lambda(x, y, \lambda) = 0$$

(d) If (x_0, y_0, λ_0) is the only critical point of F, then assume that (x_0, y_0) is the solution to the problem. If F has more than one critical point, then evaluate $z = f(x, y)$ at (x_0, y_0) for each critical point (x_0, y_0, λ_0) of F. Assume that the largest of these values is the maximum value of $f(x, y)$ subject to the constraint $g(x, y) = 0$, and the smallest is the minimum value of $f(x, y)$ subject to the constraint $g(x, y) = 0$.

---

1. Step 1. Maximize $f(x, y) = 2xy$
   Subject to: $g(x, y) = x + y - 6 = 0$

Step 2. $F(x, y, \lambda) = f(x, y) + \lambda g(x, y)$
   $$= 2xy + \lambda(x + y - 6)$$

Step 3. $F_x = 2y + \lambda = 0$ (1)
   $F_y = 2x + \lambda = 0$ (2)
   $F_\lambda = x + y - 6 = 0$ (3)

   From (1) and (2), we obtain:
   $$x = -\frac{\lambda}{2}, \; y = -\frac{\lambda}{2}$$
   Substituting these into (3), we have:
   $$-\frac{\lambda}{2} - \frac{\lambda}{2} - 6 = 0$$
   $$\lambda = -6.$$
   Thus, the critical point is $(3, 3, -6)$.

Step 4. Since $(3, 3, -6)$ is the only critical point for F, we conclude that max $f(x, y) = f(3, 3) = 2 \cdot 3 \cdot 3 = 18$.

3. Step 1. Minimize $f(x, y) = x^2 + y^2$
Subject to: $g(x, y) = 3x + 4y - 25 = 0$

Step 2. $F(x, y, \lambda) = f(x, y) + \lambda g(x, y)$
$\qquad\qquad = x^2 + y^2 + \lambda(3x + 4y - 25)$

Step 3. $F_x = 2x + 3\lambda = 0 \qquad$ (1)
$F_y = 2y + 4\lambda = 0 \qquad$ (2)
$F_\lambda = 3x + 4y - 25 = 0$ (3)

From (1) and (2), we obtain:

$x = -\dfrac{3\lambda}{2}, \; y = -2\lambda$

Substituting these into (3), we have:

$3\left(-\dfrac{3\lambda}{2}\right) + 4(-2\lambda) - 25 \; = 0$

$\qquad\qquad \dfrac{25}{2}\lambda \; = -25$

$\qquad\qquad \lambda \; = -2$

The critical point is $(3, 4, -2)$.

Step 4. Since $(3, 4, -2)$ is the only critical point for F, we conclude
that min $f(x, y) = f(3, 4) = 3^2 + 4^2 = 25$.

5. Step 1. Maximize and minimize $f(x, y) = 2xy$
Subject to: $g(x, y) = x^2 + y^2 - 18 = 0$

Step 2. $F(x, y, \lambda) = f(x, y) + \lambda g(x, y)$
$\qquad\qquad = 2xy + \lambda(x^2 + y^2 - 18)$

Step 3. $F_x = 2y + 2\lambda x = 0 \qquad$ (1)
$F_y = 2x + 2\lambda y = 0 \qquad$ (2)
$F_\lambda = x^2 + y^2 - 18 = 0 \qquad$ (3)

From (1), (2), and (3), we obtain the critical points
$(3, 3, -1)$, $(3, -3, 1)$, $(-3, 3, 1)$ and $(-3, -3, -1)$.

Step 4. $f(3, 3) = 2\cdot3\cdot3 = 18$
$f(3, -3) = 2\cdot3(-3) = -18$
$f(-3, 3) = 2(-3)\cdot3 = -18$
$f(-3, -3) = 2(-3)(-3) = 18$
Thus, max $f(x, y) = f\,(3, 3) = f(-3, -3) = 18$;
min $f(x, y) = f(3, -3) = f(-3, 3) = -18.$

7. Let x and y be the required numbers.

Step 1. Maximize $f(x, y) = xy$
Subject to: $x + y = 10$ or $g(x, y) = x + y - 10 = 0$

Step 2. $F(x, y, \lambda) = xy + \lambda(x + y - 10)$

Step 3. $F_x = y + \lambda = 0 \qquad$ (1)
$F_y = x + \lambda = 0 \qquad$ (2)
$F_\lambda = x + y - 10 = 0 \qquad$ (3)

From (1) and (2), we obtain:

$x = -\lambda$, $y = -\lambda$

Substituting these into (3), we have:

$\lambda = -5$

The critical point is $(5, 5, -5)$.

Step 4. Since $(5, 5, -5)$ is the only critical point for F, we conclude that max $f(x, y) = f(5, 5) = 5 \cdot 5 = 25$. Thus, the maximum product is 25 when $x = 5$ and $y = 5$.

9. Step 1. Minimize $f(x, y, z) = x^2 + y^2 + z^2$

Subject to: $g(x, y) = 2x - y + 3z + 28 = 0$

Step 2. $F(x, y, z, \lambda) = x^2 + y^2 + z^2 + \lambda(2x - y + 3z + 28)$

Step 3. $F_x = 2x + 2\lambda = 0$ (1)

$F_y = 2y - \lambda = 0$ (2)

$F_z = 2z + 3\lambda = 0$ (3)

$F_\lambda = 2x - y + 3z + 28 = 0$ (4)

From (1), (2), and (3), we obtain:

$x = -\lambda$, $y = \dfrac{\lambda}{2}$, $z = -\dfrac{3}{2}\lambda$

Substituting these into (4), we have:

$$2(-\lambda) - \frac{\lambda}{2} + 3\left(-\frac{3}{2}\lambda\right) + 28 \;\; = 0$$

$$-\frac{14}{2}\lambda + 28 \;\; = 0$$

$$\lambda = 4$$

The critical point is $(-4, 2, -6, 4)$.

Step 4. Since $(-4, 2, -6, 4)$ is the only critical point for F, we conclude that min $f(x, y, z) = f(-4, 2, -6) = 56$.

11. Step 1. Maximize and minimize $f(x, y, z) = x + y + z$

Subject to: $g(x, y, z) = x^2 + y^2 + z^2 - 12 = 0$

Step 2. $F(x, y, z, \lambda) = f(x, y, z) + \lambda g(x, y, z)$

$$= x + y + z + \lambda(x^2 + y^2 + z^2 - 12)$$

Step 3. $F_x = 1 + 2x\lambda = 0$ (1)

$F_y = 1 + 2y\lambda = 0$ (2)

$F_z = 1 + 2z\lambda = 0$ (3)

$F_\lambda = x^2 + y^2 + z^2 - 12 = 0$ (4)

From (1), (2), and (3), we obtain:

$x = -\dfrac{1}{2\lambda}$, $y = -\dfrac{1}{2\lambda}$, $z = -\dfrac{1}{2\lambda}$

Substituting these into (4), we have:

$$\left(-\frac{1}{2\lambda}\right)^2 + \left(-\frac{1}{2\lambda}\right)^2 + \left(-\frac{1}{2\lambda}\right)^2 - 12 \quad = 0$$

$$\frac{3}{4\lambda^2} - 12 = 0$$

$$1 - 16\lambda^2 = 0$$

$$\lambda = \pm\frac{1}{4}$$

Thus, the critical points are $\left(2,\ 2,\ 2,\ -\frac{1}{4}\right)$ and $\left(-2,\ -2,\ -2,\ \frac{1}{4}\right)$.

Step 4. $f(2,\ 2,\ 2) = 2 + 2 + 2 = 6$
$f(-2,\ -2,\ -2) = -2 - 2 - 2 = -6$
Thus, max $f(x,\ y,\ z) = f(2,\ 2,\ 2) = 6$;
   min $f(x,\ y,\ z) = f(-2,\ -2,\ -2) = -6$.

13. Step 1.   Maximize $f(x,\ y) = y + xy^2$
Subject to: $x + y^2 = 1$ or $g(x,\ y) = x + y^2 - 1 = 0$

Step 2. $F(x,\ y,\ \lambda) = y + xy^2 + \lambda(x + y^2 - 1)$

Step 3. $F_x = y^2 + \lambda = 0$ (1)
$F_y = 1 + 2xy + 2y\lambda = 0$ (2)
$F_\lambda = x + y^2 - 1 = 0$ (3)

From (1), $\lambda = -y^2$ and from (3), $x = 1 - y^2$. Substituting these values into (2), we have

$$1 + 2(1 - y^2)y - 2y^3 = 0$$
or $$4y^3 - 2y - 1 = 0$$

Using a graphing utility to solve this equation, we get $y \approx 0.885$. Then $x \approx 0.217$ and max $f(x,\ y) = f(0.217,\ 0.885) \approx 1.055$.

15. The constraint $g(x,\ y) = y - 5 = 0$ implies $y = 5$. Replacing y by 5 in the function f, the problem reduces to maximizing the function $h(x) = f(x,\ 5)$, a function of one independent variable.

17. Step 1.   Minimize cost function $C(x,\ y) = 6x^2 + 12y^2$
Subject to: $x + y = 90$ or $g(x,\ y) = x + y - 90 = 0$

Step 2. $F(x,\ y,\ \lambda) = 6x^2 + 12y^2 + \lambda(x + y - 90)$

Step 3. $F_x = 12x + \lambda = 0$ (1)
$F_y = 24y + \lambda = 0$ (2)
$F_\lambda = x + y - 90 = 0$ (3)
From (1) and (2), we obtain
$$x = -\frac{\lambda}{12},\ y = -\frac{\lambda}{24}$$

Substituting these into (3), we have:

$$-\frac{\lambda}{12} - \frac{\lambda}{24} - 90 = 0$$

$$\frac{3\lambda}{24} = -90$$

$$\lambda = -720$$

The critical point is (60, 30, -720).

Step 4. Since (60, 30, -720) is the only critical point for F, we conclude that:

$$\min C(x, y) = C(60, 30) = 6 \cdot 60^2 + 12 \cdot 30^2$$
$$= 21,600 + 10,800$$
$$= \$32,400$$

Thus, 60 of model A and 30 of model B will yield a minimum cost of \$32,400 per week.

19. (A) Step 1. Maximize the production function $N(x, y) = 50x^{0.8}y^{0.2}$
Subject to the constraint: $C(x, y) = 40x + 80y = 400,000$
i.e., $g(x, y) = 40x + 80y - 400,000 = 0$

Step 2. $F(x, y, \lambda) = 50x^{0.8}y^{0.2} + \lambda(40x + 80y - 400,000)$

Step 3. $F_x = 40x^{-0.2}y^{0.2} + 40\lambda = 0$ (1)
$F_y = 10x^{0.8}y^{-0.8} + 80\lambda = 0$ (2)
$F_\lambda = 40x + 80y - 400,000 = 0$ (3)

From (1), $\lambda = -\dfrac{y^{0.2}}{x^{0.2}}$. From (2), $\lambda = -\dfrac{x^{0.8}}{8y^{0.8}}$.

Thus, we obtain
$$-\frac{y^{0.2}}{x^{0.2}} = -\frac{x^{0.8}}{8y^{0.8}} \quad \text{or} \quad x = 8y$$

Substituting into (3), we have:
$$320y + 80y - 400,000 = 0$$
$$y = 1000$$

Therefore, $x = 8000$, $\lambda \approx -0.6598$, and the critical point is (8000, 1000, -0.6598). Thus, we conclude that:
$$\max N(x, y) = N(8000, 1000) = 50(8000)^{0.8}(1000)^{0.2}$$
$$\approx 263,902 \text{ units}$$

and production is maximized when 8000 labor units and 1000 capital units are used.

(B) The marginal productivity of money is $-\lambda \approx 0.6598$. The increase in production if an additional \$50,000 is budgeted for production is:
$0.6598(50,000) = 32,990$ units

21. Let x = length, y = width, and z = height.

Step 1.   Maximize volume $V = xyz$
                Subject to: $S(x, y, z) = xy + 3xz + 3yz - 192 = 0$

Step 2. $F(x, y, z, \lambda) = xyz + \lambda(xy + 3xz + 3yz - 192)$

Step 3. $F_x = yz + \lambda(y + 3z) = 0 \qquad (1)$
               $F_y = xz + \lambda(x + 3z) = 0 \qquad (2)$
               $F_z = xy + \lambda(3x + 3y) = 0 \qquad (3)$
               $F_\lambda = xy + 3xz + 3yz - 192 = 0 \qquad (4)$
               Solving this system of equations, (1)-(4), simultaneously, yields:
               $x = 8, \ y = 8, \ z = \dfrac{8}{3}, \ \lambda = -\dfrac{4}{3}$
               Thus, the critical point is $\left(8, \ 8, \ \dfrac{8}{3}, \ -\dfrac{4}{3}\right)$.

Step 4.   Since $\left(8, \ 8, \ \dfrac{8}{3}, \ -\dfrac{4}{3}\right)$ is the only critical point for F:

               $\max V(x, y, z) = V\left(8, \ 8, \ \dfrac{8}{3}\right) = \dfrac{512}{3} \approx 170.67$
               Thus, the dimensions that will maximize the volume of the box are: Length $x = 8$ inches; Width $y = 8$ inches; Height $z = \dfrac{8}{3}$ inches.

23. Step 1.   Maximize $A = xy$
                Subject to: $P(x, y) = y + 4x - 400 = 0$

Step 2. $F(x, y, \lambda) = xy + \lambda(y + 4x - 400)$

Step 3. $F_x = y + 4\lambda = 0 \qquad (1)$
               $F_y = x + \lambda = 0 \qquad (2)$
               $F_\lambda = y + 4x - 400 = 0 \qquad (3)$
               From (1) and (2), we have:
               $y = -4\lambda$ and $x = -\lambda$
               Substituting these into (3), we obtain:
               $-4\lambda - 4\lambda - 400 = 0$
               Thus, $\lambda = -50$ and the critical point is $(50, 200, -50)$.

Step 4.   Since $(50, 200, -50)$ is the only critical point for F,
               $\max A(x, y) = A(50, 200) = 10,000$.
               Therefore, $x = 50$ feet, $y = 200$ feet will produce the maximum area $A(50, 200) = 10,000$ square feet.

Things to remember:

<u>1</u>. For a set of n points (x_1, y_1), (x_2, y_2), $\ldots$, (x_n, y_n), the coefficients m and d of the least squares line
   $$y = mx + d$$
   are the solutions of the system of NORMAL EQUATIONS

   $$\left(\sum_{k=1}^{n} x_k\right)m + nd = \sum_{k=1}^{n} y_k \tag{1}$$

   $$\left(\sum_{k=1}^{n} x_k^2\right)m + \left(\sum_{k=1}^{n} x_k\right)d = \sum_{k=1}^{n} x_k y_k$$

   and are given by the formulas

   $$m = \frac{n\left(\sum\limits_{k=1}^{n} x_k y_k\right) - \left(\sum\limits_{k=1}^{n} x_k\right)\left(\sum\limits_{k=1}^{n} y_k\right)}{n\left(\sum\limits_{k=1}^{n} x_k^2\right) - \left(\sum\limits_{k=1}^{n} x_k\right)^2} \tag{2}$$

   and

   $$d = \frac{\sum\limits_{k=1}^{n} y_k - m\left(\sum\limits_{k=1}^{n} x_k\right)}{n} \tag{3}$$

   [<u>Note</u>: To find m and d, either solve system (1) directly, or use formulas (2) and (3). If the formulas are used, the value of m must be calculated first since it is used in formula (3).

1.

x_k	y_k	$x_k y_k$	x_k^2
1	1	1	1
2	3	6	4
3	4	12	9
4	3	12	16
Totals 10	11	31	30

Thus, $\sum\limits_{k=1}^{4} x_k = 10$, $\sum\limits_{k=1}^{4} y_k = 11$, $\sum\limits_{k=1}^{4} x_k y_k = 31$, $\sum\limits_{k=1}^{4} x_k^2 = 30$.

Substituting these values into formulas (2) and (3) for m and d, respectively, we have:

$$m = \frac{n\left(\sum\limits_{k=1}^{n} x_k y_k\right) - \left(\sum\limits_{k=1}^{n} x_k\right)\left(\sum\limits_{k=1}^{n} y_k\right)}{n\left(\sum\limits_{k=1}^{n} x_k^2\right) - \left(\sum\limits_{k=1}^{n} x_k\right)^2} = \frac{4(31) - (10)(11)}{4(30) - (10)^2} = \frac{14}{20} = 0.7$$

$$d = \frac{\sum\limits_{k=1}^{n} y_k - m\left(\sum\limits_{k=1}^{n} x_k\right)}{n} = \frac{11 - 0.7(10)}{4} = 1$$

Thus, the least squares line is $y = mx + d = 0.7x + 1$. Refer to the graph at the right.

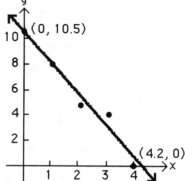

3.

x_k	y_k	$x_k y_k$	x_k^2
1	8	8	1
2	5	10	4
3	4	12	9
4	0	0	16
Totals 10	17	30	30

Thus, $\sum\limits_{k=1}^{4} x_k = 10$, $\sum\limits_{k=1}^{4} y_k = 17$, $\sum\limits_{k=1}^{4} x_k y_k = 30$, $\sum\limits_{k=1}^{4} x_k^2 = 30$.

Substituting these values into system (1), we have:

$10m + 4d = 17$
$30m + 10d = 30$

The solution of this system is $m = -2.5$, $d = 10.5$. Thus, the least squares line is $y = mx + d = -2.5x + 10.5$. Refer to the graph at the right.

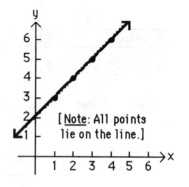

5.

x_k	y_k	$x_k y_k$	x_k^2
1	3	3	1
2	4	8	4
3	5	15	9
4	6	24	16
Totals 10	18	50	30

Thus, $\sum\limits_{k=1}^{4} x_k = 10$, $\sum\limits_{k=1}^{4} y_k = 18$, $\sum\limits_{k=1}^{4} x_k y_k = 50$, $\sum\limits_{k=1}^{4} x_k^2 = 30$.

Substituting these values into the formulas for m and d [formulas (2) and (3)], we have:

$$m = \frac{4(50) - (10)(18)}{4(30) - (10)^2} = \frac{20}{20} = 1$$

$$d = \frac{18 - 1(10)}{4} = \frac{8}{4} = 2$$

Thus, the least squares line is $y = mx + d = x + 2$. Refer to the graph at the right.

7.

	x_k	y_k	$x_k y_k$	x_k^2
	0	10	0	0
	5	22	110	25
	10	31	310	100
	15	46	690	225
	20	51	1020	400
Totals	50	160	2130	750

Thus, $\sum\limits_{k=1}^{5} x_k = 50$, $\sum\limits_{k=1}^{5} y_k = 160$, $\sum\limits_{k=1}^{5} x_k y_k = 2130$, $\sum\limits_{k=1}^{5} x_k^2 = 750$.

Substituting these values into formulas (2) and (3) for m and d, respectively, we have:

$$m = \frac{5(2130) - (50)(160)}{5(750) - (50)^2} = \frac{2650}{1250} = 2.12$$

$$d = \frac{160 - 2.12(50)}{5} = \frac{54}{5} = 10.8$$

Thus, the least squares line is $y = 2.12x + 10.8$.
When $x = 25$, $y = 2.12(25) + 10.8 = 63.8$.

9.

	x_k	y_k	$x_k y_k$	x_k^2
	−1	14	−14	1
	1	12	12	1
	3	8	24	9
	5	6	30	25
	7	5	35	49
Totals	15	45	87	85

Thus, $\sum\limits_{k=1}^{5} x_k = 15$, $\sum\limits_{k=1}^{5} y_k = 45$, $\sum\limits_{k=1}^{5} x_k y_k = 87$, $\sum\limits_{k=1}^{5} x_k^2 = 85$.

Substituting these values into formulas (2) and (3) for m and d, respectively, we have:

$$m = \frac{5(87) - (15)(45)}{5(85) - (15)^2} = \frac{-240}{200} = -1.2$$

$$d = \frac{45 - (-1.2)(15)}{5} = 12.6$$

Thus, the least squares line is
$y = -1.2x + 12.6$.
When $x = 2$, $y = -1.2(2) + 12.6 = 10.2$.

11.

x_k	y_k	$x_k y_k$	x_k^2
0.5	25	12.5	0.25
2.0	22	44.0	4.00
3.5	21	73.5	12.25
5.0	21	105.0	25.00
6.5	18	117.0	42.25
9.5	12	114.0	90.25
11.0	11	121.0	121.00
12.5	8	100.0	156.25
14.0	5	70.0	196.00
15.5	1	15.5	240.25
Totals 80.0	144	772.5	887.50

Thus, $\sum\limits_{k=1}^{10} x_k = 80$, $\sum\limits_{k=1}^{10} y_k = 144$, $\sum\limits_{k=1}^{10} x_k y_k = 772.5$, $\sum\limits_{k=1}^{10} x_k^2 = 887.5$.

Substituting these values into formulas (2) and (3) for m and d, respectively, we have:

$$m = \frac{10(772.5) - (80)(144)}{10(887.5) - (80)^2} = \frac{-3795}{2475} \approx -1.53$$

$$d = \frac{144 - (-1.53)(80)}{10} = \frac{266.4}{10} = 26.64$$

Thus, the least squares line is
$y = -1.53x + 26.64$.
When $x = 8$, $y = -1.53(8) + 26.64 = 14.4$.

13. Minimize
$$F(a, b, c) = (a + b + c - 2)^2 + (4a + 2b + c - 1)^2$$
$$+ (9a + 3b + c - 1)^2 + (16a + 4b + c - 3)^2$$

$$F_a(a, b, c) = 2(a + b + c - 2) + 8(4a + 2b + c - 1)$$
$$+ 18(9a + 3b + c - 1) + 32(16a + 4b + c - 3)$$
$$= 708a + 200b + 60c - 126$$

$$F_b(a, b, c) = 2(a + b + c - 2) + 4(4a + 2b + c - 1)$$
$$+ 6(9a + 3b + c - 1) + 8(16a + 4b + c - 3)$$
$$= 200a + 60b + 20c - 38$$

$$F_c(a, b, c) = 2(a + b + c - 2) + 2(4a + 2b + c - 1)$$
$$+ 2(9a + 3b + c - 1) + 2(16a + 4b + c - 3)$$
$$= 60a + 20b + 8c - 14$$

The system is: $F_a(a, b, c) = 0$
$F_b(a, b, c) = 0$
$F_c(a, b, c) = 0$

or: $708a + 200b + 60c = 126$
$200a + 60b + 20c = 38$
$60a + 20b + 8c = 14$

The solution is $(a, b, c) = (0.75, -3.45, 4.75)$, which gives us the equation for the parabola shown at the right:
$$y = ax^2 + bx + c$$
or
$$y = 0.75x^2 - 3.45x + 4.75$$
The given points: $(1, 2)$, $(2, 1)$, $(3, 1)$, $(4, 3)$ also appear on the graph.

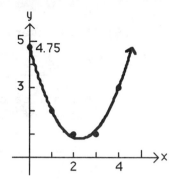

15. System (1) is:

$$\left(\sum_{k=1}^{n} x_k\right)m + nd = \sum_{k=1}^{n} y_k \qquad \text{(a)}$$

$$\left(\sum_{k=1}^{n} x_k^2\right)m + \left(\sum_{k=1}^{n} x_k\right)d = \sum_{k=1}^{n} x_k y_k \qquad \text{(b)}$$

Multiply equation (a) by $-\left(\sum_{k=1}^{n} x_k\right)$, equation (b) by n, and add the resulting equations. This will eliminate d from the system.

$$\left[-\left(\sum_{k=1}^{n} x_k\right)^2 + n\sum_{k=1}^{n} x_k^2\right]m = -\left(\sum_{k=1}^{n} x_k\right)\left(\sum_{k=1}^{n} y_k\right) + n\sum_{k=1}^{n} x_k y_k$$

Thus,

$$m = \frac{n\left(\sum_{k=1}^{n} x_k y_k\right) - \left(\sum_{k=1}^{n} x_k\right)\left(\sum_{k=1}^{n} y_k\right)}{n\left(\sum_{k=1}^{n} x_k^2\right) - \left(\sum_{k=1}^{n} x_k\right)^2}$$

which is equation (2). Solving equation (a) for d, we have

$$d = \frac{\sum_{k=1}^{n} y_k - m\left(\sum_{k=1}^{n} x_k\right)}{n}$$

which is equation (3).

17. (A) Suppose that $n = 5$ and $x_1 = -2$, $x_2 = -1$, $x_3 = 0$, $x_4 = 1$, $x_5 = 2$.

Then $\sum_{k=1}^{5} x_k = -2 - 1 + 0 + 1 + 2 = 0$. Therefore, from formula (2),

$$m = \frac{5\sum_{k=1}^{5} x_k y_k}{5\sum_{k=1}^{5} x_k^2} = \frac{\sum x_k y_k}{\sum x_k^2}$$

From formula (3), $\quad d = \dfrac{\sum_{k=1}^{5} y_k}{5}$,

which is the average of y_1, y_2, y_3, y_4, and y_5.

(B) If the average of the x-coordinates is 0, then

$$\frac{\sum_{k=1}^{n} x_k}{n} = 0$$

Then all calculations will be the same as in part (A) with "n" instead of 5.

19. (A)

x_k	y_k	$x_k y_k$	x_k^2
0	23.8	0	0
1	16.5	16.5	1
2	19.0	38.0	4
3	29.0	87.0	9
4	37.9	151.6	16
5	51.2	256.0	25
6	61.1	366.6	36
Totals 21	238.5	915.7	91

Thus, $\sum_{k=1}^{7} x_k = 21$, $\sum_{k=1}^{7} y_k = 238.5$, $\sum_{k=1}^{7} x_k y_k = 915.7$, $\sum_{k=1}^{7} x_k^2 = 91$.

Substituting these values into the formulas for m and d, we have:

$$m = \frac{7(915.7) - 21(238.5)}{7(91) - (21)^2} = \frac{1401.4}{196} = 7.15$$

$$d = \frac{238.5 - 7.15(21)}{7} = \frac{88.35}{7} \approx 12.62$$

Thus, the least squares line is $y = 7.15x + 12.62$.

(B) 1998 corresponds to $x = 13$. The monthly production in the 13th year will be $y \approx 7.15(13) + 12.62 = 105.57$ or 105.57 thousand per month.

21. (A)

x_k	y_k	$x_k y_k$	x_k^2
5.0	2.0	10.0	25.00
5.5	1.8	9.9	30.25
6.0	1.4	8.4	36.00
6.5	1.2	7.8	42.25
7.0	1.1	7.7	49.00
Totals 30.0	7.5	43.8	182.50

Thus, $\sum_{k=1}^{5} x_k = 30$, $\sum_{k=1}^{5} y_k = 7.5$, $\sum_{k=1}^{5} x_k y_k = 43.8$, $\sum_{k=1}^{5} x_k^2 = 182.5$.

Substituting these values into the formulas for m and d, we have:

$$m = \frac{5(43.8) - (30)(7.5)}{5(182.5) - (30)^2} = \frac{-6}{12.5} = -0.48$$

$$d = \frac{7.5 - (-0.48)(30)}{5} = 4.38$$

Thus, a demand equation is $y = -0.48x + 4.38$.

(B) Cost: $C = 4y$

Revenue: $R = xy = -0.48x^2 + 4.38x$

Profit: $P = R - C = -0.48x^2 + 4.38x - 4(-0.48x + 4.38)$

or $P(x) = -0.48x^2 + 6.3x - 17.52$

Now, $P'(x) = -0.96x + 6.3$.

Critical value: $P'(x) = -0.96x + 6.3 = 0$

$$x = \frac{6.3}{0.96} \approx 6.56$$

$P''(x) = -0.96$ and $P''(6.56) = -0.96 < 0$

Thus, $P(x)$ has a maximum at $x = 6.56$; the price per bottle should be $6.56 to maximize the monthly profit.

23.

x_k	y_k	$x_k y_k$	x_k^2
50	15	750	2500
55	13	715	3025
60	10	600	3600
65	6	390	4225
70	2	140	4900
Totals 300	46	2595	18,250

Thus, $\sum_{k=1}^{5} x_k = 300$, $\sum_{k=1}^{5} y_k = 46$, $\sum_{k=1}^{5} x_k y_k = 2595$, $\sum_{k=1}^{5} x_k^2 = 18,250$.

Substitituting these values into the formulas for m and d, we have:

$$m = \frac{5(2595) - (300)(46)}{5(18,250) - (300)^2} = \frac{-825}{1250} = -0.66$$

$$d = \frac{46 - (-0.66)300}{5} = 48.8$$

(A) The least squares line for the data is $P = -0.66T + 48.8$.

(B) $P(57) = -0.66(57) + 48.8 = 11.18$ beats per minute.

25.

x_k	y_k	$x_k y_k$	x_k^2
1.7	51	86.7	2.89
2.1	49	102.9	4.41
2.3	53	121.9	5.29
2.4	36	86.4	5.76
3.6	65	234.0	12.96
3.7	35	129.5	13.69
4.7	29	136.3	22.09
6.2	40	248.0	38.44
7.1	34	241.4	50.41
7.4	29	214.6	54.76
8.7	20	174.0	75.69
11.9	23	273.7	141.61
Totals 61.8	464	2049.4	428.00

Thus, $\sum_{k=1}^{12} x_k = 61.8$, $\sum_{k=1}^{12} y_k = 464$, $\sum_{k=1}^{12} x_k y_k = 2049.4$, $\sum_{k=1}^{12} x_k^2 = 428$.

Substitituting these values into the formulas for m and d, we have:

$$m = \frac{12(2049.4) - (61.8)(464)}{12(428) - (61.8)^2} = \frac{-4082.4}{1316.76} = -3.1$$

$$d = \frac{464 - (-3.1)(61.8)}{12} = \frac{655.58}{12} = 54.6$$

(A) The least squares line for the data is $D = -3.1A + 54.6$.

(B) If $A = 3.0$, then $D = -3.1(3.0) + 54.6 \approx 45$ or 45%.

27. (A) Enter the data in a calculator or computer. (We used a TI-85.) The totals are:

$$n = 23, \qquad \sum x = 1098, \qquad \sum y = 343.61,$$
$$\sum x^2 = 73,860, \qquad \sum xy = 18,259.08$$

Now, the least squares line can be calculated either by using formulas (2) and (3), or by using the linear regression feature. We used the latter to get

$$m = 0.08653 \quad \text{and} \quad b = 10.81$$

Therefore, the least squares line is:
$$y = 0.08653x + 10.81$$

(B) Using the result in (A), an estimate for the winning height in the pole vault in the Olympic games of 2008 is:

$$y = 0.08653(112) + 10.81 \approx 20.50 \text{ feet}$$

EXERCISE 8-6

Things to remember:

GIVEN A FUNCTION $z = f(x, y)$:

1. $\int f(x, y)\,dx$ means antidifferentiate $f(x, y)$ with respect to x, holding y fixed.

   $\int f(x, y)\,dy$ means antidifferentiate $f(x, y)$ with respect to y, holding x fixed.

2. The DOUBLE INTEGRAL of $f(x, y)$ over the rectangle $R = \{(x, y) \mid a \le x \le b, c \le y \le d\}$ is:

$$\iint\limits_{R} f(x, y)\,dA = \int_a^b \left[\int_c^d f(x, y)\,dy \right] dx$$

$$= \int_c^d \left[\int_a^b f(x, y)\,dx \right] dy$$

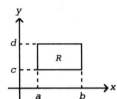

3. The AVERAGE VALUE of $f(x, y)$ over the rectangle
$R = \{(x, y) \mid a \le x \le b, c \le y \le d\}$ is:

$$\frac{1}{(b - a)(d - c)} \iint\limits_R f(x, y)\,dA$$

4. If $f(x, y) \ge 0$ over a rectangle
$R = \{(x, y) \mid a \le x \le b, c \le y \le d\}$,

then the VOLUME of the solid formed by graphing f over R is given by:

$$V = \iint\limits_R f(x, y)\,dA$$

---

1. (A) $\displaystyle\int 12x^2y^3\,dy = 12x^2\int y^3\,dy$ (x is treated as a constant.)

$$= 12x^2\,\frac{y^4}{4} + C(x) \quad \begin{array}{l}\text{(The "constant" of integration}\\\text{is a function of } x.)\end{array}$$

$$= 3x^2y^4 + C(x)$$

(B) $\displaystyle\int_0^1 12x^2y^3\,dy = 3x^2y^4 \Big|_0^1 = 3x^2$

3. (A) $\displaystyle\int (4x + 6y + 5)\,dx$

$$= \int 4x\,dx + \int (6y + 5)\,dx \quad \text{(y is treated as a constant.)}$$

$$= 2x^2 + (6y + 5)x + E(y) \quad \begin{array}{l}\text{(The "constant" of integration}\\\text{is a function of } y.)\end{array}$$

$$= 2x^2 + 6xy + 5x + E(y)$$

(B) $\displaystyle\int_{-2}^3 (4x + 6y + 5)\,dx = (2x^2 + 6xy + 5x)\Big|_{-2}^3$

$$= 2\cdot3^2 + 6\cdot3y + 5\cdot3 - [2(-2)^2 + 6(-2)y + 5(-2)]$$
$$= 30y + 35$$

5. (A) $\displaystyle\int \frac{x}{\sqrt{y + x^2}}\,dx = \int (y + x^2)^{-1/2}x\,dx = \frac{1}{2}\int (y + x^2)^{-1/2}2x\,dx$

Let $u = y + x^2$,

then $du = 2x\,dx$.

$$= \frac{1}{2}\int u^{-1/2}\,du$$

$$= u^{1/2} + E(y) = \sqrt{y + x^2} + E(y)$$

(B) $\displaystyle\int_0^2 \frac{x}{\sqrt{y + x^2}}\,dx = \sqrt{y + x^2}\Big|_0^2 = \sqrt{y + 4} - \sqrt{y}$

7. $\displaystyle\int_{-1}^{2}\int_{0}^{1} 12x^2y^3\,dy\ dx = \int_{-1}^{2}\left[\int_{0}^{1} 12x^2y^3\,dy\right]dx = \int_{-1}^{2} 3x^2\,dx$ (see Problem 1)

$$= x^3\Big|_{-1}^{2} = 8 + 1 = 9$$

9. $\displaystyle\int_{1}^{4}\int_{-2}^{3}(4x+6y+5)\,dx\ dy = \int_{1}^{4}\left[\int_{-2}^{3}(4x+6y+5)\,dx\right]dy$

$$= \int_{1}^{4}(30y+35)\,dy \quad \text{(see Problem 3)}$$

$$= (15y^2 + 35y)\Big|_{1}^{4}$$

$$= 15\cdot 4^2 + 35\cdot 4 - (15+35) = 330$$

11. $\displaystyle\int_{1}^{5}\int_{0}^{2}\frac{x}{\sqrt{y+x^2}}\,dx\ dy = \int_{1}^{5}\left[\int_{0}^{2}\frac{x}{\sqrt{y+x^2}}\,dx\right]dy$

$$= \int_{1}^{5}(\sqrt{4+y} - \sqrt{y})\,dy \quad \text{(see Problem 5)}$$

$$= \left[\frac{2}{3}(4+y)^{3/2} - \frac{2}{3}y^{3/2}\right]\Big|_{1}^{5}$$

$$= \frac{2}{3}(9)^{3/2} - \frac{2}{3}(5)^{3/2} - \left(\frac{2}{3}\cdot 5^{3/2} - \frac{2}{3}\cdot 1^{3/2}\right)$$

$$= 18 - \frac{4}{3}(5)^{3/2} + \frac{2}{3} = \frac{56 - 20\sqrt{5}}{3}$$

13. $\displaystyle\iint_{R} xy\ dA = \int_{0}^{2}\int_{0}^{4} xy\ dy\ dx = \int_{0}^{2}\left[\int_{0}^{4} xy\ dy\right]dx = \int_{0}^{2}\left[\frac{xy^2}{2}\Big|_{0}^{4}\right]dx$

$$= \int_{0}^{2} 8x\ dx = 4x^2\Big|_{0}^{2} = 16$$

$\displaystyle\iint_{R} xy\ dA = \int_{0}^{4}\int_{0}^{2} xy\ dy\ dx = \int_{0}^{4}\left[\int_{0}^{2} xy\ dx\right]dy = \int_{0}^{4}\left[\frac{x^2y}{2}\Big|_{0}^{2}\right]dy$

$$= \int_{0}^{4} 2y\ dy = y^2\Big|_{0}^{4} = 16$$

15. $\displaystyle\iint_{R}(x+y)^5\,dA = \int_{-1}^{1}\int_{1}^{2}(x+y)^5\,dy\ dx = \int_{-1}^{1}\left[\int_{1}^{2}(x+y)^5\,dy\right]dx$

$$= \int_{-1}^{1}\left[\frac{(x+y)^6}{6}\Big|_{1}^{2}\right]dx = \int_{-1}^{1}\left[\frac{(x+2)^6}{6} - \frac{(x+1)^6}{6}\right]dx$$

$$= \left[\frac{(x+2)^7}{42} - \frac{(x+1)^7}{42}\right]\Big|_{-1}^{1} = \frac{3^7}{42} - \frac{2^7}{42} - \frac{1}{42} = 49$$

$$\iint\limits_R (x + y)^5 dA = \int_1^2 \int_{-1}^1 (x + y)^5 dx\ dy = \int_1^2 \left[\int_{-1}^1 (x + y)^5 dx\right] dy$$

$$= \int_1^2 \left[\frac{(x + y)^6}{6}\bigg|_{-1}^1\right] dy = \int_1^2 \left[\frac{(y + 1)^6}{6} - \frac{(y - 1)^6}{6}\right] dy$$

$$= \left[\frac{(y + 1)^7}{42} - \frac{(y - 1)^7}{42}\right]\bigg|_1^2 = \frac{3^7}{42} - \frac{1}{42} - \frac{2^7}{42} = 49$$

17. Average value $= \dfrac{1}{(5 - 1)[1 - (-1)]} \iint\limits_R (x + y)^2 dA$

$$= \frac{1}{8}\int_{-1}^1 \int_1^5 (x + y)^2 dx\ dy = \frac{1}{8}\int_{-1}^1 \left[\frac{(x + y)^3}{3}\bigg|_1^5\right] dy$$

$$= \frac{1}{8}\int_{-1}^1 \left[\frac{(5 + y)^3}{3} - \frac{(1 + y)^3}{3}\right] dy = \frac{1}{8}\left[\frac{(5 + y)^4}{12} - \frac{(1 + y)^4}{12}\right]\bigg|_{-1}^1$$

$$= \frac{1}{96}[6^4 - 2^4 - 4^4] = \frac{32}{3}$$

19. Average value $= \dfrac{1}{(4 - 1)(7 - 2)} \iint\limits_R \dfrac{x}{y} dA = \dfrac{1}{15}\int_1^4 \int_2^7 \dfrac{x}{y} dy\ dx$

$$= \frac{1}{15}\int_1^4 \left[x \ln y\right]_2^7 dx = \frac{1}{15}\int_1^4 [x \ln 7 - x \ln 2] dx$$

$$= \frac{\ln 7 - \ln 2}{15}\int_1^4 x\ dx = \frac{\ln 7 - \ln 2}{15}\cdot\frac{x^2}{2}\bigg|_1^4$$

$$= \frac{\ln 7 - \ln 2}{15}\left(\frac{4^2}{2} - \frac{1^2}{2}\right) = \frac{1}{2}(\ln 7 - \ln 2)$$

$$= \frac{1}{2}\ln\left(\frac{7}{2}\right) \approx 0.626$$

21. $V = \iint\limits_R (2 - x^2 - y^2) dA = \int_0^1 \int_0^1 (2 - x^2 - y^2) dy\ dx$

$$= \int_0^1 \left[\int_0^1 (2 - x^2 - y^2) dy\right] dx = \int_0^1 \left[\left(2y - x^2 y - \frac{y^3}{3}\right)\bigg|_0^1\right] dx$$

$$= \int_0^1 \left(2 - x^2 - \frac{1}{3}\right) dx = \int_0^1 \left(\frac{5}{3} - x^2\right) dx = \left(\frac{5}{3}x - \frac{x^3}{3}\right)\bigg|_0^1 = \frac{5}{3} - \frac{1}{3} = \frac{4}{3}$$

23. $V = \iint\limits_R (4 - y^2) dA = \int_0^2 \int_0^2 (4 - y^2) dx\ dy = \int_0^2 \left[\int_0^2 (4 - y^2) dx\right] dy$

$$= \int_0^2 \left[(4x - xy^2)\bigg|_0^2\right] dy = \int_0^2 (8 - 2y^2) dy = \left(8y - \frac{2}{3}y^3\right)\bigg|_0^2 = 16 - \frac{16}{3} = \frac{32}{3}$$

25. $\displaystyle\iint\limits_{R} xe^{xy} dA = \int_0^1 \int_1^2 xe^{xy} dy\ dx = \int_0^1 \left[\int_1^2 xe^{xy} dy\right] dx$

$\displaystyle = \int_0^1 \left[x\int_1^2 e^{xy} dy\right] dx = \int_0^1 \left[x \cdot \frac{e^{xy}}{x}\Big|_1^2\right] dx = \int_0^1 \left[e^{xy}\Big|_1^2\right] dx$

$\displaystyle = \int_0^1 (e^{2x} - e^x)\ dx = \left(\frac{e^{2x}}{2} - e^x\right)\Big|_0^1 = \frac{e^2}{2} - e - \left(\frac{1}{2} - 1\right)$

$\displaystyle = \frac{e^2}{2} - e + \frac{1}{2}$

27. $\displaystyle\iint\limits_{R} \frac{2y + 3xy^2}{1 + x^2} dA = \int_0^1 \int_{-1}^1 \frac{2y + 3xy^2}{1 + x^2} dy\ dx = \int_0^1 \left[\int_{-1}^1 \frac{2y + 3xy^2}{1 + x^2} dy\right] dx$

$\displaystyle = \int_0^1 \left[\frac{1}{1 + x^2}(y^2 + xy^3)\Big|_{-1}^1\right] dx$

$\displaystyle = \int_0^1 \left[\frac{1}{1 + x^2}(1 + x - [1 - x])\right] dx$

$\displaystyle = \int_0^1 \frac{2x}{1 + x^2} dx = \ln(1 + x^2)\Big|_0^1 \qquad$ Substitution: $u = 1 + x^2$
$\qquad\qquad\qquad\qquad\qquad\qquad\qquad\qquad\qquad\qquad\qquad\qquad\qquad du = 2x\ dx$

$= \ln 2$

29. $\displaystyle\int_0^2 \int_0^2 (1 - y)\ dx\ dy = \int_0^2 \left[\int_0^2 (1 - y)\ dx\right] dy$

$\displaystyle = \int_0^2 \left[(x - xy)\Big|_0^2\right] dy$

$\displaystyle = \int_0^2 (2 - 2y)\ dy$

$\displaystyle = (2y - y^2)\Big|_0^2 = 0$

Since $f(x, y) = 1 - y$ is NOT nonnegative over the rectangle $R = \{(x, y) \mid 0 \le x \le 2,\ 0 \le y \le 2\}$ the double integral does not represent the volume of solid.

31. $f(x, y) = x^3 + y^2 - e^{-x} - 1$ on $R = \{(x, y) \mid -2 \le x \le 2,\ -2 \le y \le 2\}$.

(A) Average value of f:

$\displaystyle\frac{1}{b - a} \cdot \frac{1}{d - c}\iint\limits_{R} f(x, y)\ dA$

$\displaystyle = \frac{1}{2 - (-2)} \cdot \frac{1}{2 - (-2)}\int_{-2}^2 \int_{-2}^2 (x^3 + y^2 - e^{-x} - 1)\ dx\ dy$

$\displaystyle = \frac{1}{16}\int_{-2}^2 \left[\left(\frac{1}{4}x^4 + xy^2 + e^{-x} - x\right)\Big|_{-2}^2\right] dy$

$$= \frac{1}{16} \int_{-2}^{2} [4y^2 + e^{-2} - e^2 - 4] dy$$

$$= \frac{1}{16} \left[\frac{4}{3} y^3 + e^{-2}y - e^2 y - 4y \right] \Big|_{-2}^{2}$$

$$= \frac{1}{16} \left[\frac{64}{3} + 4e^{-2} - 4e^2 - 16 \right] = \frac{1}{3} + \frac{1}{4} e^{-2} - \frac{1}{4} e^2$$

(B)

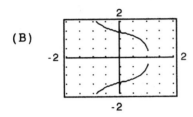

(C) $f(x, y) > 0$ at the points which lie to the right of the curve in part (B); $f(x, y) < 0$ at the points which lie to the left of the curve in part (B).

33. $S(x, y) = \frac{y}{1 - x}$, $0.6 \leq x \leq 0.8$, $5 \leq y \leq 7$.

The *average* total amount of spending is given by:

$$T = \frac{1}{(0.8 - 0.6)(7 - 5)} \iint_R \frac{y}{1 - x} dA = \frac{1}{0.4} \int_{0.6}^{0.8} \int_{5}^{7} \frac{y}{1 - x} dy \ dx$$

$$= \frac{1}{0.4} \int_{0.6}^{0.8} \left[\frac{1}{1 - x} \cdot \frac{y^2}{2} \Big|_{5}^{7} \right] dx = \frac{1}{0.4} \int_{0.6}^{0.8} \frac{1}{1 - x} \left(\frac{49}{2} - \frac{25}{2} \right) dx$$

$$= \frac{12}{0.4} \int_{0.6}^{0.8} \frac{1}{1 - x} dx = 30 \left[-\ln(1 - x) \right] \Big|_{0.6}^{0.8}$$

$$= 30[-\ln(0.2) + \ln(0.4)] = 30 \ln 2 \approx \$20.8 \text{ billion}$$

35. $N(x, y) = x^{0.75} y^{0.25}$, $10 \leq x \leq 20$, $1 \leq y \leq 2$

$$\text{Average value} = \frac{1}{(20 - 10)(2 - 1)} \int_{10}^{20} \int_{1}^{2} x^{0.75} y^{0.25} dy \ dx$$

$$= \frac{1}{10} \int_{10}^{20} \left[x^{0.75} \frac{y^{1.25}}{1.25} \Big|_{1}^{2} \right] dx = \frac{1}{10} \int_{10}^{20} \left[x^{0.75} \frac{2^{1.25} - 1}{1.25} \right] dx$$

$$= \frac{1}{12.5} (2^{1.25} - 1) \int_{10}^{20} x^{0.75} dx = \frac{1}{12.5} (2^{1.25} - 1) \frac{x^{1.75}}{1.75} \Big|_{10}^{20}$$

$$= \frac{1}{21.875} (2^{1.25} - 1)(20^{1.75} - 10^{1.75}) \approx 8.375 \text{ or } 8375 \text{ items}$$

37. $C = 10 - \frac{1}{10}d^2 = 10 - \frac{1}{10}(x^2 + y^2) = C(x, y)$, $-8 \leq x \leq 8$, $-6 \leq y \leq 6$

Average concentration $= \frac{1}{16(12)}\int_{-8}^{8}\int_{-6}^{6}\left[10 - \frac{1}{10}(x^2 + y^2)\right]dy\,dx$

$\qquad = \frac{1}{192}\int_{-8}^{8}\left[10y - \frac{1}{10}\left(x^2 y + \frac{y^3}{3}\right)\right]\Big|_{-6}^{6}dx$

$\qquad = \frac{1}{192}\int_{-8}^{8}\left\{60 - \frac{1}{10}\left(6x^2 + \frac{216}{3}\right) - \left[-60 - \frac{1}{10}\left(-6x^2 - \frac{216}{3}\right)\right]\right\}dx$

$\qquad = \frac{1}{192}\int_{-8}^{8}\left[120 - \frac{1}{10}(12x^2 + 144)\right]dx$

$\qquad = \frac{1}{192}\left[120x - \frac{1}{10}(4x^3 + 144x)\right]\Big|_{-8}^{8}$

$\qquad = \frac{1}{192}(1280) = \frac{20}{3} \approx 6.67$ insects per square foot

39. $C = 100 - 15d^2 = 100 - 15(x^2 + y^2) = C(x, y)$, $-2 \leq x \leq 2$, $-1 \leq y \leq 1$

Average concentration $= \frac{1}{4(2)}\int_{-2}^{2}\int_{-1}^{1}[100 - 15(x^2 + y^2)]dy\,dx$

$\qquad = \frac{1}{8}\int_{-2}^{2}(100y - 15x^2 y - 5y^3)\Big|_{-1}^{1}dx$

$\qquad = \frac{1}{8}\int_{-2}^{2}(190 - 30x^2)dx = \frac{1}{8}(190x - 10x^3)\Big|_{-2}^{2}$

$\qquad = \frac{1}{8}(600) = 75$ parts per million

41. $L = 0.0000133xy^2$, $2000 \leq x \leq 3000$, $50 \leq y \leq 60$

Average length $= \frac{1}{10,000}\int_{2000}^{3000}\int_{50}^{60}0.0000133xy^2 dy\,dx$

$\qquad = \frac{0.0000133}{10,000}\int_{2000}^{3000}\left[\frac{xy^3}{3}\Big|_{50}^{60}\right]dx$

$\qquad = \frac{0.0000133}{10,000}\int_{2000}^{3000}\frac{91,000}{3}x\,dx = \frac{1.2103}{30,000}\cdot\frac{x^2}{2}\Big|_{2000}^{3000}$

$\qquad = \frac{1.2103}{60,000}(5,000,000) \approx 100.86$ feet

43. $Q(x, y) = 100\left(\frac{x}{y}\right)$, $8 \leq x \leq 16$, $10 \leq y \leq 12$

Average intelligence $= \frac{1}{16}\int_{8}^{16}\int_{10}^{12}100\left(\frac{x}{y}\right)dy\,dx = \frac{100}{16}\int_{8}^{16}\left[x\ln y\Big|_{10}^{12}\right]dx$

$\qquad = \frac{100}{16}\int_{8}^{16}x(\ln 12 - \ln 10)dx$

$\qquad = \frac{100(\ln 12 - \ln 10)}{16}\cdot\frac{x^2}{2}\Big|_{8}^{16}$

$\qquad = \frac{100(\ln 12 - \ln 10)}{32}(192)$

$\qquad = 600\ln(1.2) \approx 109.4$

1. $f(x, y) = 2000 + 40x + 70y$

$f(5, 10) = 2000 + 40 \cdot 5 + 70 \cdot 10 = 2900$

$f_x(x, y) = 40$

$f_y(x, y) = 70$ $\hspace{3cm}$ (8-1, 8-2)

2. $z = x^3 y^2$

$\dfrac{\partial z}{\partial x} = 3x^2 y^2$

$\dfrac{\partial^2 z}{\partial x^2} = \dfrac{\partial \left(\frac{\partial z}{\partial x} \right)}{\partial x} = \dfrac{\partial (3x^2 y^2)}{\partial x} = 6xy^2$

$\dfrac{\partial z}{\partial y} = 2x^3 y$

$\dfrac{\partial^2 z}{\partial x \partial y} = \dfrac{\partial \left(\frac{\partial z}{\partial y} \right)}{\partial x} = \dfrac{\partial (2x^3 y)}{\partial x} = 6x^2 y$ $\hspace{2cm}$ (8-2)

3. $\displaystyle \int (6xy^2 + 4y)\, dy = 6x \int y^2 dy + 4 \int y\, dy = 6x \cdot \dfrac{y^3}{3} + 4 \cdot \dfrac{y^2}{2} + C(x)$

$\hspace{5cm} = 2xy^3 + 2y^2 + C(x)$ $\hspace{2cm}$ (8-6)

4. $\displaystyle \int (6xy^2 + 4y)\, dx = 6y^2 \int x\, dx + 4y \int dx = 6y^2 \cdot \dfrac{x^2}{2} + 4yx + E(y)$

$\hspace{5cm} = 3x^2 y^2 + 4xy + E(y)$ $\hspace{2cm}$ (8-6)

5. $\displaystyle \int_0^1 \int_0^1 4xy\, dy\, dx = \int_0^1 \left[\int_0^1 4xy\, dy \right] dx = \int_0^1 \left[2xy^2 \Big|_0^1 \right] dx$

$\hspace{4cm} = \displaystyle \int_0^1 2x\, dx = x^2 \Big|_0^1 = 1$ $\hspace{2cm}$ (8-6)

6. $f(x, y) = 3x^2 - 2xy + y^2 - 2x + 3y - 7$

$f(2, 3) = 3 \cdot 2^2 - 2 \cdot 2 \cdot 3 + 3^2 - 2 \cdot 2 + 3 \cdot 3 - 7 = 7$

$f_y(x, y) = -2x + 2y + 3$

$f_y(2, 3) = -2 \cdot 2 + 2 \cdot 3 + 3 = 5$ $\hspace{3cm}$ (8-1, 8-2)

7. $f(x, y) = -4x^2 + 4xy - 3y^2 + 4x + 10y + 81$

$f_x(x, y) = -8x + 4y + 4$ $\hspace{3cm}$ $f_y(x, y) = 4x - 6y + 10$

$f_{xx}(x, y) = -8$ $\hspace{4.5cm}$ $f_{yy}(x, y) = -6$

$f_{xy}(x, y) = 4$

Now, $f_{xx}(2, 3) \cdot f_{yy}(2, 3) - [f_{xy}(2, 3)]^2 = (-8)(-6) - 4^2 = 32$. $\hspace{1cm}$ (8-2)

8. $f(x, y) = x + 3y$ and $g(x, y) = x^2 + y^2 - 10$.

Let $F(x, y, \lambda) = f(x, y) + \lambda g(x, y) = x + 3y + \lambda(x^2 + y^2 - 10)$.

Then, we have:

$F_x = 1 + 2x\lambda$

$F_y = 3 + 2y\lambda$

$F_\lambda = x^2 + y^2 - 10$

Setting $F_x = F_y = F_\lambda = 0$, we obtain:

$1 + 2x\lambda = 0$ (1)

$3 + 2y\lambda = 0$ (2)

$x^2 + y^2 - 10 = 0$ (3)

From the first equation, $x = -\dfrac{1}{2\lambda}$; from the second equation, $y = -\dfrac{3}{2\lambda}$.

Substituting these into the third equation gives:

$$\frac{1}{4\lambda^2} + \frac{9}{4\lambda^2} - 10 = 0$$

$$40\lambda^2 = 10$$

$$\lambda^2 = \frac{1}{4}$$

$$\lambda = \pm\frac{1}{2}$$

Thus, the critical points are $\left(-1,\ -3,\ \dfrac{1}{2}\right)$ and $\left(1,\ 3,\ -\dfrac{1}{2}\right)$.   (8-4)

9.

x_k	y_k	$x_k y_k$	x_k^2
2	12	24	4
4	10	40	16
6	7	42	36
8	3	24	64
Totals 20	32	130	120

Thus, $\displaystyle\sum_{k=1}^{4} x_k = 20$, $\displaystyle\sum_{k=1}^{4} y_k = 32$, $\displaystyle\sum_{k=1}^{4} x_k y_k = 130$, $\displaystyle\sum_{k=1}^{4} x_k^2 = 120$.

Substituting these values into the formulas for m and d, we have:

$$m = \frac{4\left(\displaystyle\sum_{k=1}^{4} x_k y_k\right) - \left(\displaystyle\sum_{k=1}^{4} x_k\right)\left(\displaystyle\sum_{k=1}^{4} y_k\right)}{4\left(\displaystyle\sum_{k=1}^{4} x_k^2\right) - \left(\displaystyle\sum_{k=1}^{4} x_k\right)^2} = \frac{4(130) - (20)(32)}{4(120) - (20)^2} = \frac{-120}{80} = -1.5$$

$$d = \frac{\displaystyle\sum_{k=1}^{4} y_k - (-1.5)\displaystyle\sum_{k=1}^{4} x_k}{4} = \frac{32 + (1.5)(20)}{4} = \frac{62}{4} = 15.5$$

Thus, the least squares line is:

$y = mx + d = -1.5x + 15.5$

When $x = 10$, $y = -1.5(10) + 15.5 = 0.5$.   (8-5)

10. $\displaystyle\iint\limits_{R} (4x + 6y)\,dA = \int_{-1}^{1} \int_{1}^{2} (4x + 6y)\,dy\,dx = \int_{-1}^{1} \left[\int_{1}^{2} (4x + 6y)\,dy \right] dx$

$\displaystyle \qquad = \int_{-1}^{1} \left[(4xy + 3y^2) \Big|_{1}^{2} \right] dx = \int_{-1}^{1} (8x + 12 - 4x - 3)\,dx$

$\displaystyle \qquad = \int_{-1}^{1} (4x + 9)\,dx = (2x^2 + 9x) \Big|_{-1}^{1} = 2 + 9 - (2 - 9) = 18$

$\displaystyle\iint\limits_{R} (4x + 6y)\,dA = \int_{1}^{2} \int_{-1}^{1} (4x + 6y)\,dx\,dy = \int_{1}^{2} \left[\int_{-1}^{1} (4x + 6y)\,dx \right] dy$

$\displaystyle \qquad = \int_{1}^{2} \left[(2x^2 + 6xy) \Big|_{-1}^{1} \right] dy = \int_{1}^{2} [2 + 6y - (2 - 6y)]\,dy$

$\displaystyle \qquad = \int_{1}^{2} 12y\,dy = 6y^2 \Big|_{1}^{2} = 24 - 6 = 18 \hspace{3cm} (8\text{-}6)$

11. $f(x,\ y) = e^{x^2 + 2y}$

$f_x(x,\ y) = e^{x^2 + 2y} \cdot 2x = 2xe^{x^2 + 2y}$

$f_y(x,\ y) = e^{x^2 + 2y} \cdot 2 = 2e^{x^2 + 2y}$

$f_{xy}(x,\ y) = 2xe^{x^2 + 2y} \cdot 2 = 4xe^{x^2 + 2y} \hspace{3cm} (8\text{-}2)$

12. $f(x,\ y) = (x^2 + y^2)^5$

$f_x(x,\ y) = 5(x^2 + y^2)^4 \cdot 2x = 10x(x^2 + y^2)^4$

$f_{xy}(x,\ y) = 10x(4)(x^2 + y^2)^3 \cdot 2y = 80xy(x^2 + y^2)^3 \hspace{2cm} (8\text{-}2)$

13. $f(x,\ y) = x^3 - 12x + y^2 - 6y$

$f_x(x,\ y) = 3x^2 - 12 \hspace{3cm} f_y(x,\ y) = 2y - 6$

$3x^2 - 12 = 0 \hspace{4cm} 2y - 6 = 0$

$\qquad x^2 = 4 \hspace{5cm} y = 3$

$\qquad x = \pm 2$

Thus, the critical points are $(2,\ 3)$ an $(-2,\ 3)$.

$f_{xx}(x,\ y) = 6x \hspace{1.5cm} f_{xy}(x,\ y) = 0 \hspace{1.5cm} f_{yy}(x,\ y) = 2$

For the critical point $(2,\ 3)$:

$f_{xx}(2,\ 3) = 12 > 0$

$f_{xy}(2,\ 3) = 0$

$f_{yy}(2,\ 3) = 2$

$f_{xx}(2,\ 3) \cdot f_{yy}(2,\ 3) - [f_{xy}(2,\ 3)]^2 = 12 \cdot 2 = 24 > 0$

Therefore, $f(2,\ 3) = 2^3 - 12 \cdot 2 + 3^2 - 6 \cdot 3 = -25$ is a local minimum.

For the critical point $(-2,\ 3)$:

$f_{xx}(-2,\ 3) = -12 < 0$

$f_{xy}(-2,\ 3) = 0$

$f_{yy}(-2,\ 3) = 2$

$f_{xx}(-2,\ 3) \cdot f_{yy}(-2,\ 3) - [f_{xy}(-2,\ 3)]^2 = -12 \cdot 2 - 0 = -24 < 0$

Thus, f has a saddle point at $(-2,\ 3)$. $\hspace{3cm} (8\text{-}3)$

14. <u>Step 1.</u> Maximize $f(x, y) = xy$

                   Subject to: $g(x, y) = 2x + 3y - 24 = 0$

<u>Step 2.</u> $F(x, y, \lambda) = f(x, y) + \lambda g(x, y) = xy + \lambda(2x + 3y - 24)$

<u>Step 3.</u> $F_x = y + 2\lambda = 0$ (1)

           $F_y = x + 3\lambda = 0$ (2)

           $F_\lambda = 2x + 3y - 24 = 0$ (3)

From (1) and (2), we obtain:

$y = -2\lambda$ and $x = -3\lambda$

Substituting these into (3), we have:

$-6\lambda - 6\lambda - 24 = 0$

             $\lambda = -2$

Thus, the critical point is $(6, 4, -2)$.

<u>Step 4.</u> Since $(6, 4, -2)$ is the only critical point for F, we conclude that max $f(x, y) = f(6, 4) = 6 \cdot 4 = 24$.   **(8-4)**

15. <u>Step 1.</u> Minimize $f(x, y, z) = x^2 + y^2 + z^2$

                   Subject to: $2x + y + 2z = 9$ or $g(x, y, z) = 2x + y + 2z - 9 = 0$

<u>Step 2.</u> $F(x, y, z, \lambda) = x^2 + y^2 + z^2 + \lambda(2x + y + 2z - 9)$

<u>Step 3.</u> $F_x = 2x + 2\lambda = 0$ (1)

           $F_y = 2y + \lambda = 0$ (2)

           $F_z = 2z + 2\lambda = 0$ (3)

           $F_\lambda = 2x + y + 2z - 9 = 0$ (4)

From equations (1), (2), and (3), we have:

$x = -\lambda,\ y = -\dfrac{\lambda}{2},$ and $z = -\lambda$

Substituting these into (4), we obtain:

$-2\lambda - \dfrac{\lambda}{2} - 2\lambda - 9 = 0$

            $\dfrac{9}{2}\lambda = -9$

            $\lambda = -2$

The critical point is: $(2, 1, 2, -2)$

<u>Step 4.</u> Since $(2, 1, 2, -2)$ is the only critical point for F, we conclude that min $f(x, y, z) = f(2, 1, 2) = 2^2 + 1^2 + 2^2 = 9$.   **(8-4)**

16.

x_k	y_k	$x_k y_k$	x_k^2
10	50	500	100
20	45	900	400
30	50	1,500	900
40	55	2,200	1,600
50	65	3,250	2,500
60	80	4,800	3,600
70	85	5,950	4,900
80	90	7,200	6,400
90	90	8,100	8,100
100	110	11,000	10,000
Totals 550	720	45,400	38,500

Thus, $\sum_{k=1}^{10} x_k = 550$, $\sum_{k=1}^{10} y_k = 720$, $\sum_{k=1}^{10} x_k y_k = 45,400$, $\sum_{k=1}^{10} x_k^2 = 38,500$.

Substituting these values into the formulas for m and d, we have:

$$m = \frac{10(45,400) - (550)(720)}{10(38,500) - (550)^2} = \frac{58,000}{82,500} = \frac{116}{165}$$

$$d = \frac{720 - \left(\frac{116}{165}\right)550}{10} = \frac{100}{3}$$

Therefore, the least squares line is:

$$y = \frac{116}{165}x + \frac{100}{3} \approx 0.703x + 33.33 \tag{8-5}$$

17. $\dfrac{1}{(b-a)(d-c)} \iint\limits_{R} f(x,\ y)\,dA = \dfrac{1}{[8-(-8)](27-0)} \displaystyle\int_{-8}^{8} \int_{0}^{27} x^{2/3} y^{1/3}\,dy\ dx$

$$= \frac{1}{16 \cdot 27} \int_{-8}^{8} \left(\frac{3}{4} x^{2/3} y^{4/3} \bigg|_{y=0}^{y=27}\right) dx$$

$$= \frac{1}{16 \cdot 27} \int_{-8}^{8} \frac{3^5}{4} x^{2/3}\,dx = \frac{9}{64} \int_{-8}^{8} x^{2/3}\,dx$$

$$= \frac{9}{64} \cdot \frac{3}{5} x^{5/3} \bigg|_{-8}^{8} = \frac{9}{64} \cdot \frac{3}{5} [2^5 - (-2)^5]$$

$$= \frac{9}{64} \cdot \frac{3}{5} \cdot 2^6 = \frac{27}{5} \tag{8-6}$$

18. $V = \displaystyle\iint\limits_{R} (3x^2 + 3y^2)\,dA = \int_0^1 \int_{-1}^1 (3x^2 + 3y^2)\,dy\ dx = \int_0^1 \left[\int_{-1}^1 (3x^2 + 3y^2)\,dy\right] dx$

$$= \int_0^1 \left[(3x^2 y + y^3) \bigg|_{-1}^1\right] dx = \int_0^1 [3x^2 + 1 - (-3x^2 - 1)]\,dx$$

$$= \int_0^1 (6x^2 + 2)\,dx = (2x^3 + 2x) \bigg|_0^1 = 4 \text{ cubic units} \tag{8-6}$$

19. $f(x, y) = x + y$; $-10 \le x \le 10$, $-10 \le y \le 10$

Prediction: average value = $f(0, 0) = 0$.

Verification:

$$\text{average value} = \frac{1}{[10 - (-10)][10 - (-10)]} \int_{-10}^{10} \int_{-10}^{10} (x + y)\, dy\, dx$$

$$= \frac{1}{400} \int_{-10}^{10} \left[\left(xy + \frac{1}{2}y^2 \right) \Big|_{-10}^{10} \right] dy$$

$$= \frac{1}{400} \int_{-10}^{10} 20x\, dx$$

$$= \frac{1}{400} (10x^2) \Big|_{-10}^{10} = 0 \qquad\qquad (8\text{-}6)$$

20. $f(x, y) = \dfrac{e^x}{y + 10}$

(A) $S = \{x, y) \mid -a \le x \le a, \ -a \le y \le a\}$

The average value of f over S is given by:

$$\frac{1}{[a - (-a)][a - (-a)]} \int_{-a}^{a} \int_{-a}^{a} \frac{e^x}{y + 10}\, dx\, dy$$

$$= \frac{1}{4a^2} \int_{-a}^{a} \left[\frac{e^x}{y + 10} \Big|_{-a}^{a} \right] dy$$

$$= \frac{1}{4a^2} \int_{-a}^{a} \left(\frac{e^a}{y + 10} - \frac{e^{-a}}{y + 10} \right) dy$$

$$= \frac{e^a - e^{-a}}{4a^2} \int_{-a}^{a} \frac{1}{y + 10}\, dy$$

$$= \frac{e^a - e^{-a}}{4a^2} (\ln |y + 10|) \Big|_{-a}^{a}$$

$$= \frac{e^a - e^{-a}}{4a^2} [\ln(10 + a) - \ln(10 - a)]$$

$$= \frac{e^a - e^{-a}}{4a^2} \ln\left(\frac{10 + a}{10 - a} \right)$$

Now, $\dfrac{e^a - e^{-a}}{4a^2} \ln\left(\dfrac{10 + a}{10 - a} \right) = 5$

is equivalent to

$$(e^a - e^{-a}) \ln\left(\frac{10 + a}{10 - a} \right) - 20a^2 = 0.$$

Using a graphing utility, the graph of

$$f(x) = (e^x - e^{-x}) \ln\left(\frac{10 + x}{10 - x} \right) - 20x^2$$

is shown at the right and $f(x) = 0$ at $x \approx \pm 6.28$.

The dimensions of the square are: 12.56×12.56.

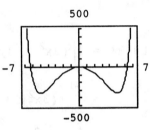

(B) To determine whether there is a square centered at $(0, 0)$ such that
$$\frac{e^a - e^{-a}}{4a^2} \ln\left(\frac{10 + a}{10 - a}\right) = 0.05,$$
graph,
$$f(x) = (e^x - e^{-x}) \ln\left(\frac{10 + x}{10 - x}\right) - 0.20x^2$$
The result is shown at the right
and $f(x) = 0$ only at $x = 0$.

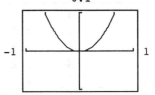

Thus, there does not exist a square centered
at $(0, 0)$ such that the average value of $f = 0.05$.   (8–6)

21. $P(x, y) = -4x^2 + 4xy - 3y^2 + 4x + 10y + 81$

    (A) $P_x(x, y) = -8x + 4y + 4$
        $P_x(1, 3) = -8 \cdot 1 + 4 \cdot 3 + 4 = 8$

    At the output level $(1, 3)$, profit will increase by \$8000 for 100
    units increase in product A if the production of product B is held
    fixed.

    (B) $P_x = -8x + 4y + 4 = 0$ (1)
        $P_y = 4x - 6y + 10 = 0$ (2)
    Solving (1) and (2) for x and y, we obtain $x = 2$, $y = 3$.
    Thus, $(2, 3)$ is a critical point.
$$P_{xx} = -8 \qquad\qquad P_{yy} = -6 \qquad\qquad P_{xy} = 4$$
$$P_{xx}(2, 3) = -8 < 0 \qquad P_{yy}(2, 3) = -6 \qquad P_{xy}(2, 3) = 4$$
$$P_{xx}(2, 3) \cdot P_{yy}(2, 3) - [P_{xy}(2, 3)]^2 = (-8)(-6) - 4^2 = 32 > 0$$

    Thus, $P(2, 3)$ is a maximum and
$$\begin{aligned} \max P(x, y) = P(2, 3) &= -4 \cdot 2^2 + 4 \cdot 2 \cdot 3 - 3 \cdot 3^2 + 4 \cdot 2 + 10 \cdot 3 + 81 \\ &= -16 + 24 - 27 + 8 + 30 + 81 \\ &= 100. \end{aligned}$$
    Thus, the maximum profit is \$100,000. This is obtained when 200
    units of A and 300 units of B are produced per month.   (8–2, 8–3)

22. Minimize $S(x, y, z) = xy + 4yz + 3xz$
    Subject to: $V(x, y, z) = xyz - 96 = 0$
    Put $F(x, y, z, \lambda) = S(x, y, z) + \lambda V(x, y, z) = xy + 4yz + 3xz + \lambda(xyz - 96)$.
    Then, we have:
$$F_x = y + 3z + \lambda yz = 0 \qquad (1)$$
$$F_y = x + 4z + \lambda xz = 0 \qquad (2)$$
$$F_z = 4y + 3x + \lambda xy = 0 \qquad (3)$$
$$F_\lambda = xyz - 96 = 0 \qquad (4)$$

    Solving the system of equations, (1)–(4), simultaneously, yields $x = 8$,
    $y = 6$, $z = 2$, and $\lambda = -1$. Thus, the critical point is $(8, 6, 2, -1)$ and
    $S(8, 6, 2) = 8 \cdot 6 + 4 \cdot 6 \cdot 2 + 3 \cdot 8 \cdot 2 = 144$
    is the minimum value of S subject to the constraint $V = xyz - 96 = 0$.

    The dimensions of the box that will require the minimum amount of
    material are:
    Length $x = 8$ inches; Width $y = 6$ inches; Height $z = 2$ inches   (8–3)

23.

	x_k	y_k	$x_k y_k$	x_k^2
	1	2.0	2.0	1
	2	2.5	5.0	4
	3	3.1	9.3	9
	4	4.2	16.8	16
	5	4.3	21.5	25
Totals	15	16.1	54.6	55

Thus, $\sum_{k=1}^{5} x_k = 15$, $\sum_{k=1}^{5} y_k = 16.1$, $\sum_{k=1}^{5} x_k y_k = 54.6$, $\sum_{k=1}^{5} x_k^2 = 55$.

Substituting these values into the formulas for m and d, we have:

$$m = \frac{5(54.6) - (15)(16.1)}{5(55) - (15)^2} = \frac{31.5}{50} \approx 0.63$$

$$d = \frac{16.1 - (0.63)(15)}{5} = 1.33$$

Therefore, the least squares line is:
$y = 0.63x + 1.33$
When $x = 6$, $y = 0.63(6) + 1.33 = 5.11$, and the profit for the sixth year
is estimated to be \$5.11 million. $\qquad$ (8-4)

24. $N(x, y) = 10x^{0.8}y^{0.2}$

(A) $N_x(x, y) = 8x^{-0.2}y^{0.2}$

$N_x(40, 50) = 8(40)^{-0.2}(50)^{0.2} \approx 8.36$

$N_y(x, y) = 2x^{0.8}y^{-0.8}$

$N_y(40, 50) = 2(40)^{0.8}(50)^{-0.8} \approx 1.67$

Thus, at the level of 40 units of labor and 50 units of capital, the
marginal productivity of labor is approximately 8.36 and the
marginal productivity of capital is approximately 1.67. Management
should encourage increased use of labor.

(B) <u>Step 1.</u> Maximize the production function $N(x, y) = 10x^{0.8}y^{0.2}$

Subject to the constraint: $C(x, y) = 100x + 50y = 10,000$

i.e., $g(x, y) = 100x + 50y - 10,000 = 0$

<u>Step 2.</u> $F(x, y, \lambda) = 10x^{0.8}y^{0.2} + \lambda(100x + 50y - 10,000)$

<u>Step 3.</u> $F_x = 8x^{-0.2}y^{0.2} + 100\lambda = 0 \qquad (1)$

$F_y = 2x^{0.8}y^{-0.8} + 50\lambda = 0 \qquad (2)$

$F_\lambda = 100x + 50y - 10,000 = 0 \qquad (3)$

From equation (1), $\lambda = \dfrac{-0.08y^{0.2}}{x^{0.2}}$, and from (2),

$\lambda = \dfrac{-0.04x^{0.8}}{y^{0.8}}$. Thus, $\dfrac{0.08y^{0.2}}{x^{0.2}} = \dfrac{0.04x^{0.8}}{y^{0.8}}$ and $x = 2y$.

Substituting into (3) yields:

$$200y + 50y = 10,000$$
$$250y = 10,000$$
$$y = 40$$

Therefore, $x = 80$ and $\lambda \approx -0.0696$. The critical point is $(80, 40, -0.0696)$. Thus, we conclude that max $N(x, y) = N(80, 40) = 10(80)^{0.8}(40)^{0.2} \approx 696$ units.

Production is maximized when 80 units of labor and 40 units of capital are used.

The marginal productivity of money is $-\lambda \approx 0.0696$. The increase in production resulting from an increase of \$2000 in the budget is:

$$0.0696(2000) \approx 139 \text{ units}$$

(C) Average number of units

$$= \frac{1}{(100 - 50)(40 - 20)}\int_{50}^{100}\int_{20}^{40} 10x^{0.8}y^{0.2}dy\ dx$$

$$= \frac{1}{(50)(20)}\int_{50}^{100}\left[\frac{10x^{0.8}y^{1.2}}{1.2}\bigg|_{20}^{40}\right]dx = \frac{1}{1000}\int_{50}^{100}\frac{10}{1.2}x^{0.8}(40^{1.2} - 20^{1.2})dx$$

$$= \frac{40^{1.2} - 20^{1.2}}{120}\int_{50}^{100} x^{0.8}dx = \frac{40^{1.2} - 20^{1.2}}{120}\cdot\frac{x^{1.8}}{1.8}\bigg|_{50}^{100}$$

$$= \frac{(40^{1.2} - 20^{1.2})(100^{1.8} - 50^{1.8})}{216} \approx \frac{(47.24)(2837.81)}{216} \approx 621$$

Thus, the average number of units produced is approximately 621.

(8-4)

25. $T(V, x) = \dfrac{33V}{x + 33} = 33V(x + 33)^{-1}$

$T_x(V, x) = -33V(x + 33)^{-2} = \dfrac{-33V}{(x + 33)^2}$

$T_x(70, 17) = \dfrac{-33(70)}{(17 + 33)^2} = \dfrac{-33(70)}{2500}$

$\qquad\qquad\qquad = -0.924$ minutes per unit increase in depth when $V = 70$ cubic feet and $x = 17$ feet

(8-2)

26. $C = 100 - 24d^2 = 100 - 24(x^2 + y^2)$

$C(x, y) = 100 - 24(x^2 + y^2),\ -2 \leq x \leq 2,\ -2 \leq y \leq 2$

Average concentration $= \dfrac{1}{4(4)}\int_{-2}^{2}\int_{-2}^{2} [100 - 24(x^2 + y^2)]dy\ dx$

$$= \frac{1}{16}\int_{-2}^{2} [100y - 24x^2y - 8y^3]\bigg|_{-2}^{2} dx$$

$$= \frac{1}{16}\int_{-2}^{2} [400 - 96x^2 - 128]dx = \frac{1}{16}\int_{-2}^{2} (272 - 96x^2)dx$$

$$= \frac{1}{16}[272x - 32x^3]\bigg|_{-2}^{2} = \frac{1}{16}(544 - 256) - \frac{1}{16}(-544 + 256)$$

$$= 18 + 18 = 36 \text{ parts per million}$$

(8-6)

27. $n(P_1, P_2, d) = 0.001 \dfrac{P_1 P_2}{d}$

$n(100,000, 50,000, 100) = 0.001 \dfrac{100,000 \times 50,000}{100} = 50,000$ (8-1)

28.

x_k	y_k	$x_k y_k$	x_k^2
30	60	1,800	900
50	75	3,750	2,500
60	80	4,800	3,600
70	85	5,950	4,900
90	90	8,100	8,100
Totals 300	390	24,400	20,000

Thus, $\displaystyle\sum_{k=1}^{5} x_k = 300,\ \sum_{k=1}^{5} y_k = 390,\ \sum_{k=1}^{5} x_k y_k = 24,400,\ \sum_{k=1}^{5} x_k^2 = 20,000.$

Substituting these values into the formulas for m and d, we have:

$m = \dfrac{5(24,400) - (300)(390)}{5(20,000) - (300)^2} = \dfrac{5000}{10,000} = 0.5$

$d = \dfrac{390 - 0.5(300)}{5} = \dfrac{240}{5} = 48$

Therefore, the least squares line is:
$y = 0.5x + 48$
When $x = 40,\ y = 0.5(40) + 48 = 68.$ (8-5)

29. (A) Enter the data in a calculator or computer. (We used a TI-85.)
The totals are:
$n = 10,\quad \sum x = 450,\quad \sum y = 470.6,$
$\sum x^2 = 28,500,\quad \sum xy = 25,062$

Now, the least squares line can be calculated either by using formulas (2) and (3) in Section 8-5, or by using the linear regression feature. We used the latter to get

$m = 0.4709 \quad \text{and} \quad b = 25.87$

Therefore, the least squares line is:
$y = 0.4709x + 25.87$

(B) Using the result in (A), an estimate for the population density in the year 2000 is:

$y = 0.4709(100) + 25.87 \approx 72.96$ persons/square mile   (8-5)

30. (A) Enter the data in a calculator or computer. (We used a TI-85.)
The totals are:
$$n = 9, \qquad \sum x = 589.44, \qquad \sum y = 634.97,$$
$$\sum x^2 = 38{,}819.351, \qquad \sum xy = 41{,}816.153$$

Now, the least squares line can be calculated either by using
formulas (2) and (3) in Section 8-5, or by using the linear
regression feature. We used the latter to get
$$m = 1.069 \quad \text{and} \quad b = 0.522$$

Therefore, the least squares line is:
$$y = 1.069x + 0.522$$

(B) Using the result in (A), an estimate for the life expectancy of a
female corresponding to a life expectancy of a male of 60 years is:
$$y = 1.069(60) + 0.522 \approx 64.66 \text{ years} \tag{8-5}$$

CHAPTER 9   TRIGONOMETRIC FUNCTIONS

Things to remember:

<u>**1**</u>. θ_{rad} = radian measure of θ

$\qquad = \dfrac{\text{Arc length}}{\text{Radius}}$

$\qquad = \dfrac{s}{R} \qquad$ [<u>Note</u>: If $R = 1$, then $\theta_{rad} = s$.]

<u>**2**</u>. DEGREE-RADIAN CONVERSION FORMULA

$$\dfrac{\theta_{deg}}{180°} = \dfrac{\theta_{rad}}{\pi_{rad}}$$

<u>**3**</u>. On a unit circle, where the origin is at the center, cos θ and sin θ are measured by the abscissa and ordinate of point P, respectively, as shown in the figure.

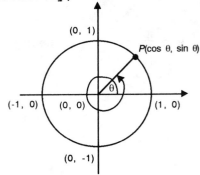

For any real number x, sin x = sin(x radians), cos x = cos(x radians).

<u>**4**</u>. $\tan x = \dfrac{\sin x}{\cos x}$, $\cos x \neq 0$

<u>**5**</u>. $\cot x = \dfrac{\cos x}{\sin x}$, $\sin x \neq 0$

<u>**6**</u>. $\sec x = \dfrac{1}{\cos x}$, $\cos x \neq 0$

<u>**7**</u>. $\csc x = \dfrac{1}{\sin x}$, $\sin x \neq 0$

1. $60° = \frac{\pi}{3}$ radians   **3.** $45° = \frac{\pi}{4}$ radians   **5.** $30° = \frac{\pi}{6}$ radians

7. II   **9.** IV   **11.** I

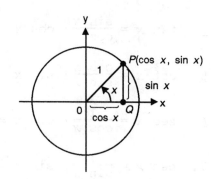

13. $\cos 0 = 1$ **15.** $\sin \pi = 0$
(see Figure 5)   (see Figure 5)

17. $\cos(-\pi) = -1$ **19.** $\frac{\pi}{3}$ radians $= 60°$

21. $\frac{\pi}{4}$ radians $= 45°$ **23.** $\frac{\pi}{6}$ radians $= 30°$

25. $\cos 30° = \frac{\sqrt{3}}{2}$ **27.** $\sin\left(\frac{\pi}{6} \text{ radians}\right) = \frac{1}{2}$ **29.** $\cos \frac{5\pi}{6} = -\frac{\sqrt{3}}{2}$
(see Figure 6)   (see Figure 6)   (see Figure 6)

31. $\sin 3 = 0.1411$ **33.** $\cos 33.74 = -0.6840$ **35.** $\sin(-43.06) = 0.7970$

37. $\theta° = 27°$

Using $\underline{2}$, $\frac{27°}{180°} = \frac{\theta}{\pi \text{ radians}}$

$\qquad \theta = \frac{27\pi}{180}$ radians

$\qquad = \frac{3\pi}{20}$ radians

39. Using $\underline{2}$, $\frac{\theta°}{180°} = \frac{\frac{\pi}{12}}{\pi}$

Thus, $\frac{\pi}{12}$ radians $= \frac{180°}{12} = 15°$

41. From Figure 6 and using $\underline{4}$,
$\tan 45° = \frac{\sin 45°}{\cos 45°}$

$\qquad = \frac{\frac{1}{\sqrt{2}}}{\frac{1}{\sqrt{2}}} = 1$

43. $\sec \frac{\pi}{3} = \frac{1}{\cos \frac{\pi}{3}}$ (using $\underline{6}$)

$\qquad = \frac{1}{\frac{1}{2}} = 2$

45. $\cot \frac{\pi}{3} = \frac{\cos \frac{\pi}{3}}{\sin \frac{\pi}{3}} = \frac{\frac{1}{2}}{\frac{\sqrt{3}}{2}} = \frac{1}{\sqrt{3}} = \frac{1}{\sqrt{3}} \cdot \frac{\sqrt{3}}{\sqrt{3}} = \frac{\sqrt{3}}{3}$

47. Applying the Pythagorean Theorem to
the triangle OPQ, we obtain:

$(\sin x)^2 + (\cos x)^2 = 1$

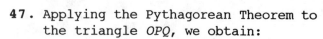

49. $y = 2 \sin \pi x$;
$0 \le x \le 2, \; -2 \le y \le 2$

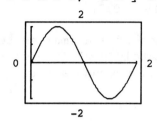

51. $y = 4 - 4 \cos \dfrac{\pi x}{2}$;
$0 \le x \le 8, \; 0 \le y \le 8$

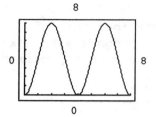

53. (A) $P(t) = 5 - 5 \cos\left(\dfrac{\pi}{26}\right)t$

$P(13) = 5 - 5 \cos\left(\dfrac{\pi}{26}\right)13 = 5 - 5 \cos\left(\dfrac{\pi}{2}\right) = 5 - 5(0) = 5$

$P(26) = 5 - 5 \cos\left(\dfrac{\pi}{26}\right)26 = 5 - 5 \cos(\pi) = 5 - 5(-1) = 10$

$P(39) = 5 - 5 \cos\left(\dfrac{\pi}{26}\right)39 = 5 - 5 \cos\left(\dfrac{3\pi}{2}\right) = 5 - 5(0) = 5$

$P(52) = 5 - 5 \cos\left(\dfrac{\pi}{26}\right)52 = 5 - 5 \cos(2\pi) = 5 - 5(1) = 0$

(B) $P(30) = 5 - 5 \cos\left(\dfrac{\pi}{26}\right)30 \approx 5 - 5(0.886) \approx 5 + 4.43 \approx 9.43$

$P(100) = 5 - 5 \cos\left(\dfrac{\pi}{26}\right)100 \approx 5 - 5(-0.886) \approx 5 - 4.43 \approx 0.57$

Interpretation: 30 weeks after January 1, the profit on a week's sales of bathing suits is $943; 100 weeks after January 1, the profit on a week's sales of bathing suits is $57.

(C)

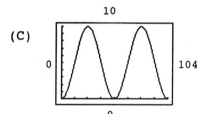

55. $V(t) = 0.45 - 0.35 \cos \dfrac{\pi t}{2}, \; 0 \le t \le 8$

(A) $V(0) = 0.45 - 0.35 \cos(0) = 0.45 - 0.35 = 0.10$

$V(1) = 0.45 - 0.35 \cos\left(\dfrac{\pi}{2}\right) = 0.45 - 0 = 0.45$

$V(2) = 0.45 - 0.35 \cos(\pi) = 0.45 + 0.35 = 0.80$

$V(3) = 0.45 - 0.35 \cos\left(\dfrac{3\pi}{2}\right) = 0.45 - 0 = 0.45$

$V(7) = 0.45 - 0.35 \cos\left(\dfrac{7\pi}{2}\right) = 0.45 - 0 = 0.45$

(B) $V(3.5) = 0.45 - 0.35 \cos\left(\dfrac{3.5\pi}{2}\right) \approx 0.45 - 0.2475 \approx 0.20$

$V(5.7) = 0.45 - 0.35 \cos\left(\dfrac{5.7\pi}{2}\right) \approx 0.45 + 0.3119 \approx 0.76$

Interpretation: The volume of air in the lungs of a normal seated adult 3.5 seconds after exhaling is approximately 0.20 liters; the volume of air is approximately 0.76 liters 5.7 seconds after exhaling.

(C)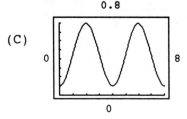

57. $d = -2.1 - 4 \sin 4\theta$

(A) $\theta = 30°$
$d = -2.1 - 4 \sin 4(30°)$
$= -2.1 - 4 \sin 120°$
$= -2.1 - 4 \cdot \dfrac{\sqrt{3}}{2} \approx -5.6°$

(B) $\theta = 10°$
$d = -2.1 - 4 \sin 4(10°)$
$= -2.1 - 4 \sin 40°$
$-2.1 - 4(0.6428) \approx -4.7°$

EXERCISE 9-2

Things to remember:

$\underline{1}$. $\dfrac{d}{dx} \sin x = \cos x$

$\underline{2}$. $\dfrac{d}{dx} \cos x = -\sin x$

$\underline{3}$. If f is a differentiable function of x, then

$\dfrac{d}{dx} \sin f(x) = [\cos f(x)]f'(x)$

$\underline{4}$. If f is a differentiable function of x, then

$\dfrac{d}{dx} \cos f(x) = [-\sin f(x)]f'(x)$

1. $\dfrac{d}{dt} \cos t = -\sin t$ (using $\underline{2}$)

3. $\dfrac{d}{dx} \sin x^3 = \cos x^3 \dfrac{d}{dx} x^3$ (using $\underline{3}$)
$= \cos x^3 (3x^2)$
$= 3x^2 \cos x^3$

5. $\dfrac{d}{dt} t \sin t = t \dfrac{d}{dt} \sin t + \sin t \dfrac{d}{dt} t$

$\qquad\qquad = t \cos t + (\sin t)1$

$\qquad\qquad = t \cos t + \sin t$

7. $\dfrac{d}{dx} \sin x \cos x = \sin x \dfrac{d}{dx} \cos x + \cos x \dfrac{d}{dx} \sin x$

$\qquad\qquad\qquad = -\sin x \sin x + \cos x \cos x$

$\qquad\qquad\qquad = (\cos x)^2 - (\sin x)^2$

9. $\dfrac{d}{dx}(\sin x)^5 = 5(\sin x)^4 \dfrac{d}{dx} \sin x = 5(\sin x)^4 \cos x$

11. $\dfrac{d}{dx}\sqrt{\sin x} = \dfrac{d}{dx}(\sin x)^{1/2} = \dfrac{1}{2}(\sin x)^{-1/2}\dfrac{d}{dx}\sin x$

$\qquad\qquad\qquad = \dfrac{1}{2}(\sin x)^{-1/2}\cos x = \dfrac{\cos x}{2\sqrt{\sin x}}$

13. $\dfrac{d}{dx}\cos\sqrt{x} = -\sin\sqrt{x}\dfrac{d}{dx}\sqrt{x}$ (using $\underline{4}$)

$\qquad\qquad = -\dfrac{1}{2}x^{-1/2}\sin\sqrt{x} = -\dfrac{\sin\sqrt{x}}{2\sqrt{x}}$

15. $f(x) = \sin x$

$\quad f'(x) = \cos x$

The slope of the graph of f at $x = \dfrac{\pi}{6}$ is: $f'\!\left(\dfrac{\pi}{6}\right) = \cos\dfrac{\pi}{6} = \dfrac{\sqrt{3}}{2} \approx 0.866$

17. f is increasing on $[-\pi,\ 0]$ $(f'(x) > 0)$; f is decreasing on $[0,\ \pi]$

$(f'(x) < 0)$; f has a local maximum at $x = 0$;

the graph of f is concave upward on $\left(-\pi,\ -\dfrac{\pi}{2}\right)$ and

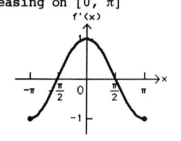

on $\left(\dfrac{\pi}{2},\ \pi\right)$ $(f'$ is increasing on these intervals$)$;

the graph of f is concave downward on $\left(-\dfrac{\pi}{2},\ \dfrac{\pi}{2}\right)$

$(f'$ is decreasing on this interval$)$; $f(x) = \cos x$, $f'(x) = -\sin x$.

19. $\dfrac{d}{dx}\tan x = \dfrac{d}{dx}\dfrac{\sin x}{\cos x} = \dfrac{\cos x \dfrac{d}{dx}\sin x - \sin x \dfrac{d}{dx}\cos x}{(\cos x)^2}$

$\qquad = \dfrac{\cos x(\cos x) - \sin x(-\sin x)}{(\cos x)^2} = \dfrac{\cos x \cos x + \sin x \sin x}{(\cos x)^2}$

$\qquad = \dfrac{(\cos x)^2 + (\sin x)^2}{(\cos x)^2} = \dfrac{1}{(\cos x)^2} = (\sec x)^2$

21. $\dfrac{d}{dx}\sin\sqrt{x^2 - 1} = \cos\sqrt{x^2 - 1}\dfrac{d}{dx}\sqrt{x^2 - 1}$

$\qquad\qquad = \cos\sqrt{x^2 - 1}\dfrac{1}{2}(x^2 - 1)^{-1/2}\dfrac{d}{dx}(x^2 - 1)$

$\qquad\qquad = \dfrac{2x \cos\sqrt{x^2 - 1}}{2\sqrt{x^2 - 1}} = \dfrac{x \cos\sqrt{x^2 - 1}}{\sqrt{x^2 - 1}}$

23. $f(x) = e^x \sin x$

$f'(x) = e^x(\cos x) + (\sin x)e^x = e^x(\sin x + \cos x)$

$f''(x) = e^x(\cos x - \sin x) + (\sin x + \cos x)e^x = 2e^x \cos x$

25. $y = x \sin \pi x$;

$0 \le x \le 9, \; -9 \le y \le 9$

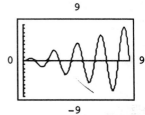

27. $y = \dfrac{\cos \pi x}{x}$;

$0 \le x \le 8, \; -2 \le y \le 3$

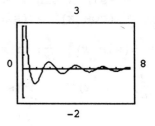

29. $y = e^{-0.3x} \sin \pi x$,

$0 \le x \le 10, \; -1 \le y \le 1$

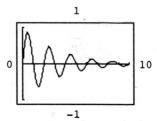

31. $P(t) = 5 - 5 \cos\left(\dfrac{\pi t}{26}\right), \; 0 \le t \le 104$

(A) $P'(t) = -5\left[-\sin\left(\dfrac{\pi t}{26}\right)\left(\dfrac{\pi}{26}\right)\right] = \dfrac{5\pi}{26} \sin\left(\dfrac{\pi t}{26}\right), \; 0 \le t \le 104$

(B) $P'(8) = \dfrac{5\pi}{26} \sin\left(\dfrac{8\pi}{26}\right) \approx 0.50$ (hundred) or \$50 per week

$P'(26) = \dfrac{5\pi}{26} \sin\left(\dfrac{26\pi}{26}\right) = 0$ or \$0 per week

$P'(50) = \dfrac{5\pi}{26} \sin\left(\dfrac{50\pi}{26}\right) \approx -0.14$ (hundred) or -\$14 per week

(C) $P'(t) = \dfrac{5\pi}{26} \sin\left(\dfrac{\pi t}{26}\right) = 0, \; 0 < t < 104$

$$\sin\left(\dfrac{\pi t}{26}\right) = 0$$

Therefore, the critical values are:

$\dfrac{\pi t}{26} = \pi$, or $t = 26$; $\qquad \dfrac{\pi t}{26} = 2\pi$ or $t = 52$; $\qquad \dfrac{\pi t}{26} = 3\pi$ or $t = 78$.

Now,

$P''(t) = \dfrac{5\pi^2}{676} \cos\left(\dfrac{\pi t}{26}\right)$

$P''(26) = \dfrac{5\pi^2}{676} \cos(\pi) = -\dfrac{5\pi^2}{676} < 0$

$$P''(52) = \frac{5\pi^2}{676} \cos(2\pi) = \frac{5\pi^2}{676} > 0$$

$$P''(78) = \frac{5\pi^2}{676} \cos(3\pi) = -\frac{5\pi^2}{676} < 0$$

Thus,

t	$P(t)$	
26	$1000	local maximum
52	$0	local minimum
78	$1000	local maximum

(D)

t	$P(t)$	
0	$0	absolute minimum
26	$1000	absolute maximum
52	$0	absolute minimum
78	$1000	absolute maximum
104	$0	absolute minimum

(E) The results in part (C) are illustrated by the graph of f shown at the right.

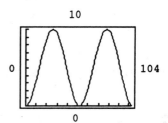

33. $V(t) = 0.45 - 0.35 \cos\left(\frac{\pi t}{2}\right),\ 0 \le t \le 8$

(A) $V'(t) = -0.35\left[-\sin\left(\frac{\pi t}{2}\right)\right]\left(\frac{\pi}{2}\right) = \frac{0.35\pi}{2} \sin\left(\frac{\pi t}{2}\right),\ 0 \le t \le 8$

(B) $V'(3) = \frac{0.35\pi}{2} \sin\left(\frac{3\pi}{2}\right) = -\frac{0.35\pi}{2} \approx -0.55$ liters per second

$V'(4) = \frac{0.35\pi}{2} \sin\left(\frac{4\pi}{2}\right) = 0.00$ liters per second

$V'(5) = \frac{0.35\pi}{2} \sin\left(\frac{5\pi}{2}\right) = \frac{0.35\pi}{2} \approx 0.55$ liters per second

(C) $V'(t) = \frac{0.35\pi}{2} \sin\left(\frac{\pi t}{2}\right) = 0,\ 0 < t < 8$

$$\sin\left(\frac{\pi t}{2}\right) = 0$$

Therefore, the critical values are:

$\frac{\pi t}{2} = \pi$ or $t = 2$; $\quad \frac{\pi t}{2} = 2\pi$ or $t = 4$; $\quad \frac{\pi t}{2} = 3\pi$ or $t = 6$.

Now,

$V''(t) = \frac{0.35\pi}{2} \cos\left(\frac{\pi t}{2}\right)\left(\frac{\pi}{2}\right) = \frac{0.35\pi^2}{4} \cos\left(\frac{\pi t}{2}\right)$

$V''(2) = \frac{0.35\pi^2}{4} \cos(\pi) = -\frac{0.35\pi^2}{4} < 0$

$$V''(4) = \frac{0.35\pi^2}{4} \cos(2\pi) = \frac{0.35\pi^2}{4} > 0$$

$$V''(6) = \frac{0.35\pi^2}{4} \cos(3\pi) = -\frac{0.35\pi^2}{4} < 0$$

Thus,

t	$V(t)$	
2	0.80	local maximum
4	0.10	local minimum
6	0.80	local maximum

(D)

t	$V(t)$	
0	0.10	absolute minimum
2	0.80	absolute maximum
4	0.10	absolute minimum
6	0.80	absolute maximum
8	0.10	absolute minimum

Thus, 0.10 liters is the absolute minimum and 0.80 liters is the absolute maximum of V for $0 \le t \le 8$.

(E) The results in part (C) are illustrated by the graph of f shown at the right.

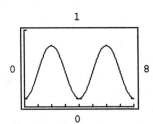

EXERCISE 9-3

Things to remember:

<u>1</u>. $\int \cos x \, dx = \sin x + C$

<u>2</u>. $\int \sin x \, dx = -\cos x + C$

<u>3</u>. $\int \sin u \, du = -\cos u + C$

<u>4</u>. $\int \cos u \, du = \sin u + C$ $\left. \begin{array}{c} \\ \\ \end{array} \right\}$ where $u = u(x)$

1. $\int \sin t \, dt = -\cos t + C$ (using <u>2</u>)

3. $\int \cos 3x \, dx = \frac{1}{3} \int \cos 3x(3 \, dx)$ [Let $u = 3x$, then $du = 3 \, dx$.]

$$= \frac{1}{3} \sin 3x + C \quad \text{(using } \underline{4})$$

5. $\int (\sin x)^{12} \cos x \, dx = \int u^{12} \, du$ [Let $u = \sin x$, then $du = \cos x \, dx$.]

$$= \frac{u^{13}}{13} + C = \frac{(\sin x)^{13}}{13} + C$$

7. $\int \sqrt[3]{\cos x} \sin x \, dx = -\int \sqrt[3]{\cos x}(-\sin x) \, dx$ [Let $u = \cos x$, then $du = -\sin x \, dx$.]

$$= -\int u^{1/3} \, du = -\frac{u^{4/3}}{\frac{4}{3}} + C = -\frac{3}{4}(\cos x)^{4/3} + C$$

9. $\int x^2 \cos x^3 \, dx = \frac{1}{3} \int \cos x^3 (3x^2) \, dx$ [Let $u = x^3$, then $du = 3x^2 \, dx$.]

$$= \frac{1}{3} \sin x^3 + C$$

11. $\displaystyle\int_0^{\pi/2} \cos x \, dx = \sin x \Big|_0^{\pi/2} = \sin \frac{\pi}{2} - \sin 0 = 1 - 0 = 1$

13. $\displaystyle\int_{\pi/2}^{\pi} \sin x \, dx = -\cos x \Big|_{\pi/2}^{\pi} = -\left(\cos \pi - \cos \frac{\pi}{2}\right) = -(-1 - 0) = 1$

15. The shaded area $= \displaystyle\int_{\pi/6}^{\pi/3} \cos x \, dx = \sin x \Big|_{\pi/6}^{\pi/3} = \sin \frac{\pi}{3} - \sin \frac{\pi}{6}$

$$= \frac{\sqrt{3}}{2} - \frac{1}{2} \approx 0.866 - 0.5 \approx 0.366$$

17. $\displaystyle\int_0^2 \sin x \, dx = -\cos x \Big|_0^2 = -(\cos 2 - \cos 0) \approx -(-0.4161 - 1) \approx 1.4161$

19. $\displaystyle\int_1^2 \cos x \, dx = \sin x \Big|_1^2 = \sin 2 - \sin 1 \approx 0.9093 - 0.8415 \approx 0.0678$

21. $\int e^{\sin x} \cos x \, dx$ [Let $u = \sin x$, then $du = \cos x \, dx$.]

$$= \int e^u \, du = e^u + C = e^{\sin x} + C$$

23. $\int \dfrac{\cos x}{\sin x} dx = \int \dfrac{du}{u}$ [Let $u = \sin x$, then $du = \cos x \, dx$.]

$$= \ln|u| + C = \ln|\sin x| + C$$

25. $\int \tan x \, dx = \int \dfrac{\sin x}{\cos x} \, dx$ [Let $u = \cos x$, then $du = -\sin x \, dx$.]

$$= -\int \dfrac{du}{u} = -\ln|u| + C = -\ln|\cos x| + C$$

27. $f(x) = e^{-x} \sin x, \ 0 \le x \le 3$

(A)

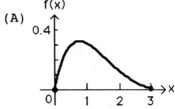

(B) $a = 0, \ b = 3, \ n = 6, \ \Delta x = \dfrac{3 - 0}{6} = \dfrac{1}{2}$

$$M_6 = f\left(\tfrac{1}{4}\right)\Delta x + f\left(\tfrac{3}{4}\right)\Delta x + f\left(\tfrac{5}{4}\right)\Delta x + f\left(\tfrac{7}{4}\right)\Delta x + f\left(\tfrac{9}{4}\right)\Delta x + f\left(\tfrac{11}{4}\right)\Delta x$$

$$= \tfrac{1}{2}(0.1927 + 0.3220 + 0.2719 + 0.1710 + 0.0820 + 0.0244)$$

$$= 0.532$$

(C) $f'(x) = -e^{-x}\sin x + e^{-x}\cos x; \ f''(x) = -2e^{-x}\cos x$

The graph of $|f''(x)| = |-2e^{-x}\cos x|$ on $[0, 3]$ is

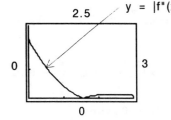

(D) $|I - M_6| \le \dfrac{\max|f''(x)|(b-a)^3}{24n^2} = \dfrac{2(3-0)^3}{24(6)^2} = \dfrac{1}{16} = 0.0625$

29. $P(t) = 5 - 5\cos\left(\dfrac{\pi t}{26}\right), \ 0 \le t \le 104$

(A) Total profit during the two-year period:

$$T = \int_0^{104}\left[5 - 5\cos\left(\dfrac{\pi t}{26}\right)\right]dt = \int_0^{104} 5 \, dt - 5\int_0^{104}\cos\left(\dfrac{\pi t}{26}\right)dt$$

$$= 5t \Big|_0^{104} - 5\left(\dfrac{26}{\pi}\right)\int_0^{104}\cos\left(\dfrac{\pi t}{26}\right)\left(\dfrac{\pi}{26}\right)dt = 520 - \dfrac{130}{\pi}\sin\left(\dfrac{\pi t}{26}\right)\Big|_0^{104} = 520$$

Thus, $T = \$520$ hundred or $\$52,000$.

(B) Total profit earned from $t = 13$ to $t = 26$:

$$T = \int_{13}^{26}\left[5 - 5\,\cos\!\left(\frac{\pi t}{26}\right)\right]dt = \int_{13}^{26} 5\,dt - 5\int_{13}^{26}\cos\!\left(\frac{\pi t}{26}\right)dt$$

$$= 5t\Big|_{13}^{26} - \frac{5(26)}{\pi}\int_{13}^{26}\cos\!\left(\frac{\pi t}{26}\right)\!\left(\frac{\pi}{26}\right)dt$$

$$= 65 - \frac{130}{\pi}\,\sin\!\left(\frac{\pi t}{26}\right)\Big|_{13}^{26} = 65 + \frac{130}{\pi} \approx 106.38$$

Thus, $T = \$106.38$ hundred or $\$10,638$.

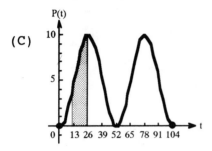

(C)

31. $P(n) = 1 + \cos\!\left(\dfrac{\pi n}{26}\right),\ \ 0 \le n \le 104$

(A) Total amount of pollutants during the two-year period:

$$T = \int_{0}^{104}\left[1 + \cos\!\left(\frac{\pi n}{26}\right)\right]dn = \int_{0}^{104} dn + \int_{0}^{104}\cos\!\left(\frac{\pi n}{26}\right)dn$$

$$= n\Big|_{0}^{104} + \frac{26}{\pi}\int_{0}^{104}\cos\!\left(\frac{\pi n}{26}\right)\!\left(\frac{\pi}{26}\right)dn = 104 + \frac{26}{\pi}\,\sin\!\left(\frac{\pi n}{26}\right)\Big|_{0}^{104} = 104 \text{ tons}$$

(B) Total amount of pollutants from the 13th week to the 52nd week:

$$T = \int_{13}^{52}\left[1 + \cos\!\left(\frac{\pi n}{26}\right)\right]dn = \int_{13}^{52} dn + \int_{13}^{52}\cos\!\left(\frac{\pi n}{26}\right)dn$$

$$= n\Big|_{13}^{52} + \frac{26}{\pi}\int_{13}^{52}\cos\!\left(\frac{\pi n}{26}\right)\!\left(\frac{\pi}{26}\right)dn = 39 + \frac{26}{\pi}\,\sin\!\left(\frac{\pi n}{26}\right)\Big|_{13}^{52}$$

$$= 39 - \frac{26}{\pi} \approx 31 \text{ tons}$$

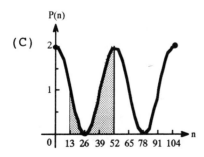

(C)

1. (A) $\dfrac{\theta°}{180°} = \dfrac{\theta_{rad}}{\pi}$

   $\theta_{rad} = \dfrac{\theta°\pi}{180°}$ [Note: θ_{rad} = radian measure of θ.]

   $\theta° = 30°$

   $\theta_{rad} = \dfrac{30°\pi}{180°} = \dfrac{\pi}{6}$

   (B) $\theta° = 45°$ $\qquad\qquad$ (C) $\theta° = 60°$ $\qquad\qquad$ (D) $\theta° = 90°$

   $\quad\theta_{rad} = \dfrac{45°\pi}{180°} = \dfrac{\pi}{4}$ $\qquad$ $\theta_{rad} = \dfrac{60°\pi}{180°} = \dfrac{\pi}{3}$ $\qquad$ $\theta_{rad} = \dfrac{90°\pi}{180°} = \dfrac{\pi}{2}$

   $\hfill$ (9-1)

2. (A) $\cos \pi = -1$ $\qquad$ (B) $\sin 0 = 0$ $\qquad$ (C) $\sin \dfrac{\pi}{2} = 1$ $\hfill$ (9-1)

3. $\dfrac{d}{dm}\cos m = -\sin m$ $\qquad$ (9-2) $\qquad$ 4. $\dfrac{d}{du}\sin u = \cos u$ $\qquad$ (9-2)

5. $\dfrac{d}{dx}\sin(x^2 - 2x + 1) = \cos(x^2 - 2x + 1)\dfrac{d}{dx}(x^2 - 2x + 1)$

   $\qquad\qquad\qquad\qquad = (2x - 2)\cos(x^2 - 2x + 1)$ $\hfill$ (9-2)

6. $\displaystyle\int \sin 3t\, dt = \dfrac{1}{3}\int \sin(3t)3\, dt = \dfrac{1}{3}\int \sin u\, du$ [Let $u = 3t$, then $du = 3\, dt$.]

   $\qquad\qquad = -\dfrac{1}{3}\cos u + C = -\dfrac{1}{3}\cos 3t + C$ $\hfill$ (9-3)

7. (A) $\theta° = \dfrac{180°\theta_{rad}}{\pi}$ $\qquad\qquad\qquad$ (B) $\theta_{rad} = \dfrac{\pi}{4}$

   $\quad\theta_{rad} = \dfrac{\pi}{6}$

   $\qquad\qquad\qquad\qquad\qquad\qquad\qquad\theta° = \dfrac{180°\left(\dfrac{\pi}{4}\right)}{\pi} = 45°$

   $\quad\theta° = \dfrac{180°\left(\dfrac{\pi}{6}\right)}{\pi} = 30°$

   (C) $\theta_{rad} = \dfrac{\pi}{3}$ $\qquad\qquad\qquad\qquad$ (D) $\theta_{rad} = \dfrac{\pi}{2}$

   $\quad\theta° = \dfrac{180°\left(\dfrac{\pi}{3}\right)}{\pi} = 60°$ $\qquad\qquad$ $\theta° = \dfrac{180°\left(\dfrac{\pi}{2}\right)}{\pi} = 90°$ $\hfill$ (9-1)

8. (A) $\sin \dfrac{\pi}{6} = \dfrac{1}{2}$ $\quad$ (B) $\cos \dfrac{\pi}{4} = \dfrac{\sqrt{2}}{2}$ $\quad$ (C) $\sin \dfrac{\pi}{3} = \dfrac{\sqrt{3}}{2}$ $\hfill$ (9-1)

9. (A) $\cos 33.7 \approx -0.6543$ $\qquad$ (B) $\sin(-118.4) \approx 0.8308$ $\hfill$ (9-1)

10. $\dfrac{d}{dx}(x^2 - 1)\sin x = (x^2 - 1)\dfrac{d}{dx}\sin x + \sin x \dfrac{d}{dx}(x^2 - 1)$

    $\qquad\qquad\qquad\qquad = (x^2 - 1)\cos x + 2x \sin x$ $\hfill$ (9-2)

11. $\frac{d}{dx}(\sin x)^6 = 6(\sin x)^5 \frac{d}{dx}\sin x = 6(\sin x)^5 \cos x$ (9-2)

12. $\frac{d}{dx}\sqrt[3]{\sin x} = \frac{d}{dx}(\sin x)^{1/3} = \frac{1}{3}(\sin x)^{-2/3}\frac{d}{dx}\sin x = \frac{1}{3}(\sin x)^{-2/3}\cos x$

(9-2)

13. $\int t \cos(t^2 - 1)\,dt = \frac{1}{2}\int \cos(t^2 - 1)2t\,dt$ [Let $u = (t^2 - 1)$, then $du = 2t\,dt$.]

$$= \frac{1}{2}\int \cos u\,du = \frac{1}{2}\sin u + C = \frac{1}{2}\sin(t^2 - 1) + C \quad (9\text{-}3)$$

14. $\displaystyle\int_0^\pi \sin u\,du = -\cos u\Big|_0^\pi = -[\cos \pi - \cos 0] = 2$ (9-3)

15. $\displaystyle\int_0^{\pi/3} \cos x\,dx = \sin x\Big|_0^{\pi/3} = \sin \frac{\pi}{3} - \sin 0 = \frac{\sqrt{3}}{2}$ (9-3)

16. $\displaystyle\int_1^{2.5} \cos x\,dx = \sin x\Big|_1^{2.5} = \sin(2.5) - \sin(1) \approx 0.5985 - 0.8415 = -0.2430$

(9-3)

17. $y = \cos x,\ y' = -\sin x$

$y'\Big|_{x=\pi/4} = -\sin\left(\frac{\pi}{4}\right) = -\frac{\sqrt{2}}{2}$ (9-2)

18. $A = \displaystyle\int_{\pi/4}^{3\pi/4} \sin x\,dx = -\cos x\Big|_{\pi/4}^{3\pi/4} = \left(-\cos\frac{3\pi}{4}\right) - \left(-\cos\frac{\pi}{4}\right) = \sqrt{2}$ (9-3)

19. $f(x) = \dfrac{\sin x}{x},\ 1 \le x \le 5$

(A)

(B) $a = 1,\ b = 5,\ n = 4,\ \Delta x = \dfrac{5 - 1}{4} = 1$

$M_4 = f\left(\frac{3}{2}\right)\Delta x + f\left(\frac{5}{2}\right)\Delta x + f\left(\frac{7}{2}\right)\Delta x + f\left(\frac{9}{2}\right)\Delta x$

$\quad = 1[0.6650 + 0.2394 + (-0.1002) + (-0.2172)]$

$\quad = 0.587$

(C) $f'(x) = \dfrac{x \cos x - \sin x}{x^2}$;

$f''(x) = \dfrac{-x^2 \sin x - 2x \cos x + 2 \sin x}{x^3}$

The graph of $|f''(x)|$ is shown at the right.

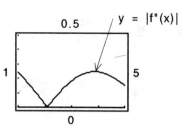

(D) $|I - M_4| \leq \dfrac{\max |f''(x)|(b - a)^3}{24n^2} = \dfrac{0.3(5 - 1)^3}{24(4)^2} = 0.05$ (9-3)

20. $\theta_{rad} = \dfrac{\theta° \pi}{180°}$

Set $\theta° = 15°$. Then $\theta_{rad} = \dfrac{15° \pi}{180°} = \dfrac{\pi}{12}$ (9-1)

21. (A) $\sin\left(\dfrac{3\pi}{2}\right) = -1$ (B) $\cos\left(\dfrac{5\pi}{6}\right) = -\dfrac{\sqrt{3}}{2}$ (C) $\sin\left(-\dfrac{\pi}{6}\right) = -\dfrac{1}{2}$ (9-1)

22. $\dfrac{d}{du} \tan u = \dfrac{d}{du} \dfrac{\sin u}{\cos u} = \dfrac{\cos u \dfrac{d}{du} \sin u - \sin u \dfrac{d}{du} \cos u}{[\cos u]^2}$

$= \dfrac{\cos u(\cos u) - \sin u(-\sin u)}{\cos^2 u}$

$= \dfrac{[\cos u]^2 + [\sin u]^2}{[\cos u]^2} = \dfrac{1}{[\cos u]^2} = [\sec u]^2$ (9-2)

23. $\dfrac{d}{dx} e^{\cos x^2} = e^{\cos x^2} \dfrac{d}{dx} \cos x^2 = e^{\cos x^2}(-\sin x^2) \dfrac{d}{dx} x^2$

$= e^{\cos x^2}(-\sin x^2)2x = -2x \sin x^2 e^{\cos x^2}$ (9-2)

24. $\displaystyle\int e^{\sin x} \cos x \, dx = \int e^u \, du$ [Let $u = \sin x$, then $du = \cos x \, dx$.]

$= e^u + C = e^{\sin x} + C$ (9-3)

25. $\displaystyle\int \tan x \, dx = \int \dfrac{\sin x}{\cos x} dx = -\int \dfrac{1}{\cos x}(-\sin x) \, dx = -\int \dfrac{1}{u} du$

[Let $u = \cos x$, $= -\ln|u| + C$
then $du = -\sin x \, dx$.] $= -\ln|\cos x| + C$ (9-3)

26. $\displaystyle\int_2^5 (5 + 2 \cos 2x) \, dx = \int_2^5 5 \, dx + \int_2^5 \cos(2x)2 \, dx = 5x \Big|_2^5 + \sin 2x \Big|_2^5$

[<u>Note</u>: In the second integral,
we let $u = 2x$ and $du = 2 \, dx$.]

$= 25 - 10 + \sin 10 - \sin 4$
$= 15 - 0.5440 + 0.7568$
$= 15.2128$ (9-3)

27. $y = \dfrac{\sin \pi x}{0.2x}$;

$1 \le x \le 8,\ -4 \le y \le 4$

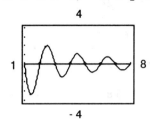

4

1 ⊢ 8

-4

(9-2, 9-3)

28. $y = 0.5x \cos \pi x$;

$0 \le x \le 8,\ -5 \le y \le 5$

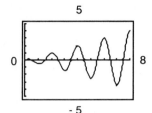

5

0 ⊢ 8

-5

(9-2, 9-3)

29. $y = 3 - 2 \cos \pi x$;

$0 \le x \le 6,$

$0 \le y \le 5$

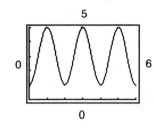

5

0 ⊢ 6

0

(9-2, 9-3)

30. $R(t) = 3 + 2 \cos\left(\dfrac{\pi t}{6}\right),\ 0 \le t \le 24$

(A) $R(0) = 3 + 2 \cos(0) = 3 + 2(1) = 5$

$R(2) = 3 + 2 \cos\left(\dfrac{2\pi}{6}\right) = 3 + 2\left(\dfrac{1}{2}\right) = 4$

$R(3) = 3 + 2 \cos\left(\dfrac{3\pi}{6}\right) = 3 + 2(0) = 3$

$R(6) = 3 + 2 \cos\left(\dfrac{6\pi}{6}\right) = 3 + 2(-1) = 1$

(B) $R(1) = 3 + 2 \cos\left(\dfrac{\pi}{6}\right) = 3 + 2\left(\dfrac{\sqrt{3}}{2}\right) \approx 4.732$

$R(22) = 3 + 2 \cos\left(\dfrac{22\pi}{6}\right) = 3 + 2\left(\dfrac{1}{2}\right) = 4$

Interpretation: The revenue is \$4,732 for a month of sweater sales 1 month after January 1; the revenue is \$4,000 for a month of sweater sales 22 months after January 1. (9-1)

31. (A) $R'(t) = 2\left[-\sin\left(\dfrac{\pi t}{6}\right)\left(\dfrac{\pi}{6}\right)\right] = -\dfrac{\pi}{3} \sin\left(\dfrac{\pi t}{6}\right),\ 0 \le t \le 24$

(B) $R'(3) = -\dfrac{\pi}{3} \sin\left(\dfrac{3\pi}{6}\right) = -\dfrac{\pi}{3} \approx -\$1.047$ thousand or $-\$1047$ per month

$R'(10) = -\dfrac{\pi}{3} \sin\left(\dfrac{10\pi}{6}\right) = -\dfrac{\pi}{3}\left(-\dfrac{\sqrt{3}}{2}\right) \approx \$0.907$ thousand or \$907 per month

$R'(18) = -\dfrac{\pi}{3} \sin\left(\dfrac{18\pi}{6}\right) = -\dfrac{\pi}{3}(0) = \$0.00$

(C) Critical values: $R'(t) = -\dfrac{\pi}{3} \sin\left(\dfrac{\pi t}{6}\right) = 0$

$$\sin\left(\dfrac{\pi t}{6}\right) = 0, \quad 0 < t < 24$$

$$\dfrac{\pi t}{6} = \pi \text{ or } t = 6$$

$$\dfrac{\pi t}{6} = 2\pi \text{ or } t = 12$$

$$\dfrac{\pi t}{6} = 3\pi \text{ or } t = 18$$

$$R''(t) = -\dfrac{\pi}{3} \cos\left(\dfrac{\pi t}{6}\right)\left(\dfrac{\pi}{6}\right) = \dfrac{-\pi^2}{18} \cos\left(\dfrac{\pi t}{6}\right)$$

$$R''(6) = \dfrac{-\pi^2}{18} \cos \pi = \dfrac{\pi^2}{18} > 0$$

$$R''(12) = \dfrac{-\pi^2}{18} \cos(2\pi) = \dfrac{-\pi^2}{18} < 0$$

$$R''(18) = \dfrac{-\pi^2}{18} \cos(3\pi) = \dfrac{\pi^2}{18} > 0$$

Thus,

t	$R(t)$	
6	\$1000	local minimum
12	\$5000	local maximum
18	\$1000	local minimum

(D)

t	$R(t)$	
0	\$5000	absolute maximum
6	\$1000	absolute minimum
12	\$5000	absolute maximum
18	\$1000	absolute minimum
24	\$5000	absolute maximum

(E) The results in part (C) are illustrated
by the graph of R shown at the right.

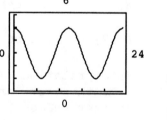

(9-2)

32. (A) Total revenue: $T = \displaystyle\int_0^{24}\left[3 + 2\cos\left(\dfrac{\pi t}{6}\right)\right]dt = \int_0^{24} 3\ dt + 2\int_0^{24}\cos\left(\dfrac{\pi t}{6}\right)dt$

$$= 3t\Big|_0^{24} + \dfrac{12}{\pi}\int_0^{24}\cos\left(\dfrac{\pi t}{6}\right)\left(\dfrac{\pi}{6}\right)dt = 72 + \dfrac{12}{\pi}\sin\left(\dfrac{\pi t}{6}\right)\Big|_0^{24}$$

$$= 72 + \dfrac{12}{\pi}[\sin 4\pi - \sin 0] = \$72 \text{ thousand or } \$72,000$$

(B) Total revenue: $T = \int_5^9 \left[3 + 2\cos\left(\frac{\pi t}{6}\right)\right] dt = \int_5^9 3\, dt + \frac{12}{\pi}\int_5^9 \cos\left(\frac{\pi t}{6}\right)\left(\frac{\pi}{6}\right) dt$

$\qquad\qquad\qquad = 3t\Big|_5^9 + \frac{12}{\pi}\sin\left(\frac{\pi t}{6}\right)\Big|_5^9$

$\qquad\qquad\qquad = 3(9 - 5) + \frac{12}{\pi}\left[\sin\left(\frac{9\pi}{6}\right) - \sin\left(\frac{5\pi}{6}\right)\right]$

$\qquad\qquad\qquad = 12 + \frac{12}{\pi}\left(-1 - \frac{1}{2}\right) = 12 - \frac{18}{\pi}$ $6.270 thousand or $6270

(C)

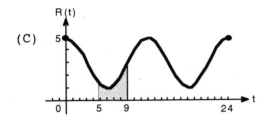

(9-3)

CHAPTER A   BASIC ALGEBRA REVIEW

1. $(3x - 4) + (x + 2) + (3x^2 + x - 8) + (x^3 + 8)$
$= 3x - 4 + x + 2 + 3x^2 + x - 8 + x^3 + 8 = x^3 + 3x^2 + 5x - 2$ (A-1)

2. $[(x + 2) + (x^3 + 8)] - [(3x - 4) + (3x^2 + x - 8)]$
$= x^3 + x + 10 - [3x^2 + 4x - 12]$
$= x^3 + x + 10 - 3x^2 - 4x + 12 = x^3 - 3x^2 - 3x + 22$ (A-1)

3. $(x^3 + 8)(3x^2 + x - 8) = x^3(3x^2 + x - 8) + 8(3x^2 + x - 8)$
$= 3x^5 + x^4 - 8x^3 + 24x^2 + 8x - 64$ (A-1)

4. $x^3 + 8$ has degree $\underline{\underline{3}}$ (A-1)

5. The coefficient of the second term in $3x^2 + x - 8$ is $\underline{\underline{1}}$ (A-1)

6. $5x^2 - 3x[4 - 3(x - 2)] = 5x^2 - 3x[4 - 3x + 6]$
$= 5x^2 - 3x(-3x + 10)$
$= 5x^2 + 9x^2 - 30x$
$= 14x^2 - 30x$ (A-1)

7. $(2x + y)(3x - 4y) = 6x^2 - 8xy + 3xy - 4y^2$
$= 6x^2 - 5xy - 4y^2$ (A-1)

8. $(2a - 3b)^2 = (2a)^2 - 2(2a)(3b) + (3b)^2$
$= 4a^2 - 12ab + 9b^2$ (A-1)

9. $(2x - y)(2x + y) - (2x - y)^2 = (2x)^2 - y^2 - (4x^2 - 4xy + y^2)$
$= 4x^2 - y^2 - 4x^2 + 4xy - y^2$
$= 4xy - 2y^2$ (A-1)

10. $(m^2 + 2mn - n^2)(m^2 - 2mn - n^2)$
$= m^2(m^2 - 2mn - n^2) + 2mn(m^2 - 2mn - n^2) - n^2(m^2 - 2mn - n^2)$
$= m^4 - 2m^3n - m^2n^2 + 2m^3n - 4m^2n^2 - 2mn^3 - m^2n^2 + 2mn^3 + n^4$
$= m^4 - 6m^2n^2 + n^4$ (A-1)

11. $(x - 2y)^3 = (x - 2y)(x - 2y)^2$
$= (x - 2y)(x^2 - 4xy + 4y^2)$
$= x(x^2 - 4xy + 4y^2) - 2y(x^2 - 4xy + 4y^2)$
$= x^3 - 4x^2y + 4xy^2 - 2x^2y + 8xy^2 - 8y^3$
$= x^3 - 6x^2y + 12xy^2 - 8y^3$ (A-1)

12. (A) $4,065,000,000,000 = 4.065 \times 10^{12}$

(B) $0.0073 = 7.3 \times 10^{-3}$ (A-4)

13. (A) $2.55 \times 10^8 = 255,000,000$ (B) $4.06 \times 10^{-4} = 0.000\,406$ (A-4)

14. $6(xy^3)^5 = 6x^5y^{15}$ (A-4)   **15.** $\dfrac{9u^8v^6}{3u^4v^8} = \dfrac{3^2u^8v^6}{3u^4v^8} = \dfrac{3^{2-1}u^{8-4}}{v^{8-6}} = \dfrac{3u^4}{v^2}$ (A-4)

16. $(2 \times 10^5)(3 \times 10^{-3}) = 6 \times 10^{5-3} = 6 \times 10^2 = 600$ (A-4)

17. $(x^{-3}y^2)^{-2} = x^6y^{-4} = \dfrac{x^6}{y^4}$ (A-4)   **18.** $u^{5/3}u^{2/3} = u^{5/3+2/3} = u^{7/3}$ (A-5)

19. $(9a^4b^{-2})^{1/2} = (3^2a^4b^{-2})^{1/2} = (3^2)^{1/2}(a^4)^{1/2}(b^{-2})^{1/2} = 3a^2b^{-1} = \dfrac{3a^2}{b}$ (A-5)

20. $\dfrac{5^0}{3^2} + \dfrac{3^{-2}}{2^{-2}} = \dfrac{1}{3^2} + \dfrac{\frac{1}{3^2}}{\frac{1}{2^2}} = \dfrac{1}{9} + \dfrac{1}{9} \cdot \dfrac{4}{1} = \dfrac{5}{9}$ (A-4)

21. $(x^{1/2} + y^{1/2})^2 = (x^{1/2})^2 + 2x^{1/2}y^{1/2} + (y^{1/2})^2 = x + 2x^{1/2}y^{1/2} + y$ (A-5)

22. $(3x^{1/2} - y^{1/2})(2x^{1/2} + 3y^{1/2}) = 6x + 9x^{1/2}y^{1/2} - 2x^{1/2}y^{1/2} - 3y$
$$= 6x + 7x^{1/2}y^{1/2} - 3y \qquad \text{(A-5)}$$

23. $12x^2 + 5x - 3$
    $a = 12, \ b = 5, \ c = -3$
    <u>Step 1</u>. Use the ac-test
            $ac = 12(-3) = -36$

pq
$(1)(-36)$
$(-1)(36)$
$(2)(-18)$
$(-2)(18)$
$(3)(-12)$
$(-3)(12)$
$(4)(-9)$
$\boxed{(-4)(9)}$
$\vdots$

        $-4 + 9 = 5 = b$
        $12x^2 + 5x - 3 = 12x^2 - 4x + 9x - 3$
                          $= (12x^2 - 4x) + (9x - 3)$
                          $= 4x(3x - 1) + 3(3x - 1)$
                          $= (3x - 1)(4x + 3)$ (A-2)

24. $8x^2 - 18xy + 9y^2$
    $a = 8, \ b = -18, \ c = 9$
    $ac = (8)(9) = 72$
    Note that $(-6)(-12) = 72$ and $-12 - 6 = -18$. Thus
    $8x^2 - 18xy + 9y^2 = 8x^2 - 12xy - 6xy + 9y^2$
                        $= (8x^2 - 12xy) - (6xy - 9y^2)$
                        $= 4x(2x - 3y) - 3y(2x - 3y)$
                        $= (2x - 3y)(4x - 3y)$ (A-2)

25. $t^2 - 4t - 6$ This polynomial cannot be factored further.   (A-2)

26. $6n^3 - 9n^2 - 15n = 3n(2n^2 - 3n - 5) = 3n(2n - 5)(n + 1)$ (A-2)

27. $(4x - y)^2 - 9x^2 = (4x - y - 3x)(4x - y + 3x) = (x - y)(7x - y)$ (A-2)

28. $2x^2 + 4xy - 5y^2$ cannot be factored further. (A-2)

29. $\dfrac{2}{5b} - \dfrac{4}{3a^3} - \dfrac{1}{6a^2b^2}$ LCD $= 30a^3b^2$

$= \dfrac{6a^3b}{6a^3b} \cdot \dfrac{2}{5b} - \dfrac{10b^2}{10b^2} \cdot \dfrac{4}{3a^3} - \dfrac{5a}{5a} \cdot \dfrac{1}{6a^2b^2}$

$= \dfrac{12a^3b}{30a^3b^2} - \dfrac{40b^2}{30a^3b^2} - \dfrac{5a}{30a^3b^2} = \dfrac{12a^3b - 40b^2 - 5a}{30a^3b^2}$ (A-3)

30. $\dfrac{3x}{3x^2 - 12x} + \dfrac{1}{6x} = \dfrac{3x}{3x(x - 4)} + \dfrac{1}{6x} = \dfrac{1}{x - 4} + \dfrac{1}{6x}$

$= \dfrac{6x}{6x(x - 4)} + \dfrac{x - 4}{6x(x - 4)} = \dfrac{6x + x - 4}{6x(x - 4)} = \dfrac{7x - 4}{6x(x - 4)}$ (A-3)

31. $\dfrac{x}{x^2 - 16} - \dfrac{x + 4}{x^2 - 4x} = \dfrac{x}{(x - 4)(x + 4)} - \dfrac{x + 4}{x(x - 4)}$

[LCD $= x(x - 4)(x + 4)$]

$= \dfrac{x^2}{x(x - 4)(x + 4)} - \dfrac{(x + 4)^2}{x(x - 4)(x + 4)}$

$= \dfrac{x^2 - (x + 4)^2}{x(x - 4)(x + 4)}$

$= \dfrac{x^2 - x^2 - 8x - 16}{x(x - 4)(x + 4)} = \dfrac{-8(x + 2)}{x(x - 4)(x + 4)}$ (A-3)

32. $\dfrac{y - 2}{y^2 - 4y + 4} \div \dfrac{y^2 + 2y}{y^2 + 4y + 4} = \dfrac{y - 2}{(y - 2)^2} \cdot \dfrac{(y + 2)^2}{y(y + 2)} = \dfrac{y + 2}{y(y - 2)}$ (A-3)

33. $\dfrac{\dfrac{1}{7 + h} - \dfrac{1}{7}}{h} = \dfrac{\dfrac{7 - (7 + h)}{7(7 + h)}}{h} = \dfrac{\dfrac{-h}{7(7 + h)}}{\dfrac{h}{1}} = \dfrac{-h}{7(7 + h)} \cdot \dfrac{1}{h} = \dfrac{-1}{7(7 + h)}$ (A-3)

34. $\dfrac{x^{-1} + y^{-1}}{x^{-2} - y^{-2}} = \dfrac{\dfrac{1}{x} + \dfrac{1}{y}}{\dfrac{1}{x^2} - \dfrac{1}{y^2}} = \dfrac{\dfrac{x + y}{xy}}{\dfrac{y^2 - x^2}{x^2y^2}} = \dfrac{x + y}{xy} \cdot \dfrac{x^2y^2}{(y - x)(y + x)} = \dfrac{xy}{y - x}$ (A-5)

35. $6\sqrt[5]{x^2} - 7\sqrt[4]{(x - 1)^3} = 6x^{2/5} - 7(x - 1)^{3/4}$ (A-5)

36. $2x^{1/2} - 3x^{2/3} = 2\sqrt{x} - 3\sqrt[3]{x^2}$ (A-5)

37. $\dfrac{4\sqrt{x} - 3}{2\sqrt{x}} = \dfrac{4x^{1/2}}{2x^{1/2}} - \dfrac{3}{2x^{1/2}} = 2 - \dfrac{3}{2}x^{-1/2}$ (A-5)

38. $\dfrac{3x}{\sqrt{3x}} = \dfrac{3x}{\sqrt{3x}} \cdot \dfrac{\sqrt{3x}}{\sqrt{3x}} = \dfrac{3x\sqrt{3x}}{3x} = \sqrt{3x}$ (A-5)

39. $\dfrac{x - 5}{\sqrt{x} - \sqrt{5}} = \dfrac{x - 5}{\sqrt{x} - \sqrt{5}} \cdot \dfrac{\sqrt{x} + \sqrt{5}}{\sqrt{x} + \sqrt{5}} = \dfrac{\cancel{(x - 5)}(\sqrt{x} + \sqrt{5})}{\cancel{x - 5}} = \sqrt{x} + \sqrt{5}$ (A-5)

40. $\dfrac{\sqrt{x - 5}}{x - 5} = \dfrac{\sqrt{x - 5}}{x - 5} \cdot \dfrac{\sqrt{x - 5}}{\sqrt{x - 5}} = \dfrac{\cancel{x - 5}}{\cancel{(x - 5)}\sqrt{x - 5}} = \dfrac{1}{\sqrt{x - 5}}$ (A-5)

41. $\dfrac{\sqrt{u + h} - \sqrt{u}}{h} = \dfrac{\sqrt{u + h} - \sqrt{u}}{h} \cdot \dfrac{\sqrt{u + h} + \sqrt{u}}{\sqrt{u + h} + \sqrt{u}}$

$\qquad = \dfrac{u + h - u}{h(\sqrt{u + h} + \sqrt{u})}$

$\qquad = \dfrac{\cancel{h}}{\cancel{h}(\sqrt{u + h} + \sqrt{u})} = \dfrac{1}{\sqrt{u + h} + \sqrt{u}}$ (A-5)

42. $\dfrac{x}{12} - \dfrac{x - 3}{3} = \dfrac{1}{2}$

Multiply each term by 12: $x - 4(x - 3) = 6$

$\qquad\qquad\qquad\qquad\quad x - 4x + 12 = 6$

$\qquad\qquad\qquad\qquad\qquad\quad -3x = 6 - 12$

$\qquad\qquad\qquad\qquad\qquad\quad -3x = -6$

$\qquad\qquad\qquad\qquad\qquad\quad\;\; x = 2$ (A-6)

43. $\qquad x^2 = 5x$

$x^2 - 5x = 0$ (solve by factoring)

$x(x - 5) = 0$

$x = 0$ or $x - 5 = 0$

$\qquad\qquad\qquad x = 5$ (A-7)

44. $3x^2 - 21 = 0$

$\quad x^2 - 7 = 0$ (solve by the square root method)

$\qquad\quad x^2 = 7$

$\qquad\quad\; x = \pm\sqrt{7}$ (A-7)

45. $\qquad x^2 - x - 20 = 0$ (solve by factoring)

$(x - 5)(x + 4) = 0$

$x - 5 = 0$ or $x + 4 = 0$

$\qquad x = 5 \qquad\qquad x = -4$ (A-7)

46. $2x = 3 + \dfrac{1}{x}$ ($x \neq 0$)

$2x^2 = 3x + 1$ (multiply by x)

$2x^2 - 3x - 1 = 0$

$x = \dfrac{-(-3) \pm \sqrt{(-3)^2 - (4)(2)(-1)}}{2(2)}$

$\quad = \dfrac{3 \pm \sqrt{9 + 8}}{4} = \dfrac{3 \pm \sqrt{17}}{4}$ (A-7)

47. $2(x + 4) > 5x - 4$

$\quad 2x + 8 > 5x - 4$

$\; 2x - 5x > -4 - 8$

$\qquad -3x > -12$ (Divide both sides by -3 and reverse the inequality)

$\qquad\qquad x < 4$ or $(-\infty, 4)$

   (A-6)

48. $1 - \dfrac{x-3}{3} \leq \dfrac{1}{2}$

Multiply both sides of the inequality by 6. We do not reverse the direction of the inequality, since $6 > 0$.

$6 - 2(x-3) \leq 3$

$\quad 6 - 2x + 6 \leq 3$

$\qquad\qquad -2x \leq 3 - 12$

$\qquad\qquad -2x \leq -9$

Divide both sides by -2 and reverse the direction of the inequality, since $-2 < 0$.

$x \geq \dfrac{9}{2}$ or $\left[\dfrac{9}{2},\ \infty\right)$

(A-6)

49. $\qquad -2 \leq \dfrac{x}{2} - 3 < 3$

$-2 + 3 \leq \dfrac{x}{2} < 3 + 3$

$\qquad 1 \leq \dfrac{x}{2} < 6$

$\qquad 2 \leq x < 12$ or $[2, 12)$

(A-6)

50. $2x - 3y = 6$

$\quad -3y = -2x + 6$

$\qquad y = \dfrac{2}{3}x - 2$ (A-6)

51. $\quad xy - y = 3$

$\quad y(x-1) = 3$

$\qquad y = \dfrac{3}{x-1}$ (A-6)

52. The GDP per person is given by:

$$\dfrac{5{,}951{,}000{,}000{,}000}{255{,}100{,}000} = \dfrac{5.951 \times 10^{12}}{2.551 \times 10^{8}} \approx 2.3328 \times 10^{4} = \$23{,}328 \qquad \text{(A-4)}$$

53. Let x = the amount invested at 8%. Then $60{,}000 - x$ = amount invested at 14%. The interest on \$60,000 at 12% for one year is:

$0.12(60{,}000) = 7200$

Thus, we want:

$0.08x + 0.14(60{,}000 - x) = 7200$

$\quad 0.08x + 8400 - 0.14x = 7200$

$\qquad\qquad\qquad\quad -0.06x = -1200$

$\qquad\qquad\qquad\qquad\quad x = 20{,}000$

Therefore, \$20,000 should be invested at 8% and \$40,000 should be invested at 14%.

(A-6)

54. Let x = number of tapes produced

$\quad$ Cost: $C = 12x + 72{,}000$

Revenue: $R = 30x$

To find the break-even point, set $R = C$.

$\qquad 30x = 12x + 72{,}000$

$\qquad 18x = 72{,}000$

$\qquad\quad x = 4000$

Thus, 4000 tapes must be sold for the producer to break even.   (A-6)

Things to remember:

1. **NATURAL NUMBER* EXPONENT**

   For n a natural number and b any real number,

   $b^n = b \cdot b \cdot \dots \cdot b$, n factors of b.

   For example, $2^3 = 2 \cdot 2 \cdot 2 \ (= 8)$,
   $3^5 = 3 \cdot 3 \cdot 3 \cdot 3 \cdot 3 \ (= 243)$.

   In the expression b^n, n is called the EXPONENT or POWER, and b is called the BASE.

   * The natural numbers are the counting numbers: 1, 2, 3, 4, ….

2. **FIRST PROPERTY OF EXPONENTS**

   For any natural numbers m and n, and any real number b,

   $b^m \cdot b^n = b^{m+n}$.

   For example, $3^3 \cdot 3^4 = 3^{3+4} = 3^7$.

3. **POLYNOMIALS**

   a. A POLYNOMIAL IN ONE VARIABLE x is constructed by adding or subtracting constants and terms of the form ax^n, where a is a real number, called the COEFFICIENT of the term, and n is a natural number.

   b. A POLYNOMIAL IN TWO VARIABLES x AND y is constructed by adding or subtracting constants and terms of the form $ax^m y^n$, bx^k, cy^j, where a, b, and c are real numbers, called COEFFICIENTS and m, n, k, and j are natural numbers.

   c. Polynomials in more than two variables are defined similarly.

   d. A polynomial with only one term is called a MONOMIAL.
   A polynomial with two terms is called a BINOMIAL.
   A polynomial with three terms is called a TRINOMIAL.

4. **DEGREE OF A POLYNOMIAL**

   a. A term of the form ax^n, $a \neq 0$, has degree n. A term of the form $ax^m y^n$, $a \neq 0$, has degree $m + n$. A nonzero constant has degree 0.

b. The DEGREE OF A POLYNOMIAL is the degree of the nonzero term with the highest degree. For example, $3x^4 + \sqrt{2}\,x^3 - 2x + 7$ has degree 4; $2x^3y^2 - 3x^2y + 7x^4 - 5y^3 + 6$ has degree 5; the polynomial 4 has degree 0.

c. The constant 0 is a polynomial but it is not assigned a degree.

5. Two terms in a polynomial are called LIKE TERMS if they have exactly the same variable factors raised to the same powers. For example, in

$$7x^5y^2 - 3x^3y + 2x + 4x^3y - 1,$$

$-3x^3y$ and $4x^3y$ are like terms.

6. To multiply two polynomials, multiply each term of one by each term of the other, and then combine like terms.

7. SPECIAL PRODUCTS

   a. $(a - b)(a + b) = a^2 - b^2$

   b. $(a + b)^2 = a^2 + 2ab + b^2$

   c. $(a - b)^2 = a^2 - 2ab + b^2$

---

1. The term of highest degree in $x^3 + 2x^2 - x + 3$ is x^3 and the degree of this term is 3.

3. $(2x^2 - x + 2) + (x^3 + 2x^2 - x + 3) = x^3 + 2x^2 + 2x^2 - x - x + 2 + 3$
$$= x^3 + 4x^2 - 2x + 5$$

5. $(x^3 + 2x^2 - x + 3) - (2x^2 - x + 2) = x^3 + 2x^2 - x + 3 - 2x^2 + x - 2$
$$= x^3 + 1$$

7. Using a vertical arrangement:

$$
\begin{array}{r}
x^3 + 2x^2 - x + 3 \\
2x^2 - x + 2 \\
\hline
2x^5 + 4x^4 - 2x^3 + 6x^2 \\
- x^4 - 2x^3 + x^2 - 3x \\
2x^3 + 4x^2 - 2x + 6 \\
\hline
2x^5 + 3x^4 - 2x^3 + 11x^2 - 5x + 6
\end{array}
$$

9. $2(u - 1) - (3u + 2) - 2(2u - 3) = 2u - 2 - 3u - 2 - 4u + 6$
$$= -5u + 2$$

11. $4a - 2a[5 - 3(a + 2)] = 4a - 2a[5 - 3a - 6]$
$$= 4a - 2a[-3a - 1]$$
$$= 4a + 6a^2 + 2a$$
$$= 6a^2 + 6a$$

13. $(a + b)(a - b) = a^2 - b^2$ (Special product 7a)

15. $(3x - 5)(2x + 1) = 6x^2 + 3x - 10x - 5$
$$= 6x^2 - 7x - 5$$

17. $(2x - 3y)(x + 2y) = 2x^2 + 4xy - 3xy - 6y^2$
$$= 2x^2 + xy - 6y^2$$

19. $(3y + 2)(3y - 2) = (3y)^2 - 2^2 = 9y^2 - 4$ (Special product 7a)

21. $(3m + 7n)(2m - 5n) = 6m^2 - 15mn + 14mn - 35n^2$
$$= 6m^2 - mn - 35n^2$$

23. $(4m + 3n)(4m - 3n) = 16m^2 - 9n^2$

25. $(3u + 4v)^2 = 9u^2 + 24uv + 16v^2$ (Special product 7b)

27. $(a - b)(a^2 + ab + b^2) = a(a^2 + ab + b^2) - b(a^2 + ab + b^2)$
$$= a^3 + a^2b + ab^2 - a^2b - ab^2 - b^3$$
$$= a^3 - b^3$$

29. $(4x + 3y)^2 = 16x^2 + 24xy + 9y^2$

31. $m - \{m - [m - (m - 1)]\} = m - \{m - [m - m + 1]\}$
$$= m - \{m - 1\}$$
$$= m - m + 1$$
$$= 1$$

33. $(x^2 - 2xy + y^2)(x^2 + 2xy + y^2) = (x - y)^2(x + y)^2$
$$= [(x - y)(x + y)]^2$$
$$= [x^2 - y^2]^2$$
$$= x^4 - 2x^2y^2 + y^4$$

35. $(3a - b)(3a + b) - (2a - 3b)^2 = (9a^2 - b^2) - (4a^2 - 12ab + 9b^2)$
$$= 9a^2 - b^2 - 4a^2 + 12ab - 9b^2$$
$$= 5a^2 + 12ab - 10b^2$$

37. $(m - 2)^2 - (m - 2)(m + 2) = m^2 - 4m + 4 - [m^2 - 4]$
$$= m^2 - 4m + 4 - m^2 + 4$$
$$= -4m + 8$$

39. $(x - 2y)(2x + y) - (x + 2y)(2x - y)$
$$= 2x^2 - 4xy + xy - 2y^2 - [2x^2 + 4xy - xy - 2y^2]$$
$$= 2x^2 - 3xy - 2y^2 - 2x^2 - 3xy + 2y^2$$
$$= -6xy$$

41. $(u + v)^3 = (u + v)(u + v)^2 = (u + v)(u^2 + 2uv + v^2)$
$$= u^3 + 3u^2v + 3uv^2 + v^3$$

43. $(x - 2y)^3 = (x - 2y)(x - 2y)^2 = (x - 2y)(x^2 - 4xy + 4y^2)$
$$= x(x^2 - 4xy + 4y^2) - 2y(x^2 - 4xy + 4y^2)$$
$$= x^3 - 4x^2y + 4xy^2 - 2x^2y + 8xy^2 - 8y^3$$
$$= x^3 - 6x^2y + 12xy^2 - 8y^3$$

45. $[(2x^2 - 4xy + y^2) + (3xy - y^2)] - [(x^2 - 2xy - y^2) + (-x^2 + 3xy - 2y^2)]$
$= [2x^2 - xy] - [xy - 3y^2] = 2x^2 - 2xy + 3y^2$

47. $(2x - 1)^3 - 2(2x - 1)^2 + 3(2x - 1) + 7$
$= (2x - 1)(2x - 1)^2 - 2[4x^2 - 4x + 1] + 6x - 3 + 7$
$= (2x - 1)(4x^2 - 4x + 1) - 8x^2 + 8x - 2 + 6x + 4$
$= 8x^3 - 12x^2 + 6x - 1 - 8x^2 + 14x + 2$
$= 8x^3 - 20x^2 + 20x + 1$

49. $2\{(x - 3)(x^2 - 2x + 1) - x[3 - x(x - 2)]\}$
$= 2\{x^3 - 5x^2 + 7x - 3 - x[3 - x^2 + 2x]\}$
$= 2\{x^3 - 5x^2 + 7x - 3 + x^3 - 2x^2 - 3x\}$
$= 2\{2x^3 - 7x^2 + 4x - 3\}$
$= 4x^3 - 14x^2 + 8x - 6$

51. $m + n$

53. Let x = amount invested at 9%.
Then $10,000 - x$ = amount invested at 12%.
The total annual income I is:
$$I = 0.09x + 0.12(10,000 - x)$$
$$= 1,200 - 0.03x$$

55. Let x = number of tickets at \$10.
Then $3x$ = number of tickets at \$30 and $4,000 - x - 3x = 4,000 - 4x$
= number of tickets at \$50.
The total receipts R are:
$$R = 10x + 30(3x) + 50(4,000 - 4x)$$
$$= 10x + 90x + 200,000 - 200x = 200,000 - 100x$$

57. Let x = number of kilograms of food A.
Then $10 - x$ = number of kilograms of food B.
The total number of kilograms F of fat in the final food mix is:
$$F = 0.02x + 0.06(10 - x)$$
$$= 0.6 - 0.04x$$

EXERCISE A-2

Things to remember:

<u>1</u>. FACTORED FORMS

A polynomial is in FACTORED FORM if it is written as the
product of two or more polynomials. A polynomial with integer
coefficients is FACTORED COMPLETELY if each factor cannot be
expressed as the product of two or more polynomials with
integer coefficients, other than itself and 1.

2. METHODS

    a. Factor out all factors common to all terms, if they are present.

    b. Try grouping terms.

    c. ac-Test for polynomials of the form
$$ax^2 + bx + c \quad \text{or} \quad ax^2 + bxy + cy^2$$
If the product ac has two integer factors p and q whose sum is the coefficient b of the middle term, i.e., if integers p and q exist so that
$$pq = ac \quad \text{and} \quad p + q = b$$
then the polynomials have first-degree factors with integer coefficients. If no such integers exist then the polynomials will not have first-degree factors with integer coefficients; the polynomials are *not factorable*.

3. SPECIAL FACTORING FORMULAS

    a. $u^2 + 2uv + v^2 = (u + v)^2$ Perfect square

    b. $u^2 - 2uv + v^2 = (u - v)^2$ Perfect square

    c. $u^2 - v^2 = (u - v)(u + v)$ Difference of squares

    d. $u^3 - v^3 = (u - v)(u^2 + uv + v^2)$ Difference of cubes

    e. $u^3 + v^3 = (u + v)(u^2 - uv + v^2)$ Sum of cubes

---

1. $3m^2$ is a common factor: $6m^4 - 9m^3 - 3m^2 = 3m^2(2m^2 - 3m - 1)$

3. $2uv$ is a common factor: $8u^3v - 6u^2v^2 + 4uv^3 = 2uv(4u^2 - 3uv + 2v^2)$

5. $(2m - 3)$ is a common factor: $7m(2m - 3) + 5(2m - 3) = (7m + 5)(2m - 3)$

7. $(3c + d)$ is a common factor: $a(3c + d) - 4b(3c + d)$
$$= (a - 4b)(3c + d)$$

9. $2x^2 - x + 4x - 2 = (2x^2 - x) + (4x - 2)$
$$= x(2x - 1) + 2(2x - 1)$$
$$= (2x - 1)(x + 2)$$

11. $3y^2 - 3y + 2y - 2 = (3y^2 - 3y) + (2y - 2)$
$$= 3y(y - 1) + 2(y - 1)$$
$$= (y - 1)(3y + 2)$$

13. $2x^2 + 8x - x - 4 = (2x^2 + 8x) - (x + 4)$
$$= 2x(x + 4) - (x + 4)$$
$$= (x + 4)(2x - 1)$$

15. $wy - wz + xy - xz = (wy - wz) + (xy - xz)$
$$= w(y - z) + x(y - z)$$
$$= (y - z)(w + x)$$

$$\text{or}\quad wy - wz + xy - xz = (wy + xy) - (wz + xz)$$
$$= y(w + x) - z(w + x)$$
$$= (w + x)(y - z)$$

17. $am - bn - bm + an = (am - bm) + (an - bn)$
$$= m(a - b) + n(a - b)$$
$$= (a - b)(m + n)$$

or $am - bn - bm + an = (am + an) - (bm + bn)$
$$= a(m + n) - b(m + n)$$
$$= (m + n)(a - b)$$

19. $3y^2 - y - 2$
    $a = 3, \; b = -1, \; c = -2$

    Step 1. Use the ac-test to test for factorability

    $ac = (3)(-2) = -6$

pq
$(1)(-6)$
$(-1)(6)$
$\boxed{(2)(-3)}$
$(-2)(3)$

    Note that $2 + (-3) = -1 = b$. Thus, $3y^2 - y - 2$ has first-degree factors with integer coefficients.

    Step 2. Split the middle term using $b = p + q$ and factor by grouping.

    $-1 = -3 + 2$
    $3y^2 - y - 2 = 3y^2 - 3y + 2y - 2 = (3y^2 - 3y) + (2y - 2)$
    $$= 3y(y - 1) + 2(y - 1)$$
    $$= (y - 1)(3y + 2)$$

21. $u^2 - 2uv - 15v^2$
    $a = 1, \; b = -2, \; c = -15$

    Step 1. Use the ac-test

    $ac = 1(-15) = -15$

pq
$(1)(-15)$
$(-1)(15)$
$\boxed{(3)(-5)}$
$(-3)(5)$

    Note that $3 + (-5) = -2 = b$. Thus $u^2 - 2uv - 15v^2$ has first-degree factors with integer coefficients.

    Step 2. Factor by grouping

    $-2 = 3 + (-5)$
    $u^2 + 3uv - 5uv - 15v^2 = (u^2 + 3uv) - (5uv + 15v^2)$
    $$= u(u + 3v) - 5v(u + 3v)$$
    $$= (u + 3v)(u - 5v)$$

23. $m^2 - 6m - 3$
  $a = 1, \; b = -6, \; c = -3$
  <u>Step 1</u>. Use the ac-test
      $ac = (1)(-3) = -3$

  $\underline{pq}$
  $(1)(-3)$
  $(-1)(3)$

  None of the factors add up to $-6 = b$. Thus, this polynomial is *not factorable*.

25. $w^2x^2 - y^2 = (wx - y)(wx + y)$ (difference of squares)

27. $9m^2 - 6mn + n^2 = (3m - n)^2$ (perfect square)

29. $y^2 + 16$
  $a = 1, \; b = 0, \; c = 16$
  <u>Step 1</u>. Use the ac-test
      $ac = (1)(16)$

  $\underline{pq}$
  $(1)(16)$
  $(-1)(-16)$
  $(2)(8)$
  $(-2)(-8)$
  $(4)(4)$
  $(-4)(-4)$

  None of the factors add up to $0 = b$. Thus this polynomial is *not factorable*.

31. $4z^2 - 28z + 48 = 4(z^2 - 7z + 12) = 4(z - 3)(z - 4)$

33. $2x^4 - 24x^3 + 40x^2 = 2x^2(x^2 - 12x + 20) = 2x^2(x - 2)(x - 10)$

35. $4xy^2 - 12xy + 9x = x(4y^2 - 12y + 9) = x(2y - 3)^2$

37. $6m^2 - mn - 12n^2 = (2m - 3n)(3m + 4n)$

39. $4u^3v - uv^3 = uv(4u^2 - v^2) = uv[(2u)^2 - v^2] = uv(2u - v)(2u + v)$

41. $2x^3 - 2x^2 + 8x = 2x(x^2 - x + 4)$ [<u>Note</u>: $x^2 - x + 4$ is *not factorable*.]

43. $r^3 - t^3 = (r - t)(r^2 + rt + t^2)$ (difference of cubes)

45. $a^3 + 1 = (a + 1)(a^2 - a + 1)$ (sum of cubes)

47. $(x + 2)^2 - 9y^2 = [(x + 2) - 3y][(x + 2) + 3y]$
  $= (x + 2 - 3y)(x + 2 + 3y)$

49. $5u^2 + 4uv - 2v^2$ is *not factorable*.

51. $6(x - y)^2 + 23(x - y) - 4 = [6(x - y) - 1][(x - y) + 4]$
  $= (6x - 6y - 1)(x - y + 4)$

53. $y^4 - 3y^2 - 4 = (y^2)^2 - 3y^2 - 4 = (y^2 - 4)(y^2 + 1)$
  $= (y - 2)(y + 2)(y^2 + 1)$

55. $27a^2 + a^5b^3 = a^2(27 + a^3b^3) = a^2[3^3 + (ab)^3]$
  $= a^2(3 + ab)(9 - 3ab + a^2b^2)$

Things to remember:

<u>1</u>. FUNDAMENTAL PROPERTY OF FRACTIONS

If a, b, and k are real numbers with b, $k \neq 0$, then
$$\frac{ka}{kb} = \frac{a}{b}.$$

A fraction is in LOWEST TERMS if the numerator and denominator have no common factors other than 1 or −1.

<u>2</u>. MULTIPLICATION AND DIVISION

For a, b, c, and d real numbers:

a. $\dfrac{a}{b} \cdot \dfrac{c}{d} = \dfrac{ac}{bd}$, b, $d \neq 0$

b. $\dfrac{a}{b} \div \dfrac{c}{d} = \dfrac{\dfrac{a}{b}}{\dfrac{c}{d}} = \dfrac{a}{b} \cdot \dfrac{d}{c}$, b, c, $d \neq 0$

The same procedures are used to multiply or divide two rational expressions.

<u>3</u>. ADDITION AND SUBTRACTION

For a, b, and c real numbers:

a. $\dfrac{a}{b} + \dfrac{c}{b} = \dfrac{a + c}{b}$, $b \neq 0$

b. $\dfrac{a}{b} - \dfrac{c}{b} = \dfrac{a - c}{b}$, $b \neq 0$

The same procedures are used to add or subtract two rational expressions (with the same denominator).

<u>4</u>. THE LEAST COMMON DENOMINATOR (LCD)

The LCD of two or more rational expressions is found as follows:

a. Factor each denominator completely, including integer factors.

b. Identify each different factor from all the denominators.

c. Form a product using each different factor to the highest power that occurs in any one denominator. This product is the LCD.

The least common denominator is used to add or subtract rational expressions having different denominators.

1. $\dfrac{d^5}{3a} \div \left(\dfrac{d^2}{6a^2} \cdot \dfrac{a}{4d^3} \right) = \dfrac{d^5}{3a} \div \left(\dfrac{\cancel{a}d^2}{24\cancel{a^2}d^3} \right) = \dfrac{d^5}{3a} \div \dfrac{1}{24ad} = \dfrac{d^5}{\cancel{3a}} \cdot \dfrac{\overset{8}{\cancel{24}}\cancel{a}d}{1} = 8d^6$

$$\underset{a \quad d}{}$$

3. $\dfrac{x^2}{12} + \dfrac{x}{18} - \dfrac{1}{30} = \dfrac{15x^2}{180} + \dfrac{10x}{180} - \dfrac{6}{180}$

$\qquad \qquad = \dfrac{15x^2 + 10x - 6}{180}$

We find the LCD of 12, 18, 30:
$12 = 2^2 \cdot 3$, $18 = 2 \cdot 3^2$, $30 = 2 \cdot 3 \cdot 5$
Thus, LCD $= 2^2 \cdot 3^2 \cdot 5 = 180$.

5. $\dfrac{4m - 3}{18m^3} + \dfrac{3}{4m} - \dfrac{2m - 1}{6m^2}$

Find the LCD of $18m^3$, $4m$, $6m^2$:
$18m^3 = 2 \cdot 3^2 m^3$, $4m = 2^2 m$,
$6m^2 = 2 \cdot 3 m^2$
Thus, LCD $= 36m^3$.

$= \dfrac{2(4m - 3)}{36m^3} + \dfrac{3(9m^2)}{36m^3} - \dfrac{6m(2m - 1)}{36m^3}$

$= \dfrac{8m - 6 + 27m^2 - 6m(2m - 1)}{36m^3}$

$= \dfrac{8m - 6 + 27m^2 - 12m^2 + 6m}{36m^3} = \dfrac{15m^2 + 14m - 6}{36m^3}$

7. $\dfrac{x^2 - 9}{x^2 - 3x} \div (x^2 - x - 12) = \dfrac{\cancel{(x - 3)}(x + 3)}{x\cancel{(x - 3)}} \cdot \dfrac{1}{(x - 4)\cancel{(x + 3)}} = \dfrac{1}{x(x - 4)}$

9. $\dfrac{2}{x} - \dfrac{1}{x - 3} = \dfrac{2(x - 3)}{x(x - 3)} - \dfrac{x}{x(x - 3)}$ $\qquad$ LCD $= x(x - 3)$

$\qquad \qquad = \dfrac{2x - 6 - x}{x(x - 3)} = \dfrac{x - 6}{x(x - 3)}$

11. $\dfrac{3}{x^2 - 1} - \dfrac{2}{x^2 - 2x + 1} = \dfrac{3}{(x - 1)(x + 1)} - \dfrac{2}{(x - 1)^2}$ $\qquad$ LCD $= (x - 1)^2(x + 1)$

$\qquad \qquad = \dfrac{3(x - 1)}{(x - 1)^2(x + 1)} - \dfrac{2(x + 1)}{(x - 1)^2(x + 1)}$

$\qquad \qquad = \dfrac{3x - 3 - 2(x + 1)}{(x - 1)^2(x + 1)} = \dfrac{3x - 3 - 2x - 2}{(x - 1)^2(x + 1)}$

$\qquad \qquad = \dfrac{x - 5}{(x - 1)^2(x + 1)}$

13. $\dfrac{x + 1}{x - 1} - 1 = \dfrac{x + 1}{x - 1} - \dfrac{x - 1}{x - 1} = \dfrac{x + 1 - (x - 1)}{x - 1} = \dfrac{2}{x - 1}$

15. $\dfrac{3}{a - 1} - \dfrac{2}{1 - a} = \dfrac{3}{a - 1} - \dfrac{-2}{-(1 - a)} = \dfrac{3}{a - 1} + \dfrac{2}{a - 1} = \dfrac{5}{a - 1}$

17. $\dfrac{2x}{x^2 - 16} - \dfrac{x - 4}{x^2 + 4x} = \dfrac{2x}{(x - 4)(x + 4)} - \dfrac{x - 4}{x(x + 4)}$ LCD $= x(x - 4)(x + 4)$

$$= \dfrac{2x(x) - (x - 4)(x - 4)}{x(x - 4)(x + 4)}$$

$$= \dfrac{2x^2 - (x^2 - 8x + 16)}{x(x - 4)(x + 4)}$$

$$= \dfrac{x^2 + 8x - 16}{x(x - 4)(x + 4)}$$

19. $\dfrac{x^2}{x^2 + 2x + 1} + \dfrac{x - 1}{3x + 3} - \dfrac{1}{6} = \dfrac{x^2}{(x + 1)^2} + \dfrac{x - 1}{3(x + 1)} - \dfrac{1}{6}$

LCD $= 6(x + 1)^2$ $\qquad = \dfrac{6x^2}{6(x + 1)^2} + \dfrac{2(x + 1)(x - 1)}{6(x + 1)^2} - \dfrac{(x + 1)^2}{6(x + 1)^2}$

$$= \dfrac{6x^2 + 2(x^2 - 1) - (x^2 + 2x + 1)}{6(x + 1)^2}$$

$$= \dfrac{7x^2 - 2x - 3}{6(x + 1)^2}$$

21. $\dfrac{2 - x}{2x + x^2} \cdot \dfrac{x^2 + 4x + 4}{x^2 - 4} = \dfrac{-\cancel{(x - 2)}}{x\cancel{(x + 2)}} \cdot \dfrac{\cancel{(x + 2)}^2}{\cancel{(x + 2)}\cancel{(x - 2)}} = -\dfrac{1}{x}$

23. $\dfrac{c + 2}{5c - 5} - \dfrac{c - 2}{3c - 3} + \dfrac{c}{1 - c} = \dfrac{c + 2}{5(c - 1)} - \dfrac{c - 2}{3(c - 1)} - \dfrac{c}{c - 1}$

LCD $= 15(c - 1)$ $\qquad = \dfrac{3(c + 2)}{15(c - 1)} - \dfrac{5(c - 2)}{15(c - 1)} - \dfrac{15c}{15(c - 1)}$

$$= \dfrac{3c + 6 - 5c + 10 - 15c}{15(c - 1)} = \dfrac{-17c + 16}{15(c - 1)}$$

25. $\dfrac{1 + \dfrac{3}{x}}{x - \dfrac{9}{x}} = \dfrac{\dfrac{x + 3}{x}}{\dfrac{x^2 - 9}{x}} = \dfrac{x + 3}{x} \cdot \dfrac{x}{x^2 - 9} = \dfrac{\cancel{x + 3}}{\cancel{x}} \cdot \dfrac{\cancel{x}}{\cancel{(x + 3)}(x - 3)} = \dfrac{1}{x - 3}$

27. $\dfrac{\dfrac{1}{2(x + h)} - \dfrac{1}{2x}}{h} = \left(\dfrac{1}{2(x + h)} - \dfrac{1}{2x}\right) \div \dfrac{h}{1}$

$$= \dfrac{x - x - h}{2x(x + h)} \cdot \dfrac{1}{h}$$

$$= \dfrac{-h}{2x(x + h)h} = \dfrac{-1}{2x(x + h)}$$

29. $\dfrac{\dfrac{x}{y} - 2 + \dfrac{y}{x}}{\dfrac{x}{y} - \dfrac{y}{x}} = \dfrac{\dfrac{x^2 - 2xy + y^2}{xy}}{\dfrac{x^2 - y^2}{xy}} = \dfrac{\cancel{(x - y)}^2}{\cancel{xy}} \cdot \dfrac{\cancel{xy}}{\cancel{(x - y)}(x + y)} = \dfrac{x - y}{x + y}$

31.

$$\frac{\dfrac{1}{3(x+h)^2} - \dfrac{1}{3x^2}}{h} = \left[\frac{1}{3(x+h)^2} - \frac{1}{3x^2}\right] \div \frac{h}{1}$$

$$= \frac{x^2 - (x+h)^2}{3x^2(x+h)^2} \cdot \frac{1}{h}$$

$$= \frac{x^2 - (x^2 + 2xh + h^2)}{3x^2(x+h)^2 h}$$

$$= \frac{-2xh - h^2}{3x^2(x+h)^2 h}$$

$$= \frac{-\cancel{h}(2x+h)}{3x^2(x+h)^2 \cancel{h}}$$

$$= -\frac{(2x+h)}{3x^2(x+h)^2} = \frac{-2x-h}{3x^2(x+h)^2}$$

33.

$$1 - \frac{1}{1 - \dfrac{1}{1 - \dfrac{1}{x}}} = 1 - \frac{1}{1 - \dfrac{1}{\dfrac{x-1}{x}}} = 1 - \frac{1}{1 - \dfrac{x}{x-1}}$$

$$= 1 - \frac{1}{\dfrac{(x-1) - x}{x-1}} = 1 - \frac{1}{\dfrac{-1}{x-1}} = 1 + x - 1 = x$$

EXERCISE A-4

Things to remember:

1. DEFINITION OF a^n, where n is an integer and a is a real number:

   a. For n a positive integer,
   $$a^n = a \cdot a \cdot \,\cdots\, \cdot a, \; n \text{ factors of } a.$$

   b. For $n = 0$,
   $$a^0 = 1, \; a \neq 0, \; 0^0 \text{ is not defined.}$$

   c. For n a negative integer,
   $$a^n = \frac{1}{a^{-n}}, \; a \neq 0.$$

   [<u>Note</u>: If n is negative, then $-n$ is positive.]

2. PROPERTIES OF EXPONENTS

   GIVEN: n and m are integers and a and b are real numbers.

   a. $a^m a^n = a^{m+n}$ $\qquad\qquad$ $a^8 a^{-3} = a^{8+(-3)} = a^5$

   b. $(a^n)^m = a^{mn}$ $\qquad\qquad$ $(a^{-2})^3 = a^{3(-2)} = a^{-6}$

   c. $(ab)^m = a^m b^m$ $\qquad\qquad$ $(ab)^{-2} = a^{-2} b^{-2}$

d. $\left(\dfrac{a}{b}\right)^m = \dfrac{a^m}{b^m},\ b \neq 0$ $\qquad\qquad$ $\left(\dfrac{a}{b}\right)^5 = \dfrac{a^5}{b^5}$

e. $\dfrac{a^m}{a^n} = a^{m-n} = \dfrac{1}{a^{n-m}},\ a \neq 0$ $\qquad$ $\dfrac{a^{-3}}{a^7} = \dfrac{1}{a^{7-(-3)}} = \dfrac{1}{a^{10}}$

<u>3</u>. SCIENTIFIC NOTATION

Let r be any real number. Then r can be expressed as the product of a number between 1 and 10 and an integer power of 10, that is, r can be written

$\qquad r = a \times 10^n,\ 1 \leq a < 10,\ n$ an integer, a in decimal form

A number expressed in this form is said to be in scientific notation.

Examples:

$$7 = 7 \times 10^0 \qquad\qquad 0.5 = 5 \times 10^{-1}$$
$$67 = 6.7 \times 10 \qquad\qquad 0.45 = 4.5 \times 10^{-1}$$
$$580 = 5.8 \times 10^2 \qquad\qquad 0.0032 = 3.2 \times 10^{-3}$$
$$43{,}000 = 4.3 \times 10^4 \qquad\qquad 0.000\,045 = 4.5 \times 10^{-5}$$

---

1. $2x^{-9} = \dfrac{2}{x^9}$ $\qquad\qquad$ **3.** $\dfrac{3}{2w^{-7}} = \dfrac{3w^7}{2}$

5. $2x^{-8}x^5 = 2x^{-8+5} = 2x^{-3} = \dfrac{2}{x^3}$ $\qquad$ **7.** $\dfrac{w^{-8}}{w^{-3}} = \dfrac{1}{w^{-3+8}} = \dfrac{1}{w^5}$

9. $5v^8v^{-8} = 5v^{8-8} = 5v^0 = 5 \cdot 1 = 5$ $\qquad$ **11.** $(a^{-3})^2 = a^{-6} = \dfrac{1}{a^6}$

13. $(x^6y^{-3})^{-2} = x^{-12}y^6 = \dfrac{y^6}{x^{12}}$ $\qquad$ **15.** $82{,}300{,}000{,}000 = 8.23 \times 10^{10}$

17. $0.783 = 7.83 \times 10^{-1}$ $\qquad\qquad$ **19.** $0.000\,034 = 3.4 \times 10^{-5}$

21. $4 \times 10^4 = 40{,}000$ $\qquad\qquad$ **23.** $7 \times 10^{-3} = 0.007$

25. $6.171 \times 10^7 = 61{,}710{,}000$ $\qquad$ **27.** $8.08 \times 10^{-4} = 0.000\,808$

29. $(22 + 31)^0 = (53)^0 = 1$

31. $\dfrac{10^{-3} \times 10^4}{10^{-11} \times 10^{-2}} = \dfrac{10^{-3+4}}{10^{-11-2}} = \dfrac{10^1}{10^{-13}} = 10^{1+13} = 10^{14}$

33. $(5x^2y^{-3})^{-2} = 5^{-2}x^{-4}y^6 = \dfrac{y^6}{5^2x^4} = \dfrac{y^6}{25x^4}$

35. $\dfrac{8 \times 10^{-3}}{2 \times 10^{-5}} = \dfrac{8}{2} \times \dfrac{10^{-3}}{10^{-5}} = 4 \times 10^{-3+5} = 4 \times 10^2$

37. $\dfrac{8x^{-3}y^{-1}}{6x^2y^{-4}} = \dfrac{4y^{-1+4}}{3x^{2+3}} = \dfrac{4y^3}{3x^5}$

39. $\dfrac{7x^5 - x^2}{4x^5} = \dfrac{7x^5}{4x^5} - \dfrac{x^2}{4x^5} = \dfrac{7}{4} - \dfrac{1}{4x^3} = \dfrac{7}{4} - \dfrac{1}{4}x^{-3}$

41. $\dfrac{3x^4 - 4x^2 - 1}{4x^3} = \dfrac{3x^4}{4x^3} - \dfrac{4x^2}{4x^3} - \dfrac{1}{4x^3} = \dfrac{3}{4}x - x^{-1} - \dfrac{1}{4}x^{-3}$

43. $\dfrac{3x^2(x-1)^2 - 2x^3(x-1)}{(x-1)^4} = \dfrac{x^2(x-1)[3(x-1) - 2x]}{(x-1)^4} = \dfrac{x^2(x-3)}{(x-1)^3}$

45. $2x^{-2}(x-1) - 2x^{-3}(x-1)^2 = \dfrac{2(x-1)}{x^2} - \dfrac{2(x-1)^2}{x^3}$

$$= \dfrac{2x(x-1) - 2(x-1)^2}{x^3}$$

$$= \dfrac{2(x-1)[x - (x-1)]}{x^3}$$

$$= \dfrac{2(x-1)}{x^3}$$

47. $\dfrac{9,600,000,000}{(1,600,000)(0.000\,000\,25)} = \dfrac{9.6 \times 10^9}{(1.6 \times 10^6)(2.5 \times 10^{-7})} = \dfrac{9.6 \times 10^9}{1.6(2.5) \times 10^{6-7}}$

$$= \dfrac{9.6 \times 10^9}{4.0 \times 10^{-1}} = 2.4 \times 10^{9+1} = 2.4 \times 10^{10}$$

$$= 24,000,000,000$$

49. $\dfrac{(1,250,000)(0.000\,38)}{0.0152} = \dfrac{(1.25 \times 10^6)(3.8 \times 10^{-4})}{1.52 \times 10^{-2}} = \dfrac{1.25(3.8) \times 10^{6-4}}{1.52 \times 10^{-2}}$

$$= 3.125 \times 10^4 = 31,250$$

51. $\dfrac{u+v}{u^{-1}+v^{-1}} = \dfrac{u+v}{\dfrac{1}{u}+\dfrac{1}{v}} = \dfrac{u+v}{\dfrac{v+u}{uv}} = (u+v) \cdot \dfrac{uv}{v+u} = uv$

53. $\dfrac{b^{-2}-c^{-2}}{b^{-3}-c^{-3}} = \dfrac{\dfrac{1}{b^2}-\dfrac{1}{c^2}}{\dfrac{1}{b^3}-\dfrac{1}{c^3}} = \dfrac{\dfrac{c^2-b^2}{b^2c^2}}{\dfrac{c^3-b^3}{b^3c^3}} = \dfrac{\cancel{(c-b)}(c+b)}{\cancel{b^2c^2}} \cdot \dfrac{\overset{bc}{\cancel{b^3c^3}}}{\cancel{(c-b)}(c^2+cb+b^2)}$

$$= \dfrac{bc(c+b)}{c^2+cb+b^2}$$

55. (A) $213,701,000,000 = 2.13701 \times 10^{11}$

(B) $95,862,000,000 = 9.5862 \times 10^{10}$

$$\frac{2.13701 \times 10^{11}}{9.5682 \times 10^{10}} = \frac{2.13701}{9.5862} \times 10 \approx 2.2293$$

(C) $$\frac{9.5862 \times 10^{10}}{2.13701 \times 10^{11}} = \frac{9.5862}{2.13701} \times 10^{-1} \approx 0.4486$$

57. (A) $$\frac{4,065,000,000,000}{255,100,000} = \frac{4.065 \times 10^{12}}{2.551 \times 10^{8}} \approx 1.5935 \times 10^{4} = \$15,935$$

(B) $$\frac{292,300,000,000}{255,100,000} = \frac{2.923 \times 10^{11}}{2.551 \times 10^{8}} \approx 1.146 \times 10^{3} = \$1,146$$

(C) $$\frac{292,300,000,000}{4,065,000,000,000} = \frac{2.923 \times 10^{11}}{4.065 \times 10^{12}} \approx 0.719 \times 10^{-1} = 0.0719 \text{ or } 7.19\%$$

59. (A) $9 \text{ ppm} = \dfrac{9}{1,000,000} = \dfrac{9}{10^{6}} = 9 \times 10^{-6}$ (B) $0.000\,009$ (C) 0.0009%

61. $\dfrac{757.5}{100,000} \times 255,100,000 = \dfrac{7.575 \times 10^{2}}{10^{5}} \cdot 2.551 \times 10^{8}$

$$= \frac{19.323825 \times 10^{10}}{10^{5}}$$
$$= 19.323825 \times 10^{5}$$
$$= 1,932,382.5$$

To the nearest thousand, there were 1,932,000 violent crimes committed in 1992.

EXERCISE A-5

Things to remember:

<u>1.</u> *n*th ROOT

Let *b* be a real number. For any natural number *n*,

  r is an *n*th ROOT of *b* if $r^{n} = b$

If *n* is odd, then *b* has exactly one real *n*th root.
If *n* is even, and $b < 0$, then *b* has NO real *n*th roots.
If *n* is even, and $b > 0$, then *b* has two real *n*th roots;
   if *r* is an *n*th root, then $-r$ is also an *n*th root.
0 is an *n*th root of 0 for all *n*

2. NOTATION

Let b be a real number and let $n > 1$ be a natural number. If n is odd, then the nth root of b is denoted

$$b^{1/n} \quad \text{or} \quad \sqrt[n]{b}$$

If n is even and $b > 0$, then the PRINCIPAL nth ROOT OF b is the positive nth root; the principal nth root is denoted

$$b^{1/n} \quad \text{or} \quad \sqrt[n]{b}$$

In the $\sqrt[n]{b}$ notation, the symbol $\sqrt{}$ is called a RADICAL, n is the INDEX of the radical and b is called the RADICAND.

3. RATIONAL EXPONENTS

If m and n are natural numbers without common prime factors, b is a real number, and b is nonnegative when b is even, then

$$b^{m/n} = \begin{cases} (b^{1/n})^m = (\sqrt[n]{b})^m \\ (b^m)^{1/n} = \sqrt[n]{b^m} \end{cases}$$

and $\quad b^{-m/n} = \dfrac{1}{b^{m/n}}, \quad b \neq 0$

The two definitions of $b^{m/n}$ are equivalent under the indicated restrictions on m, n, and b.

4. PROPERTIES OF RADICALS

If m and n are natural numbers greater than 1 and x and y are positive real numbers, then

a. $\sqrt[n]{x^n} = x$ $\qquad\qquad$ $\sqrt[3]{x^3} = x$

b. $\sqrt[n]{xy} = \sqrt[n]{x}\,\sqrt[n]{y}$ $\qquad$ $\sqrt[5]{xy} = \sqrt[5]{x}\,\sqrt[5]{y}$

c. $\sqrt[n]{\dfrac{x}{y}} = \dfrac{\sqrt[n]{x}}{\sqrt[n]{y}}$ $\qquad\quad$ $\sqrt[4]{\dfrac{x}{y}} = \dfrac{\sqrt[4]{x}}{\sqrt[4]{y}}$

---

1. $6x^{3/5} = 6\sqrt[5]{x^3}$ $\qquad\qquad$ **3.** $(4xy^3)^{2/5} = \sqrt[5]{(4xy^3)^2}$

5. $(x^2 + y^2)^{1/2} = \sqrt{x^2 + y^2}$ $\qquad$ **7.** $5\sqrt[4]{x^3} = 5x^{3/4}$
[Note: $\sqrt{x^2 + y^2} \neq x + y$.]

9. $\sqrt[5]{(2x^2y)^3} = (2x^2y)^{3/5}$ $\qquad$ **11.** $\sqrt[3]{x} + \sqrt[3]{y} = x^{1/3} + y^{1/3}$

13. $25^{1/2} = (5^2)^{1/2} = 5$ $\qquad\quad$ **15.** $16^{3/2} = (4^2)^{3/2} = 4^3 = 64$

17. $-36^{1/2} = -(6^2)^{1/2} = -6$

19. $(-36)^{1/2}$ is not a rational number -36 does not have a real square root; $(-36)^{1/2}$ is not a real number.

21. $\left(\dfrac{4}{25}\right)^{3/2} = \left(\left(\dfrac{2}{5}\right)^2\right)^{3/2} = \left(\dfrac{2}{5}\right)^3 = \dfrac{2^3}{5^3} = \dfrac{8}{125}$

23. $9^{-3/2} = (3^2)^{-3/2} = 3^{-3} = \dfrac{1}{3^3} = \dfrac{1}{27}$

25. $x^{4/5}x^{-2/5} = x^{4/5-2/5} = x^{2/5}$

27. $\dfrac{m^{2/3}}{m^{-1/3}} = m^{2/3-(-1/3)} = m^1 = m$

29. $(8x^3y^{-6})^{1/3} = (2^3x^3y^{-6})^{1/3} = 2^{3/3}x^{3/3}y^{-6/3} = 2xy^{-2} = \dfrac{2x}{y^2}$

31. $\left(\dfrac{4x^{-2}}{y^4}\right)^{-1/2} = \left(\dfrac{2^2x^{-2}}{y^4}\right)^{-1/2} = \dfrac{2^{2(-1/2)}x^{-2(-1/2)}}{y^{4(-1/2)}} = \dfrac{2^{-1}x^1}{y^{-2}} = \dfrac{xy^2}{2}$

33. $\dfrac{8x^{-1/3}}{12x^{1/4}} = \dfrac{2}{3x^{1/4+1/3}} = \dfrac{2}{3x^{7/12}}$

35. $\sqrt[5]{(2x+3)^5} = [(2x+3)^5]^{1/5} = 2x+3$

37. $\sqrt{18x^3}\sqrt{2x^3} = \sqrt{36x^6} = (6^2x^6)^{1/2} = (6^2)^{1/2}(6^2)^{1/2}(x^6)^{1/2} = 6x^3$

39. $\dfrac{\sqrt{6x}\sqrt{10}}{\sqrt{15x}} = \sqrt{\dfrac{60x}{15x}} = \sqrt{4} = 2$

41. $3x^{3/4}(4x^{1/4} - 2x^8) = 12x^{3/4+1/4} - 6x^{3/4+8}$
$$= 12x - 6x^{3/4+32/4} = 12x - 6x^{35/4}$$

43. $(3u^{1/2} - v^{1/2})(u^{1/2} - 4v^{1/2}) = 3u - 12u^{1/2}v^{1/2} - u^{1/2}v^{1/2} + 4v$
$$= 3u - 13u^{1/2}v^{1/2} + 4v$$

45. $(5m^{1/2} + n^{1/2})(5m^{1/2} - n^{1/2}) = (5m^{1/2})^2 - (n^{1/2})^2 = 25m - n$

47. $(3x^{1/2} - y^{1/2})^2 = (3x^{1/2})^2 - 6x^{1/2}y^{1/2} + (y^{1/2})^2 = 9x - 6x^{1/2}y^{1/2} + y$

49. $\dfrac{\sqrt[3]{x^2} + 2}{2\sqrt[3]{x}} = \dfrac{x^{2/3} + 2}{2x^{1/3}} = \dfrac{x^{2/3}}{2x^{1/3}} + \dfrac{2}{2x^{1/3}} = \dfrac{1}{2}x^{1/3} + \dfrac{1}{x^{1/3}} = \dfrac{1}{2}x^{1/3} + x^{-1/3}$

51. $\dfrac{2\sqrt[4]{x^3} + 3\sqrt[3]{x}}{3x} = \dfrac{2x^{3/4} + 3x^{1/3}}{3x} = \dfrac{2x^{3/4}}{3x} + \dfrac{3x^{1/3}}{3x}$
$$= \dfrac{2}{3}x^{3/4-1} + x^{1/3-1} = \dfrac{2}{3}x^{-1/4} + x^{-2/3}$$

53. $\dfrac{2\sqrt[3]{x} - \sqrt{x}}{4\sqrt{x}} = \dfrac{2x^{1/3} - x^{1/2}}{4x^{1/2}} = \dfrac{2x^{1/3}}{4x^{1/2}} - \dfrac{x^{1/2}}{4x^{1/2}} = \dfrac{1}{2}x^{1/3-1/2} - \dfrac{1}{4} = \dfrac{1}{2}x^{-1/6} - \dfrac{1}{4}$

55. $\dfrac{12mn^2}{\sqrt{3mn}} = \dfrac{12mn^2}{\sqrt{3mn}} \cdot \dfrac{\sqrt{3mn}}{\sqrt{3mn}} = \dfrac{12mn^2\sqrt{3mn}}{3mn} = 4n\sqrt{3mn}$

57. $\dfrac{2}{\sqrt{x-2}} = \dfrac{2}{\sqrt{x-2}} \cdot \dfrac{\sqrt{x-2}}{\sqrt{x-2}} = \dfrac{2\sqrt{x-2}}{x-2}$

59. $\dfrac{7(x-y)^2}{\sqrt{x}-\sqrt{y}} = \dfrac{7(x-y)^2}{\sqrt{x}-\sqrt{y}} \cdot \dfrac{\sqrt{x}+\sqrt{y}}{\sqrt{x}+\sqrt{y}} = \dfrac{7(x-y)^2(\sqrt{x}+\sqrt{y})}{x-y}$

$$= 7(x-y)(\sqrt{x}+\sqrt{y})$$

61. $\dfrac{\sqrt{5xy}}{5x^2y^2} = \dfrac{\sqrt{5xy}}{5x^2y^2} \cdot \dfrac{\sqrt{5xy}}{\sqrt{5xy}} = \dfrac{5xy}{5x^2y^2\sqrt{5xy}} = \dfrac{1}{xy\sqrt{5xy}}$

63. $\dfrac{\sqrt{x+h}-\sqrt{x}}{h} = \dfrac{\sqrt{x+h}-\sqrt{x}}{h} \cdot \dfrac{\sqrt{x+h}+\sqrt{x}}{\sqrt{x+h}+\sqrt{x}}$

$$= \dfrac{x+h-x}{h(\sqrt{x+h}+\sqrt{x})} = \dfrac{h}{h(\sqrt{x+h}+\sqrt{x})}$$

$$= \dfrac{1}{\sqrt{x+h}+\sqrt{x}}$$

65. $\dfrac{\sqrt{t}-\sqrt{x}}{t-x} = \dfrac{\sqrt{t}-\sqrt{x}}{t-x} \cdot \dfrac{\sqrt{t}+\sqrt{x}}{\sqrt{t}+\sqrt{x}} = \dfrac{t-x}{(t-x)(\sqrt{t}+\sqrt{x})} = \dfrac{1}{\sqrt{t}+\sqrt{x}}$

67. $-\dfrac{1}{2}(x-2)(x+3)^{-3/2} + (x+3)^{-1/2} = \dfrac{-(x-2)}{2(x+3)^{3/2}} + \dfrac{1}{(x+3)^{1/2}}$

$$= \dfrac{-x+2+2(x+3)}{2(x+3)^{3/2}}$$

$$= \dfrac{x+8}{2(x+3)^{3/2}}$$

69. $\dfrac{(x-1)^{1/2} - x\left(\dfrac{1}{2}\right)(x-1)^{-1/2}}{x-1} = \dfrac{(x-1)^{1/2} - \dfrac{x}{2(x-1)^{1/2}}}{x-1}$

$$= \dfrac{\dfrac{2(x-1)^{1/2}(x-1)^{1/2}}{2(x-1)^{1/2}} - \dfrac{x}{2(x-1)^{1/2}}}{x-1}$$

$$= \dfrac{\dfrac{2(x-1)-x}{2(x-1)^{1/2}}}{x-1} = \dfrac{x-2}{2(x-1)^{3/2}}$$

71. $\dfrac{(x+2)^{2/3} - x\left(\dfrac{2}{3}\right)(x+2)^{-1/3}}{(x+2)^{4/3}} = \dfrac{(x+2)^{2/3} - \dfrac{2x}{3(x+2)^{1/3}}}{(x+2)^{4/3}}$

$$= \dfrac{\dfrac{3(x+2)^{1/3}(x+2)^{2/3}}{3(x+2)^{1/3}} - \dfrac{2x}{3(x+2)^{1/3}}}{(x+2)^{4/3}}$$

$$= \dfrac{\dfrac{3(x+2)-2x}{3(x+2)^{1/3}}}{(x+2)^{4/3}} = \dfrac{x+6}{3(x+2)^{5/3}}$$

73. $22^{3/2} = 22^{1.5} \approx 103.2$ or $22^{3/2} = \sqrt{(22)^3} = \sqrt{10,648} \approx 103.2$

75. $827^{-3/8} = \dfrac{1}{827^{3/8}} = \dfrac{1}{827^{0.375}} \approx \dfrac{1}{12.42} \approx 0.0805$

77. $37.09^{7/3} \approx 37.09^{2.3333} \approx 4,588$

EXERCISE A-6

Things to remember:

1. **FIRST DEGREE, OR LINEAR, EQUATIONS AND INEQUALITIES**

    A FIRST DEGREE, or LINEAR, EQUATION in one variable x is an equation that can be written in the form

    $$ax + b = 0, \qquad \text{STANDARD FORM}$$

    where a and b are constants and $a \neq 0$. If the equality symbol $=$ is replaced by $<$, $>$, $\leq$, or $\geq$, then the resulting expression is called a FIRST DEGREE, or LINEAR, INEQUALITY.

2. **SOLUTIONS**

    A SOLUTION OF AN EQUATION (or inequality) involving a single variable is a number that when substituted for the variable makes the equation (or inequality) true. The set of all solutions is called the SOLUTION SET. To SOLVE AN EQUATION (or inequality) we mean that we determine the solution set. Two equations (or inequalities) are EQUIVALENT if they have the same solution set.

3. **EQUALITY PROPERTIES**

    An equivalent equation will result if:

    a) The same quantity is added to or subtracted from each side of a given equation.

    b) Each side of a given equation is multiplied by or divided by the same nonzero quantity.

4. **INEQUALITY PROPERTIES**

    An equivalent inequality will result and the SENSE WILL REMAIN THE SAME if each side of the original inequality:

    a) Has the same real number added to or subtracted from it.

    b) Is multiplied or divided by the same positive number.

    An equivalent inequality will result and the SENSE WILL REVERSE if each side of the original inequality:

    c) Is multiplied or divided by the same negative number.

    NOTE: Multiplication and division by 0 is not permitted.

<u>5</u>. The double inequality $a \leq x \leq b$ means that $a \leq x$ and $x \leq b$. Other variations, as well as a useful interval notation, are indicated in the following table.

Interval Notation	Inequality Notation	Line Graph
$[a, b]$	$a \leq x \leq b$	
$[a, b)$	$a \leq x < b$	
$(a, b]$	$a < x \leq b$	
(a, b)	$a < x < b$	
$(-\infty, a]$	$x \leq a$	
$(-\infty, a)$	$x < a$	
$[b, \infty)$	$x \geq b$	
(b, ∞)	$x > b$	

[<u>Note</u>: An endpoint on a line graph has a square bracket through it if it is included in the inequality and a parenthesis through it if it is not.]

1.

$2m + 9 = 5m - 6$

$2m + 9 - 9 = 5m - 6 - 9$ [using <u>3</u>(a)]

$2m = 5m - 15$

$2m - 5m = 5m - 15 - 5m$ [using <u>3</u>(a)]

$-3m = -15$

$\dfrac{-3m}{-3} = \dfrac{-15}{-3}$ [using <u>3</u>(b)]

$m = 5$

3.

$x + 5 < -4$

$x + 5 - 5 < -4 - 5$ [using <u>4</u>(a)]

$x < -9$

5.

$-3x \geq -12$

$\dfrac{-3x}{-3} \leq \dfrac{-12}{-3}$ [using <u>4</u>(c)]

$x \leq 4$

7. $-4x - 7 > 5$

$-4x > 5 + 7$

$-4x > 12$

$x < -3$

Graph of $x < -3$ is:

9. $2 \leq x + 3 \leq 5$

$2 - 3 \leq x \leq 5 - 3$

$-1 \leq x \leq 2$

Graph of $-1 \leq x \leq 2$ is:

11. $\frac{y}{7} - 1 = \frac{1}{7}$

Multiply both sides of the equation by 7. We obtain:

$y - 7 = 1$ [using $\underline{3}$(b)]

$\qquad y = 8$

15. $\frac{y}{3} = 4 - \frac{y}{6}$

Multiply both sides of the equation by 6. We obtain:

$2y = 24 - y$

$3y = 24$

$\; y = 8$

19. $3 - y \le 4(y - 3)$

$\quad 3 - y \le 4y - 12$

$\quad\;\; -5y \le -15$

$\qquad\;\; y \ge 3$

[$\underline{\text{Note}}$: Division by a negative number, -3.]

23. $\frac{m}{5} - 3 < \frac{3}{5} - m$

Multiply both sides of the inequality by 5. We obtain:

$m - 15 < 3 - 5m$

$\qquad 6m < 18$

$\qquad\; m < 3$

27. $2 \le 3x - 7 < 14$

$7 + 2 \le 3x < 14 + 7$

$\quad\; 9 \le 3x < 21$

$\quad\; 3 \le x < 7$

Graph of $3 \le x < 7$ is:

31. $3x - 4y = 12$

$\qquad 3x = 12 + 4y$

$3x - 12 = 4y$

$\qquad\quad y = \frac{1}{4}(3x - 12)$

$\qquad\quad y = \frac{3}{4}x - 3$

13. $\frac{x}{3} > -2$

Multiply both sides of the inequality by 3. We obtain:

$x > -6$ [using $\underline{4}$(b)]

17. $10x + 25(x - 3) = 275$

$10x + 25x - 75 = 275$

$\qquad\qquad\; 35x = 275 + 75$

$\qquad\qquad\; 35x = 350$

$\qquad\qquad\quad x = \frac{350}{35}$

$\qquad\qquad\quad x = 10$

21. $\frac{x}{5} - \frac{x}{6} = \frac{6}{5}$

Multiply both sides of the equation by 30. We obtain:

$6x - 5x = 36$

$\qquad\;\; x = 36$

25. $0.1(x - 7) + 0.05x = 0.8$

$0.1x - 0.7 + 0.05x = 0.8$

$\qquad\qquad\quad 0.15x = 1.5$

$\qquad\qquad\qquad\; x = \frac{1.5}{0.15}$

$\qquad\qquad\qquad\; x = 10$

29. $-4 \le \frac{9}{5}C + 32 \le 68$

$\quad -36 \le \frac{9}{5}C \le 36$

$-36\left(\frac{5}{9}\right) \le C \le 36\left(\frac{5}{9}\right)$

$\quad\; -20 \le C \le 20$

Graph of $-20 \le C \le 20$ is:

33. $Ax + By = C$

$\qquad By = C - Ax$

$\qquad\; y = \frac{C}{B} - \frac{Ax}{B}, \; B \ne 0$

or $\qquad y = -\left(\frac{A}{B}\right)x + \frac{C}{B}$

35.
$$F = \frac{9}{5}C + 32$$
$$\frac{9}{5}C + 32 = F$$
$$\frac{9}{5}C = F - 32$$
$$C = \frac{5}{9}(F - 32)$$

37.
$$A = Bm - Bn$$
$$A = B(m - n)$$
$$B = \frac{A}{m - n}$$

39. $-3 \le 4 - 7x < 18$
$$-3 - 4 \le -7x < 18 - 4$$
$$-7 \le -7x < 14.$$

Dividing by -7, and recalling $\underline{4}$(c), we have
$1 \ge x > -2$ or $-2 < x \le 1$

The graph is:

41. Let x = number of $15 tickets. Then the number of $25 tickets = $8000 - x$.
Now,
$$15x + 25(8000 - x) = 165,000$$
$$15x + 200,000 - 25x = 165,000$$
$$-10x = -35,000$$
$$x = 3,500$$

Thus, 3,500 $15 tickets and $8,000 - 3,500 = 4,500$ $25 tickets were sold.

43. Let x = the amount invested at 10%. Then $12,000 - x$ is the amount invested at 15%.

Required total yield = 12% of $12,000 = $0.12 \cdot 12,000$ = $1,440. Thus,
$$0.10x + 0.15(12,000 - x) = 0.12 \cdot 12,000$$
$$10x + 15(12,000 - x) = 12 \cdot 12,000 \quad \text{(multiply both sides by 100)}$$
$$10x + 180,000 - 15x = 144,000$$
$$-5x = -36,000$$
$$x = \$7,200$$
Thus, we get $7,200 invested at 10% and $12,000 - 7,200 = \$4,800$ invested at 15%.

45. Let x be the price of the car in 1992. Then
$$\frac{x}{5,000} = \frac{140.3}{38.8} \quad \text{(refer to Table 2, Example 9)}$$
$$x = 5,000 \cdot \frac{140.3}{38.8} = 18,079.90$$

To the nearest dollar, the car would sell for $18,080.

47. Let x = number of books produced. Then

    Costs: $C = 1.60x + 55,000$
Revenue: $R = 11x$

To find the break-even point, set $R = C$:

$$11x = 1.60x + 55,000$$
$$9.40x = 55,000$$
$$x = 5851.06383$$

Thus, 5851 books will have to be sold for the publisher to break even.

49. Let x = the number of rainbow trout in the lake. Then,
$\dfrac{x}{200} = \dfrac{200}{8}$ (since proportions are the same)

$x = \dfrac{200}{8}(200)$

$x = 5,000$

51. $IQ = \dfrac{\text{Mental age}}{\text{Chronological age}}(100)$

$\dfrac{\text{Mental age}}{9}(100) = 140$

$\text{Mental age} = \dfrac{140}{100}(9)$

$\qquad\qquad\quad = 12.6$ years

EXERCISE A-7

Things to remember:

<u>1.</u> A quadratic equation in one variable is an equation of the form

(A) $ax^2 + bx + c = 0$,

where x is a variable and a, b, and c are constants, $a \neq 0$.

<u>2.</u> Quadratic equations of the form $ax^2 + c = 0$ can be solved by the SQUARE ROOT METHOD. The solutions are:

$$x = \pm\sqrt{\dfrac{-c}{a}} \text{ provided } \dfrac{-c}{a} \geq 0;$$

otherwise, the equation has no real solutions.

<u>3.</u> If the left side of the quadratic equation (A) can be FACTORED,

$$ax^2 + bx + c = (px + q)(rx + s),$$

then the solutions of (A) are

$$x = \dfrac{-q}{p} \text{ or } x = \dfrac{-s}{r}.$$

<u>4</u>. The solutions of (A) are given by the QUADRATIC FORMULA:

$$x = \frac{-b \pm \sqrt{b^2 - 4ac}}{2a}$$

The quantity $b^2 - 4ac$ under the radical is called the DISCRIMINANT and:

   (i)   (A) has two real solutions if $b^2 - 4ac > 0$;
   (ii) (A) has one real solution if $b^2 - 4ac = 0$;
   (iii) (A) has no real solution if $b^2 - 4ac < 0$.

<u>5</u>. FACTORABILITY THEOREM

The second-degree polynomial, $ax^2 + bx + c$, with integer coefficients, can be expressed as the product of two first-degree polynomials with integer coefficients if and only if $\sqrt{b^2 - 4ac}$ is an integer.

<u>6</u>. FACTOR THEOREM

If r_1 and r_2 are solutions of $ax^2 + bx + c = 0$, then $ax^2 + bx + c = a(x - r_1)(x - r_2)$.

---

1. $2x^2 - 22 = 0$
$x^2 - 11 = 0$
$\quad x^2 = 11$
$\quad\quad x = \pm\sqrt{11}$

3. $(x - 1)^2 = 4$
$x - 1 = \pm\sqrt{4} = \pm 2$
$\quad x = 1 \pm 2 = -1 \text{ or } 3$

5. $2u^2 - 8u - 24 = 0$
$u^2 - 4u - 12 = 0$
$(u - 6)(u + 2) = 0$
$u - 6 = 0 \text{ or } u + 2 = 0$
$\quad u = 6 \text{ or } \quad\quad u = -2$

7. $\quad\quad x^2 = 2x$
$\quad x^2 - 2x = 0$
$x(x - 2) = 0$
$x = 0 \text{ or } x - 2 = 0$
$\quad\quad\quad\quad x = 2$

9. $x^2 - 6x - 3 = 0$

$$x = \frac{-b \pm \sqrt{b^2 - 4ac}}{2a}, \quad a = 1, \ b = -6, \ c = -3$$

$$= \frac{-(-6) \pm \sqrt{(-6)^2 - 4(1)(-3)}}{2(1)}$$

$$= \frac{6 \pm \sqrt{48}}{2} = \frac{6 \pm 4\sqrt{3}}{2} = 3 \pm 2\sqrt{3}$$

11. $3u^2 + 12u + 6 = 0$

Since 3 is a factor of each coefficient, divide both sides by 3.
$u^2 + 4u + 2 = 0$

$$u = \frac{-b \pm \sqrt{b^2 - 4ac}}{2a}, \quad a = 1, \ b = 4, \ c = 2$$

$$= \frac{-4 \pm \sqrt{4^2 - 4(1)(2)}}{2(1)} = \frac{-4 \pm \sqrt{8}}{2} = \frac{-4 \pm 2\sqrt{2}}{2} = -2 \pm \sqrt{2}$$

13.

$$2x^2 = 4x$$

$x^2 = 2x$ (divide both sides by 2)

$x^2 - 2x = 0$ (solve by factoring)

$x(x - 2) = 0$

$x = 0$ or $x - 2 = 0$

               $x = 2$

15. $4u^2 - 9 = 0$

$4u^2 = 9$ (solve by square root method)

$u^2 = \dfrac{9}{4}$

$u = \pm\sqrt{\dfrac{9}{4}} = \pm\dfrac{3}{2}$

17.

$$8x^2 + 20x = 12$$

$8x^2 + 20x - 12 = 0$

$2x^2 + 5x - 3 = 0$

$(x + 3)(2x - 1) = 0$

$x + 3 = 0$ or $2x - 1 = 0$

$x = -3$ or $2x = 1$

                     $x = \dfrac{1}{2}$

19.

$$x^2 = 1 - x$$

$x^2 + x - 1 = 0$

$x = \dfrac{-b \pm \sqrt{b^2 - 4ac}}{2a}, \quad a = 1, \ b = 1, \ c = -1$

$= \dfrac{-1 \pm \sqrt{(1)^2 - 4(1)(-1)}}{2(1)} = \dfrac{-1 \pm \sqrt{5}}{2}$

21.

$$2x^2 = 6x - 3$$

$2x^2 - 6x + 3 = 0$

$x = \dfrac{-b \pm \sqrt{b^2 - 4ac}}{2a}, \quad a = 2, \ b = -6, \ c = 3$

$= \dfrac{-(-6) \pm \sqrt{(-6)^2 - 4(2)(3)}}{2(2)} = \dfrac{6 \pm \sqrt{12}}{4} = \dfrac{6 \pm 2\sqrt{3}}{4} = \dfrac{3 \pm \sqrt{3}}{2}$

23.

$$y^2 - 4y = -8$$

$y^2 - 4y + 8 = 0$

$y = \dfrac{-b \pm \sqrt{b^2 - 4ac}}{2a}, \quad a, = 1, \ b = -4, \ c = 8$

$= \dfrac{-(-4) \pm \sqrt{(-4)^2 - 4(1)(8)}}{2(1)} = \dfrac{4 \pm \sqrt{-16}}{2}$

Since $\sqrt{-16}$ is not a real number, there are no real solutions.

25. $(x + 4)^2 = 11$

$x + 4 = \pm\sqrt{11}$

$x = -4 \pm \sqrt{11}$

27. $\dfrac{3}{p} = p$

$p^2 = 3$

$p = \pm\sqrt{3}$

29.

$$2 - \dfrac{2}{m^2} = \dfrac{3}{m}$$

$2m^2 - 2 = 3m$

$2m^2 - 3m - 2 = 0$

$(2m + 1)(m - 2) = 0$

$m = -\dfrac{1}{2}, \ 2$

31. $x^2 + 40x - 84$

    <u>Step 1</u>. Test for factorability

$$\sqrt{b^2 - 4ac} = \sqrt{(40)^2 - 4(1)(-84)} = \sqrt{1936} = 44$$

    Since the result is an integer, the polynomial has first-degree factors with integer coefficients.

    <u>Step 2</u>. Use the factor theorem

$$x^2 + 40x - 84 = 0$$
$$x = \frac{-40 \pm 44}{2} = 2, \ -42 \quad \text{(by the quadratic formula)}$$

    Thus, $x^2 + 40x - 84 = (x - 2)(x - [-42]) = (x - 2)(x + 42)$

33. $x^2 - 32x + 144$

    <u>Step 1</u>. Test for factorability

$$\sqrt{b^2 - 4ac} = \sqrt{(-32)^2 - 4(1)(144)} = \sqrt{448} \approx 21.166$$

    Since this is not an integer, the polynomial is not factorable.

35. $2x^2 + 15x - 108$

    <u>Step 1</u>. Test for factorability

$$\sqrt{b^2 - 4ac} = \sqrt{(15)^2 - 4(2)(-108)} = \sqrt{1089} = 33$$

    Thus, the polynomial has first-degree factors with integer coefficients.

    <u>Step 2</u>. Use the factor theorem

$$2x^2 + 15x - 108$$
$$x = \frac{-15 \pm 33}{4} = \frac{9}{2}, \ -12$$

    Thus, $2x^2 + 15x - 108 = 2\left(x - \frac{9}{2}\right)(x - [-12]) = (2x - 9)(x + 12)$

37. $4x^2 + 241x - 434$

    <u>Step 1</u>. Test for factorability

$$\sqrt{b^2 - 4ac} = \sqrt{(241)^2 - 4(4)(-434)} = \sqrt{65025} = 255$$

    Thus, the polynomial has first-degree factors with integer coefficients.

    <u>Step 2</u>. Use the factor theorem

$$4x^2 + 241x - 434$$
$$x = \frac{-241 \pm 255}{8} = \frac{14}{8}, \ -\frac{496}{8} \text{ or } \frac{7}{4}, \ -62$$

    Thus, $4x^2 + 241x - 434 = 4\left(x - \frac{7}{4}\right)(x + 62) = (4x - 7)(x + 62)$

39.
$$A = P(1 + r)^2$$
$$(1 + r)^2 = \frac{A}{P}$$
$$1 + r = \sqrt{\frac{A}{P}}$$
$$r = \sqrt{\frac{A}{P}} - 1$$

41. Setting the supply equation equal to the demand equation, we have

$$\frac{x}{450} + \frac{1}{2} = \frac{6,300}{x}$$

$$\frac{1}{450}x^2 + \frac{1}{2}x = 6,300$$

$$x^2 + 225x - 2,835,000 = 0$$

$$x = \frac{-225 \pm \sqrt{(225)^2 - 4(1)(-2,835,000)}}{2} \qquad \text{(quadratic formula)}$$

$$= \frac{-225 \pm \sqrt{11,390,625}}{2}$$

$$= \frac{-225 \pm 3375}{2}$$

$$= 1,575 \text{ units}$$

Note, we discard the negative root since a negative number of units cannot be produced or sold. Substituting $x = 1,575$ into either equation (we use the demand equation), we get

$$p = \frac{6,300}{1,575} = 4$$

Supply equals demand at $4 per unit.

43. $A = P(1 + r)^2 = P(1 + 2r + r^2) = Pr^2 + 2Pr + P$
Let $A = 144$ and $P = 100$. Then,
$100r^2 + 200r + 100 = 144$
$100r^2 + 200r - 44 = 0$
Using the quadratic formula,

$$r = \frac{-200 \pm \sqrt{(200)^2 - 4(100)(-44)}}{200}$$

$$= \frac{-200 \pm 240}{200} = -2.2, \ 0.20$$

Since $r > 0$, we have $r = 0.20$ or 20%.

45. $v^2 = 64h$
For $h = 1$, $v^2 = 64(1) = 64$. Therefore, $v = 8$ ft/sec.
For $h = 0.5$, $v^2 = 64(0.5) = 32$. Therefore, $v = \sqrt{32} = 4\sqrt{2} \approx 5.66$ ft/sec.

APPENDIX B SPECIAL TOPICS

Things to remember:

<u>1</u>. $a \in A$ means "a is an element of set A."

<u>2</u>. $a \notin A$ means "a is not an element of set A."

<u>3</u>. $\varnothing$ means "the empty set" or "null set."

<u>4</u>. $S = \{x \mid P(x)\}$ means "S is the set of all x such that $P(x)$ is true."

<u>5</u>. $A \subset B$ means "A is a subset of B."

<u>6</u>. $A = B$ means "A and B have exactly the same elements."

<u>7</u>. $A \not\subset B$ means "A is not a subset of B."

<u>8</u>. $A \neq B$ means "A and B do not have exactly the same elements."

<u>9</u>. $A \cup B = A$ union $B = \{x \mid x \in A \text{ or } x \in B\}$.

<u>10</u>. $A \cap B = A$ intersection $B = \{x \mid x \in A \text{ and } x \in B\}$.

<u>11</u>. $A' =$ complement of $A = \{x \in U \mid x \notin A\}$, where U is a universal set.

---

1. T   **3.** T   **5.** T   **7.** T   **9.** $\{1, 3, 5\} \cup \{2, 3, 4\} = \{1, 2, 3, 4, 5\}$

11. $\{1, 3, 4\} \cap \{2, 3, 4\} = \{3, 4\}$ **13.** $\{1, 5, 9\} \cap \{3, 4, 6, 8\} = \varnothing$

15. $\{x \mid x - 2 = 0\}$
$x - 2 = 0$ is true for $x = 2$.
Hence, $\{x \mid x - 2 = 0\} = \{2\}$.

17. $x^2 = 49$ is true for $x = 7$ and -7.
Hence, $\{x \mid x^2 = 49\} = \{-7, 7\}$.

19. $\{x \mid x$ is an odd number between 1 and 9 inclusive$\} = \{1, 3, 5, 7, 9\}$.

21. $U = \{1, 2, 3, 4, 5\}$; $A = \{2, 3, 4\}$.   Then $A' = \{1, 5\}$.

23. From the Venn diagram, A has 40 elements.

25. A' has 60 elements.

27. $A \cup B$ has 60 elements $(35 + 5 + 20)$.

29. $A' \cap B$ has 20 elements (common elements between A' and B.)

31. $(A \cap B)'$ has 95 elements. [<u>Note</u>: $A \cap B$ has 5 elements.]

33. $A' \cap B'$ has 40 elements.

35. (A) $\{x \mid x \in R$ or $x \in T\}$.
   $= R \cup T$ ("or" translated
      as $\cup$, union)
   $= \{1, 2, 3, 4\} \cup \{2, 4, 6\}$
   $= \{1, 2, 3, 4, 6\}$

   (B) $R \cup T = \{1, 2, 3, 4, 6\}$

37. $Q \cap R = \{2, 4, 6\} \cap \{3, 4, 5, 6\}$
   $= \{4, 6\}$
   $P \cup (Q \cap R) = \{1, 2, 3, 4\} \cup \{4, 6\}$
   $= \{1, 2, 3, 4, 6\}$

39. Yes. $A \cup B = B$ can be
represented by the Venn diagram

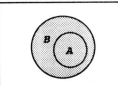

From the diagram, we see that
$A \subset B$. Thus, the given
statement is *true*.

41. Yes. The given statement is always
true. To understand this, see
the following Venn diagram.

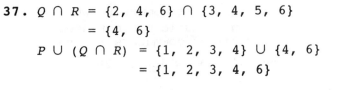

43. Yes. The given statement is
true. To understand this, see
the following Venn diagram.

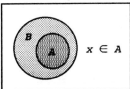

From the diagram, we conclude
that $x \in B$.

45. (A) Set $\{a\}$ has two subsets:
   $\{a\}$ and $\varnothing$

   (B) Set $\{a, b\}$ has four subsets:
   $\{a, b\}$, $\varnothing$, $\{a\}$, $\{b\}$

   (C) Set $\{a, b, c\}$ has eight subsets:
   $\{a, b, c\}$, $\varnothing$, $\{a\}$, $\{b\}$, $\{c\}$,
   $\{a, b\}$, $\{a, c\}$, $\{b, c\}$

   Parts (A), (B), and (C) suggest
   the following formula:
   The number of subsets in a set
   with n elements $= 2^n$.

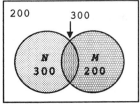

47. The Venn diagram that corresponds
to the given information is shown
at the right. We can see that
$N \cup M$ has $300 + 300 + 200 = 800$
students.

49. $(N \cup M)'$ has 200 students
[because $N \cup M$ has 800
students and $(N \cup M)'$ has $1000 - 800 = 200$].

51. $N' \cap M$ has 200 students.

53. The number of commuters who
listen to either news or music
= number of commuters in the
set $M \cup N$, which is 800.

55. The number of commuters who do
not listen to either news or
music = number of commuters in
set $(N \cup M)'$, which is
$1000 - 800 = 200$.

57. The number of commuters who listen to music but not news = number of commuters in the set $N' \cap M$, which is 200.

59. The six two-person subsets that can be formed from the given set $\{P,\ V_1,\ V_2,\ V_3\}$ are:

$\{P,\ V_1\}$ $\{P,\ V_3\}$ $\{V_1,\ V_3\}$
$\{P,\ V_2\}$ $\{V_1,\ V_2\}$ $\{V_2,\ V_3\}$

61. From the given Venn diagram $A \cap Rh = \{A+,\ AB+\}$

63. Again, from the given Venn diagram: $A \cup Rh = \{A-,\ A+,\ B+,\ AB-,\ AB+,\ O+\}$

65. From the given Venn diagram: $(A \cup B)' = \{O+,\ O-\}$

67. $A' \cap B = \{B-,\ B+\}$

69. Statement (2): for every $a,\ b \in C$, aRb and bRa means that everyone in the clique relates to one another.

EXERCISE B-2

Things to remember:

1. **THE SET OF REAL NUMBERS**

SYMBOL	NAME	DESCRIPTION	EXAMPLES
N	Natural numbers	Counting numbers (also called positive integers)	1, 2, 3, ...
Z	Integers	Natural numbers, their negatives, and 0	... −2, −1, 0, 1, 2, ...
Q	Rational numbers	Any number that can be represented as $\frac{a}{b}$, where a and b are integers and $b \neq 0$. Decimal representations are repeating or terminating.	-4; 0; 1; 25; $\frac{-3}{5}$; $\frac{2}{3}$; 3.67; $-0.333\overline{3}$; $5.272\overline{27}$
I	Irrational numbers	Any number with a decimal representation that is nonrepeating and non-terminating.	$\sqrt{2}$; π; $\sqrt[3]{7}$; $1.414213...$; $2.718281828...$
R	Real numbers	Rationals and irrationals	

<u>2</u>. BASIC PROPERTIES OF THE SET OF REAL NUMBERS

Let a, b, and c be arbitrary elements in the set of real numbers R.

ADDITION PROPERTIES

ASSOCIATIVE: $(a + b) + c = a + (b + c)$

COMMUTATIVE: $a + b = b + a$

    IDENTITY: 0 is the additive identity; that is, $0 + a = a + 0$ for all a in R, and 0 is the only element in R with this property.

    INVERSE: For each a in R, $-a$ is its unique additive inverse; that is, $a + (-a) = (-a) + a = 0$, and $-a$ is the only element in R relative to a with this property.

MULTIPLICATION PROPERTIES

ASSOCIATIVE: $(ab)c = a(bc)$

COMMUTATIVE: $ab = ba$

    IDENTITY: 1 is the multiplicative identity; that is, $1a = a1 = a$ for all a in R, and 1 is the only element in R with this property.

    INVERSE: For each a in R, $a \neq 0$, $\frac{1}{a}$ is its unique multiplicative inverse; that is, $a\left(\frac{1}{a}\right) = \left(\frac{1}{a}\right)a = 1$, and $\frac{1}{a}$ is the only element in R relative to a with this property.

DISTRIBUTIVE PROPERTIES

$$a(b + c) = ab + ac$$
$$(a + b)c = ac + bc$$

<u>3</u>. SUBTRACTION AND DIVISION

For all real numbers a and b.

SUBTRACTION: $a - b = a + (-b)$
$$7 - (-5) = 7 + [-(-5)] = 7 + 5 = 12$$

DIVISION: $a \div b = a\left(\frac{1}{b}\right), \; b \neq 0$
$$9 \div 4 = 9\left(\frac{1}{4}\right) = \frac{9}{4}$$

NOTE: 0 can never be used as a divisor!

<u>4</u>. PROPERTIES OF NEGATIVES

For all real numbers a and b.

a. $-(-a) = a$ b. $(-a)b = -(ab) = a(-b) = -ab$

c. $(-a)(-b) = ab$ d. $(-1)a = -a$

e. $\dfrac{-a}{b} = -\dfrac{a}{b} = \dfrac{a}{-b},\ \ b \neq 0$

f. $\dfrac{-a}{-b} = -\dfrac{-a}{b} = -\dfrac{a}{-b} = \dfrac{a}{b},\ \ b \neq 0$

<u>5</u>. ZERO PROPERTIES

For all real numbers a and b.

a. $a \cdot 0 = 0$

b. $ab = 0$ if and only if $a = 0$ or $b = 0$ (or both)

<u>6</u>. FRACTION PROPERTIES

For all real numbers a, b, c, d, and k (division by 0 excluded).

a. $\dfrac{a}{b} = \dfrac{c}{d}$ if and only if $ad = bc$ b. $\dfrac{ka}{kb} = \dfrac{a}{b}$

c. $\dfrac{a}{b} \cdot \dfrac{c}{d} = \dfrac{ac}{bd}$ d. $\dfrac{a}{b} \div \dfrac{c}{d} = \dfrac{a}{b} \cdot \dfrac{d}{c}$

e. $\dfrac{a}{b} + \dfrac{c}{b} = \dfrac{a + c}{b}$ f. $\dfrac{a}{b} - \dfrac{c}{b} = \dfrac{a - c}{b}$

g. $\dfrac{a}{b} + \dfrac{c}{d} = \dfrac{ad + bc}{bd}$ h. $\dfrac{a}{b} - \dfrac{c}{d} = \dfrac{ad - bc}{bd}$

---

1. $uv = vu$ **3.** $3 + (7 + y) = (3 + 7) + y$ **5.** $1(u + v) = u + v$

7. T; Associative property of multiplication

9. T; Distributive property   **11.** T; Definition of subtraction

13. T; Commutative property of addition   **15.** T; Property of negatives

17. T; Multiplicative inverse property   **19.** T; Property of negatives

21. F; $\dfrac{a}{b} + \dfrac{c}{d} = \dfrac{ad + bc}{bd}$ **23.** T; Distributive property

25. T; Zero property

27. No. For example: $2\left(\dfrac{1}{2}\right) = 1$. In general $a\left(\dfrac{1}{a}\right) = 1$ whenever $a \neq 0$.

29. (A) False. For example, -3 is an integer but not a natural number.

(B) True

(C) True. For example, for any natural number n, $n = \frac{n}{1}$.

31. $\sqrt{2}$, $\sqrt{3}$, ...; in general, the square root of any rational number that is not a perfect square; π, e.

33. (A) $8 \in N, Z, Q, R$ (B) $\sqrt{2} \in R$

(C) $-1.414 = -\frac{1414}{1000} \in Q, R$ (D) $\frac{-5}{2} \in Q, R$

35. (A) True. This is the associative property of addition.

(B) False. For example, $(3 - 7) - 4 = -4 - 4 = -8$
$$\neq 3 - (7 - 4) = 3 - 3 = 0.$$

(C) True. This is the associative property of multiplication.

(D) False. For example, $(12 \div 4) \div 2 = 3 \div 2 = \frac{3}{2}$
$$\neq 12 \div (4 \div 2) = 12 \div 2 = 6.$$

37.
$$C = 0.090909...$$
$$100C = 9.090909...$$
$$100C - C = (9.090909...) - (0.090909...)$$
$$99C = 9$$
$$C = \frac{9}{99} = \frac{1}{11}$$

39. (A) $\frac{13}{6} = 2.1666666...$ (repeating decimal)

(B) $\sqrt{21} \approx 4.5825756...$; $\sqrt{21}$ is an irrational number

(C) $\frac{7}{16} = 0.4375$ (terminating decimal)

(D) $\frac{29}{111} = 0.261261261...$ (repeating decimal)

EXERCISE B-3

Things to remember:

1. SEQUENCES

   A SEQUENCE is a function whose domain is a set of successive integers. If the domain of a given sequence is a finite set, then the sequence is called a FINITE SEQUENCE; otherwise, the sequence is an INFINITE SEQUENCE. In general, unless stated to the contrary or the context specifies otherwise, the domain of a sequence will be understood to be the set N of natural numbers.

2. NOTATION FOR SEQUENCES

Rather than function notation $f(n)$, n in the domain of a given sequence f, subscript notation a_n is normally used to denote the value in the range corresponding to n, and the sequence itself is denoted $\{a_n\}$ rather than f or $f(n)$. The elements in the range, a_n, are called the TERMS of the sequence; a_1 is the first term, a_2 is the second term, and a_n is the nth term or general term.

3. SERIES

Given a sequence $\{a_n\}$. The sum of the terms of the sequence, $a_1 + a_2 + a_3 + \cdots$ is called a SERIES. If the sequence is finite, the corresponding series is a FINITE SERIES; if the sequence is infinite, then the corresponding series is an INFINITE SERIES. Only finite series are considered in this section.

4. NOTATION FOR SERIES

Series are represented using SUMMATION NOTATION. If $\{a_k\}$, $k = 1, 2, \ldots, n$ is a finite sequence, then the series

$$a_1 + a_2 + a_3 + \cdots + a_n$$

is denoted

$$\sum_{k=1}^{n} a_k.$$

The symbol $\sum$ is called the SUMMATION SIGN and k is called the SUMMING INDEX.

5. ARITHMETIC MEAN

If $\{a_k\}$, $k = 1, 2, \ldots, n$, is a finite sequence, then the ARITHMETIC MEAN $\bar{a}$ of the sequence is defined as

$$\bar{a} = \frac{1}{n} \sum_{k=1}^{n} x_k.$$

---

1. $a_n = 2n + 3$;
$a_1 = 2\cdot 1 + 3 = 5$
$a_2 = 2\cdot 2 + 3 = 7$
$a_3 = 2\cdot 3 + 3 = 9$
$a_4 = 2\cdot 4 + 3 = 11$

5. $a_n = (-3)^{n+1}$;
$a_1 = (-3)^{1+1} = (-3)^2 = 9$
$a_2 = (-3)^{2+1} = (-3)^3 = -27$
$a_3 = (-3)^{3+1} = (-3)^4 = 81$
$a_4 = (-3)^{4+1} = (-3)^5 = -243$

3. $a_n = \dfrac{n + 2}{n + 1}$; $a_1 = \dfrac{1 + 2}{1 + 1} = \dfrac{3}{2}$
$a_2 = \dfrac{2 + 2}{2 + 1} = \dfrac{4}{3}$
$a_3 = \dfrac{3 + 2}{3 + 1} = \dfrac{5}{4}$
$a_4 = \dfrac{4 + 2}{4 + 1} = \dfrac{6}{5}$

7. $a_n = 2n + 3$; $a_{10} = 2 \cdot 10 + 3 = 23$

9. $a_n = \dfrac{n + 2}{n + 1}$; $a_{99} = \dfrac{99 + 2}{99 + 1} = \dfrac{101}{100}$

11. $\displaystyle\sum_{k=1}^{6} k = 1 + 2 + 3 + 4 + 5 + 6 = 21$

13. $\displaystyle\sum_{k=4}^{7} (2k - 3) = (2 \cdot 4 - 3) + (2 \cdot 5 - 3) + (2 \cdot 6 - 3) + (2 \cdot 7 - 3)$

$$= 5 + 7 + 9 + 11 = 32$$

15. $\displaystyle\sum_{k=0}^{3} \dfrac{1}{10^k} = \dfrac{1}{10^0} + \dfrac{1}{10^1} + \dfrac{1}{10^2} + \dfrac{1}{10^3} = 1 + \dfrac{1}{10} + \dfrac{1}{100} + \dfrac{1}{1000} = \dfrac{1111}{1000} = 1.111$

17. $a_1 = 5$, $a_2 = 4$, $a_3 = 2$, $a_4 = 1$, $a_5 = 6$. Here $n = 5$ and the arithmetic mean is given by:

$$\bar{a} = \dfrac{1}{5} \sum_{k=1}^{5} a_k = \dfrac{1}{5} (5 + 4 + 2 + 1 + 6) = \dfrac{18}{5} = 3.6$$

19. $a_1 = 96$, $a_2 = 65$, $a_3 = 82$, $a_4 = 74$, $a_5 = 91$, $a_6 = 88$, $a_7 = 87$, $a_8 = 91$, $a_9 = 77$, and $a_{10} = 74$. Here $n = 10$ and the arithmetic mean is given by:

$$\bar{a} = \dfrac{1}{10} \sum_{k=1}^{10} a_k = \dfrac{1}{10} (96 + 65 + 82 + 74 + 91 + 88 + 87 + 91 + 77 + 74)$$

$$= \dfrac{825}{10} = 82.5$$

21. $a_n = \dfrac{(-1)^{n+1}}{2^n}$; $a_1 = \dfrac{(-1)^2}{2^1} = \dfrac{1}{2}$

$$a_2 = \dfrac{(-1)^3}{2^2} = -\dfrac{1}{4}$$

$$a_3 = \dfrac{(-1)^4}{2^3} = \dfrac{1}{8}$$

$$a_4 = \dfrac{(-1)^5}{2^4} = -\dfrac{1}{16}$$

$$a_5 = \dfrac{(-1)^6}{2^5} = \dfrac{1}{32}$$

23. $a_n = n[1 + (-1)^n]$; $a_1 = 1[1 + (-1)^1] = 0$

$$a_2 = 2[1 + (-1)^2] = 4$$

$$a_3 = 3[1 + (-1)^3] = 0$$

$$a_4 = 4[1 + (-1)^4] = 8$$

$$a_5 = 5[1 + (-1)^5] = 0$$

25. $a_n = \left(-\dfrac{3}{2}\right)^{n-1}; \quad a_1 = \left(-\dfrac{3}{2}\right)^0 = 1$

$$a_2 = \left(-\dfrac{3}{2}\right)^1 = -\dfrac{3}{2}$$

$$a_3 = \left(-\dfrac{3}{2}\right)^2 = \dfrac{9}{4}$$

$$a_4 = \left(-\dfrac{3}{2}\right)^3 = -\dfrac{27}{8}$$

$$a_5 = \left(-\dfrac{3}{2}\right)^4 = \dfrac{81}{16}$$

27. Given -2, -1, 0, 1, $\ldots$ The sequence is the set of successive integers beginning with -2. Thus, $a_n = n - 3$, $n = 1, 2, 3, \ldots$.

29. Given 4, 8, 12, 16, $\ldots$ The sequence is the set of positive integer multiples of 4. Thus, $a_n = 4n$, $n = 1, 2, 3, \ldots$.

31. Given $\dfrac{1}{2}$, $\dfrac{3}{4}$, $\dfrac{5}{6}$, $\dfrac{7}{8}$, $\ldots$ The sequence is the set of fractions whose numerators are the odd positive integers and whose denominators are the even positive integers. Thus,

$$a_n = \dfrac{2n - 1}{2n}, \quad n = 1, 2, 3, \ldots .$$

33. Given 1, -2, 3, -4, $\ldots$ The sequence consists of the positive integers with alternating signs. Thus,

$$a_n = (-1)^{n+1}n, \quad n = 1, 2, 3, \ldots .$$

35. Given 1, -3, 5, -7, $\ldots$ The sequence consists of the odd positive integers with alternating signs. Thus,

$$a_n = (-1)^{n+1}(2n - 1), \quad n = 1, 2, 3, \ldots .$$

37. Given 1, $\dfrac{2}{5}$, $\dfrac{4}{25}$, $\dfrac{8}{125}$, $\ldots$ The sequence consists of the nonnegative integral powers of $\dfrac{2}{5}$. Thus,

$$a_n = \left(\dfrac{2}{5}\right)^{n-1}, \quad n = 1, 2, 3, \ldots .$$

39. Given x, x^2, x^3, x^4, $\ldots$ The sequence is the set of positive integral powers of x. Thus, $a_n = x^n$, $n = 1, 2, 3, \ldots$.

41. Given x, $-x^3$, x^5, $-x^7$, $\ldots$ The sequence is the set of positive odd integral powers of x with alternating signs. Thus,

$$a_n = (-1)^{n+1}x^{2n-1}, \quad n = 1, 2, 3, \ldots .$$

43. $\displaystyle\sum_{k=1}^{5}(-1)^{k+1}(2k-1)^2 = (-1)^2(2\cdot 1 - 1)^2 + (-1)^3(2\cdot 2 - 1)^2$
$$+ (-1)^4(2\cdot 3 - 1)^2 + (-1)^5(2\cdot 4 - 1)^2$$
$$+ (-1)^6(2\cdot 5 - 1)^2$$
$$= 1 - 9 + 25 - 49 + 81$$

45. $\displaystyle\sum_{k=2}^{5}\frac{2^k}{2k+3} = \frac{2^2}{2\cdot 2 + 3} + \frac{2^3}{2\cdot 3 + 3} + \frac{2^4}{2\cdot 4 + 3} + \frac{2^5}{2\cdot 5 + 3}$

$$= \frac{4}{7} + \frac{8}{9} + \frac{16}{11} + \frac{32}{13}$$

47. $\displaystyle\sum_{k=1}^{5}x^{k-1} = x^0 + x^1 + x^2 + x^3 + x^4 = 1 + x + x^2 + x^3 + x^4$

49. $\displaystyle\sum_{k=0}^{4}\frac{(-1)^k x^{2k+1}}{2k+1} = \frac{(-1)^0 x}{2\cdot 0 + 1} + \frac{(-1)x^3}{2\cdot 1 + 1} + \frac{(-1)^2 x^5}{2\cdot 2 + 1} + \frac{(-1)^3 x^7}{2\cdot 3 + 1} + \frac{(-1)^4 x^9}{2\cdot 4 + 1}$

$$= x - \frac{x^3}{3} + \frac{x^5}{5} - \frac{x^7}{7} + \frac{x^9}{9}$$

51. (A) $2 + 3 + 4 + 5 + 6 = \displaystyle\sum_{k=1}^{5}(k+1)$

(B) $2 + 3 + 4 + 5 + 6 = \displaystyle\sum_{j=0}^{4}(j+2)$

53. (A) $1 - \dfrac{1}{2} + \dfrac{1}{3} - \dfrac{1}{4} = \displaystyle\sum_{k=1}^{4}\frac{(-1)^{k+1}}{k}$

(B) $1 - \dfrac{1}{2} + \dfrac{1}{3} - \dfrac{1}{4} = \displaystyle\sum_{j=0}^{3}\frac{(-1)^j}{j+1}$

55. $2 + \dfrac{3}{2} + \dfrac{4}{3} + \cdots + \dfrac{n+1}{n} = \displaystyle\sum_{k=1}^{n}\frac{k+1}{k}$

57. $\dfrac{1}{2} - \dfrac{1}{4} + \dfrac{1}{8} - \cdots + \dfrac{(-1)^{n+1}}{2^n} = \displaystyle\sum_{k=1}^{n}\frac{(-1)^{k+1}}{2^k}$

59. $a_1 = 2$ and $a_n = 3a_{n-1} + 2$ for $n \geq 2$.
$a_1 = 2$
$a_2 = 3\cdot a_1 + 2 = 3\cdot 2 + 2 = 8$
$a_3 = 3\cdot a_2 + 2 = 3\cdot 8 + 2 = 26$
$a_4 = 3\cdot a_3 + 2 = 3\cdot 26 + 2 = 80$
$a_5 = 3\cdot a_4 + 2 = 3\cdot 80 + 2 = 242$

61. $a_1 = 1$ and $a_n = 2a_{n-1}$ for $n \geq 2$.

$a_1 = 1$

$a_2 = 2 \cdot a_1 = 2 \cdot 1 = 2$

$a_3 = 2 \cdot a_2 = 2 \cdot 2 = 4$

$a_4 = 2 \cdot a_3 = 2 \cdot 4 = 8$

$a_5 = 2 \cdot a_4 = 2 \cdot 8 = 16$

63. In $a_1 = \dfrac{A}{2}$, $a_n = \dfrac{1}{2}\left(a_{n-1} + \dfrac{A}{a_{n-1}}\right)$, $n \geq 2$, let $A = 2$. Then:

$a_1 = \dfrac{2}{2} = 1$

$a_2 = \dfrac{1}{2}\left(a_1 + \dfrac{A}{a_1}\right) = \dfrac{1}{2}(1 + 2) = \dfrac{3}{2}$

$a_3 = \dfrac{1}{2}\left(a_2 + \dfrac{A}{a_2}\right) = \dfrac{1}{2}\left(\dfrac{3}{2} + \dfrac{2}{3/2}\right) = \dfrac{1}{2}\left(\dfrac{3}{2} + \dfrac{4}{3}\right) = \dfrac{17}{12}$

$a_4 = \dfrac{1}{2}\left(a_3 + \dfrac{A}{a_3}\right) = \dfrac{1}{2}\left(\dfrac{17}{12} + \dfrac{2}{17/12}\right) = \dfrac{1}{2}\left(\dfrac{17}{12} + \dfrac{24}{17}\right) = \dfrac{577}{408} \approx 1.414216$

$$\text{and } \sqrt{2} \approx 1.414214$$

EXERCISE B-4

Things to remember:

<u>1</u>. A sequence of numbers a_1, a_2, a_3, ..., a_n, ..., is called an ARITHMETIC PROGRESSION if there is constant d, called the COMMON DIFFERENCE, such that

$$a_n - a_{n-1} = d,$$

that is,

$$a_n = a_{n-1} + d$$

for all $n > 1$.

<u>2</u>. If a_1, a_2, a_3, ..., a_n, ..., is an arithmetic progression with common difference d, then

$$a_n = a_1 + (n - 1)d$$

for all $n > 1$.

<u>3</u>. The sum S_n of the first n terms of an arithmetic progression a_1, a_2, a_3, ..., a_n, ..., with common difference d, is given by

(a) $S_n = \dfrac{n}{2}[2a_1 + (n - 1)d]$

or by

(b) $S_n = \dfrac{n}{2}(a_1 + a_n)$.

<u>**4.**</u> A sequence of numbers a_1, a_2, a_3, $\ldots$, a_n, $\ldots$, is called a GEOMETRIC PROGRESSION if there exists a nonzero constant r, called the COMMON RATIO, such that

$$\frac{a_n}{a_{n-1}} = r,$$

that is,

$$a_n = ra_{n-1}$$

for all $n > 1$.

<u>**5.**</u> If a_1, a_2, a_3, $\ldots$, a_n, $\ldots$, is a geometric progression with common ration r, then

$$a_n = a_1 r^{n-1}$$

for all $n > 1$.

<u>**6.**</u> The sum S_n of the first n terms of a geometric progression a_1, a_2, a_3, $\ldots$, a_n, $\ldots$, with common ration r, is given by:

$$S_n = \frac{a_1(r^n - 1)}{r - 1}, \ r \neq 1,$$

or by

$$S_n = \frac{ra_n - a_1}{r - 1}, \ r \neq 1.$$

<u>**7.**</u> If a_1, a_2, a_3, $\ldots$, a_n, $\ldots$, is an infinite geometric progression with common ratio r having the property $-1 < r < 1$, then the sum S_∞ is defined to be:

$$S_\infty = \frac{a_1}{1 - r}.$$

---

1. (A) is an arithmetic progression; $a_2 - a_1 = a_3 - a_2 = 3$. Thus, $d = 3$, $a_4 = 14$, and $a_5 = 17$.

(B) is not an arithmetic progression, since $a_2 - a_1 = 8 - 4 = 4 \neq a_3 - a_2 = 16 - 8 = 8$.

(C) is not an arithmetic progression, since $-4 - (-2) = -2 \neq -8 - (-4) = -4$.

(D) is an arithmetic progression; $a_2 - a_1 = a_3 - a_2 = -10$. Thus, $d = -10$, $a_4 = -22$, and $a_5 = -32$.

3. (A) is a geometric progression; $\dfrac{a_2}{a_1} = \dfrac{a_3}{a_2} = -2$. Thus, $r = -2$, $a_4 = -8$, $a_5 = 16$.

   (B) is not a geometric progression, since $\dfrac{a_2}{a_1} = \dfrac{6}{7} \neq \dfrac{a_3}{a_2} = \dfrac{5}{6}$.

   (C) is a geometric progression; $\dfrac{a_2}{a_1} = \dfrac{a_3}{a_2} = \dfrac{1}{2}$. Thus, $r = \dfrac{1}{2}$, $a_4 = \dfrac{1}{4}$, $a_5 = \dfrac{1}{8}$.

   (D) is not a geometric progression, since $\dfrac{a_2}{a_1} = \dfrac{-4}{2} = -2 \neq \dfrac{a_3}{a_2} = \dfrac{-3}{2}$

5. $a_2 = a_1 + d = 7 + 4 = 11$
   $a_3 = a_2 + d = 11 + 4 = 15$ (using $\underline{1}$)

7. $a_{21} = a_1 + (21 - 1)d = 2 + 20 \cdot 4 = 82$ (using $\underline{2}$)
   $S_{31} = \dfrac{31}{2}[2a_1 + (31 - 1)d] = \dfrac{31}{2}[2 \cdot 2 + 30 \cdot 4] = \dfrac{31}{2} \cdot 124 = 1922$ [using $\underline{3}$(a)]

9. Using $\underline{3}$(b), $S_{20} = \dfrac{20}{2}(a_1 + a_{20}) = 10(18 + 75) = 930$

11. $a_2 = a_1 r = 3(-2) = -6$
    $a_3 = a_2 r = -6(-2) = 12$
    $a_4 = a_3 r = 12(-2) = -24$ (using $\underline{4}$)

13. Using $\underline{6}$, $S_7 = \dfrac{-3 \cdot 729 - 1}{-3 - 1} = \dfrac{-2188}{-4} = 547$.

15. Using $\underline{5}$, $a_{10} = 100(1.08)^9 = 199.90$.

17. Using $\underline{5}$, $200 = 100r^8$. Thus, $r^8 = 2$ and $r = \sqrt[8]{2} \approx 1.09$.

19. Using $\underline{6}$, $S_{10} = \dfrac{500[(0.6)^{10} - 1]}{0.6 - 1} \approx 1242$,
    $S_\infty = \dfrac{500}{1 - 0.6} = 1250$.

21. Let $a_1 = 13$, $d = 2$. Then, using $\underline{2}$, we can find n:
    $67 = 13 + (n - 1)2$ or $2(n - 1) = 54$
    $n - 1 = 27$
    $n = 28$
    Therefore, using $\underline{3}$(b), $S_{28} = \dfrac{28}{2}[13 + 67] = 14 \cdot 80 = 1120$.

23. (A) $2 + 4 + 8 + \cdots$. Since $r = \dfrac{4}{2} = \dfrac{8}{4} = \cdots = 2$ and $|2| = 2 > 1$, the sum does not exist.

(B) $2, -\frac{1}{2}, \frac{1}{8}, \ldots$. In this case, $r = \dfrac{-\frac{1}{2}}{2} = \dfrac{\frac{1}{8}}{-\frac{1}{2}} = \ldots = -\frac{1}{4}$.

Since $|r| < 1$,

$$S_\infty = \frac{2}{1 - \left(-\frac{1}{4}\right)} = \frac{2}{\frac{5}{4}} = \frac{8}{5} = 1.6.$$

25. $f(1) = -1$, $f(2) = 1$, $f(3) = 3$, ... This is an arithmetic progression with $a_1 = -1$, $d = 2$. Thus, using $\underline{3}(a)$,

$$f(1) + f(2) + f(3) + \cdots + f(50) = \frac{50}{2}[2(-1) + 49 \cdot 2] = 25 \cdot 96 = 2400$$

27. $f(1) = \frac{1}{2}$, $f(2) = \left(\frac{1}{2}\right)^2 = \frac{1}{4}$, $f(3) = \left(\frac{1}{2}\right)^3 = \frac{1}{8}$, ... This is a geometric progression with $a_1 = \frac{1}{2}$ and $r = \frac{1}{2}$. Thus, using $\underline{6}$:

$$f(1) + f(2) + \cdots + f(10) = S_{10} = \frac{\frac{1}{2}\left[\left(\frac{1}{2}\right)^{10} - 1\right]}{\frac{1}{2} - 1} \approx 0.999$$

29. Consider the arithmetic progression with $a_1 = 1$, $d = 2$. This progression is the sequence of odd positive integers. Now, using $\underline{3}(a)$, the sum of the first n odd positive integers is:

$$S_n = \frac{n}{2}[2 \cdot 1 + (n - 1)2] = \frac{n}{2}(2 + 2n - 2) = \frac{n}{2} \cdot 2n = n^2$$

31. Consider the time line:

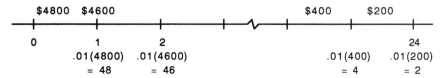

The total cost of the loan is $2 + 4 + 6 + \cdots + 46 + 48$. The terms form an arithmetic progression with $n = 24$, $a_1 = 2$, and $a_{24} = 48$. Thus, using $\underline{3}(b)$:

$$S_{24} = \frac{24}{2}(2 + 48) = 24 \cdot 25 = \$600$$

33. This is a geometric progression with $a_1 = 3,500,000$ and $r = 0.7$. Thus, using $\underline{7}$:

$$S_\infty = \frac{3,500,000}{1 - 0.7} \approx \$11,670,000$$

35.

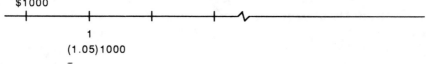

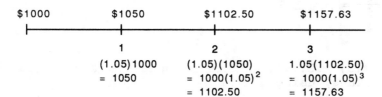

In general, after n years, the amount A_n in the account is:

$$A_n = 1000(1.05)^n$$

Thus, $A_{10} = 1000(1.05)^{10} \approx \$1628.89$

and $A_{20} = 1000(1.05)^{20} \approx \$2653.30$

EXERCISE B-5

Things to remember:

<u>1</u>. If n is a positive integer, then n FACTORIAL, denoted $n!$, is the product of the integers from 1 to n; that is,

$$n! = n \cdot (n - 1) \cdot \ldots \cdot 3 \cdot 2 \cdot 1 = n(n - 1)!$$

Also, $1! = 1$ and $0! = 1$.

<u>2</u>. If n and r are nonnegative integers and $r \leq n$, then:

$$C_{n,r} = \frac{n!}{r!(n - r)!}$$

<u>3</u>. BINOMIAL THEOREM

For all natural numbers n:

$$(a + b)^n = C_{n,0}a^n + C_{n,1}a^{n-1}b + C_{n,2}a^{n-2}b^2$$
$$+ \cdots + C_{n,n-1}ab^{n-1} + C_{n,n}b^n.$$

---

1. $6! = 6 \cdot 5 \cdot 4 \cdot 3 \cdot 2 \cdot 1 = 720$

3. $\dfrac{10!}{9!} = \dfrac{10 \cdot 9!}{9!} = 10$

5. $\dfrac{12!}{9!} = \dfrac{12 \cdot 11 \cdot 10 \cdot 9!}{9!} = 1320$

7. $\dfrac{5!}{2!3!} = \dfrac{5 \cdot 4 \cdot 3!}{2 \cdot 1 \cdot 3!} = 10$

9. $\dfrac{6!}{5!(6 - 5)!} = \dfrac{6 \cdot 5!}{5!1!} = 6$

11. $\dfrac{20!}{3!17!} = \dfrac{20 \cdot 19 \cdot 18 \cdot 17!}{3!17!}$
$= \dfrac{20 \cdot 19 \cdot 18}{3 \cdot 2 \cdot 1} = 1140$

13. $C_{5,3} = \dfrac{5!}{3!(5-3)!} = \dfrac{5!}{3!2!} = 10$ (see Problem 7)

15. $C_{6,5} = \dfrac{6!}{5!(6-5)!} = 6$ (see Problem 9)

17. $C_{5,0} = \dfrac{5!}{0!(5-0)!} = \dfrac{5!}{1 \cdot 5!} = 1$

19. $C_{18,15} = \dfrac{18!}{15!(18-15)!} = \dfrac{18 \cdot 17 \cdot 16 \cdot 15!}{15!3!} = \dfrac{18 \cdot 17 \cdot 16}{3 \cdot 2 \cdot 1} = 816$

21. Using $\underline{3}$,
$$(a+b)^4 = C_{4,0}a^4 + C_{4,1}a^3b + C_{4,2}a^2b^2 + C_{4,3}ab^3 + C_{4,4}b^4$$
$$= a^4 + 4a^3b + 6a^2b^2 + 4ab^3 + b^4$$

23. Using $\underline{3}$,
$$(x-1)^6 = [x+(-1)]^6$$
$$= C_{6,0}x^6 + C_{6,1}x^5(-1) + C_{6,2}x^4(-1)^2 + C_{6,3}x^3(-1)^3$$
$$\quad + C_{6,4}x^2(-1)^4 + C_{6,5}x(-1)^5 + C_{6,6}(-1)^6$$
$$= x^6 - 6x^5 + 15x^4 - 20x^3 + 15x^2 - 6x + 1$$

25. $(2a-b)^5 = [2a+(-b)]^5$
$$= C_{5,0}(2a)^5 + C_{5,1}(2a)^4(-b) + C_{5,2}(2a)^3(-b)^2 + C_{5,3}(2a)^2(-b)^3$$
$$\quad + C_{5,4}(2a)(-b)^4 + C_{5,5}(-b)^5$$
$$= 32a^5 - 80a^4b + 80a^3b^2 - 40a^2b^3 + 10ab^4 - b^5$$

27. The fifth term in the expansion of $(x-1)^{18}$ is:
$$C_{18,4}x^{14}(-1)^4 = \dfrac{18 \cdot 17 \cdot 16 \cdot 15}{4 \cdot 3 \cdot 2 \cdot 1}x^{14} = 3060x^{14}$$

29. The seventh term in the expansion of $(p+q)^{15}$ is:
$$C_{15,6}p^9q^6 = \dfrac{15 \cdot 14 \cdot 13 \cdot 12 \cdot 11 \cdot 10}{6 \cdot 5 \cdot 4 \cdot 3 \cdot 2 \cdot 1}p^9q^6 = 5005p^9q^6$$

31. The eleventh term in the expansion of $(2x+y)^{12}$ is:
$$C_{12,10}(2x)^2y^{10} = \dfrac{12 \cdot 11}{2 \cdot 1}4x^2y^{10} = 264x^2y^{10}$$

33. $C_{n,0} = \dfrac{n!}{0!(n-0)!} = \dfrac{n!}{1 \cdot n!} = 1$ $\qquad$ $C_{n,n} = \dfrac{n!}{n!(n-n)!} = \dfrac{n!}{n!0!} = 1$

35. The next two rows are:

1 5 10 10 5 1   and   1 6 15 20 15 6 1,

respectively. These are the coefficients in the binomial expansions of $(a+b)^5$ and $(a+b)^6$.